x6. £16-95

SECOND EDITION

The Geography Teacher's Guide to the Classroom

SECOND EDITION

The Geography Teacher's Guide to the Classroom

Edited by
John Fien Rodney Gerber Peter Wilson
Brisbane College of Advanced Education, Australia

M

© John Fien, Rodney Gerber, Peter Wilson, 1989

Artwork by Brian Pieper and Macmillan India Ltd.

All rights reserved.
No part of this publication
may be reproduced or transmitted
in any form or by any means,
without permission.

First published 1984
Second edition published 1989 by
THE MACMILLAN COMPANY OF AUSTRALIA PTY LTD
107 Moray Street, South Melbourne 3205
6 Clarke Street, Crows Nest 2065

Associated companies and representatives
throughout the world

National Library of Australia
cataloguing in publication data

The Geography teacher's guide to the classroom.

   2nd ed.
   Includes index.
   ISBN 0 333 47870 3.

   1. Geography — Study and teaching. I. Fien, John,
   1951– . II. Gerber, Rod. III. Wilson, Peter,
   1941– .

910'.7'1

Set in Baskerville by Macmillan India Ltd., Bangalore 25.
Printed in Hong Kong

Contents

Contributors   vi

Introduction to the Second Edition   vii

1. Geography: A Medium for Education
   John Fien, *Bernard Cox* and *Wayne Fossey*   1
2. Knowledge and Teaching Styles in the Geography Classroom
   David Hall   10
3. Critical Inquiry: The Emerging Perspective in Geography Teaching
   V. Leo Bartlett   22
4. Expository Teaching for Meaningful Learning in Geography
   F. Geoffrey Jones   35
5. Teaching for Thinking in the Geography Classroom
   Peter Wilson   44
6. Making Inquiry Learning Work in the Geography Classroom
   Bernard Cox   64
7. Designing Worksheets to Promote Student Inquiry in Geography
   Brian Hoepper   75
8. Learning Geography through Classroom and Library Research
   Louis Murray   85
9. Models and Reality: Integrating Practical Work and Fieldwork in Geography
   Barrie McElroy   95
10. Learning Geography through Fieldwork
    Kevin Laws   104
11. Developing Valuing and Decision-making Skills in the Geography Classroom
    Brian Maye   118
12. Promoting Social and Political Education in Geography Teaching
    Rob Gilbert   131
13. Look into my Mind: Qualitative Inquiry in Teaching Geography
    V. Leo Bartlett   141
14. Developing Environmental Awareness and Appreciation
    Robin Hall   153
15. Teaching Skills in Geography
    Bernard Cox   169
16. Teaching Graphics in Geography Lessons
    Rod Gerber   179
17. Using Maps Well in the Geography Classroom
    Rod Gerber and *Peter Wilson*   197
18. Language in the Geography Classroom
    Bryan Stephenson   210
19. Using Textbooks and Reading for Understanding in Geography
    John Lidstone   225
20. Using Games and Simulations in the Geography Classroom
    John Fien, *Robert Herschell* and *John Hodgkinson*   251
21. Using Computers in Geography Teaching
    John Lidstone   261
22. Geographical Facts from Geographical Figures: Turning Students on with 'Stats' in Geography
    John Wolforth   286
23. The Diagnosis of Student Learning in Geography
    Rod Gerber   301
24. Individualising Learning in Geography
    John Lidstone   312
25. Teaching the Less Able Students in Geography
    William Pick and *Malcolm Renwick*   333
26. Planning and Teaching a Geography Curriculum unit
    John Fien   346
27. School-based Curriculum Development in Geography
    John Fien   359
28. Planning a School-based Assessment Program
    Warren Halloway   371
29. Selecting and Evaluating Resources for Geography Teaching
    Peter Maccoll   384
30. Evaluating Your Geography Courses
    Barrie McElroy   398
31. On Being a Geography Teacher in the 1990s and Beyond
    William Romey and *William Elberty Jnr*   407

Acknowledgements   418

Contributors

John Fien, *Rodney Gerber* and *Peter Wilson* Lecturers in Geographical Education, Brisbane College of Advanced Education, Kelvin Grove Campus, Brisbane, Australia

V. Leo Bartlett Lecturer in Geographical Education, University of Queensland, Brisbane, Australia

Bernard Cox Lecturer in Geographical Education, University of Queensland, Brisbane, Australia

William Elberty Jnr Lecturer in Geography, St Lawrence University, Canton, New York, USA

Wayne Fossey Deputy Principal, Benowa High School, Gold Coast, Australia

Rob Gilbert Dean of Education, James Cook University of North Queensland, Townsville, Australia

David Hall Lecturer in Geographical Education, Bristol University, Bristol, England

Robin Hall Lecturer in Geography, Mitchell College of Advanced Education, Bathurst, Australia

Warren Halloway Head of the Centre for Social Science Education, Armidale College of Advanced Education, Armidale, Australia

Robert Herschell Curriculum Officer, Queensland Department of Education, Brisbane, Australia

John Hodgkinson Principal, Murgon High School, Murgon, Australia

Brian Hoepper Lecturer in Social Science Education, Brisbane College of Advanced Education, Kelvin Grove Campus, Brisbane, Australia

F. Geoffrey Jones Formerly Lecturer in Geographical Education, University of New England, Armidale, Australia

Kevin Laws Lecturer in Geographical Education, University of Sydney, Sydney, Australia

John Lidstone Lecturer in Geographical Education, Brisbane College of Advanced Education, Kelvin Grove Campus, Brisbane, Australia

Barrie McElroy Lecturer in Geographical Education, South Australian College of Advanced Education, Salisbury Campus, Adelaide, Australia

Peter Maccoll Curriculum Officer, Queensland Department of Education, Brisbane, Australia

Brian Maye Lecturer in Social Science Education, Armidale College of Advanced Education, Armidale, Australia

Louis Murray Formerly Lecturer in Geography, Bedford College of Higher Education, Bedford, England

William Pick Head Teacher, Seacroft Park Middle School, Leeds. Formerly Senior Lecturer in Curriculum Development, Leeds Polytechnic, Leeds, England

Malcolm Renwick Freelance Educational Consultant. Formerly Senior Lecturer in Geography, Leeds Polytechnic, Leeds, England

William Romey Lecturer in Geography, St Lawrence University, Canton, New York, USA

Bryan Stephenson Lecturer in Geographical Education, University of Exeter, Exeter, England

John Wolforth Lecturer in Geographical Education, McGill University, Montreal, Canada

Introduction to the Second Edition

The title for this book was difficult to select because all the ones that originally came to mind had been used for other books and journals already. We would have been quite happy with *Geography In and Out of School, Classroom Geographer, Teaching Geography, A Guide to Modern Geography Teaching* or many of the other titles that earlier writers had used. All of these would have been appropriate illustrations of the practical advice on geography teaching that this book provides.

We finally decided on our title, *The Geography Teacher's Guide to the Classroom*, during a discussion about the scope and purpose of this book with a group of experienced teachers. One of them, who was perhaps a little daunted by the broad implications for the geography classroom contained in the list of contents, suggested, only half jokingly, that catching a flying saucer to Saturn would be easier than doing all the preparation he thought was intended for busy teachers. He wanted to know if we had forgotten the heavy day-in day-out schedule of teaching that characterises today's secondary schools, where most preparation is done at night in a teacher's own time.

Consequently, the various chapters of *The Geography Teacher's Guide to the Classroom* have been written with the needs of busy experienced teachers in mind. Additionally, the book has been written as a guide to students in colleges and universities preparing for a career teaching geography. We believe the advice of the 26 geography teachers and curriculum specialists from Australia, England, Canada and the United States who wrote the various chapters of this book will go a considerable way in developing and maintaining the skills, insights and enthusiasm so necessary for good geography teaching.

Readers who are familiar with Douglas Adams' radio play and book, *The Hitchhiker's Guide to the Galaxy* (Adams 1979), may have realised the origin of the title *The Geography Teacher's Guide to the Classroom* from the earlier reference to 'catching a flying saucer to Saturn'. Incidentally, there must have been something of a geographer in Adams for in a second book he claimed that all civilisations go through three stages of development, those of Survival, Inquiry and Sophistication. These he describes as the *How*, *Why* and *Where* eras. The sophisticated *Where* era was, of course, the ultimate that any civilisation could reach (Adams 1980).

It is our hope that teachers following some of the ideas in this book will create parallels in their geography classrooms to the excitement of new things discovered, interest in their surroundings, widened horizons and adventure that the two space age heroes, Arthur Dent and Ford Prefect, found in their hyperspatial explorations in Adam's books.

Geographical Education into the 1990s

A variety of developments over the last two decades has helped make geographical education one of the most theoretically developed and active of curriculum studies areas. The more important developments included the application of educational theory to geography teaching, developments in the discipline of geography, and the rise of applied curriculum studies. Each chapter in this book seeks to integrate aspects of these developments and provide practical classroom examples of ways through which the benefits of improved educational and geographical understandings may be realised.

Educational Theory and Geographical Education

Several collections of readings on educational philosophy, psychology and theory within the context of the geography curriculum were published in England, Australia and North America in the 1970s. These included: *New Movements in the Study and Teaching of Geography* (Graves 1972); *New Perspectives on Geographical Education* (Bale, Graves and Walford 1973); the two volumes of *Readings in Geographical Education* (Biddle 1968 and Biddle and Deer 1973); and *The Social Sciences and Geographic Education: A Reader* (Ball, Steinbrink and Stoltman 1971). Graves in *Geography in Education* (1975) and Hall in *Geography and the Geography Teacher* (1976) integrated and further developed the ideas in these collections of readings to provide a theoretic foundation for geographical education. These books have been very useful for tertiary studies in geographical education and for reflecting on classroom practices.

Books that teachers have described as more pragmatically related to their day to day activities include *Handbook for Geography Teachers* (Long 1974), *New Directions in Geography Teaching* (Walford 1973), *Teaching Geography* (Bailey 1974), *A Guide to Modern Geography Teaching* (Dunlop 1976), *New Perspectives on*

Geographic Education (Manson and Ridd 1977), *Learning through Geography* (Slater 1982) and *Handbook for Geography Teachers* (Boardman 1986). These books focus directly on curriculum planning and teaching methods for secondary school geography courses. Each contains much practical advice and many detailed lesson ideas. Unfortunately, most of them have a distinct British or American bias and some are rapidly dating. Only Manson and Ridd (1977) and Slater (1982) give attention to classroom issues or strategies such as values education, inquiry learning, the language across the curriculum movement, catering for individual differences, school-based curriculum development and course evaluation. However, specialised articles and pamphlets have been prepared on these topics for geography teachers as have books on the classroom use of simulation, quantitative techniques, fieldwork and computer-assisted learning. *The Geography Teacher's Guide to the Classroom* provides insights into many such issues and strategies in one volume.

Each chapter contains a common structure, starting with a scene-setting introduction, and followed by practical 'how-to-do-it' advice on the various topics. Problems that need to be considered and ways of overcoming them are outlined. Frequent examples of classroom ideas, lesson samples, short exercises, transcripts of students talking, etc. are provided. Most often, up to a half of each chapter is 'straight from the chalk-face'. The examples used are not overtly parochial and teachers in many countries should find them relevant. Each chapter also suggests references for further reading on the topic.

The Discipline of Geography: Expanding Perspectives

Few subject areas have mushroomed as much as geography in the last two decades. Geographers of the past would be delighted with the growth in student numbers in secondary and tertiary geography courses. Matching this growth, and in many cases its cause, has been an expansion of the applied side of geography, its usefulness in people's lives and its role in government, corporate policy making, community action and planning. This has brought new goals and perspectives to the operation of the discipline, including the development of spatial theories and models, Marxist analyses of people–environment situations, the promotion of social justice and the search for existential understandings of the human experience of particular environments. These new goals in geography have broadened the range of possible objectives, content and relevance of school geography courses. They have brought an increased concern for the social context of learning, the environmental perceptions and experiences of students in course development and the development of teaching strategies to promote the clarification and analysis of values, effective decision-making skills and participatory citizenship.

These ideas have been introduced to teachers at workshops and conferences and through articles in journals such as *Geography*, *Geographical Education*, *Journal of Geography*, *Teaching Geography* and *Bulletin of Environmental Education*. The books *Signposts for Geography Teaching* (Walford 1981), *Geographical Education: Reflection and Action* (Huckle 1983) and *Teaching Geography for a Better World* (Fien and Gerber 1988) have summarised these developments in volumes that have contained an introductory overview followed by examples of their educational implications. These books provide a guide to worthwhile aims and content for geography teaching in the 1980s. This book should be seen as complementary to them through its focus on the development of teaching approaches through which the aims of geographical education may be achieved.

Applied Curriculum Studies

Curriculum theory and practice has been a major area of research in education over recent years. Formal curriculum studies seek to relate appropriate ideas from philosophy, sociology and psychology to educational processes in order to provide frameworks for the development, implementation and evaluation of teaching programs. Curriculum theory has provided many new ideas and practices for teachers to consider and has enabled them to question many taken for granted assumptions about the nature of knowledge, the arrangement of subjects within a total curriculum, the needs of learners and appropriate methods of organising learning.

Various geography curriculum projects have reflected this problematic nature of much curriculum decision making. For instance, the High School Geography Project brought a renewed emphasis on the conceptual structure of geography as a discrete discipline of study in American schools, while the British Place, Time and Society and Environmental Studies Projects have shown how geography may be integrated with other subjects as a medium for social and environmental education. The Victorian Secondary Geography Education Project was an early guide to strategies for school-based curriculum development. The British Schools' Council's Geography for the Young School Leaver and Geography 14–18 Projects illustrate contrasting approaches to the classroom through their use of the objectives and process models of curriculum development, respectively, while the Geography 16–19 Project has pioneered new approaches to issue-based learning, syllabus development and assessment (Naish et al. 1987).

In effect, these curriculum development projects have fused selected trends in educational theory and the discipline of geography to produce programs more attuned to the needs of students and society than were the rather static regional geography courses of not so very long ago. However, the reforms that such developments have brought to geography classrooms in many parts of the world could easily stagnate as curriculum initiatives and inservice education opportunities slow down in the financially stringent world of education today. The focus on curriculum development and teaching strategies in this book represents an attempt to provide student teachers and experienced ones too with a guide to establishing and maintaining a modern geography classroom.

The Scope of this Book

Knowledge and Teaching Styles

The first edition was considered unusual for books of its type as there was no attempt to provide an explicit analysis of the aims or objectives of geography teaching. Neither was the range of possible content for geography courses surveyed. This second edition recognises the need for such an analysis of the nature of geography and its educational role, especially for new teachers, and so begins with three introductory chapters. The first chapter canvasses the role of geography as a medium for education in the secondary school.

The partner chapters to this introduction, written by David Hall and Leo Bartlett, address the assumptions behind various teaching styles. These authors expose various traditions of education and geography teaching—their philosophical derivation, the aims and knowledge they deem important and related teaching strategies. Hall does this by describing and reflecting on lessons he has seen taught in various schools by student teachers he has supervised. Bartlett, on the other hand, explores the various types of 'geographical' explanations various teachers may choose to use to help students understand a particular problem—in this case: Why did Port Arthur crumble? Geography teaching is as diverse an enterprise as there are geographers and geography teachers. Each of us has our own emphases and, intentionally or not, they place us within one of various traditions that Hall and Bartlett describe. You will recognise yourself in at least one of the vignettes they describe. Once you have done that, it will be relatively easy to select teaching methods that reflect your style of teaching and the aims and content for geography courses you espouse. This will provide general guidance to the variety of chapters of this book into which you should 'dip' for advice and teaching examples. However, do not leave your reading there. Please consider the other traditions Hall and Bartlett have described. Think about them and perhaps experiment with some of the ideas and teaching activities they suggest. In this way, your own experiences may provide new insights into the scope and purposes of geography teaching for you.

Teaching Strategies

By far, the largest portion of this book is devoted to providing practical advice on a variety of teaching approaches and methods. Assumptions about a great many aims of geographical education are explicit in many chapters in this teaching strategies section. All of the strategies see geography as a *medium* for education and not as a body of content or skills to be learnt for its own sake.

Each one has the development of specific competencies and insights that students will be able to use in their possible leisure, citizenship, occupational and conservationist life roles outside the classroom. Thus, there are chapters on inquiry, logical thinking and research skills (Chapters 5–8), on the development of language skills (Chapter 18), valuing processes and decision-making skills (Chapter 11), social and political literacy (Chapter 12), heightened sensory awareness and aesthetic appreciation (Chapter 14), and skills for graphic communication (Chapters 16 and 17) and numeracy (Chapter 22). Other teaching strategies that are considered include: qualitative inquiry which seeks to raise student consciousness and sensitivity to the people and places around them (Chapter 13), teaching through exposition (Chapter 4), the use of simulations and games in the classroom (Chapter 20), practical work (Chapter 9), fieldwork (Chapter 10), and computer-assisted instruction (Chapter 21). This list of teaching strategies represents many that have been used well by geography teachers for many years, some that have sometimes been used relatively unsuccessfully in many classrooms, and others that will be new to some teachers. This blend of the old and the new provides practical advice for geography teaching based upon both many years of teaching experience and the results of recent research on classroom practices.

Catering for Individual Differences in Students

The third section of the book is concerned with the task of catering for specific individual and group learning needs. The section begins with a chapter on strategies for diagnosing the learning strengths and weaknesses of students in geographical understandings and skills (Chapter 23). The consequent need to cater for a wide range of abilities in any class of students is addressed in Chapter 24. This chapter needs to be read in conjunction with the chapter on teaching the less able (Chapter 25). Much debate exists over the question of student streaming and mixed-ability teaching. Arguments also exist about the sociology of knowledge and the devaluation of individuals in catering for low achievers in both streamed and mixed-ability classes. The chapters in this section touch only briefly on these issues. Instead, they focus on providing practical advice on the learning characteristics and needs of students, and the teaching strategies which cater for these amidst the reality of the range of student groupings that different schools and school policies provide.

School-Based Curriculum Development

The theme of the fourth section of this book is provided by the needs of geography teachers in relation to school-based curriculum development. Chapter 27 addresses the broad issues involved in this task. It is linked to chapters which provide detailed suggestions on course implementation through unit and lesson planning strategies (Chapter 26), selecting and effectively using teaching materials (Chapter 29), student assessment (Chapter 28) and course evaluation (Chapter 30).

On Being a Geography Teacher in the 1990s

The final chapter of *The Geography Teacher's Guide to the Classroom* explores the personal and professional traits needed by geography teachers to meet the needs of students in the 1990s and beyond. The authors of this chapter show how a concern for student growth can be met through empathetic teaching involving open-ended learning experiences, simulation, fieldwork and many of the other strategies described in this book.

Conclusion

Each of the chapters of *The Geography Teacher's Guide to the Classroom* has been written by people committed to the potential of geographical education in the lives of individuals and communities. Each has been written to guide student teachers of geography and to be a source of information to which experienced teachers may go for practical advice on specific issues and teaching strategies. Together, the chapters provide several principles for good geography teaching. Specifically, they stress:

1. the importance of various basic modes of communication—mapping and graphics, speaking and reading, and statistics;
2. the importance of student field, classroom and library research;
3. the importance of student thinking and inquiry;
4. the importance of values and understanding social and political processes in resolving geographical problems;
5. the need for a variety of teaching approaches;
6. the importance of diagnosing students' abilities and developing appropriate learning programs;
7. the need for school-based curriculum development;
8. the interrelationship between the evaluation of courses, student achievement and our teaching;
9. the central role in the classroom of the personal interest, commitment and enthusiasm of the teacher.

Consider these principles, add to them as you find others that suit your teaching style, and use them as your *Geography Teacher's Guide to the Classroom*. With them, you will have no need for the advice, 'Don't Panic', found on the front cover of *The Hitchhiker's Guide to the Galaxy*.

References

Adams, D. (1979) *The Hitchhiker's Guide to the Galaxy*, London: Pan Books.
Adams, D. (1980) *The Restaurant at the End of the Universe*, London: Pan Books.
Bailey, P. (1974) *Teaching Geography*, Newton Abbot: David and Charles.
Bale, J., Graves, N. and Walford, R. (eds) (1973) *New*

Perspectives in Geographical Education, Edinburgh: Oliver and Boyd.

Ball, J., Steinbrink, J. and Stoltman, J. (eds) (1971) *The Social Sciences and Geographic Education: A Reader*, New York: John Wiley and Sons.

Biddle, D. (ed.) (1968) *Readings in Geographical Education*, Volume 1, Sydney: Whitcombe and Tombs.

Biddle, D. and Deer, C. (eds) (1973) *Readings in Geographical Education*, Volume 2, Sydney: Whitcombe and Tombs.

Boardman, D. (ed.) (1986) *Handbook for Geography Teachers*, Sheffield: Geographic Association.

Dunlop, S. (1976) *A Guide to Modern Geography Teaching*, London: Heinemann.

Fien, J. and Gerber, R. (eds) (1988) *Teaching Geography for a Better World*, 2nd edition, Edinburgh; Oliver and Boyd.

Graves, N. (ed.) (1972) *New Movements in the Study and Teaching of Geography*, London: Temple Smith. (Also Melbourne: Cheshire.)

Graves, N. (1975) *Geography in Education*, London: Heinemann.

Hall, D. (1976) *Geography and the Geography Teacher*, London: George Allen and Unwin.

Huckle, J. (ed.) (1983) *Geographical Education: Reflection and Action*, Oxford: Oxford University Press.

Long, M. (ed.) (1974) *Handbook for Geography Teachers*, 6th edition, London: Methuen.

Manson, G. and Ridd, M. (eds) (1977) *New Perspectives in Geographic Education: Putting Theory into Practice*, Dubuque: Kendall/Hunt.

Naish, M. et al. (1987) *Geography 16–19: The Contribution of a Curriculum Development Project to 16–19 Education*, London: Longman.

Slater, F. (1982) *Learning Through Geography*, London: Heinemann.

Walford, R. (ed.) (1973) *New Directions in Geography Teaching*, London: Longman.

Walford, R. (ed.) (1981) *Signposts for Geography Teaching*, London: Longman.

1

Geography: A Medium for Education[1]

John Fien, Bernard Cox and Wayne Fossey

There are two purposes to this chapter, both of which are introductory. Firstly, the chapter introduces a view of geography as a way of looking at and studying the world rather than as a body of set information to be memorised. This view of geography, which might be described more accurately as a *geographical perspective*, is based on the key questions that guide geographers in their investigations into the relationships between people and the natural and social environments. These key questions give rise to a particular range of concepts, values and investigation skills which are useful not just to professional geographers but also to every one of us in our daily lives. Gradually acquiring an understanding of these concepts, evaluating the importance of these values and developing competence in these skills are important parts of the education process. The second purpose of this chapter is to introduce these understandings, skills and values as objectives of geographical education in order to illustrate the ways in which geography may be an important medium for the education of young people.

Geography is included in the school curriculum because it can make a valuable contribution to the education of young people. When it first entered the curriculum in the nineteenth century in Britain, Germany and France, it was because a knowledge of other countries and their products was considered an important aspect of the education of future colonial administrators and traders and would thus serve 'the needs of empire'. In addition, in the days before computer mapping and satellite imagery, the development of mapping and sketching skills was important in the training of military officers who were much more dependent on an understanding of terrain than are our present-day, high-technology warriors. Today, the educational purposes of geography are not so utilitarian. Today, geography teachers talk of the broad contributions of geography in terms of the understandings, skills and values that geography promotes in order to help students better explore the environmental, social, economic and political aspects of the world. A focus on these educational purposes of geography is always necessary to ensure that the full potential of geography as *a medium for education* is being realised.

Sadly, however, even in the not too distant past, not all geography teachers kept the wide educational potential of geography in mind. This led one student to write this poem called 'The Geography Demon':

> I hate my geography lesson!
> It's nothing but nonsense and names.
> To torture me so every morning
> I think it the greatest of shames.
>
> The brooks, they flow into rivers,
> And the rivers flow into the sea;
> For my part I hope they enjoy it,
> But what does it matter to me?

Sue Townsend who wrote that famous book, *The Secret Diary of Adrian Mole* (Townsend 1982), must have had a similar sort of geography teacher. On Friday, 3 April, she had Adrian recording in his diary:

Got full marks in the geography test, today. Yes! I am proud to report I got twenty out of twenty! I was also complimented on the neat presentation of my work. There is nothing I don't know about the Norwegian leather industry.
(p. 60)

The following Sunday, his diary tells how he tried to console his friend, Nigel, who had been jilted by Pandora, with some of this new knowledge:

Nigel came round this morning. He is still mad about Pandora. I tried to take his mind off her by talking about the Norwegian leather industry but he couldn't get interested somehow.
(p. 61)

Many of us may be able to tell classroom horror stories like these. However, it is the purpose of this chapter to provide the other side of the picture by outlining some of the purposes that geography may serve as a useful medium for the education of young people.

Well taught, geography involves the education of young people in a number of important ways which might be usefully categorised as education *about*, *in*, and *for* the environment and society in which we all live. These three categories arise from three assumptions about the purposes of education and the most appropriate ways of planning learning experiences for students:

Assumptions
1. Knowledge *about* society and the environment within areal units, ranging through local, regional, national and global scales, and the development of inquiry skills are essential to the development of informed and active citizens, and a further desire to know about people and places, a love of learning, and a commitment to life-long education.

2. Experiences *in* society and in the environment are a major source of understanding in geography. Therefore, students already have a range of geographical knowledge, skills and values simply because of their daily experiences of people and environments. Geography as a medium for education develops these social and environmental understandings and abilities to enable students to fulfil their potentials as human beings and citizens.

3. Education *for* the society and environments in which students live entails knowledge, thinking process, skill and values objectives for geography teaching which will enable students to participate in and seek to improve their society and environment.

   There are many global dimensions to life today. While, students may be members of particular cultural groups and inhabit specific local environments, learning geography expands students' horizons so that they appreciate the network of interactions between societies and environments around the world. This helps them to 'think globally but act locally' in making decisions about everyday life.

Appropriate learning experiences
Many learning experiences may be organised to assist students to acquire knowledge about society and the environment and to develop their inquiry skills. These include learning by:

- Listening to the teacher in expository lessons and from reading printed material.
- Research in libraries.
- Practicals based on maps, photographs, statistical information, graphs and diagrams.
- Programmed learning, simulation game or computer program.
- Problem solving and discovery learning.
- Vicarious learning through listening to or watching audio-visual materials, and analysing artistic expresions such as landscape poetry or paintings.

Geography teachers can help students develop their social and environmental understandings by organising learning experiences in the community outside the school. These include:

- Field work and study camps.
- Community-based learning.
- Personal perceptions of environments and mental mapping.
- Work experience.
- Learing to work constructively with other people in problem-solving groups.

Education for the betterment of society and environments can take place when students engage in learning experiences that involve:

- Analysing the viewpoints and underlying values of the various people and groups involved in a social or environmental controversy.
- Investigations into past and present power relationships at work in any social or environmental situation.
- The development of empathy with people in different social and environmental settings. Sensitive audio-visuals, guest speakers and role plays can be excellent aids in the development of empathy.
- Clarification and justification of one's own views on a controversy or problem.
- Reflection on a problem followed by an appraisal of alternative solutions and decisions about appropriate actions that could be taken. Such action could involve environmental improvement projects, work in community groups or raising money to help finance an aid project.
- Learning through providing service to others.

The Nature and Purpose of Geography *for* Education

A recurring theme in geography is the way people organise and use the environment. Thus geography considers how, in certain situations, people may be constrained by environmental circumstances, for example when Australian grain farmers suffer from recurring cycles of drought, flood and locust plagues, or when the pattern of social and farming life in India is dominated by the monsoons. However, geography is concerned also with the way people have sought to control environmental constraints and with the social and environmental effects of human actions. An example of this is the construction of dams to control flood waters and the consequent effects, good and bad, of expanded agriculture through irrigation, reduced sediment load in the river, decreased siltation at the river mouth, decreased deposition on the flood plain, and a decreased supply of sand for beach deposition. Geography is especially concerned with the contrasting patterns of people–environment relationships, contrasting patterns of human use of the earth, and the different environmental, social, economic and political processes that led to the differences. Such contrasts, patterns and processes give distinctive character to places around the world and provide much of the fascination for geographers and travellers, alike.

In studying the character of places and patterns of human use of the earth, geographers place great importance on explaining the causes and processes involved, the social and environmental consequences of any pattern or event, and making reasoned decisions to improve the quality of society and environments. Thus most geographical investigations take place in four stages:

1. The observation, recording and description of a spatial or environmental pattern, activity, question, issue or problem.
2. Explaining the causes and processes involved in producing the matters under investigation.
3. Exploring and evaluating all the likely social and environmental outcomes and impacts.
4. Making decisions on the best way to conserve and/or improve the lives of people and the quality of the environment after a careful analysis of all possible alternatives.

Thus geography is about:

- the description of places and patterns of human use of the earth;
- the pattern of locations, distributions and other impacts that result from environmental decisions and the natural and social processes that cause them;
- the ways the environment influences human decisions;
- the ways people modify natural and human environments;

Inquiry procedures	Key questions	Guiding concepts
Perception, definition and classification using skills of observation, recording and description	**What and where?** 1. What are the questions, issues and problems being studied? What is their scale, appearance and character? 2. Where are the various elements under investigation located? What patterns of distribution do they reflect/or are they part of? 3. How are they perceived by people from a variety of societal backgrounds?	1. Location 2. Distribution 3. Pattern 4. Landscape 5. Region 6. Spatial association 7. Culture and society 8. Perception
Explanation and prediction using skills of application, analysis and synthesis	**How? Why?** 1. How are the issues, problems and patterns structured? How are they related in natural and/or social systems? 2. What natural and social processes are operating to cause and change locations, patterns and systems? **What impact?** 1. What are the effects of these processes on people and environments? 2. How may these effects be evaluated?	1. People–environment relationships 2. Natural processes 3. Social/economic processes 4. System 5. Energy 6. Flow of goods, people and ideas 7. Change through time 8. Power
Evaluation and decision making using skills of values analysis and problem solving	**How ought?** 1. What criteria may be used to evaluate the appropriateness of locations, patterns and systems? 2. What alternatives should be considered in making decisions about changes to patterns, structures and systems? 3. Who decides and for whom? 4. Who gains and who loses as a result of the decision?	1. Social justice 2. Quality of life 3. Quality of the environment 4. Economic profit 5. Conflict/harmony 6. Planning 7. Decision making

Figure 1.1 A framework of inquiry procedures, key questions and guiding concepts in geography (after Naish et al. 1987)

- how individual and group perceptions, values and actions influence environmental decision making;
- how disagreements about decisions on environmental management may be resolved;
- how environments ought to be managed to attempt to redress social inequalities and ensure future social and environmental well-being for all.

This view of geography is far removed from the older traditions of regional description and 'capes and bays'. The emphasis now is on concepts and principles, on the key questions posed by geographers for gathering and organising data, and on the investigation and decision-making skills they use. It is a geography concerned with the active investigation of issues relevant to the lives of students. It offers them opportunities to explore questions and issues that arise in their society and environment and to seek answers in local and broader contexts.

In both the classroom and in fieldwork, students should be called on to use the skills of geographers as they learn. For example, they can be provided with a range of information, data and evidence relating to a topic, or shown how to obtain it first hand through fieldwork or from secondary sources such as books and magazines, maps, photographs and statistics. Such data can be analysed and evaluated using a variety of critical thinking, statistical and cartographic skills. Students can be required also to consider the values position of all people involved in an issue and to clarify their own attitudes and values in relation to the issue before proceeding to a decision and the presentation of the results of their investigation.

This view of the educational value of geography is based upon the key questions guiding it as a discipline of knowledge. These key questions fit into four categories:

What and where?
1. What are the questions, issues or problems being studied? What is their scale, appearance and character?
2. Where are the various elements under investigation located? What patterns do they reflect or are part of?
3. How are they perceived by people from a variety of societal backgrounds?

How? Why?
1. How are the issues, problems and patterns structured? How are they related in natural and/or social systems?
2. Why do natural and social processes operate to cause and change locations, patterns and systems?

What impact?
1. What are the effects of these processes on people and environments?

How ought?
1. What criteria may be used to evaluate the appropriateness of locations, patterns and systems?
2. What alternatives should be considered in making decisions about changes to patterns, structures and systems?
3. Who decides and for whom?
4. Who gains and who loses as a result of the decision?

These four categories of key questions match the four steps in a geographical investigation previously outlined. Together, the key questions and steps in an investigation provide a conceptual and methodological unity to the study of geography which is illustrated in Figure 1.1.

Aims of Geographical Education

The study of geography can thus serve as a medium for the attainment of a broad range of educational aims and objectives. All the syllabus documents you use will outline the particular aims of the program of studies that you and your students are pursuing. There may be a degree of variety between the aims of different syllabuses depending on the approach to geography taken in particular syllabuses and the age of the students for whom they have been planned. However, there will be a very high degree of commonality between the aims, also. For example, all will suggest that geography can or should help students in the four general areas of knowledge and understanding, thinking skills, investigation skills, and attitudes and values. Very often you will even find common wording and phrases in different lists of aims.

As an exercise, you might care to compare the following two expressions of aims of geographical education. The first comes from the British Department of Education and Science (DES) and lists the ten aims that provide direction for geography teaching for students in Years 5 to 10 of their education. The second list was prepared by the Schools Council Geography 16–19 Project for Years 11–12 geography to assist teachers plan courses for students in the higher year levels. As you compare the two sets of aims, looking for similarities and any differences, you might find it useful to classify each aim according to whether it indicates a knowledge, thinking skill, investigation skill or attitudinal objective.

DES Aims

1. To develop students' understanding of their surroundings and extend their interest in, and knowledge and understanding of other places.

2. To gain a perspective within which they can place local, national and international events.

3. To learn about the variety of physical and human conditions on the earth's surface; the different ways in which people have reacted to, modified and shaped environments; and the influence of environmental conditions (physical and human) on social, political and economic activities.

4. To appreciate more fully the significance in human affairs of the location of places and of the links between places, and develop understanding of the spatial organisation of human activities.

5. To gain understanding of the processes which have produced pattern and variety on the earth's surface and which bring about change.

6. To develop a sensitive awareness of the contrasting opportunities and constraints facing different peoples living in different places under different economic, social, political and physical conditions.

7. To develop an understanding of the nature of multicultural and multi-ethnic societies and a sensitivity to cultural and racial prejudice and injustice.

8. To gain a fuller understanding of some controversial social, economic, political and environmental issues which have a geographical dimension, reflect on their own and other people's attitudes to these issues, and make their own informed judgements.

9. To develop a wide range of skills and competencies that are required for geographical enquiry and are widely applicable in other contexts.

10. To act more effectively in their environment as individuals and as members of society.
(After DES 1986)

Geography 16–19 Project Aims

1. To develop an awareness of the geographer's contribution to understanding and attempting to resolve man–environment questions, issues and problems at different scales, and so an understanding of:
   (a) the key questions and guiding concepts of geography;
   (b) the functioning and characteristics of both natural and human systems and their interrelationships;
   (c) methods of recognising, describing and analysing the spatial consequences of man–environment interrelationships; and
   (d) processes operating to produce spatial patterns and structures.

2. To develop knowledge of some regional and systematic aspects of the geography of selected parts of both the developed and less developed worlds.

3. To develop knowledge of the global implications of some important man–environment issues in the modern world.

4. To develop a degree of competence in practising a range of intellectual, social, communication, practical and study skills, including particularly the ability to use and prepare maps of different types and scales.

5. To develop the ability to use such skills in following through logical steps in geographical enquiry and the clarification of values.

6. To develop an attitude of concern for the quality of the environment, for the condition of human life and for the biosphere as a life support system.

7. To develop the ability to relate to the environment and to sense conditions which either enhance or threaten survival of living things.

8. To develop an approach to learning which will facilitate awareness of the nature and significance of attitudes and values in environmental questions, issues and problems.
(After Naish 1987)

The two sets of aims are remarkably similar, aren't they? What you probably also noticed was that each statement generally involved a combination of various types of educational intent. This is due to the level of generality at which they have been expressed. Figure 1.2 shows that as teachers translate these broad statements of educational aims into particular objectives for their own schools, you will find that they become more specific and that objectives are often then categorised into four categories of: knowledge, investigation skills, critical thinking skills and values objectives. This is in order to make the objectives more useful in the planning of actual units of work and lessons for students, something which the rather broad statements in the two lists cannot do. However, for our intentions of introducing the purposes of geographical education, an understanding of general aims is sufficient at this point.

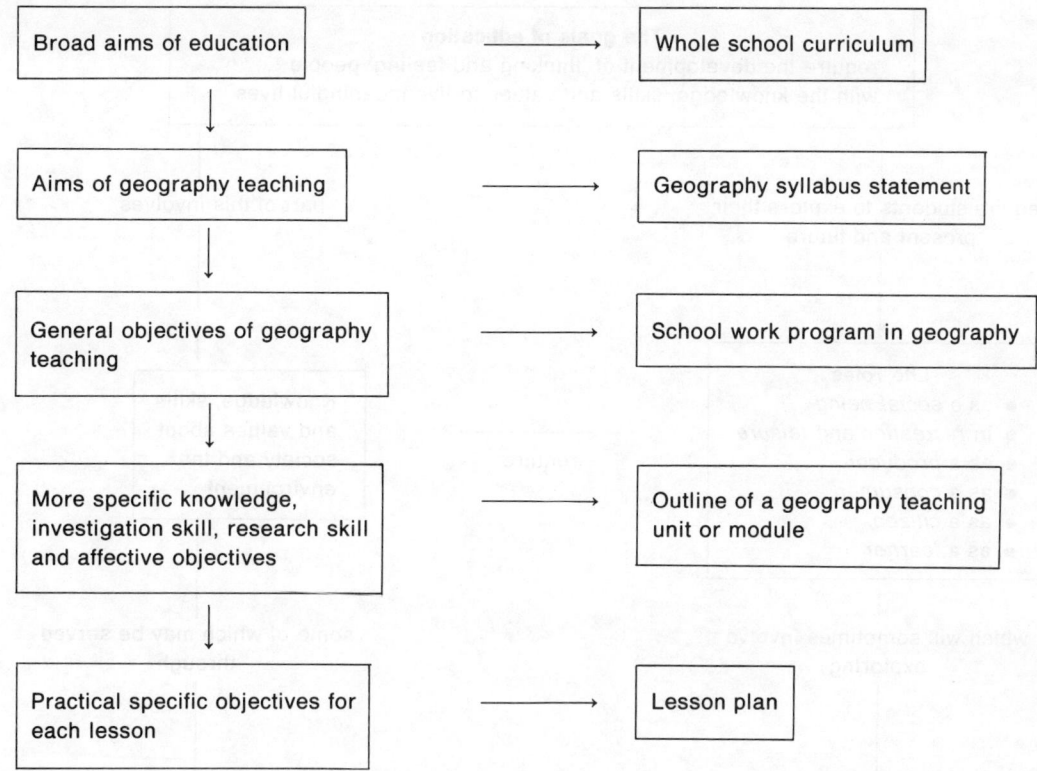

Figure 1.2 The translation of general aims into more specific objectives in the teaching process (after Biddle 1976)

The general aims of geography that you have examined indicate how geography may serve as a medium for the attainment of a broad range of educational goals. Figure 1.3 indicates *how* this may occur. The *goals* of education support the development of thinking, feeling and involve people with the knowledge, values and thinking and investigation skills necessary to live meaningful, independent and socially responsible lives. Any education with this as its goal must help students explore their present and future life roles—and many of the roles we play in life often demand decisions about geographical questions, issues and problems, such as:

Life role	**Geographical relevance**
Social	Where to rent or buy a house in relation to family, work and recreation patterns.
Recreation	Where and when to take a holiday and how to use leisure time for the benefit of self, the community and the environment.
Producer	Where to find work and the costs and benefits of environmental controls on industry.
Consumer	Where to shop, alternative transport patterns, and whether to purchase goods that might be overpackaged or produced by socially or environmentally unjust means.
Citizen	How to encourage others to conserve the environment and analyse the social and environmental implications of political issues.
Learner	How to acquire information, evaluate and interpret information, and to make decisions.

Decisions on questions and issues such as these are based upon the geographical knowledge, values and skills acquired through life's experiences, some of which may be school studies in geography. Such knowledge, values and skills may be considered a *personal geography* (Fien 1983). It should be the aim of school programs in geography to provide learning experiences which extend and refine the personal geographies of students. These experiences which include studies at local, national and global scales enhance all life roles by encouraging and preparing students to become informed, reflective and involved. In this way, they may become more effective in meeting the social and environmental questions,

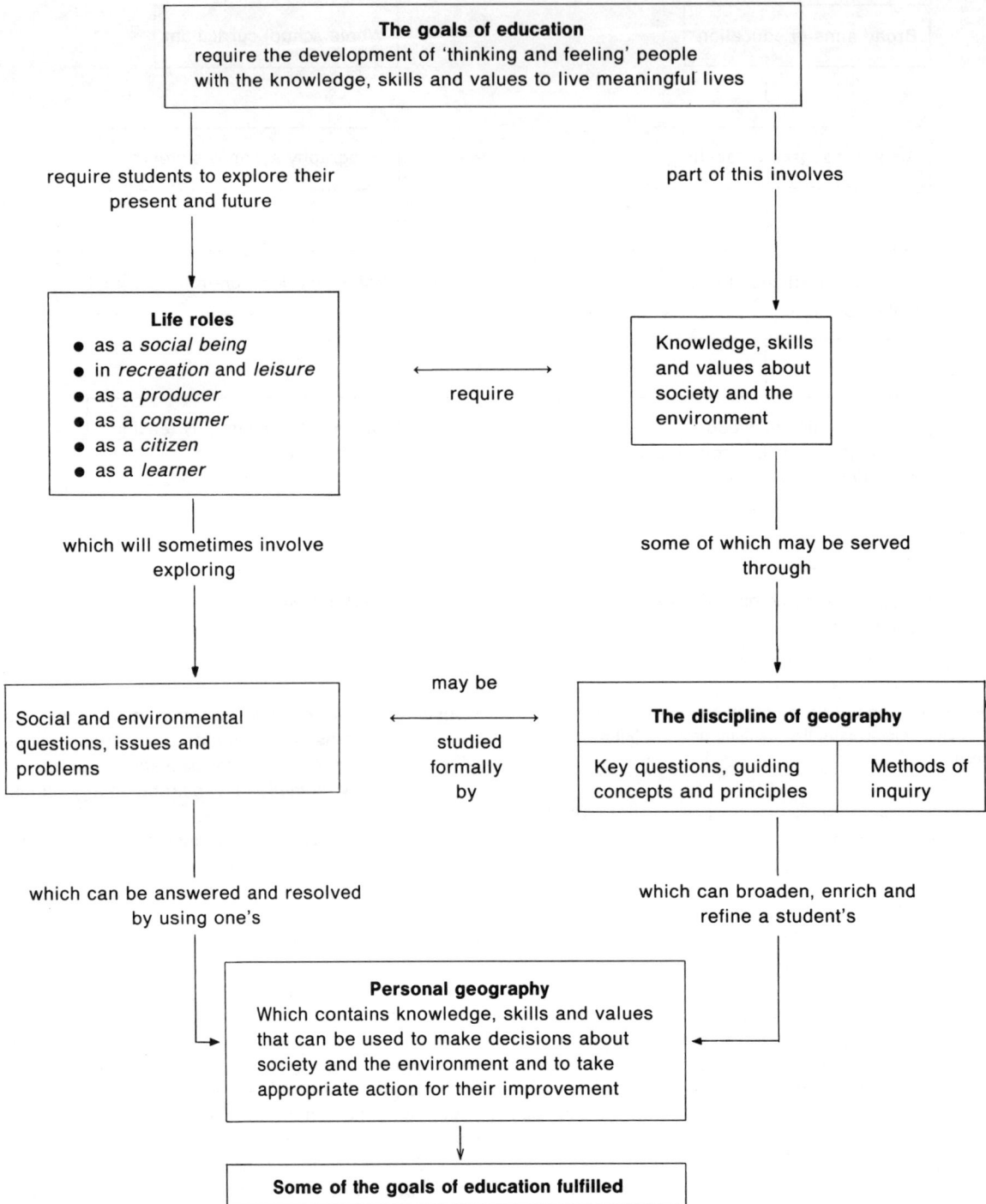

Figure 1.3 Geography's role in fulfilling some of the aims of education

issues and problems encountered in daily life and, thus, more skilled in their life roles.

Conclusion

This overview of the nature of geography and its educational potential has illustrated how learning geography can be a meaningful and enjoyable experience for students. Teachers are the key to success in the geography classroom for it is their role to match the educational and life role needs and interests of students with appropriate aspects of geography. They do this by identifying educational goals that can be served by geography, and selecting appropriate knowledge, skill, thinking and attitudinal objectives supported by appropriate activities and resources for learning.

The best way to be sure that the learning experiences being planned are capable of achieving the full range of objectives inherent in geographical education is to organise the teaching units around the key questions asked by geographers as outlined in Figure 1.1. It is not essential or even possible for students to explore every one of the questions in the middle column of Figure 1.1 in every lesson. However, over the two–four-week period usually devoted to a teaching unit in geography, students should encounter ideas and inquiries related to the four broad steps in the geographical study of a topic. That is, in the study of any topic students should investigate the following four geographical aspects of the topic:

Key question 1: What and where?
The nature of the pattern, activity or event being studied and its location.

Key question 2: How? Why?
The causes and processes that brought it about.

Key question 3: What impact?
The social and environmental consequences of the pattern, activity or event and any changes that are occurring.

Key question 4: How ought?
The range of management strategies available to conserve or improve the situation and thus make the earth a better place in which to live.

No school-level study of any topic in geography is complete unless these four aspects of it have been explored. This is necessary for the full educational potential of geography to be achieved. Indeed, as Figure 1.1 shows, particular geographical concepts, skills and processes are related to the different stages of a geographical inquiry. For example, unless 'How ought?' type questions are explored to help students analyse, evaluate and decide on different management strategies for the topic under investigation, students will not get much practice in developing values analysis, problem-solving or decision-making skills, nor an understanding of concepts such as quality of life, quality of the environment, conflict/harmony and planning. And if this were to happen, the geographical integrity of the work being taught would be doubtful and students deprived of the valuable educational objectives that geography teaching has the potential to provide.

Note

1. Parts of this chapter were first published in an article in *Queensland Geographer*, the journal of the Geography Teachers' Association of Queensland, and are reproduced with permission.

References

Biddle, D.S. (1976) *Translating Curriculum Theory into Practice in Geographical Education: A Systems Approach*, Geographical Education Monograph Series No. 1, Melbourne: Australian Geography Teachers' Association.

DES (1986) *Geography from 5 to 16*, Curriculum Matters 7, London: HMSO.

Fien, J. (1983) 'Humanistic Geography', in J. Huckle (ed.) *Geographical Education: Reflection and Action*, Oxford: Oxford University Press.

Townsend, S. (1982) *The Secret Diary of Adrian Mole Aged 13$\frac{1}{4}$*, London: Methuen.

Naish, M. et al. (1987) *Geography 16–19: The Contribution of a Curriculum Development Project to 16–19 Education*, London: Longman.

2

Knowledge and Teaching Styles in the Geography Classroom

David Hall

This is a chapter about teaching styles and approaches to objectives in geographical education. Geography teaching is as diverse an enterprise as there are geographers and geography teachers. Each of us has our own emphases which are derived (often unconsciously) from a variety of philosophical traditions or approaches to knowledge. These control what is taught and how it is taught in the geography classroom. David Hall identifies three approaches to knowledge—empiricism, rationalism and humanism—and illustrates these with vignettes of classroom activities he has observed. He shows how some of the educationally beneficial features of each may be absorbed into the geography classroom through the adoption of an issues-based people–environment approach to the selection of knowledge which he calls an 'ecological' perspective.

You will recognise yourself in at least one of the vignettes. Use this information to identify the types of objectives, content and teaching methods that match your philosophical orientations and teaching styles.

Today, Monday, wonderfully Spring has arrived! A light frost on the lawn as I left home, the journey to work unexpectedly completed in 20 minutes despite my late departure in the office rush hour under brilliant sunshine and a cloudless ice-blue sky. The clothes on the line should dry well: we are on the fringe of a high-pressure system centred over the Baltic with high winds from the southeast of cP origin. The Avon Gorge reflected whites, browns and greens from its craggy sunlit slopes, with the river at low tide seemingly impotent ever to have etched successfully its antecedent course vertically through those highly resistant strata. The phasing of the lights at the busy Hotwells intersection must have been adjusted at last for me to have filtered into the mainstream of traffic so quickly.

Today, my mind is tense and troubled. Late last night my fifteen-year-old daughter bounced home resplendent with three times the volume of hair that I recollect was in her possession last Friday. She has spent £40 on 'extenders': a new concept for my impoverished vocabulary by which simulated hair material is platted and stitched onto the natural strands in the name of fashion. The money earned by shelf and rack tidying at a boutique on Saturday has been passed back into circulation. No appreciation of how this personal act in support of the Chancellor of the Exchequer might affect her father's job performance! I rang my ex-wife and suggested she might come round early *en route* to her work so that we could review the situation collectively. Many headteachers would surely draw the line the other side of this latest bid for beautification; and with a new assessment system in operation, our daughter can scarcely afford to be gated. My own life is unstable enough without these machinations. That is why I departed later than planned this morning.

Life experience, by such events, confirms *and confronts* our own habits, assumptions and feelings, forcing us in the most powerful way possible to reflect, review and evaluate our views about the kinds of knowledge which are educationally worthwhile, and the ways in which schools in general and geography in particular might be organised effectively to help it.

What are we trying to achieve in geography lessons? Does the opening paragraph summarise it all? Is it (i) to convey some sense of real places by making quite specific factual references at different levels of scale (the Avon Gorge, the Baltic, Hotwells); or (ii) to acquire a broad vocabulary of nouns and adjectives so that a real landscape can be communicated (clear blue cloudless sky, craggy sunlit slopes)? Or do we seek (iii) to transmit deeper meaning, where words function not merely at the surface of description, but at a conceptual level where an 'educated' readership possesses a shared understanding of the meaning of terms such as 'antecedence', 'cP air', and 'resistant strata' which they can handle intellectually through a familiarity with theories of rejuvenated erosion cycles, of air masses and source regions, and the reasoning upon which a classification of rock types is based according to their mode of origin? Equally one might (iv) appreciate the quantitative analysis of traffic flows through periodicity values in converging network patterns, or assimilate readily, by cross-reference to one's own field survey data sets, the manipulation of traffic movement to reduce congestion at places such as the Hotwells intersection.

Here, then, are all of the ingredients of geography: the understanding of the nature of the earth's surface by particular attention to the character of places, the complex nature of people's relationships and interactions with the environment, and the importance of human affairs in the location and spatial organisation of human activities (Department of Education and Science 1986). Yet each specialist geographer and geography teacher handles these ingredients according to a particular viewpoint about the nature of the subject and its function in the school curriculum. For many, the outlook is inherited: a matter of insemination of the knowledge prevalent in one's formative years as a student, or one which has been shaped by the prevailing ethos and expectation of a particular school or education system.

In the succeeding sections of this chapter I wish to consider the three major theories of knowledge (empiricism, rationalism and humanism) and examine three further approaches to geography teaching which are derived from these foundations (the positivist, the realist and the ecological approaches). The listing of subject topics in a work program does not tell us with any certainty about the approach that will be adopted in a particular classroom. This comes from the assumptions about knowledge made by teachers in their everyday work in the classroom. However, unless these assumptions are made explicit and open to criticism and review, the professionalism of a teacher ossifies and one effectively becomes a prisoner of the visions of others (Berlin 1978).

Empiricism: Knowledge as the Transmission of Fact

Until the 1960s, learning in the geography classroom was the memorisation of facts for their own sake. The function of the teacher was to devise methods whereby factual material could be efficiently memorised either by constant reinforcement through memory tests, or by skilful use of visual aids: maps on the blackboard in coloured chalk, 35-mm slides, graphs/diagrams and photographs to enrich verbal descriptions and note taking. As the mind was seen as a bucket, students were the passive recipients of second-hand information. Some buckets might be bigger that others and less prone to leakage (memory failure): the best could not only store large quantities of it but sort and retrieve it on demand in conversation or on quiz questioning.

Even today, this form of teaching is common in many classrooms, and student teachers either willingly or unconsciously adopt it.

Andrew was a hard-working student teacher who prepared his lessons diligently and systematically. For this lesson with a mixed-ability class of 13 year olds he had carefully prepared on an overhead transparency three cross-sections of a glaciated valley: before, during and after the ice age. Each cross-section had three arrows pointing to particular features which remained unlabelled: the class copied each diagram into the exercise book adding a selected descriptor at each arrowhead: valley floor, concave slope, river, glacier, moraine, steep slope, mountain peak, shoulder, scree. For homework three words from a previous lesson were to be designed as cartoon pictures ('pictoons') where the letters could be deformed topologically to depict visually their physical shapes: plateau, gorge, meander.

Andrew's lesson highlights the advantages and problems of learning contents which are energised by facts. The delivery was clear and unambiguous, the annotations either correct or incorrect. There were no discipline problems for the students understood and accepted their passive role. The able could pass time at leisure, and the less able copy out diagrams without much threat of error or confusion. The homework was simplistic, but as original as it was undemanding. The lesson conformed with school norms, in that what was accomplished could be measured in the exercise book, like the payment of a compositor by the measurement of lineage in a newspaper. The geography department too valued the display of good homework, and endorsed the knowledge objectives with its own view that the content of geography is, at root, factual and that to know about glaciated landscapes was an essential component of the special knowledge that geography contributed to the curriculum.

The intensely formal atmosphere of the classroom, with its rows of desks and quiet indifference, was in harmony with the conservative ideology that geography as a subject required both graphicacy and language, but no personal or social engagement with issues of landscape other than to name them.

The most obvious reason for the survival of a factual descriptive view of geographical knowledge is that it makes good commonsense. You are better at geography if you know that Cairns is north of Townsville and that neither are in Western Australia; or that the Suez Canal connects the Mediterranean with the Red Sea and not the Atlantic with the Pacific Ocean. This is the vulgar view of geography, with the geographer seen as the walking atlas.

Of course it is highly desirable for an individual to have some means of placing the brute facts of the distributions of land, air and ocean in their spatial context, just as in knowing the basic verbs in the study of a language, or the names of individuals in a class to be taught. Automatic recall for some immediate purpose is valuable, but it is hardly a sufficient criterion for an educated community.

The empirical view also draws its strength from its close connection with the *empirical* mode of experience. As a theory of knowledge its assumptions are that what we know is the product of received sensations of hearing, taste, touch, smell and especially vision. The image of the mind is that of a tape recorder with blank tapes upon which sensations can be registered. Imprints constitute the core of what we come to know; the mind is characteristically neutral and the facts speak for themselves and ultimately define what we understand.

Historically, it is hardly surprising that geography has remained saturated with empiricist assumptions about its content. The exploration of the unknown world as it exploded in the late fifteenth century was devoted to the collection of such factual knowledge. The need was to know what was out there, without any necessary need to search for principles or cause/effect relationships which might bring pattern, order and regularity to the recorded facts. Theory was weak, and often it was quite misleading, as in the hypothesis that land and sea should be globally distributed in equal proportion, or that the torrid zones were uninhabitable, so that explorers had to overcome the superstitious fantasies encouraged by defective theorising.

The high status accorded to facts registered by the senses leads, however, to the view that events and phenomena are to be regarded as discrete entities. Facts are ultimately real and exist independently of other facts; they are the 'objects' of thinking. Real knowledge is tangible, clear and distinct, consisting of an assemblage of objects which, like billiard balls, are real and independent of each other. Indeed it is a billiard table view of reality, for the patterns we see (on the landscape as on the billiard table) are to be explained by the external structures which are imposed upon them (the specific geology or the rectangular shape of the table) and the external processes or forces (demand for water supply, the billiard cue) which operate through them. This ignores the idea that what constitutes the reality of any object is conditional upon activity and inter-relationship, what Whitehead (1933) called the Laws of Nature as immanent or intrinsic to the character of things themselves rather than Laws of Nature as mechanically imposed from without.

For example, in settlement studies, villages are seen as discrete objects and the reasons for their locations sought in factors unique to that site (water supply, aspect, height, etc.). Empiricism rejects the idea that village locations might be related through a set of general principles and that the character of a village cannot be understood if considered only in relation to its particular site and situation. Such a shortcoming penalises the educational potential of the subject, relegating any treatment of relationships and patterns to little more than an assemblage of mechanical descriptions.

In sum, the emphasis upon empirical description produces at best the illustrated lecture and the transmission style of teaching with the students accepting the vulgar knowledge objectives as central to the subject, and acquiescent to be ranked one against the other on an achievement scale based upon a correct response to questions of detail and location. At worst, it is the *Gradgrind curriculum* so aptly described by Charles Dickens in *Hard Times* where poor Toby chanted the characteristics of the horse as a catechism at the command of the inspector. Learning is then, in Sir Alec Clegg's (1967) apt phrase, 'a burden on the memory rather than a light in the mind'.

Rationalism and Concept-Based Knowledge

Although the empiricist viewpoint was the dominant rationale of geography teaching well into the 1960s, its limitations had long been acknowledged. Developments in the natural sciences throughout the nineteenth century diffused ideas about processes, cause/effect relationships and classifactory methods which helped make sense of discrete data by the use of rules and general explanatory statements. In addition, mathematical geography challenged the empiricist claim that sense perception and inductive procedures were the fundamental routes of knowledge. Consider again the explorers of the sixteenth century. Their success was based upon ideas about the regularity of winds and the use of latitude and longitude to plot position and bearing. Yet, meridians and parallels are not observed through the senses, nor are contours, natural regions or the boundary between urban and rural areas. And what of the conflict between appearance and reality? Our senses tell us that it is the sun which moves whilst the earth remains still! Only by the exercise of reason do we discover that a wave only *appears* to move, and that in *reality* it is only the wave form which is in forward linear movement. These examples illustrate the function of the mind in not just recording external reality, but in actively *constructing* it.

Rationalism is the name given to this view of knowledge. It is a tradition of thought which has been a counterweight to empiricism since Aristotle and has dominated the general cultural outlook of many historical periods. Rationalism holds that knowledge is founded on the innate power of the mind to organise, interpret, relate and anticipate events. Without the mind as a searchlight, external experience is confusing, ephemeral and chaotic. It affirms the primacy of ideas, theories and concepts which make sense of the objective world.

Mike was an able and lively student teacher, with a commitment to make students think for themselves. The group of 16 students settled down, once the late arrivals had straggled in from the upper school study centre, in a brainstorming session to a lively analysis and discussion of 'Unemployment and Industry in Britain'. Mike distributed a spread sheet of information about the per cent unemployed by region, and a graph plotting trends in manufacturing employment in the past decade. Concepts and indices of structural unemployment, footloose industries and spatial concentration of sunrise industries along the Reading–Bristol axis, in particular, were analysed and discussed, and the level of thinking switched from inter-regional imbalance to intra-regional variation and even local imbalances within growth areas investigated. Mismatch features (e.g. York as a growth centre; tin mining in the southwest in a regional growth area) were evaluated.

I think that readers would agree that this lesson, even with an age difference, was pitched at a level which was much more stimulating than the copying and sorting of factual information in Andrew's lesson. In terms of Bloom's taxonomy of mental processes, Mike's class is operating above the level of recall, and mostly at the levels of comprehension and analysis. There is a genuine desire to treat knowledge more actively with an emphasis upon discussion and higher-order thinking. The focus is upon the understanding of concepts through established procedures of analysis and reflection.

The topic is fresh and the ideas contemporary in form, so that it might be said a 'modern geography' is being practised. But the concepts are still being transmitted in that the brainstorming session called back into discussion the concepts that had already been studied formally on a previous occasion. Intellectually, there is little difference between the cognitive behaviours required of students in Mike's lesson and those in the early part of the century which might have discussed Mackinder's theory of the Heartland or a decade later concepts such as sequent occupance and industrial inertia. It is a context in which many teachers flourished at selective schools and at university, and even in schools today it offers emotional and personal rewards where there are niches for it to be practised.

Many syllabuses which are now concept based have this academic and 'high-culture' view of the subject (Lawton 1975). It replicates the 'meritocratic/bourgeois' model of education where subjects are hard, demanding but, with effort and sound intellect, ultimately rewarding. Consequently, teaching styles which promote knowledge, as in initiation into the particular practices and perspectives of an academic discipline, are usually conservative and as equally open to formal lecturing procedures as is the learning of factual content. The big problem with the rationalist approach in education is its stress on intellectualism, with a heavy emphasis on rigour and the elimination of feeling. It was invaluable in an era when nations and colonies required civil servants and administrators who could do their duty in a detached manner, but who should be discouraged from questioning the political structures which legitimated their authority. Thus it still remains attractive to inspectors and many middle class parents who see education as the professionalisation of the young. And, as with Mike's approach, the teaching style can veer away from formal initiation by the analysis of the texts, to a liberal outlook which encourages the progressive refining and reformulation of ideas. But the teaching style is never radical either in its attitude to socioeconomic structures which control personal behaviour and opportunity or to the encouragement of alternatives which might deeply transform them.

For the classroom student the immediate problem with a rationalist approach is not the exclusion of other possible modes of learning, but that heavy stress on intellectualism, often in an overloaded teaching/examination syllabus, causes compression of knowledge in which concepts are quickly reduced to labels, and abstract terms to proper nouns. Without reference to the modes of inquiry which produced the conceptual structures in the first place, rationalism can breed a 'jabberwocky curriculum' (Hall 1980). Learning in such classrooms is then not 'theory soaked', it is suffocated.

Figure 2.1 Positivism at the intersection of empiricism and rationalism

Positivism: Theoretical Knowledge

Having considered teaching approaches drawn from empiricism and rationalism, their combination into a single system of inquiry is the characteristic feature of the 'new geography' which emerged in the 1960s and 1970s. This fusion is achieved through three steps:

1. Postulating a set of principles or assumptions about phenomena, patterns, or relationships.

2. Using analytical reasoning to deduce a sequence of logical consequences to explain it.

3. Searching for, recording and classifying information in accordance with the implications of the principles and the dictates of the exploratory model.

There is an interaction between ideas and factual detail using accepted analytical techniques: thus positivism has the advantage of being at the interface between the rational/analytic mode of experience and the empirical/intuitive mode (Figure 2.1). It is easily transferred into the classroom, and is attractive because it develops certain skills and mental processes such as data collection, sorting and classification, and hypothesis formation and testing. Where empiricism provided factual knowledge which was processed to form generalisations through rationalism, positivism provides knowledge in the form of generalisation, principles, theories and laws which can be applied in many different circumstances.

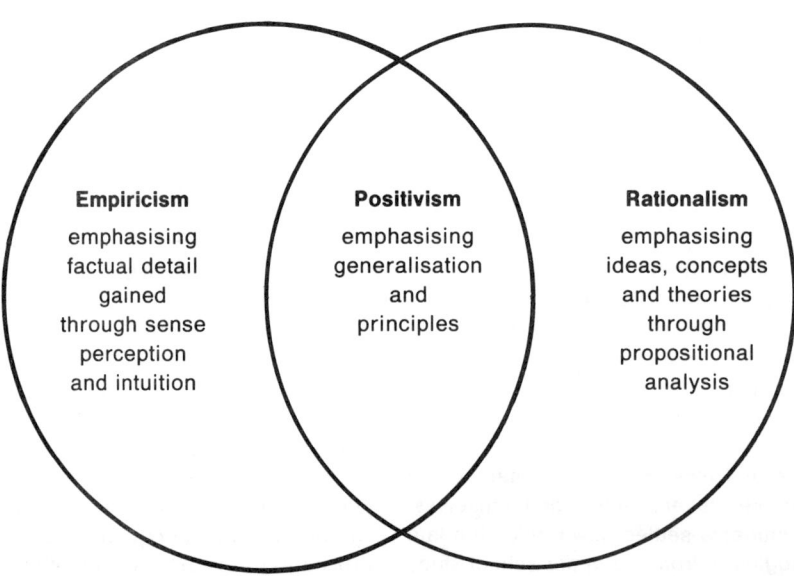

Positivism was strongly represented in many units in the American High School Geography Project. For example, in the 'Geography of Cities' unit, an activity on the growth of Chicago required students to develop models of the city based upon two given principles (the isotropic surface and the continuous outward urban growth from a point of origin). Students then derived census tract data and used the degree of correspondence as a basis for the generation of 'positive' knowledge about the processes of city growth.

Sarah distributed two worksheets to a class of 30 thirteen year olds. On one sheet was printed data of the mean maximum and minimum temperature, monthly rainfall and relative humidity of In Salah. On the second sheet there was an explanation of the atmospheric circulation, and the linkage between rainfall and the two controls of uplift and continentality. The students plotted climographs, calculated the total rainfall for the year, and the months when the average temperature dropped below 20°C. A few slides were shown of deserts, and the students asked, 'What things were needed to grow crops?'. The notes suggested the criterion of at least 6 mm of water per year was needed for plants to survive: some students thought acid rain might help things along. Sarah was visibly annoyed that such a suggestion could be made in a non-industrial region of the globe.

Apart from the lack of a well-defined structure, which inhibited learning, there was another problem with this lesson. Sarah was attempting to analyse data through climographs which seemed as inert to her pupils as the concepts of instability, relative and absolute humidity. These were weakly treated with no linkage established between the data and the physical principles. The high valuation that Sarah placed on knowledge as an assemblage of abstract concepts and second-hand facts did not help her cope flexibly with the ignorance displayed by pupils, but merely created antipathy.

It is always possible to increase the value of positivism by adding a fourth step to those stated earlier: the formulation of changes based on the 'evidence' of the first three steps and working out the way of putting them into practice. In this context, positivistic knowledge can become pragmatic: it can be used as a mechanism to institute or to control change. For example, network theory can be applied to the road networks of Bristol I use on my journey to work and, combined with data from research into traffic flow, analysed in terms of periodicity to generate working models of ways in which the flow problems could be eased.

If this step is taken, liberal attitudes can be adopted either by conscious decision by the teacher in the preparation of his scheme of work and lesson plans or, informally by students, if the topic is handled as an investigation where fieldwork is encouraged and the knowledge is utilised in some way. In positivism, however, conflict about policies or priorities in handling change is not central to the inquiry. Even the liberal assumes that we are seeking to facilitate and to ease the pressure of congestion, by working out ways this can be done by minimising the cost: benefit ratio in cash terms, rather than considering individual human rights and minority rights as central to the treatment of the topic. Therefore, there has been a strong reaction against positivism in the 1980s by a growing band of humanistic educators and geographers. And it is precisely the tightly geared combination of empiricism and rationalism which treats people as aggregated populations, and manipulates them in contexts void of values, feelings and empathy which is the object of the criticism.

For statistics do not bleed. Positivism emerged at a moment in history of economic expansion, the application of technology not only to consumerism but also to political ends, and the ingress of the mass media as mechanisms of social control over thought and action. It became connected with the manipulation in a Kafkaesque world of decision making behind closed doors, of directive styles of management in government, industry and even by extension to schools. And its benefits in making society more efficient at a mechanical level when set in an operative pragmatic mode can also be reduced, as in Sarah's lesson, to the level of a formal lesson of disjointed fact and inert principles, lacking in any sense of relevance or application to the life world.

Humanism: Affective Knowledge

An alternative approach to knowledge has gathered momentum in the 1980s which not only differs in its assumptions from empiricism and rationalism, but also avoids the trap of dissociated learning which so easily occurs when they are combined in a positivist perspective. Humanism has a long history but, in geography today, it has a distinctive outlook by appealing directly to personal knowledge and experience (Buttimer and Seamon 1980; Fien 1983). Knowledge is regarded as the product of assimilated experience based upon the interaction of the inner self with the external world in the process of living. Thus the nature of the child's contribution to the learning process in geography becomes central to its education in the subject: ideas, feelings, attitudes and values are equally significant in the appreciation of issues from a local to a global scale as are the quantities, the descriptions and the concepts used to make sense of them.

Mark had selected Ghana as a case study for discussing the prospects of introducing further technology in agriculture with a very low ability group of 14 year old pupils. They had been discussing the difficulties of getting farmers to accept even such a simple ploughing machine as the 'snail' which is three times as powerful as a pair of oxen and allows the land through early ploughing to grow an extra crop. But to indicate that everyone (not just farmers or blacks) has attitudes which hinder change, Mark asked everyone to fill in a list on a yes/no basis about items they would eat if served for dinner: edible snails, octopus, snake, raw whale blubber, sheep's eyes, horse steaks, monkey, rabbit.

This revealed that no one was in anyway prepared to eat anything, except possibly rabbit, other than the two most 'difficult' boys who had scored 8 and were therefore 'very adventurous'. The whole class then accused them of being economical with the truth and that they would never eat sheep's eyes. 'Bring 'em in, bring 'em in' retorted one in defiant tones. I imagined, at this point, the entry of a waiter in black bow tie and tails holding a tray with two steaming dishes high above his head to settle this argument once and for all.

This example highlights the way in which an appeal to the self binds the life world of the individual to outside issues which is the essence of humanism: the sense that one is not alone and that a study of the real world is not disembodied from the person who experiences it. Other examples of this approach to geography would include role play of interest groups involved in the production, distribution and marketing of world crops such as bananas, of manufacturing and trading where groups have unequal access to technology and energy, of an inquiry into proposals to turn an area of urban wildscape into a large commercial leisure park; lessons requiring pupils to judge the sense of landscape offered by a Lowry painting of industrial Manchester, or asking pupils to lie outstretched on the school field and collect all the events that they can feel, smell, or hear. The humanistic approach aims to activate student's feelings and involvement, and to promote a critical and personal appreciation of the decision-making process whose effects are as visible in the landscape as are the processes of geomorphology. In such a context a study of the rights and responsibilities of individuals and of social groups are as significant as the understanding of skills, concepts and of facts. David Hicks has summarised this as thinking of tasks for our pupils in terms of the three Ps: the *personal*, the *political*, and the *planetary* (Hicks and Fisher 1985).

With its respect for the interlocking web of human perceptions of reality, and conflicts of interest, humanism is rooted in the sense of the personal and not in the empirical or rational traditions which regard subjective feeling as a probable source of error. When Scott said of the South Pole, 'This is an awful place', he was not making an empirical statement but he was talking geography. If such feelings are denied legitimacy in the geography classroom, our sense of place must be impoverished. It would admit into the classroom the study of the landscape of Sugar Loaf Mountain, using an oblique air photograph, but deny the legitimacy of one taken of the centre of Hiroshima in August 1945.

The growth of humanism will surely continue for two major reasons. The first is the convincing argument advanced, for example, by Seymour Papert (1980) when, in discussing the significance of LOGO for children, he emphasises the crucial importance of syntonic learning—the linkage of the body in its movements, intentions, goals and desires to the powerful ideas such as a mathematical variable, differential calculus and recursive procedures in programmable statements as a key for entry into 'mathland'. The second is the accelerating pace of change in many people's lives, not just in terms of journey to work, landscape modification, and housing and eating habits, but in terms of job security, mobility and even in family life and norms. Humanism provides the means for the children of every ability to enter the world of powerful ideas in geography by preventing their presentation in an inert and formalised structure and supporting children as they build up their own structures with materials drawn from the surrounding culture. It is on this basis that the second paragraph in the introduction to this chapter is as much a part of it as the first, and it may be the one that presses on us with persistent ferocity. This is the real personal world in which the child lives equally with adults on the journey through adolescence. What proportion of the readership will remember it long after the first has been forgotten? Can geography help us with an understanding of personal experiences by reaching out and sharing inner thoughts and feelings of others? Does an understanding, be it regional, ecological or spatial, of the world 'out there' disembodied from our own immediate concerns help us reach outwards from the cramping parochialism of subjective worlds, or is there some way in which geographical knowledge can effectively integrate them?

'I am a teacher more than a geographer, and a person more than a teacher' writes Robin Richardson (1983) in daring us to be a teacher. Humanism argues that it is possible to construct a course in geography in such a way that pupils are offered insights into the human condition, both in personal and in social relationships, and from this dignified foundation come also to know and care for the necessary yet fragile relationships of life on earth at local, national and global levels of scale and perhaps see 'geography land' as one both of personal delight and, through active personal concern and responsible social effort, of becoming a better world (Fien and Gerber 1988).

In this sense, humanism does not lend itself to a conservative outlook or a didactic style of teaching. It cannot be concerned with grading of ability by norm referencing or seeing the class as a basic teaching unit. Because its direction of growth is towards the future, rather than a knowledge of the present for its own sake, it will emphasise movement and growth, social and personal criticism and development and a caring and compassionate as well as a competent society. Individualism is valued, but only within limits of the total well-being of the planet earth. It is therefore attractive to teachers whose ideology is concerned with being socially critical of existing conditions in the appraisal of landscape, rather than leaving more general issues of the relationship between such landscapes and the structure of society outside their scheme of work.

Of course, it is possible to adopt a humanistic approach without undertaking a radical critique, of expecting the 'invisible hand' of the market place and floating currency exchange rates to function in a liberal and progressive sense. But it is also likely that teachers who think of justice more as equality of opportunity and the reward of merit will gravitate towards an updated version of 'real geography' at the intersection of empiricism and humanism: realism.

Realism

Realism has long been popular in geography and social studies courses. It has empirical roots and emphasises the lives of people, places and events for their own sake without much concern with concepts, principles or political issues or decisions taking. Popular culture values descriptions of people and places under condition of tension or stress: the 'blood and thunder' topics of earthquakes, volcanoes and typhoons; eye witness accounts of daring scientists collecting lava samples from inside a volcano or of a first mate on a cargo vessel anchored offshore at the moment of the Mont Pelée eruption. Topics less turbulent than thunder, ash and asphyxiating gas would be the old 'curiosity and wonder' descriptions ranging from 'redskins and rattlesnakes' in the Mohave Desert to cheese and clogs in the Dutch polders.

However, such an approach can excite a class and be highly motivating:

Barbara was quite excited. A lot of hard work had gone into preparing the India day, and the regular timetable had been suspended so that the whole of Year 9 could take part. Five events had been planned, with each class of students visiting each one on the merry-go-round principle. In one classroom, students drew roles by lot as families (landless peasant, rich peasant, etc.) in an Indian village being visited by a government official seeking to withdraw part of the land from cultivation to develop a small textile factory. In another, students were examining different forms of dress. A third event was in the gymnasium with an expert in yoga with everyone taking part. The fourth was an excerpt from an Indian film on Calcutta. Finally, the home economics department had loaned rooms and equipment for an expert on Indian cooking to demonstrate the preparation of staple dishes, with each student being given a range of food and drink to eat at the end of the demonstration.

In this case, the link with humanism was very strong and included many different modes of learning experience. It is an example of realism cast in a modern rather than conventional form, and follow-up work included the investigation of social issues at a deeper and more critical level. Certainly, it is a progressive and liberal outlook, encouraging an educational aim generally proposed for schools: 'to help pupils understand the world in which they live, and the interdependence of individuals, groups, and nations' (Department of Education and Science 1985.

It may be easy to dismiss this type of knowledge as 'vulgar' (see Lawton 1975) because it is readily associated with tabloid newspapers and cheap television advertising. To overemphasise this type of knowledge, or to see it as the core of geography would be to exploit the ephemeral at the neglect of more fundamental issues, when education has a responsibility to raise student awareness above the level of bread and circuses. At its best, however, it is a link with the emotions: the excitement, curiosity and risk which drove the explorers forward, or the tragic account which calls for the powers of empathy with human tragedy. But the penalty of taking this approach too far is the subordination of the intellect to emotion and the opportunity for fantasy to infect a world ruled by objective ideologies. Perhaps it was not only for lack of data that the cartographers of the Age of Discovery added demons and monsters to the vast areas of the map beyond the boundaries of *terra cognita*!

The Ecological Perspective

In Figure 2.2 a Venn diagram represents the merging of the approaches to knowledge examined. A focus on the area of overlap of the three circles provides an interactive approach to knowledge in geographical education which might be termed an ecological view.

Consider Marion's class, for example:

Marion had joined a team of six teachers responsible for a program common to five Year 9 classes studying geography on a block timetable. At this point, the students were examining the patterns of shopping in the local town. Already, teams of students had interviewed shoppers at different precincts using a standard questionnaire, and the pool of information was now available as a data base to test a series of hypotheses about shopping habits, range of goods, hierarchies in retailing, minimum population requirements, and the mismatch between perception and reality. Each person had chosen or suggested an hypothesis to be tested, and agreed to a 'contract'

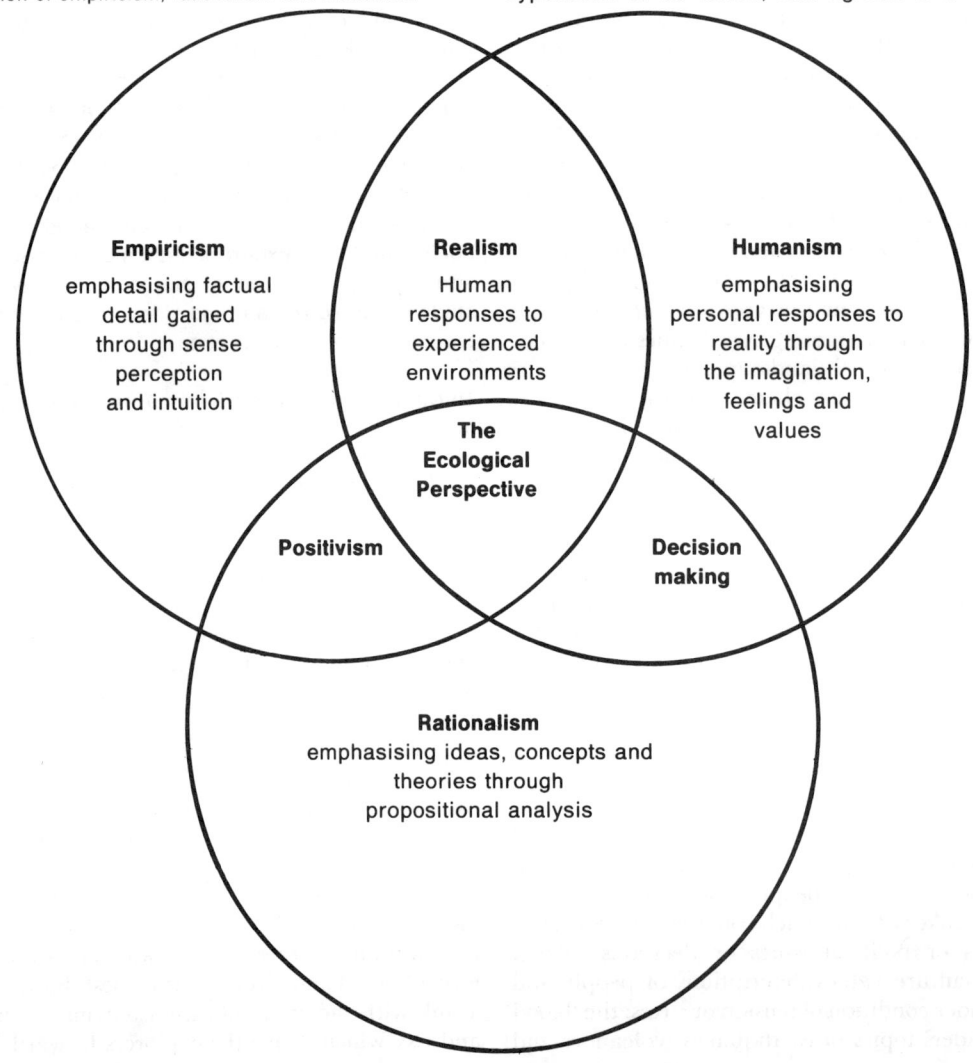

Figure 2.2 The ecological approach to knowledge at the intersection of empiricism, rationalism and humanism

drawn up with a teacher to indicate the scope and the method of inquiry.

The teachers had drawn up a series of skill cards which were available on open access for self-instruction on a given technique (i.e. how to draw proportional circles, to calculate median or quartiles, to correlate graphically two variables, etc.). After three years at the school, the students were familiar with these methods, and for Marion it seemed that teaching was a matter of making flexible responses to diverse circumstances: suggesting ideas, demonstrating a skill, and criticising inaccurate or poorly done work. Her students were extremely interested in this work, for the shopping patterns of the town were rich and varied. It had a large retired population, a high seasonal influx of tourists, and an industrial sector based upon high-technology products. Seasonal rhythms, the growth of cash-and-carry wholesale outlets and the effects of hyper-markets and the problems they posed for the elderly and those on small pensions were just some of the issues being investigated.

In this approach to teaching, knowledge is not seen as information, nor as the acquisition of abstract concepts which are part of an elitist subculture, but as an activity in which these elements combine with feeling and imagination to conduct an inquiry into an issue which lies across the human and physical domains of geography. It is an approach which employs a dynamic and flexible process framework to meet the dilemmas with coping in a world where events and issues are in ceaseless change, where there is an explosion in information, and where the present is exposed to continuous qualification, adjustment and repair.

As we have seen in the discussion on humanism, teachers are educators of children: their function is cultural. The criteria for judging what is worthwhile knowledge in schools has changed from that formally defined by an elitist academic subculture, based on research in higher education usually in terms of established subjects, to a knowledge of methodologies universally desirable for an open, fluid, dynamic and technologically based society. It is a change from 'knowing that' to 'knowing how', from a concern with the cognitive repertoires of scholars to a practical concern with people's competencies manifest in ordinary life (Ryle 1949). This implies a move away from a curriculum underpinned by a 'collection code' of objectives to an 'integrated code' where what is relevant is judged against the aims and objectives of the school as an educational unit in a given culture (Bernstein 1975).

Marion's methodology contrasts with the conventional positivistic blueprint for inquiry which reduces scientific investigation to a sterile routine adopted in many classrooms and laboratories. It is an 'open' approach to knowledge as advocated by Karl Popper (1959, 1971) and offers wonder, romance and excitement in exploration as well as rigour in the collection and processing of data. Thus humanism is an endemic perspective in Popper's formulation, as the person constantly refines observations (empiricism) and interpretations (rationalism) against broad social and person-centred values (humanism), thus creating an open, self-critical and evolving system of knowledge. Because of this broader remit, content will sustain cross-curricular issues including a sustained attention to multicultural aspects of geography (as with Barbara's event) and the prevention of stereotyped images or bias in looking at human activities particularly with regard to gender and race. Open inquiry also encourages economic awareness and the placing of political events and decisions in the understanding of landscape. The ecological approach to knowledge makes contemporary people–environment issues the focus of course construction. In this way, particular phenomena or events (such as a landslip, a flood or the production and marketing of a commodity) can be from a widened systems analysis into a dynamic inquiry which explores the impact of observed instabilities or proposed changes on landscapes and habitats.

Consider, for example, the classroom study of stream flow and discharge. A purely positivist treatment would consider the ratio of the wetted perimeter to the area cross-section and register the variations in hydraulic radius in different hypothetical situations. The next step would be to consider discharge fluctuations and the effect of these on traction load and on river banks when the 'overbankful' stage occurs. By the collection of empirical data in real streams, the rational model can be geared to the real world, and predictions made about the behaviour of these real streams under varying conditions of discharge.

However, important people–environment issues are excluded from this study. Add a human component, such as stream dredging, into the lessons and the knowledge emphasis is transformed. Values and decision making are now involved because a feedback loop has been introduced into the system. Gravel removal alters the geometry of the cross-section, and has a cascading effect beyond the original theme. The changed situation will require some 'valves' to prevent runway feedback occurring, the restoration of balance by the engineering of a negative feedback loop, or by a refusal to allow excavation to take place at all. Values and decision

Themes	Issues*
Ecosystem management	Local area microclimates* Wildlife management in East Africa Forest management in New Zealand
Coping with natural hazards	Flooding in the local area The effect of blizzards on isolated farmers Tidal surge dangers in London
Mineral development	Gravel dredging in a local stream Sand mining on Moreton Island, Australia The effect of mining operations on the indigenous people of Bougainville and Mappoon
Rural management	Soil erosion in the 'Dust Bowl' Herbicides and human health The Green Revolution
Migrations of people	Refugee camps in Africa and Asia 'Guest workers' in Asia and Europe Retirement resorts for the wealthy
Changing urban environments	The preservation of local heritage districts and buildings The redevelopment of Dockland, London Squatter settlements in Amsterdam and Hong Kong

*Select local issues as often as possible to illustrate the relevance of the general principles, to activate fully the feelings and values of students, and to help develop a concern for local community affairs

Figure 2.3 Sample themes and issues for study in geography according to the ecological view of knowledge (after Naish et al. 1987)

making are involved now. The issues of preservation (no dredging), conservation (controlled dredging) or exploitation (uncontrolled dredging) force a consideration of values and ideologies. The morality of change, a perspective of humanism, has been admitted into the knowledge component of the study without any loss of traditional academic concerns.

Therefore methodology and knowledge acquisition are interlocked like two faces of a coin. Syllabus planning is pragmatic in its selection of issues, and through the avoidance of formally separated study into 'physical', 'human', 'regional', or 'systematic' units responds constructively to the need for structural and conceptual unity in a geography course. It is a perspective adopted by the British Schools Council Geography 16–19 Project in the development of courses and teaching materials with its stress on the people–environment approach (Naish et al. 1987). The project suggests a list of over 20 people–environment themes for study, all of which give rise to contemporary issues of importance. Some of the themes it identifies and some appropriate issues for case study are outlined in Figure 2.3. Such themes, issues and case studies may be handled by students from the beginning of their studies if they are presented to them at the appropriate level.

Conclusion

This chapter began with a personal sketch of a journey to work. The writer was sustained by the stimulus of the physical details of landscape which endure during a human lifetime, and yet whose recurrence is always subtly different from day to day in light, tone and resonance. There is reassurance to be gained from the elemental rhythms of day and of season, and pleasure from the evocation of concepts and of intellectual speculations which come to mind bringing further order and sensitivity to the rites of passage. Conscience is also disturbed by the manifest differences in social opportunity and advantage reflected uneasily in the transect through the changing landscape of an urban city complex. Our geography is therefore as much an essay of personal awareness as a transmitted experience learnt in the artificial confines of a classroom. Yet even within it, the way in which learning experiences are presented are of the greatest significance. As teachers, our views of knowledge determine our styles of teaching, and they need to respond to the emerging cultural outlooks of the late twentieth century. Rex Walford (1985) recalls the graffiti he observed recorded on a bridge near Bedworth in Warwickshire: 'Geography is everywhere'. It is vital that we treat the knowledge that it offers in a way that pupils' thoughts and actions are engaged and which appeals to curiosity, conscience and intellectual competence.

References

Berlin, I. (1978) in B. Magee (ed.) *Men of Ideas*, London: BBC.

Bernstein, B. (1975) *Class, Codes and Control*, Volume 3, London: Routledge and Kegan Paul.

Buttimer, A. and Seamon, D. (1980) *The Human Experience of Space and Place*, London: Croom Helm.

Clegg, A. (1967) in J.S. Maclure (ed.) *Curriculum Innovation in Practice*, London: HMSO.

Department of Education and Science (1985) *The Curriculum from 5 to 16*, London: HMSO.

Department of Education and Science (1986) *Geography from 5 to 16*, London: HMSO.

Fien, J. (1983) 'Humanistic Geography' in J. Huckle (ed.) *Geographical Education: Reflection and Action*, Oxford: Oxford University Press.

Fien, J. and Gerber, R. (eds) (1988) *Teaching Geography for a Better World*, 2nd edition, Edinburgh: Oliver and Boyd.

Hall, D.B. (1980) 'Changing Outlooks in Geography', in P. Weigand and K. Orrel (eds) *New Leads in Geographical Education*, Sheffield: Geographical Associaton.

Hicks, D. and Fisher, S. (1985) *World Studies 8–13 Handbook*, London: Oliver and Boyd.

Huckle, J. (ed.) (1983) *Geographical Education: Reflection and Action*, Oxford: Oxford University Press.

Lawton, D. (1975) *Class, Culture and the Curriculum*, London: Routledge and Kegan Paul.

Naish, M. et al. (1987) *Geography 16–19: The Contribution of a Curriculum Development Project to 16–19 Education*, London: Longman.

Papert, S. (1980) *Mindstorms*, New York: Basic Books.

Popper, K.R. (1959) *The Logic of Scientific Discovery*, London: Routledge and Kegan Paul.

Popper, K.R. (1971) *Objective Knowledge*, Oxford: Oxford University Press.

Richardson, R. (1983) 'Daring to be a Teacher', in J. Huckle (ed.) *Geographical Education: Reflection and Action*, Oxford: Oxford University Press.

Ryle, G. (1949) *The Concept of Mind*, London: Hutchinson.

Walford, R. (1985) in R. King (ed.) *Geographical Futures*, Sheffield: Geographical Association.

Whitehead, A.N. (1933) *Adventures of Ideas*, Cambridge: Cambridge University Press.

3

Critical Inquiry: The Emerging Perspective in Geography Teaching

V. Leo Bartlett

No teaching is value-free. As David Hall illustrated in Chapter 2, teachers select an approach to objectives, content and teaching style that matches their particular philosophy of life and education. In this chapter, Leo Bartlett intensifies the points made in Chapter 2 by showing that the philosophical stances we choose have an ideological base.

There are three ideological views of geography teaching—the classical or conservative view, the liberal or interpretive view and the socially critical view. The attributes of these three views are illustrated and evaluated through an examination of how each one would approach the teaching of the inquiry, 'Why did Port Arthur crumble?'.

The socially critical view is shown to be the one that gives greatest scope to the inquiry-, issues- and values-orientated approaches of the ecological perspective.

Many teachers of geography have found that inquiry teaching is often both challenging and frustrating. What they may not be sufficiently sensitive to, perhaps, is how an inquiry is influenced by their perspective on teaching (their *conduct* knowledge) and their perspective on knowledge which relates to what they knowingly or unknowingly intend their students to learn (the *content* knowledge of the inquiry). Their perspective (or frame of reference) on both teaching and learning results in a view or orientation to inquiry.

In this chapter, we shall see how both conduct and content forms of knowledge are important in inquiry teaching in geography. This will be done in three ways. First, we shall discuss a view of inquiry called *critical inquiry*. This is the principal theme of the chapter. The aim will be to see how the degree of autonomy we have over our inquiry teaching is determined by the level of control we have over the two forms of knowledge upon which we base our teaching and our method of inquiry. Furthermore, we want to suggest that taking control of this knowledge and our teaching actions is the basis for what we may call a critical perspective in inquiry.

While a critical perspective with its underlying view on inquiry in geography is the principal interest of this chapter, we want to show that there are other perspectives representing different views of knowledge. Hence, the second purpose of the chapter is to show that there is not one but three views of inquiry which we can tentatively name the classical, the interpretive and the critical. Third, we want to see how these three views are different in kind, that is, how they produce different kinds of knowledge content through the exercise of different forms of craft (conduct) knowledge. We want to see that when we teach, our view of inquiry will lead to specific kinds of questions which in turn will result in a specific knowledge product; and that our teaching may encompass these views in a complementary way.

Critical Inquiry: The Emerging Perspective in Geography Teaching 23

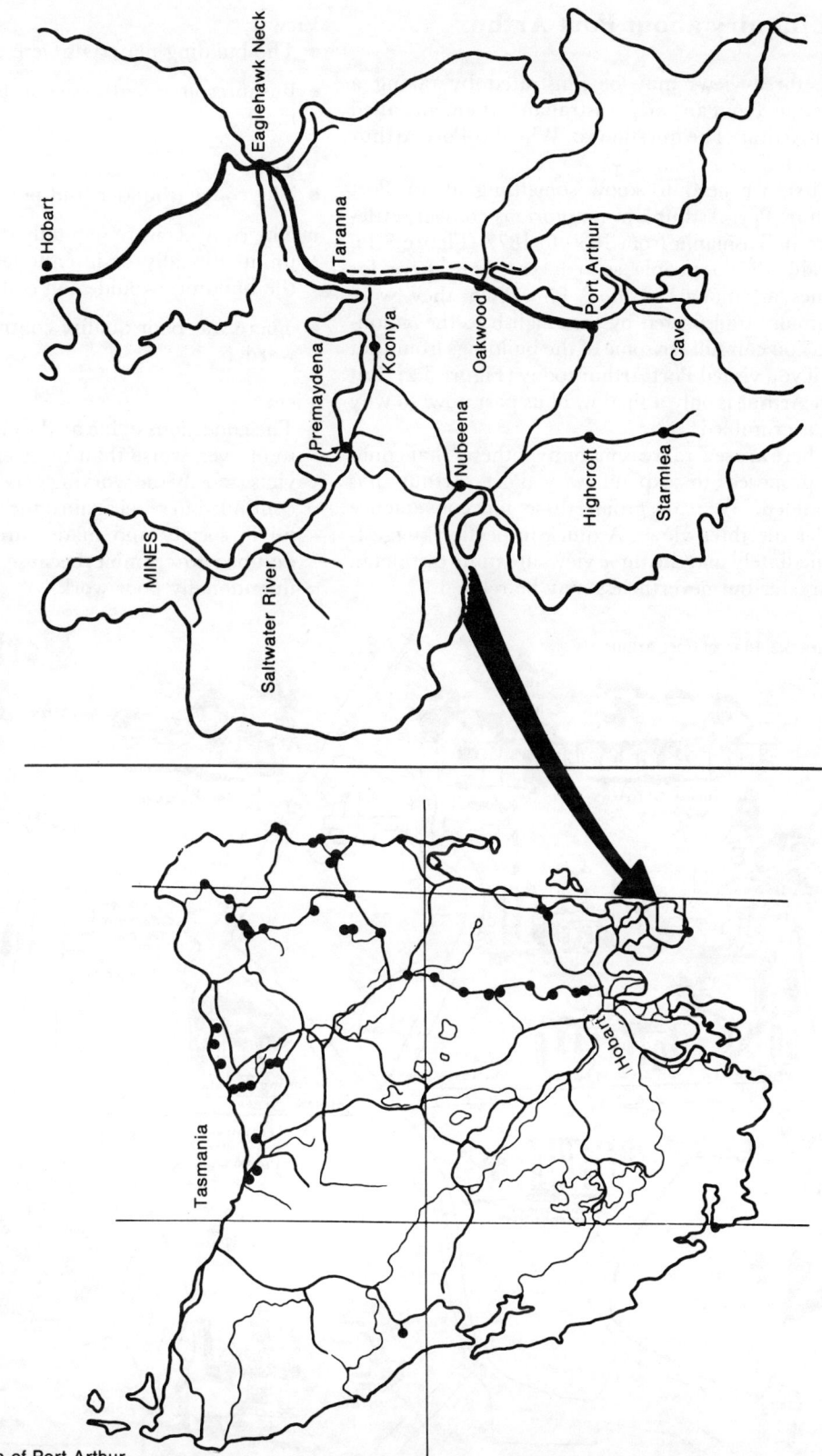

Figure 3.1 The location of Port Arthur

An Inquiry about Port Arthur

The three views may be illustrated by taking a question about an early Australian settlement called Port Arthur. The question is: 'Why did Port Arthur crumble?'.

First, we need to know something about Port Arthur. Port Arthur was a notorious convict settlement in Tasmania from 1830 to 1877 (Figure 3.1). In all, 12 500 people served sentences there for crimes additional to those for which they were originally transported by the English to the 'colonies'. You can still see some of the buildings from that era if you visited Port Arthur today (Figure 3.2). But Port Arthur is only a shadow of its past now: so why has it crumbled?

There are several reasons or hypotheses that could be proposed to explain why Port Arthur has crumbled. We can group these for convenience under the three views. A quick inspection suggests immediately that the three views are quite distinct in character but nevertheless may be related.

View 1:
- The building materials were of poor quality.
- Bushfires frequently swept through the settlement.

View 2:
- The convict builders did not know their job.
- The convicts were mentally deranged and either unintentionally or intentionally worked so that the buildings would soon collapse.
- There was poor quality control by unscrupulous warders.

View 3:
- The conditions of life at Port Arthur were similar to or even worse than those experienced by convicts (mostly the working class) in England. They continued to rebel against the few rich and powerful in society and made sure Port Arthur was destined to crumble because of their shoddy or intentionally poor work.

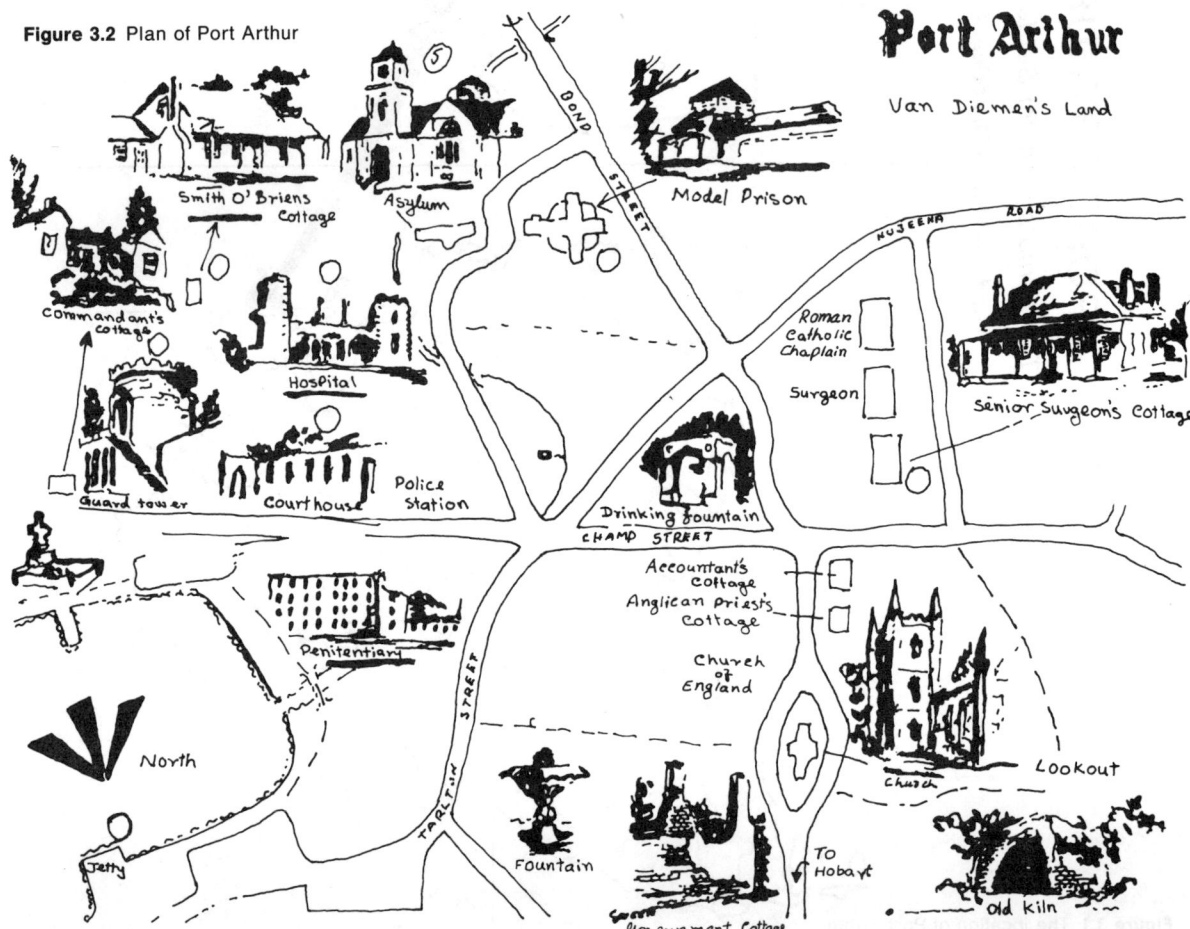

Figure 3.2 Plan of Port Arthur

These various reasons represent the 'knowledge product' of three quite different forms of inquiry. All the reasons and the views (or perspectives) within which they are lodged are reasonable explanations; but which view with its reasons provides the 'best' explanation? To answer this question we have to search for valid evidence, most of it secondary evidence from indirect observation, to begin to understand and finally explain why Port Arthur crumbled.

Here is an account about Port Arthur:

The convicts of the penal settlement of Port Arthur built their own prison, even down to making the bricks themselves. Indeed, brick-making was a major activity. They were used in large numbers and sent to Hobart and beyond. Port Arthur is now in ruins due in part to the poor quality of the bricks. The evidence shows that the bricks were underfired and that some, at least, were 'puddled' with seawater. ('Puddling' was the action of mixing the clay used to make the bricks with seawater.) After the bricks were cemented into place, rainwater entered the bricks and carried the soluble seawater salts through them. Evaporation left the salts behind at exposed surfaces. As the salts concentrated, the bricks disintegrated with the relentless work of rain, wind and sun.

So there is good evidence to suggest that physical reasons are the cause for Port Arthur's crumbling. This evidence also shows *that* the buildings have indeed crumbled. It seems to suggest that view 1 is correct; and it is. But it is an interpretation based on a view of Port Arthur as an assemblage of physical objects rather than an analysis of how the convicts who constructed the decaying buildings many years ago interacted with their environment. It is interesting to note that the physical interpretation was written by a soil scientist.

We need to demonstrate that Port Arthur has crumbled when we teach about the geography of place. If it has not physically deteriorated, our attempts to find reasons (that are related to human agents) for its crumbling seem quite a waste of time. But somehow we have forgotten or omitted the fact that people built the settlement at the 'bidding' of their masters; and that people, such as the convicts, had motivations, feelings, values and intentions that need to be taken into account. Exactly why did they 'puddle' and underfire the bricks? This brings us to view 2 in the search for 'good' or 'best' explanations.

View 2 includes people in its interpretation—the convicts and the warders. It suggests that they were motivated to puddle the bricks. If only we could get inside the minds of these people, we might understand better why Port Arthur is in ruins! The best we can do is to look at the evidence about life in old Port Arthur. Here are some sources.

1. An original document:
. . . they (the bricks brought from Port Arthur) are of such a wretched and worthless description that I cannot consent to their being used for any building purpose.
(Henry Hunter, builder of a Hobart gaol in 1876)

2. A history book:
O'Hara Booth was the greatest commandant Port Arthur saw during its operation. For eleven years he held a rigid grip on the reins of control.
(B. Smith, an historian, 1947)

3. Text from a novel:
So Rufus Dawes was relegated to his old life again and came back to his prison with the hatred of his kind that his prison had bred in him increased a hundred-fold. He was unable at first to apprehend the details of his misery . . . all hope of justice and mercy had gone from him forever.
(Marcus Clarke, a novelist, 1874)

4. Verses from a poem:

>   I was the conscript
>   Sent to hell
>   To make the desert
>   The luring well.
>
>   I split the rock
>   I bent the tree
>   The nation was—
>   Because of me.
>
>   Shame on the mouth
>   That would deny
>   The knotted hands
>   That set us high
>   (Mary Gilmore, a poet)

5. Text extract from a newspaper:
Mr Wiltshire said there were many Australians who shunned their convict past. The [Heritage] Commission found it difficult to convince people at Port Arthur, Tasmania, of the need to preserve convict ruins there. 'Some people still feel a bit ashamed', he said.
(*Courier Mail*, 8 February, 1982)

There are many other sources such as these which support view-2-style explanations. All have one thing in common—they emphasise the understandings of individuals: understandings from which we can 'filter' the meanings of and answers to the question, 'Why did Port Arthur crumble?'. View 2 helps us search for the 'truth' through our individual human understandings of places and events, themselves. This is the view best represented in Chapter 13, on qualitative inquiry. It is also known by the terms interpretive, humanistic and person-centered inquiry.

But there is other evidence about Port Arthur that we have not yet considered. This evidence includes the things we have heard, read about or even seen, all of which adds to our experience and understandings of the place. Let's look at the society from which the convicts at Port Arthur came.

Robert Hughes (1987) has described the early days of white Australia as 'the fatal shores' because English lawmakers, usually the rich, used the new colony of Australia as a dumping ground to get rid of their 'criminal class', usually the poor. Hughes says there were four main periods of convict transportation during which 160 000 men and women were sent in bondage to Australia.

Phase	Events
1787–1810	About 9300 men and 2500 women were cleared from English hulks and jails: about 7% of the total number of convicts transported.
1811–1830	Australia established as a criminal waste-disposal system. After Waterloo, England experienced a number of minor crises: unemployment, industrialisation and loss of hand trades for the workers. Police established, resulting in an increase in the number of felons. Sharp rise in number of convicts transported.
1831–1840	Peak and then decline in transportation. Penitentiaries (similar to the one at Port Arthur) founded. Transportation to NSW ceases.
1840–1860	Convicts no longer used as pioneers. From 1841–1850, 26 000 convicts sent to Van Diemen's Land (Tasmania). Ceased 1851. All transportation to Australia ceased by 1868.

It seems quite clear that many convicts were sent to Van Diemen's Land after 1840 and the system became overcrowded and broke down. Many of the convicts at Port Arthur were two (or more) time offenders. But there is no doubting the severe brutality of their imprisonment. So it is reasonable to inquire 'Why did Port Arthur crumble?' by asking the following types of questions:

1. Were convicts the victims of an unjust society in England?
2. Were most convicts from the working class and if so how did this come about?
3. Did the British 'dump' their criminals on the far-away 'fatal' shore in order to protect their property?
4. Were the convicts treated as slaves in the new colony of Australia?
5. Were women who were made to work in the factories maltreated in the early days of settlement?
6. Were the Irish convicts (many convicts came from Ireland which had been 'taken over' by the British) treated with special harshness?
7. Was the convict better off in the new land even though he/she was still captive?

These questions require an understanding not only of individual convicts and their treatment but also of what it was like to live in English society and then to be a part of the newly emerging society of what we now know as Australia.

Here are the conclusions of two writers.

1. The English wished not only to get rid of the 'criminal class' but if possible to forget about it. Australia was 'invisible', its contents filthy and unnameable. The English at the time firmly believed in a criminal class and illustrators depicted the 'criminal type' as a mask of low cunning, stunted but alert. (Summary of Hughes 1987)

2. The idea that transportation offered criminals the opportunity to reform was as much a part of British penal thinking as were the desires to protect property and to punish [the criminal convict]. There is no doubting the system's brutalities but it is important to understand that many harshly treated convicts broke the laws again in the new colonies. For every convict who felt the lash or endured the miseries of . . . Port Arthur, there were numbers who were never flogged, who never saw a place of secondary punishment, who made something of life in the new land. (Review by Alan Frost, La Trobe University, in *The Australian*, 21 February 1987)

Whichever you select you should realise that:

1. The evidence used needs to be tested for its 'truth'.
2. Differences of interpretation arise from the way the evidence is treated. The questions a teacher asks influence the conclusions that are made. For

Critical Inquiry: The Emerging Perspective in Geography Teaching

Purpose of the inquiry

View 1	View 2	View 3
Knowledge about the physical condition of the bricks	Awareness about how the convicts felt, saw themselves and were motivated to act	Knowledge about individuals and groups; and how they changed or could be changed in society

Understanding of the base inquiry question: 'Why did Port Arthur crumble?'

What was the physical condition of the bricks? What other conditions (bushfires) could be responsible for Port Arthur's crumbling?	What 'motivated' the convicts? How did they understand their plight?	Who had the power in English society? Did all people have equal opportunity to share this power?

Nature of the question being understood

Physical	Moral	Political
Natural	Intellectual	Societal
'Scientific'		

Source of evidence in understanding the question

Soil analyst	Historian	Political scientist
Meteorologist	Archivist	Historian
Bushfire expert	Planner	Political economist
	Psychologist	

Character of the inquiry

Conservative classical	Liberal or interpretive	Socially critical

Figure 3.3 Approaches to understanding and explanation in geography

example, the questions in view 1 are about physical objects such as bricks. The questions in view 2 are about the understanding of individual people. Questions in view 3 are about the understandings of individual people but they also ask further questions about how individuals (for example the convicts) relate to their society.

3. Ideas about justice and social fairness are raised in view 3.

4. Our conclusions are in part influenced by how well we handle the available evidence, our inquiry skills, *and* what we *believe* are important in our treatment of evidence.

Our interpretations or conclusions depend on a number of key elements related to our view of inquiry. These elements are listed in Figure 3.3.

The way we inquired into the convict settlement at Port Arthur can be applied to any way we might like to teach geography, with the consequence that specific knowledge content will be produced.

The Nature of Critical Inquiry

The inquiry into Port Arthur attempted to demonstrate that there are three main views that give rise to knowledge and that form the basis of our teaching. It might be unkind to suggest that many of us tend to be oriented to the first view. In the past decade we have seen the emergence of the second view which is well represented in many of the chapters in this book. But there is only a scant literature about the third view, which appeals to the critical social sciences and to a notion of critical inquiry. What then, is the nature of critical inquiry?

In the introduction it was claimed that the degree of autonomy and responsibility that teachers have over their work is determined by the level of control that they can exercise over their actions; and that this process of control is called critical inquiry. The word 'critical' does not mean criticising or being negative; it refers to the stance of enabling teachers to see their classroom actions in relation to the historical, social and cultural context in which their teaching is actually embedded. This is intended to allow them to develop themselves individually and collectively; to deal with contemporary events and structures (for example, the attitudes of others or the bureaucratic thinking of institutions) and not to take these structures for granted.

Another name we might use for critical inquiry is to talk about the *socially critical* orientation to teaching. Apple (1975) sums up this orientation in the following way:

It requires a painful process of radically examining current positions and asking pointed questions about the relationship that exists between these positions and the social structure from which they arise. It also necessitates a serious in-depth search for alternatives to those almost unconscious lenses we employ and an ability to cope with an ambiguous situation for which answers can now be only dimly seen and will not be easy to come by.

Becoming critical involves the realisation that, as teachers of geography, we are both the producers *and* creators of our own history. In practical terms, this means we will engage in systematic and social forms of inquiry that examine the origin and consequences of everyday teaching so that we come to see the factors that impede changes for its improvement. If, for example, we concluded our social inquiry and teaching of Port Arthur with view 1, then we might ask ourselves whether we derived the best explanation and whether the kind of questions we asked to get the explanation were the best ones we could have asked. We could be assisted in this process by asking ourselves the following:

1. What counts as knowledge in geography?
2. How is the school knowledge in geography organised?
3. What are the underlying 'codes' that structure school knowledge in geography?
4. How is what counts as knowledge transmitted?
5. How is access to such knowledge determined?
6. What kind of cultural system does knowledge legitimate?
7. Whose interests are being served by the production and legitimation of this knowledge?

If you look again at Figure 3.3, you would probably come to the conclusion that the explanation proposed for Port Arthur's crumbling in view 1 is a very 'scientific' one, and that the kind of thinking (assumptions underlying the questions) is similarly very 'scientific' or technical. Alternatively, reasons postulated in view 3 begin to reclaim critically some of the reasons in societal terms (without denying the reasons associated with the actions of individual persons such as the convicts) and the possible distortions in our language and thinking about Port Arthur's decay.

There may be distortions or contradictions in the evidence upon which we base our judgements of course. Both evidence and judgements need to be truth-tested. For example, we could examine the claims of the *evidence* used to judge why Port Arthur is crumbling by asking the following kinds of questions.

1. Who stands to gain most from the author's texts?
2. What are his or her interests in the issue? (Find phrases, sentences . . .).
3. What image of self does he or she wish to present?
4. Who is the intended audience?
5. What are their interests in the issue?
6. Of what does the author wish to persuade his or her readers?
7. What content does the author focus on?
8. What is omitted?
9. Are biases or prejudices apparent (such as racism, scientism, sexism, ageism)?
10. What alternative viewpoints or arguments exist that are not mentioned or acknowledged?
11. What meaning is produced by the interaction of the formal, photo/sketches and written content?

The questions provide specific means that can be used to perform a critique on the concrete materials. They also suggest that teachers can examine and raise questions about how the content of the materials evidence might be related to specific sociopolitical contexts.

The general aim is to unveil critically these contradictions in our thinking and our actions as teachers: when we do this we take control over our own knowledge. The notion of *power to take control* is 'up front' in this idea of control; as is the aim to empower and emancipate ourselves as teachers from false knowledge that guides our teaching. We need to consider further this idea of conduct knowledge in geographical inquiry.

Conduct Knowledge and Inquiry

Our teaching is guided by our ideas and actions and it is the relationship between idea and action that is central to our thinking about critical inquiry. If we examined our teaching in the classroom, its conduct knowledge basis could be described under the two categories of *craft* and *professional*.

Craft knowledge is skills based and product oriented; it is essentially technical in character. Hence, technical actions in teaching rely more on skills and following rules, for example for managing the class and for structuring a lesson. Teaching actions based on craft knowledge do not imply the same degree of autonomy and responsibility as do actions based on committed choice.

Professional knowledge, on the other hand, is based on the interactive social life of the geography classroom. Any action that is based on professional knowledge is said to require teacher judgement that always seeks 'the good', the right or best action. When teachers exercise their personal judgements they are exercising a degree of autonomy and responsibility in their teaching.

Examples of craft and professional knowledge which underlie inquiry might be illustrated with reference to the Port Arthur example. They are discussed under the three views.

View 1

We made the claim that our inquiry involves the relationship between our ideas and actions. In a technical orientation to inquiry, there may be no relationship between what is done and the idea behind the action. The idea is replaced with a form of rule following and the implementation of a set of skills. For example, one teaching myth concerning management of a class at the beginning of the school year is that the teacher must begin 'hard' and then relax as the year progresses; or that the golden rule for good management is 'no talk, when the teacher talks'.

This technical approach is analogous to the form of analysis and orientation to inquiry that would have to be followed in view 1. The explanation according to this view required a 'deterministic', 'objective', lawlike view of a 'real' world. To actually provide the reasons, you would need a set of instruments that could measure the physical properties of buildings crumbling at the site of the former penal settlement; and this would involve skills and a knowledge of rules-to-be-followed more so than personal or professional knowledge.

View 2

Sometimes teaching actions are characterised as 'craft' with the teacher's skill being very important to achieve the desired end. There are some aspects of teaching about maps that require the teacher to exercise 'working' knowledge and skills. Quite often in using 'craft' knowledge, there are pre-specified ends. For example, it is relatively easy to write skill objectives with a high degree of explicitness.

The same cannot be said about writing cognitive or affective objectives which more frequently involve student–teacher interaction and teacher actions that are judgement guided. This is the interpretive inquiry of view 2. It requires the teacher to explore the relationships between our ideas and actions. The view assumes that, as teachers, we are able to know what is the best course of action; that we are unimpeded in our teaching; that we are free to act rightly if we will to do so. Notice in the Port Arthur example, we used multiple forms of evidence and multiple realities that were derived from the lived experience of students (hence the term person-centred) or from the writing and thinking of a past and contemporary tradition and history (hence the term humanistic).

View 3

Now it may be true of *some* of our teaching actions that they are guided unhindered by our ideas and intentions, but it may be too much to assume that *all* of our actions are unimpeded by distortions to our ideas. These distortions or contradictions are ideological. Ideology is a difficult word. Here it means the beliefs, values and strongly held positions that characterise our patterns of thinking. Our 'codes' of conduct and the way we think about knowledge content in geography (well exemplified in the three views) are ideological to the extent that they represent our self-interests or the interests of the class group. Our knowledge and actions are rendered partially distorted (at least) by these interests.

There are two further aspects of ideology that should be mentioned. First, many of our teaching ideas and actions may not be distorted because of our individual selves. Institutional interests may be the basis of contradiction in how we teach or act. For example, many discipline problems in the classroom may not be due to or be changed through any action we might take. The contradictions may be in the wider community's or society's attitudes to discipline. This is referred to (along with other elements) as the *social structure* that constrains, limits or governs our teaching in many ways. The idea of 'hidden curriculum' necessarily conveys the notion that many aspects of the curriculum that are not apparent to us are a distortion of 'the good' at work in schools. A second source of contradictions lies in *tradition*. Many of our teaching actions perpetuate myths about teaching and favour the status quo. We may reproduce these myths and rituals and produce them anew in our daily teaching without being aware of them. At the beginning of a school year, many teachers exhort their students to work 'hard' so that they will be rewarded with academic success. This myth of invested effort is a contradiction because it assumes that *all* students have equal access to resources including the teacher when more often than not they don't; that all students have an equal opportunity to succeed in a grading system that *must* reward a few or in an employment setting where only few can be employed in a situation of economic stringency; and so on. Tradition and knowledge about professional conduct that has been 'handed down' to the teacher often goes unchallenged.

The aim then of this third view is to liberate our thought and action from these kinds of distortions. To be autonomous in inquiry that we use in the classroom, it is necessary to gain control of the ideas that control our actions. It is this *disposition* to emancipate our *praxis* (practice) that characterises a critical perspective. In the Port Arthur example, you would have noted a further aspect of this critical perspective. The questions and subsequent reasons appealed to a sense of the group or, more explicitly, society. The disposition not to rely solely on personal and individual judgement but to engage in what is called *social critique*, which unveils contradictions, requires a collective effort between teachers and students. This is necessary because the source of distortion resides in communication between and among individual people. Hence, to criticise is to interact collectively and to emancipate ourselves as an interactive group from contradictions.

Becoming Critical in Teaching Geography

Becoming critical involves the examination of the taken for granted and the assumed in our practice. This means we will want to ask ourselves about the meanings, values and motives we hold in relation to our *conduct knowledge* as teachers. In addition, we shall need to examine the form of analysis or inquiry we implement in teaching the *knowledge content* of geography. Some of the following kinds of questions will help us in this process:

1. What was it that caused me to want to become a teacher of geography?
2. Do these reasons still exist for me now?
3. What does it mean to be a teacher?
4. Is the teacher I am the person I am?
5. Where did the ideas I embody in the teaching of geography come from historically?
6. How did I come to appropriate them?
7. Why do I continue to endorse them now in my teaching?
8. Whose interests are being served?
9. Who has power in my classroom and how is it expressed?
10. How do the power relationships influence my interactions with students?
11. How might I teach differently?
12. What is the nature of the knowledge that guides my teaching of content described as geography?
13. Who creates this knowledge?
14. How was it possible for it to emerge during the evolution of the classroom system?
15. Whose interests does this knowledge about teaching in geography serve?
16. How do/can I personally work to uncover the contradictions in my teaching?
17. How does what I do affect the life chances of students?
18. What connections do I make (outside the associations in geography) with organisations outside the school to demonstrate my active role in society?
19. Do I work to uncover the 'hidden curriculum' (the ideological assumptions and distortions) in my teaching, my subject department and my school?

You might have realised that there is a present and a past notion in some of these questions. That is, we may need to link contemporary and historical influences when we attempt to unravel the contradictions in our teaching of geography.

One of the principal ideas we might have in our minds in becoming critical is the idea of *problematising* our teaching. This does not mean problem solving, for example as it is proposed in Chapter 5. It consists of 'problem posing' rather than problem solving. It focuses on the problem setting and the relevant questions that will critically investigate that setting. You might compare this critical stance with a more traditional problem solving which in many ways is more like the kind of questions and their knowledge basis occurring in view 1 of the Port Arthur example.

Paolo Freire contrasts a critical view (which he calls critical pedagogy) with the traditional view of teaching as problem solving when he talks about the *banking* concept in education, a kind of 'money-in, money-out' idea. The assumptions behind the banking model of teaching are:

1. The teacher teaches and the students are taught.
2. The teacher knows everything and the students know nothing.
3. The teacher thinks and the students are thought about.
4. The teacher talks and the students listen—meekly.
5. The teacher disciplines and the students are disciplined.
6. The teacher chooses and enforces his or her choice, and the students comply.
7. The teacher acts and the students have the illusion of acting through the action of the teacher.
8. The teacher chooses the program content, and the students (who were not consulted) adapt to it.
9. The teacher confuses the authority of knowledge with his or her own professional authority, which he or she sets in opposition to the freedom of the students.
10. The teacher is the subject of the learning process, while the students are mere objects.

These kinds of statements do three things to us as teachers of geography because we have an inbuilt ideology that distorts our teaching. For example, most statements affirm a view of teaching without examining the legitimacy of the assumptions upon which they are based. If you accept the dictum that 'the teacher talks and the students must listen' then you are affirming uncritically that students must think in order to learn. In legitimating this ideology, you may be sanctioning, in effect, the fact that the students will not need to think at all; that is, the teacher is the 'star' in the learning process. Fortunately, we are becoming more aware that this may not be the case; that students, like teachers, have their own ideology about listening; and that this ideology which gives rise to their resistance is more powerful in learning than anything the teacher might do or say.

The three things that Freire's statements about traditional inquiry do to us are:

1. As a form of ideology, they maintain the status quo or the existing order. They ensure 'no change,' stability and predictability in the classroom by making sure the rights of students are repressed.
2. They engender a 'false consciousness'. As teachers we often attempt to establish routines not for any educational reasons related to the rights or good of students but because it is the most effective means of avoiding conflict and suppressing (some might call it oppressing) student will.
3. They often mask contradictions by publicly misinterpreting them or excluding them. For example, the way teachers often label students as 'good' or 'bad' in the first lessons of the school year is at variance with their professed wish that all students should achieve academically.

The problematising of inquiry is probably best summed up by Tom (1985):

To make teaching problematic is to raise doubts about what, under ordinary circumstances, appears to be effective or wise practice. The object of our doubts might be accepted principles of good pedagogy, typical ways teachers respond to classroom management issues, customary beliefs about the relationships of schooling and society, or ordinary definitions of teacher authority—both in the classroom and in the broader school context.

Problematising Questions in Geography

The main thrust of the previous discussion was the need to problematise our conduct knowledge as teachers of geography. How then is this exemplified in the kind of questions we shall ask in teaching specific geography content knowledge? We have seen one example in the inquiry about Port Arthur. Below are three other examples with the three views included for comparison.

Example 1: The World and its Populations

View 1:
- How many people are there in the world? Where do most people live in the world?
- How many people are there in the older age groups?
- Is there a relationship between the age of people and the places where they live?

View 2:
- What do we mean by the word 'race'?
- How did the word appear in the English language?
- How is it used by most people?
- What are people's attitude to white races? Black races?

View 3:
- Do the actions of some powerful groups or countries in the world cause the death of people in other parts of the world?
- Do all people have access to the same life chances, finance and power?

Example 2: Natural Resources in the World

View 1:
- What is a resource?
- Where are most of the world's mineral resources?
- Is there a relationship between the mineral wealth and financial wealth in different countries?

View 2:
- What is your attitude to uranium mining?
- How do people share the world's resources?

View 3:
- Are resources distributed fairly among southern and northern countries?
- Are there examples of exploitation of individual resource-rich but money-poor nations in the world today?

Example 3: Refugee People

View 1:
- How many refugees have moved into the Sudan in recent years?
- What has been the influence of drought on the movements of these peoples?
- What are the 'push' factors (e.g. drought) and the 'pull' factors (e.g. hope to get food) that have forced people to migrate?

View 2:
- What influences people to decide to migrate?
- Do people lose their identity in a new land to which they migrate? How does this influence their lifestyle?

View 3:
- Are the refugees of Ethiopia victims of cruel oppression between rival groups with different ideas about society?
- How has the exploitation of colonial nations (e.g. England in the Sudan in the nineteenth century) led to the present suffering of people in this part of Africa?

What questions do you believe students should be investigating? Which ones will be of most benefit to them and to other people in making the world a better place in which to live?

Signposts of Critical Inquiry

The questions in view 3 in these examples reflect a critical social science perspective but, of themselves, do not constitute necessarily critical inquiry. Together, critical social science and critical inquiry form what we might call 'educational science'. The nature of a critical educational science is not yet known clearly and is the task for educators for the twenty-first century. We can, however, indicate the character and the signposts of a critical inquiry that is tied to the kinds of questions that may be derived from view 3.

One word of warning before we suggest the signposts for a critical inquiry. The concepts we have used in the discussion of critical inquiry may be new to many teachers; concepts such as critical, ideology, control, professional and others. There is a conceptual overload and an apparent dressing up of old ideas in big words and a new language. These words do not mean anything unless they are related to and derived from teachers' own practice (we made the statement that a critical inquiry must begin with the categories teachers themselves understand). The main task of critical inquiry is to unmask the contradictions in our practice and the understandings upon which it is based.

The rudiments or signposts of critical inquiry in geography include the following:

1. Inquiry must begin with student experiences. Experiences are the subjective meanings students give to what they observe and do.
2. Students must be assisted to analyse these experiences carefully. This will not be easy if only because we seem to know so little about the psychology of inquiry learning. It is important that we begin the analysis with students' subjective categories as the beginning point of inquiry.
3. No attempt should be made to reaffirm experiences that are based on uncritical categories, for example sexual stereotyping and racism.
4. The teacher must attempt to develop an understanding of the various ways in which students' perceptions and identities are produced through their cultural environment including their family, school and community.
5. Teachers must explore ways to self-evaluate their communication with students; to examine how communication in the geography classroom is used to legitimate or silence students' attempts to participate actively in dialogue.
6. Geography classrooms must be opened to diverse resources and traditions by creating active links with the community. This will necessitate the need for teachers to familiarise themselves with the culture, economy and historical traditions that belong to the surrounding community.
7. One immediate consequence is that teachers need to develop inquiries around these traditions, histories and forms of knowledge that are often ignored within school curriculums. What is to be learned and how it is to be learned involve very different approaches to the selection, organisation and transmission of geography in a remote rural school compared to a high-tech metropolitan school.
8. Inquiry that links student experiences with aspects of community life that sustain and inform such experiences needs to be implemented. The geography of cemeteries in which the study of community history, ethnic groupings, etc. takes on added meaning when related to critical inquiry practice.
9. Inquiry involving critical thinking, ethical decision making and social participation needs to be implemented according to the levels of maturity of students.
10. Finally, a critical perspective involves the analysis of social agencies other than the students' school and community. How these social agencies work to produce, distribute and legitimate particular forms of knowledge and social relations needs to be unveiled so that student understanding can be broadened to institutions other than their school. The geography classroom might then become a public place for critical debate about issues such as the role of corporate states, agribusiness, information production and exchange, the relation between northern and southern countries and others.

The signposts that might be employed to assess whether we are engaging in critical inquiry in our classrooms might be summarised in the following questions (Habermas in McCarthy 1978).

1. How much do we talk about our teaching with *comprehensiveness?* That is, do we speak in such a way that we understand what is happening in our classrooms *now* and how our practice is limited and governed from without?
2. How strongly do we seek to democratise our communication, to speak with *sincerity?* When we speak to students and other colleagues in our school, the honesty with which we communicate and the intentions we have must be explicit.
3. How do we build the *legitimation* of our communication with others so that positions of authority are not being used or abused to exert power over (as opposed to give power to) others?
4. Does the element of *truth* characterise our actions (what we do, *in* what we say, *by* what we say, and *through* what we say) so that they are credible; so that they can be justified as committed, enlightened and free; and so that alternative actions have been considered for their truth?

The path to a critical perspective is not an easy one. Maxine Greene (1980), writing about the search for a critical pedagogy, makes the following comment:

To engage with our students as persons is to affirm our own incompleteness, our consciousness of spaces still to be explored, desires still to be tapped, possibilities still to be opened and pursued. At once, it is to rediscover the value of care, to reach back to experiences of caring and being cared for as sources of an ethical ideal. We have to find out how to open spaces . . . where a better state of things can be imagined; because it is only through the projection of a better social order that we can perceive the gaps in what exists and try to transform and repair. I would like to think that this can happen in classrooms, in corridors, in schoolyards, in the streets around.

If we, as teachers of geography for the 1990s, can say that our inquiry is characterised by these signposts, then we can say that we are taking control of the knowledge that influences our teaching as well as the content knowledge of what is taught. Taking this kind of control and responsibility is the basis for a critical perspective in inquiry in geography classrooms.

References

Apple, M. (1975) 'Scientific Interests and the Nature of Educational Institutions', in W. Pinar (ed.) *Curriculum Theorizing*, Berkeley CA: McCutcheon.

Aronowitz, S. and Giroux, H.A. (1985) *Education Under Siege*, Massachusetts: Bergin and Garvey.

Bartlett, V.L. (1983) 'Questions and Viewpoints: The Art of Interpretation in Geography', *New Zealand Journal of Geography*, April, 7–13.

Brenkman, J.L. (1983) 'Seeing Beyond the Interests of Industry: Teaching Critical Thinking', *Journal of Education*, Vol. 165(3), 283–294.

Giroux, H. and McLaren, P. (1987) 'Teacher Education and the Politics of Engagement: The Case for Democratic Schooling', *Harvard Educational Review*, Vol. 56(4), 213–238.

Greene, M. (1980) 'In Search of a Critical Pedagogy', *Harvard Educational Review*, Vol. 56(4), 427–441.

Grundy, S. (1987) 'Critical Pedagogy and the Control of Professional Knowledge', *Discourse: The Australian Journal of Educational Studies*, Vol. 7(2), 21–36.

Habermas, J. (1979) *Communication and the Evolution of Society*, Boston: Beacon Press.

Hughes, R. (1987) *The Fatal Shore*, Melbourne: Penguin Books.

Kemmis, S., Cole, P. and Suggett, D. (1983) *Orientations to Curriculum and Transition: Toward the Socially Critical School*, Melbourne: The Victorian Institute of Education.

McCarthy, T. (1978) *The Critical Theory of Jurgen Habermas*, Boston: MIT Press.

Smyth, W.J. (1985) 'An Alternative and Critical Perspective for Clinical Supervision in Schools', in K. Sirotnik and J. Oakes (eds) *Critical Perspectives on the Organisation and Improvement of Schooling*, Massachusetts: Kluwer-Nijhoff.

Tom, A. (1985) 'Inquiry into Inquiry-Orientated Teacher Education', *Journal of Teacher Education*, Vol. 36(5), 35–44.

4

Expository Teaching for Meaningful Learning in Geography

F. Geoffrey Jones

The previous three chapters have concentrated on the ideologies and goals of geography teaching. This chapter focuses on the goal of helping students acquire the geographical knowledge to accompany the values important for responsible citizenship. Direct teacher exposition is one of the most common methods used to impart such knowledge. Students have to be able to relate new information to the knowledge they already possess and to apply it to new situations for it to be learnt effectively. Geoff Jones advocates the use of 'advance organisers', as developed by Ausubel, as a means of ensuring that expository teaching is meaningful to students. This requires more than the rote recall of information. A three-stage model for preparing and teaching a unit on manufacturing cities based upon the advance organisers of 'raw materials', 'transformation' and 'producer and consumer goods' is used as an example of using expository methods well in geography.

David P. Ausubel (1968) described expository learning as meaningful verbal learning. Such learning is considered expository because the entire subject matter to be mastered is presented to students, either orally or in writing, *in its final form*. Meaningful verbal learning requires students to relate new or unfamiliar ideas to knowledge and most cognitive structures that they already possess. This form of expository learning should not be confused with rote learning where the learner may be able to recall information, but not necessarily understand it. Ausubel claims, with justification, that exposition is the form of teaching and learning most commonly used in schools. A survey on teaching methods conducted by the Schools Council Geography 16-19 Project indicated that this is certainly true in matriculation geography classes in England (Schools Council Geography 16-19 Project 1978). Studies reported on by Bartlett and Hull (1978) revealed similar results in Queensland. This chapter outlines a model to make such teaching and learning more effective and efficient than might normally be the case with less well organised teacher 'chalk and talk'.

Expository learning is dependent upon the reception of material that either enlarges existing concepts or establishes new ones where once knowledge was lacking. The act of reception forms an integral part of Ausubel's expository learning model and has been referred to simply as *reception learning*. The key feature is that the entire content of what is to be learned is presented to the learner in its final form. The learner's task is to receive, internalise and be able to recall and apply the material learned.

There are five variables to consider when structuring material to be internalised by the learner. These are: advance organiser, generality, subsumption, progressive differentiation and integrative reconciliation.

1. *Advance organiser*. An advance organiser is a set of ideas or concepts given to the learner prior to the presentation of material to be learned. The advance organiser provides introductory material at a higher level of abstraction, generality, and inclusiveness than the subsequent learning tasks (Ausubel 1963). Advance organisers should provide a stable cognitive structure upon which new learning can be anchored. Some of the advantages of using advance organisers are the provision of inclusive organising elements for material which follows and an increase in memory and recall. The use of advance organisers is appropriate in at least two situations. The first is when the learner has no relevant information to which new learning can be related, while the second is when relevant information is present but is not recognised as relevant by the learner. These two situations necessitate two different types of advance organisers, the expository organiser and the comparative organiser. The *expository organiser* presents a description or exposition of relevant concepts and is appropriately employed when the learner has no relevant information to which he/she can relate new material. The *comparative organiser* employs similarities and differences between new material and the student's existing cognitive structure. Comparative organisers are used when relevant information is presented but may not be recognised as relevant by the learner.

2. *Generality*. Generality refers to the fact that in the presentation of material, the most inclusive and general concepts are presented first. Related materials are then subsumed under the more general concepts.

3. *Subsumption*. Ausubel uses the term 'subsumption' to refer to a concept or idea which incorporates or subsumes other concepts. Subsumers arranged in hierarchical fashion assist in developing cognitive structures. Subsumption then is the incorporation of meaningful material into existing cognitive structures.

4. *Progressive differentiation*. Progressive differentiation is the process through which attributes of subconcepts and subcategories of an advance organiser are identified and distinguished from one another. Subsumption and progressive differentiation form the process from whole to part (i.e. the deductive approach) in which the most general and inclusive ideas precede more detailed and specific explanation.

5. *Integrative reconciliation*. Integrative reconciliation is the process of exploring relationships between new and previously learned concepts.

An Expository Learning Model

There are many times, even when working independently, when it is necessary for students to acquire specific information as meaningfully and efficiently as possible. An important aspect of expository learning is the way in which material is structured for the learners to allow for reception. Learners may use this structure as part of their own knowledge categories and thus expand their cognitive frameworks by incorporating new learning. The sequence of phases that can be used provides a structured way of thinking which students can adopt in learning situations outside the classroom. The meaningful way in which new ideas and information are linked with patterns of thinking can stimulate students to learn about the unfamiliar and the unknown.

Implicit in this discussion is the idea that there is a model that can be used to guide both the teacher and the learner. Joyce and Weil (1980) identified a three-phase expository learning model based upon Ausubel's ideas. It is illustrated in Figure 4.1.

The major purposes of this model include:

1. developing students' conceptual structures;
2. assisting in a student's meaningful assimilation of information and ideas;
3. stimulating students' interest in inquiry; and
4. developing students' abilities for precise thinking.

Where the objectives of a lesson or curriculum unit include any such purposes, this expository model is well worth adopting.

Explanation of the Model

There are three phases of operation associated with the model. Phase 1 clarifies the objectives of the lesson and presents the advance organiser. Phase 2 presents the material to be learned in a structured logical manner. Phase 3 emphasises the relationship between the new learning material and the existing knowledge of the students.

Phase 1. Presentation of the advance organiser. Present students with the lesson objectives at the start of the

Expository Teaching for Meaningful Learning in Geography 37

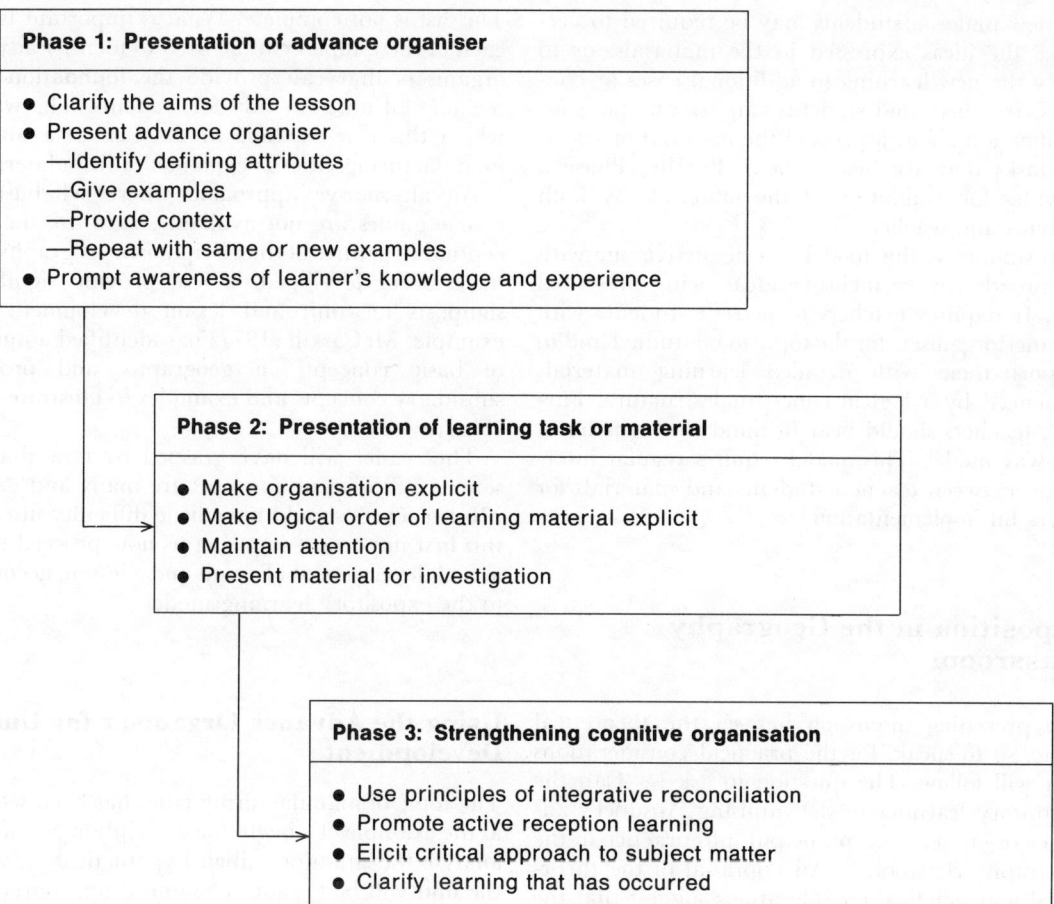

Figure 4.1 The three phases in Ausubel's advance organiser model (adapted from Joyce and Weil 1980)

lesson. This provides both teacher and students with a built-in evaluation device that can be used to determine the degree of success of the materials used. The advance organiser can then be presented to the students. An advance organiser is a more abstract set of ideas or concepts given to the learner prior to the presentation of the material to be learned. The advance organiser provides introductory material at a higher level of abstraction, generality, and inclusiveness than subsequent learning tasks. It should provide a cognitive skeleton upon which new learning can adhere. Remember that either an expository organiser or a comparative organiser may be used depending upon the familiarity of students with the learning material.

Phase 2: Presentation of the learning task or materials. Once students have been provided with the advance organiser, preferably in written form (which can be used on later occasions), new material may be presented. Often it is assumed that this material can be presented by teacher lecture only. While this approach is quite acceptable, other approaches including films, reading, audio and video recordings can also be used. Whichever approach is used, the important thing is that the material should be structured in such a way that its logical sequence is made obvious to students. This can often be done through the development of sequenced learning tasks that highlight the conceptual structure of the topic being studied.

Phase 3: Strengthening cognitive organisation. The new material presented in Phase 2 is now related to the students' existing cognitive structures. This involves activities in which they may be asked to provide a summary of the major elements of the new learnings, or state differences or similarities between the parts of the material. Students should be referred back to the advance organiser and asked to identify aspects of the material that relate to propositions evident in the advance organiser. To grasp fully the significant relationships between the organiser and the new material students may be required to verbalise the ideas expressed in the materials, or to apply the new learning to additional cases or concepts. It is here that students can ask questions for clarification about aspects of the material or learning tasks that are before them. Finally, Phase 3 provides for evaluation of the materials by both students and teacher.

In summary, this model is a deductive one with the broader, more inclusive ideas being presented first. It requires teachers to present students with advance organisers for the topic to be studied and to support them with detailed learning materials sequenced by a logical conceptual structure. Further, teachers should bear in mind that it is not a one-way model. This model requires regular interaction between teacher, students and materials for successful implementation.

Exposition in the Geography Classroom

The preceding discussion has set the theoretical scene, so to speak, for the practical considerations that will follow. The question to ask is, 'Can the expository learning model, utilising Ausubel's advance organiser concept, be put into practice in the geography classroom?'. An appraisal of the theoretical and practical considerations suggest that the approach is most amenable to the classroom where meaningful and efficient learning is planned. How, then, can it take place?

Using the Expository Model in the Classroom

The first task of the teacher is to identify a satisfactory structure around which the geography course can develop for example:

This list is not complete. What is important is that each of these topics can be used to identify advance organisers that can provide the foundation of a meaningful unit of study for students. The way in which this can be done in the case of a unit on manufacturing cities is explained a little later.

An alternative approach, where syllabuses or course guides are not available, is to use the conceptual structures of the discipline of geography that have been developed. These provide significant signposts for unit and lesson development. For example, McCaskill (1977) has identified a number of basic concepts in geography and provides subsidiary concepts and examples to illustrate them (Figure 4.2).

The reader will have grasped by now that the sources of topics or concepts are many and varied. Most teachers should have little difficulty in taking this first important step. Let us now proceed to the actual development of a unit and a lesson, according to the expository learning model.

Using the Advance Organiser for Unit Development

The topic of manufacturing cities has been selected as the example. To begin the description, we need to analyse the situation, albeit hypothetical, in which the unit will be taught. The unit is appropriate for Year 10 students who are being introduced to the topic. They have received little instruction on this topic in earlier grades.

The teacher's first task is to develop from the topic a conceptual structure in which the advance organisers are identified. The identification and construction of a conceptual structure of a topic is not an easy task for the teacher. However, Brown and Macaulay (1971) have identified a *topic analysis* approach which can be of considerable assistance to the teacher in developing a suitable conceptual

Landscape geography	Human geography	People–environment geography
Monsoon lands	Manufacturing cities	Climate and people
River studies	Mining towns	Rural life and landscapes
Forest lands	Political boundaries	Soil erosion
Mountain lands	Population growth	Rainforest conservation
Grasslands	An Asian city	Endangered species
		Floods and drought

Expository Teaching for Meaningful Learning in Geography 39

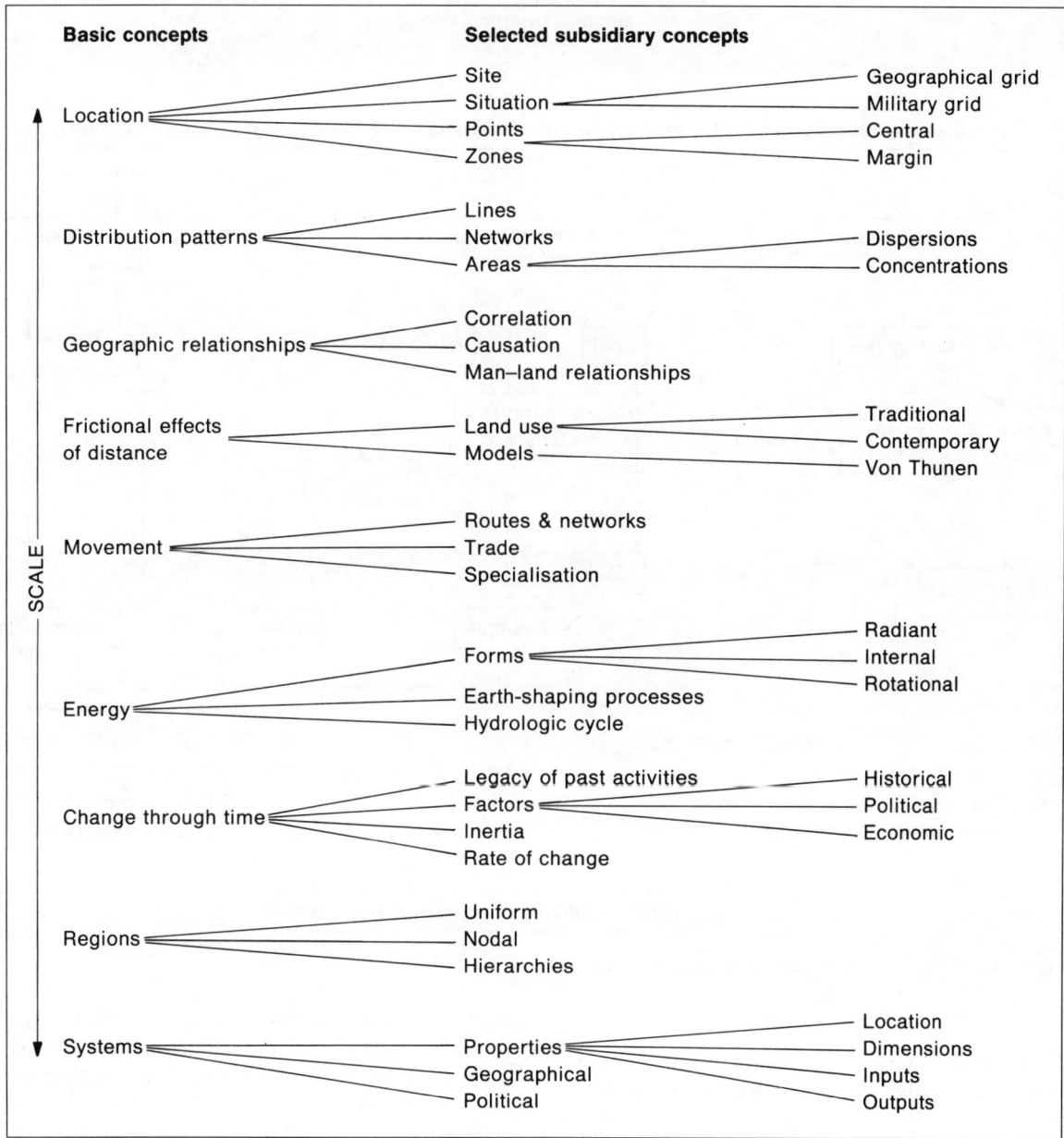

Figure 4.2 A conceptual structure of geography (after McCaskill 1977)

structure. Topic analysis may be of assistance to the teacher because it allows the development of a detailed analysis of the main ideas embedded in a topic, and requires that these ideas be presented in a structured (hierarchical) manner. It could further be said that topic analysis should be one of the first steps in unit planning, used both to organise necessary information and to establish criteria for the selection of appropriate material. Brown and Macaulay identified four major characteristics of topic analysis which could be useful. They are:

1. The topic can be related to an idea context appropriate to the geography syllabus.

2. The organisation of the analysis is hierarchical and in a deductive format.

3. The content must be in an explanatory mode of exposition.

4. Ideas must be expressed in single sentences.

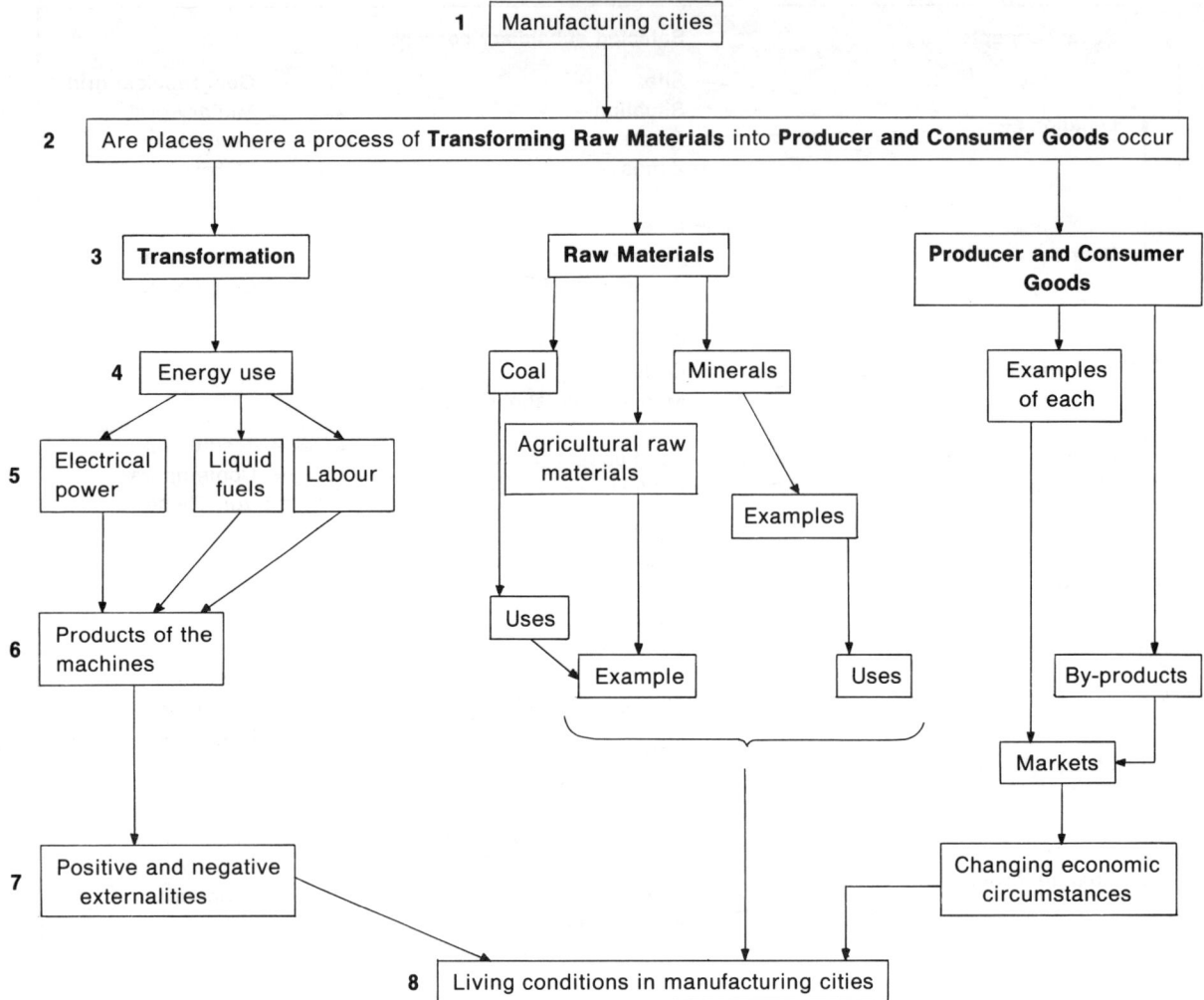

Figure 4.3 A structure of the topic 'Manufacturing Cites'

In order to acquire a more complete understanding of the topic analysis approach the reader should seek Brown and Macaulay's (1971) article for an indepth explanation.

For our purpose here, Figure 4.3 outlines a possible structure that could be used for teaching the topic 'Manufacturing Cities'. The reader will note that

1. transformation
2. raw materials
3. producer and consumer goods

have been identified as advance organisers for the topic. These concepts provide the structure through which the advance organisers can be applied to the various examples of manufacturing cities. For example: *raw materials* can be studied by using the more concrete concepts of *types*, *sources* and *uses* of raw materials which can be applied to *types of manufacturing cities*.

As the conceptual structure is now arranged, there remains one further task, and that is selecting a number and variety of manufacturing cities to which the structure can be applied. Some examples which illustrate the variety of industrial landscapes in manufacturing cities are:

1. An iron and steel industry city—Pittsburgh, USA.
2. A shipbuilding industry city—Göteborg, Sweden.

3. A car-manufacturing city—Detroit, USA.
4. A mixed-manufacturing city—Osaka, Japan.
5. A textile-manufacturing city—Bombay, India.

These are a few examples of a more complete list of manufacturing cities that students could study. Time and emphasis in the curriculum will determine how far the teacher and students will or can pursue each example.

Advance Organisers and Lesson Development

After completing the organisation of the unit the teacher can now turn to planning lessons using the model in Figure 4.1. The first series of lessons should concentrate on presenting the expository advance organisers to the class. Definitions, descriptions and then examples of their meaning should be given. Movie film, slides, photograph and map interpretation exercises, fieldwork, readings and vivid teacher descriptions are all useful. Questions, discussion and teacher-directed student activities which focus upon the abstract terms should dominate this initial stage of instruction. Students must grasp the abstract concepts fully in order to apply them meaningfully to the examples that will follow in future lessons. Once students have grasped their meanings, the advance organisers can be applied to the examples or case studies through the subset of concepts developed for the unit. The following outline provides one possible way of developing a series of lessons using an iron and steel manufacturing city as an example. It is based upon the three-phase model in Figure 4.1.

Phase 1: Present the advance organiser. The three advance organisers—transformation, raw materials and producer and consumer goods—should be presented at the beginning of the lesson. Students should read the information in Figure 4.4 in preparation for a brief discussion that focuses upon the definition of an iron and steel city. Special characteristics that would differentiate an iron and steel city from other manufacturing cities could be identified and used as a basis of comparison at a later time. The iron and steel city of Pittsburgh might then be identified as the example and the special characteristics of an iron and steel city would be discussed in relationship to the example, Pittsburgh. Films, slides, graphs, tables, maps and other pictorial and statistical materials should be used to locate the example and to help students obtain a mental image of the physical and human environment of the city.

Phase 2: Identify the specific concepts of each advance organiser. Give the students a handout based on Figure 4.5. Once this has been done, apply the concepts to the example in such a way that students will understand how the concept takes on meaning within the context of the example. To do this, the teacher could begin with the questions: What three attributes might determine where an iron and steel plant would locate? What do these attributes have in common? How is each important to its location? Display some pictures showing iron and steel city characteristics. What do they tell us about the site of iron and steel mills? How do each of the three

Figure 4.4 Introducing the advance organiser

An iron and steel manufacturing city—Pittsburgh

A *manufacturing city* processes *raw materials* and manufactures goods in *factories*. Factories that process raw materials and produce small amounts of a product are called heavy industries. Light industries produce high-value finished goods, e.g. cars, televisions, toasters and the like. Processing and manufacturing provide the *economic base* for a manufacturing city.

A manufacturing city is an *assembly* point for raw materials. Large quantities of raw materials must be transported cheaply if not located close to factories. Railways and water transport are often used. Factories also require space for buildings and for storage of raw materials and manufactured products. Flat land close to water transportation or a railway is ideal. Large amounts of energy are consumed. Manufacturing industries use *energy* to convert raw materials to products. Electricity, oil and natural gas are used as energy sources.

A manufacturing city contains offices and headquarters generally in or near the *Central Business District* (CBD). Offices arrange for the sale and delivery of products to customers. However, factories do not locate near the CBD or residential districts. They locate in environments that try to diminish the effects of noise, smoke and fumes on the rest of the city. The movement of raw materials into a city and the movement of finished products out of the city provide the *function* for a manufacturing city.

Manufacturing and the city: a case study of Pittsburgh

Pittsburgh, the heart of the iron and steel industry in the United States, is an example of a manufacturing city. It is *located* in Pennsylvania at the junction of the Allegheny and Monongahela Rivers which then form the Ohio River. (Use a map of the USA to locate the state, the city and the rivers.)

The city of Pittsburgh and environs is the largest iron and steel producer in Pennsylvania. It produces around 20 million tonnes per year. The Pittsburgh area has a population of about three million *people* of which approximately one quarter of a million work in the iron and steel industry (use a map of the city and environs of Pittsburgh).

Pittsburgh has many *heavy industries.* They use large volumes of *raw material. Coal* deposits are mined close to the rivers where *iron ore* is moved by barge and train from the Mesabi Range in Minnesota. *Limestone* is available locally. Most industries are spread along the banks of rivers as are the rail lines. Water and rail are the most *cost efficient forms of transport* for raw materials.

Electricity from oil provides *energy* to industries. Finished iron and steel is generally *stored* before being *transported* to *customers* in the United States.

Light industries have also developed. They use iron and steel to manufacture high-value products which are *distributed* all over the United States.

Away from the manufacturing arena, Pittsburgh is similar to other cities. It contains a *commercial centre* for offices of big companies as well as restaurants, retail stores, hotels and residential areas which are located on the hilly slopes along and overlooking the rivers.

Figure 4.5 Identifying the specific concepts of each advance organiser

Pittsburgh and other manufacturing cities

Pittsburgh is located at the centre of a vast iron and steel market. It is an assembly point for bulky raw materials which are transported by train or barge. Factories process the raw materials to produce finished products which are then delivered to customers.

While there are a large number of new industries developed around iron and steel now, iron and steel production still provides Pittsburgh's economic base.

Pittsburgh is a large iron and steel manufacturing city with heavy and light industries which provide activities and facilities to support its large population. Its geographic location keeps it at the heart of the iron and steel region of the northern United States.

Applications to other case studies

1. How does the industrial landscape of the shipbuilding city of Göteborg, Sweden, differ from the industrial scene in Pittsburgh?
2. How does the availability of different raw materials differ for a textile-manufacturing city such as Bombay when compared to the iron and steel industry of Pittsburgh?
3. Which Australian and English cities have industrial landscapes like that of Pittsburgh?
4. Are industrial cities in Third World countries any different from Pittsburgh?
5. What other industrial activities or functions apart from those illustrated in Pittsburgh are found in specific cities around the world?

Figure 4.6 Strengthening cognitive organisation

attributes relate to Pittsburgh as an iron and steel producing city? The teacher can now assist students to understand how the three concepts provide an understanding of the advance organiser within the context of the example. Each advance organiser can be treated in the same way. A variety of possible techniques to reinforce the content can be used here. They include: questions, pictures, photographs, slides, film strips, films, supplementary reading materials, fieldwork, simulation games, library research and statistical exercises.

Phase 3: Strengthening cognitive organisation. Students should now be able to relate the newly acquired material to material learned previously. This involves *student activity*. They may be asked to link the subconcepts by devising propositions or generalisations to show how they are related to each other (see Figure 4.6). However, to make the learning meaningful, the concepts must be referred back to the advance organisers so that learning is founded upon the structure initially established. To enhance their learning further, students may be asked to apply their newly acquired knowledge to other examples or cases as illustrated below. Students should be encouraged to ask questions about the content studied and the materials used to support it. Finally, the conceptual structure and objectives of the unit can provide a basis for *formative* evaluation which will allow the teacher to determine, firstly, how well students have understood the material and, secondly, how successful instruction has been.

Conclusion

This chapter has described the theoretical considerations for an expository instructional mode using David P. Ausubel's 'advance organiser'. Joyce and Weil (1980) operationalised an instructional model that is both sound and efficient for the teacher and learner. The key to the successful use of the model is to develop a suitable structure built around the advance organiser(s) and the subconcepts. A content outline has been included to illustrate how the conceptual structure can be developed and (*ipso facto*) evaluated. Finally, a lesson has been included that illustrates how the advance organisers and the subconcepts interrelate to form a substantial cognitive learning package complete with supporting visuals.

If teachers were to assess honestly their approaches to content development and teaching most would confirm Ausubel's observation that 'expository teaching is the most frequently used form of instruction'. The model described in this chapter outlines one way that such instruction can be made more effective, efficient and enjoyable.

References

Ausubel, D.P. (1963) *The Psychology of Meaningful Verbal Learning*, New York: Grune and Stratton.

Ausubel, D.P. (1968) *Educational Psychology: A Cognitive View*, New York: Holt, Rinehart and Winston.

Bartlett, V.L. and Hull, C. (1978) 'The Geography Classroom as Centre for Research', *Journal of Geography Teachers' Association of Queensland*, Vol. 13(2), 7–13.

Brown, D.W.F. and Macaulay, J.U. (1971) 'Topic Analysis in the Teaching of Geography', *Geography and Education*, Proceedings of the Sixth New Zealand Geography Conference, Vol. 2, 83–9. Reprinted in D. Biddle and C. Deer (eds) (1973) *Readings in Geographical Education, Volume 2*, Sydney: Whitcombe and Tombs.

Jones, F.G. (1974) *Functions of Cities*, Geography Curriculum Project, Publication No. 74-1. Athens, Georgia: University of Georgia.

Joyce, B. and Weil, M. (1980) *Models of Teaching*, 2nd edition, Englewood Cliffs, New Jersey: Prentice Hall.

McCaskill, M. (1977) *Patterns on the Land: Basic Concepts in Geography*, Melbourne: Longman.

Schools Council Geography 16–19 Project (1978). *A Summary Report of the Teacher's Questionnaire Survey*, London: Schools Council.

5

Teaching for Thinking in the Geography Classroom

Peter Wilson

Teaching for thinking should be regarded as one of the fundamental educational goals because many societies are moving from an industrial age to an information age. To attain this goal, there must be a reconceptualisation of curriculum and instruction with emphasis on a learner-centred classroom concentrating on teaching for thinking, based on different levels of complexity of content and the types of problems students must solve. This chapter outlines a carefully planned sequence for teaching thinking skills and a range of techniques to help the teacher evaluate the quality (as opposed to quantity) of students' thinking. These methods embody many of the curriculum and instructional suggestions outlined in the other chapters. Students can learn to think more effectively if the teacher concentrates on teaching them how to do so.

The teacher gave out the map of a small Japanese rural village (Figure 5.1) and asked her class to study the map and the question below it. She was a young teacher and regularly took note of the types of oral and written answers her students gave in class. She was just finishing a small section of work on farming around the world and comparing various practices with those of Australian farmers. She was keen to ensure that her students understood the reasons for Asian village farming compared to large scale farming in Australia. She asked for a volunteer to read the question below the map aloud for the class:

The Australian farmer normally lives away from the village with his land grouped around his home in a single piece. Why do you think the Japanese farmer lives in the village with his land scattered in a number of places?

She gave the students time to write their answers and then asked four of them to read their answers to the class

Konai: 'The farmer has his rice fields next to the drainage so he can easily drain the fields. He has his mulberry and his vegetables growing on dry crop area, because they need a dryer area to grow. Each farmer probably has a bit of dry land and a bit of land near the drainage systems. This would make it fair.'

'That's good, Konai', she said aloud while in her mind she noted that Konai was usually able to weave several pieces of information together in most of her work.

Ted: 'All Japanese have huge families'.

The teacher laughed along with the rest of the class and wondered why Ted always gave such silly answers. His thinking rarely seemed related to the problem on hand.

Figure 5.1 Map of a small Japanese rural village (after Rhys 1966)

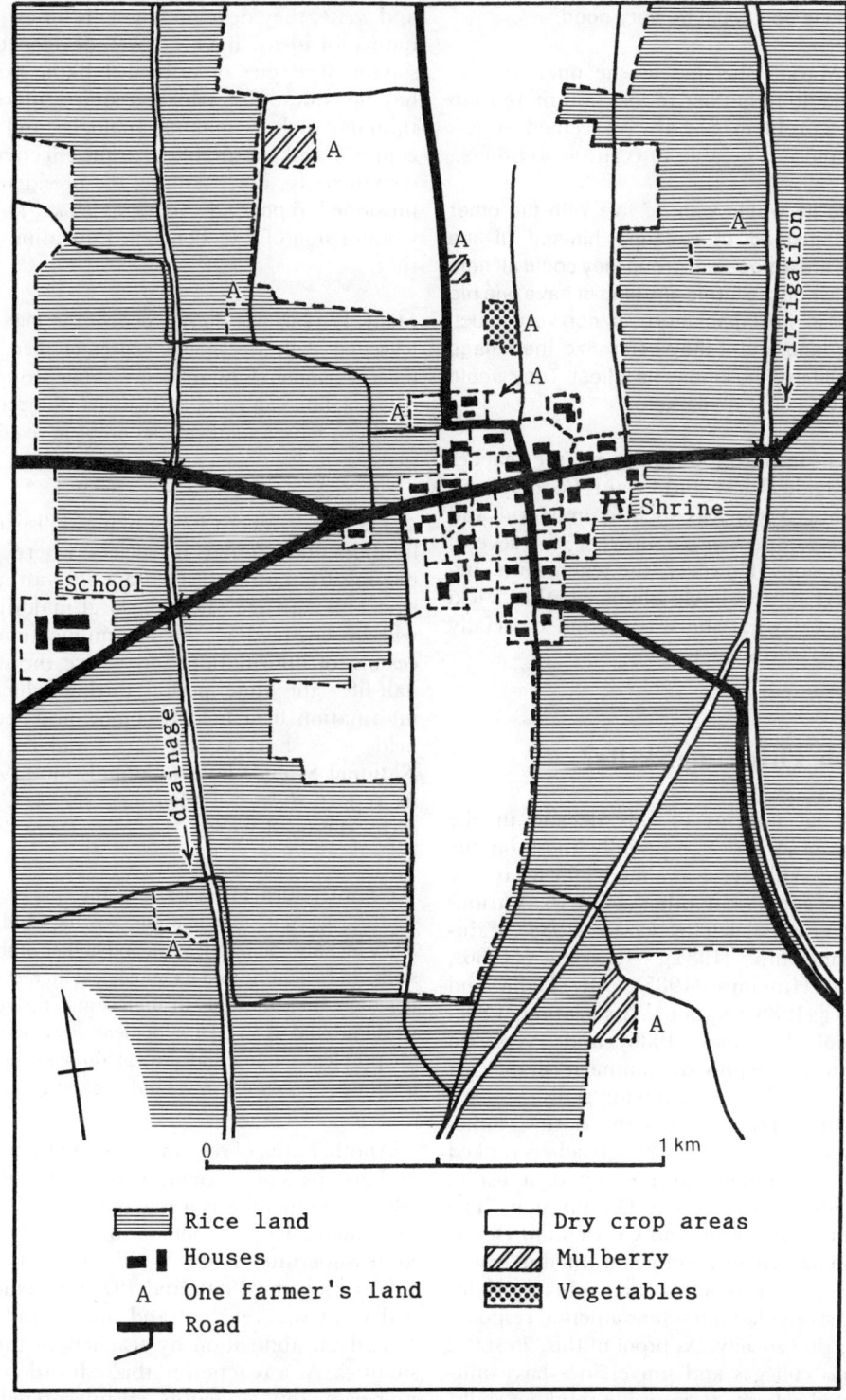

Question: The Australian farmer normally lives away from the village with his land grouped around his home in a single piece. Why do you think the Japanese farmer lives in the village with his land scattered in a number of separate places?

Maria: Because some of the other land might not have good enough soil to grow rice. And the land he picked the soil could be very good.

'Well done, Maria', she said as she puzzled over what she could do to get her to focus on more than one aspect of a problem. She always seemed to pick on one issue and not be able to relate it to others.

Vicki: Because he would want to live with the other farmers instead of isolating himself. If the farmers are a close knit group they could all help each other on the farms and kind of have one big farm. Also in Japan there is not very much farming land and if they centralize the village they could use the land to its fullest. They would also have a central market.

'Did you all hear that good answer?' the teacher said. 'Read it again, Vicki, and I want the rest of you to listen to this.' Vicki was such a smart kid and generally gave comprehensive answers with several possible conclusions. However, the teacher was beginning to have doubts lately about whether Vicki really understood what she was saying, especially some of the big words she used.

Why Teach Thinking Skills?

The educational reforms of this decade in the United States of America, especially those on the area of thinking skills, serve as a model for what is or what should be going on in other countries. Various authors and groups such as Costa (1985), Educational Leadership (1984), Kearney (1986), Marzano and Hutchins (1985), Mc Tighe and Schollenberger (1985), National Association of Secondary School Principles (1986) and Presseisen (1986) have started to provide summaries of the vast array of writings related to thinking skills.

Teaching for thinking is one of the 'hottest' topics in American education today where teachers ranked improvement in thinking at the top of a list of educational goals in a recent Gallup poll. The Association for Supervision and Curriculum Development found 82% of its members wanted information and inservice on how to teach thinking skills. This is not a passing fad but a fundamental response to changes in that society. As proof of this, 26 states and numerous colleges and universities have initiated mandated reforms to improve thinking skills.

Various reports and studies are now highlighting what teachers, parents, employers and others have been saying about the quality of thinking skills in students for sometime. While students learn to read and write, they develop few skills for examining the nature of ideas, have no well-developed problem-solving strategies or critical-thinking skills, cannot handle questions which require more complex thinking such as making analogies and organising concepts and are unable to argue effectively or write convincingly. For example, the presidentially commissioned report, *A Nation at Risk* (The National Commission of Excellence in Education 1983), said that:

Many 17-year-olds do not possess the higher order intellectual skills we should expect of them. Nearly 40 percent cannot draw inferences from written material; only one fifth can write a persuasive essay; and only one third can solve a mathematics problem requiring several steps . . .

The new concern for thinking skills stems from a fundamental change in society where people are moving from an industrial age into an information age. This is an age where the large majority of people will be engaged in the communication and processing of information and where the information half-life (the time period during which half the information in a field becomes outdated) of some fields is as short as six years—and reducing. The National Science Board Commission (1983, p.5) in its report, *Educating Americans for the 21st Century*, states the need well:

We must return to basics, but the 'basics' of the 21st century are not only reading, writing, and arithmetic. They include communication and higher problem-solving skills, and scientific and technological literacy—the thinking tools that allow us to understand the technological world around us . . . Development of students' capacities for problem-solving and critical thinking in all areas of learning is presented as a fundamental goal.

Another area of reform suggested by these authors and groups is in curriculum and instruction. Part of what is happening is a reaction to the *back-to-basics* movement. Kean (1986, p. 6) cites the California State Superintendent of Public Instruction who said that during the 1960s and 1970s the students decided what was relevant and fun to study, and this caused an abdication by teachers of their responsibilities. As a reaction to this, education went back to basics, but educators misinterpreted what the public wanted and did not go back to such things as geography, history, science, writing, high expectations, homework and discipline in the classroom. Educators instead narrowed the curriculum to basic skills. Kean also refers to the Superintendant of the San Jose School District who puts the problem well:

With the return to basics, we screwed off the kids' heads, poured in the information, and asked them to regurgitate the information by asking questions at the end of the week. But we didn't teach them how to use that information.

The back-to-basics movement created discrete add-on programs for selected groups and generally concentrated on the bottom 25%. Now people are asking for programs for *all* students—below average, average, above average and gifted.

What is needed is a reconceptualisation of curriculum and instruction to integrate thinking skills into all aspects of school experience especially the daily interaction between teacher, students and curriculum. This will mean a change in instructional style. Thinking must be seen as essential to all school subjects and its development should be considered as a means as well as an end. Many studies are critical of current instructional style. For example, John Goodlad in his book *A Place Called School* reports on the observations in over 1000 classrooms which showed that about 75% of time was on instruction, and 70% of this was verbal interaction with the teacher out-talking students by three to one. Less than 1% of teacher talk invited students to give more than mere recall.

Thus, there is a need to move from the teacher-dominated classroom to a learner-centred classroom based on instruction in thinking skills. This does not imply a move away from content. Quite the opposite, as improving thinking skills will emphasise mastery of content and allow students to develop strategies for processing information in a variety of forms. As students learn to process information, teachers will need to move away from *objectivity* in teaching and testing. Teachers must no longer assert that there is only one *right* answer and only one *right* way to arrive at their answer. This will mean a fundamental change in teaching and learning style. However, Kearney (1986) sounds a warning that many teachers, education systems and communities may not accept these changes. He warns that once students begin to think they will not limit their thinking to prescribed subjects. They will demand the right to question, to challenge and to demand. They will question, challenge and demand teachers to justify their positions on any number of issues in curriculum, content, school rules, society and the environment. In fact 'Thinking students can, in short, be very inconvenient students' (Kearney 1986, p. 10).

What are Thinking Skills?

Most teachers would agree that helping students to develop and increase their ability to think is a major educational goal. However, while this may be an accepted goal, there is little agreement among psychologists, philosophers and educationalists on a clear definition and specification of thinking skills. Also, most teachers, when they are pressed, have trouble in clearly outlining what are thinking skills and how they can be taught in the classroom. This chapter will not get into the debate among and between psychologists and philosophers with regards to what is thinking, but when the words 'think' and 'thinking' are used in this chapter this will mean that a number of processes are being used, such as knowing, inferring, analysing, judging, generalising, predicting or decision making, etc. Mayer (1977, p. 6) has suggested that a general definition of thinking involves three basic ideas:

1. Thinking is cognitive (i.e. involves knowing, perceiving and conceiving), occurs internally in the mind or cognitive system and is inferred indirectly from behaviour.

2. Thinking is a process which involves some manipulation of, or set of operations on, knowledge in the cognitive system.

3. Thinking is directed and results in behaviour which solves or is directed towards the solution of a problem.

There is little agreement on a definition of thinking and a range of adjectives and terms have been used to describe thinking, such as convergent thinking, divergent thinking, critical or reflective thinking, creative thinking, lateral thinking, problem solving and decision making. Glatthorn and Baron (1985, pp. 49–53) provide a useful summary and suggest that there are at least seven types of thinking which are:

1. Diagnosis—or trouble-shooting where a person uses hypotheses about the source of a problem and the evidence may consist of the results of a test that is performed, for example *my car won't start*.

2. Hypothesis testing—which is the process of forming and testing ideas and theories.

3. Reflection—which is the search for general principles or rules based on evidence which is largely gathered from memory.

4. Insight—or the *eureka* phenomenon where solutions come suddenly and with certainty.

5. Artistic creation—as an important type of conscious thinking where the artists' critical reaction to the evidence, such as the movement in a ballet, are most important.

6. Prediction—which is similar to reflection where the search for goals is not as controlled and may rely on past situations or similar cases.

7. Decision making—which is a type of thinking in which the possibilities are courses of actions or plans.

Also, there is no agreement on how many thinking skills and attributes there are. Most teachers are familiar with the long-standing categories of skills suggested by Bloom and his associates (1956) of knowledge, comprehension, application, analysis, synthesis and evaluation. Yet, other people such as Kearney (1986) list over 40 thinking skills and suggest there may be almost 100. He provides a 'brief' list (Figure 5.2) of 'some' of the skills and attributes various authors have identified as constituting higher-order thinking skills.

However, there is an assumption that these categories of thinking skills are based on levels which are both cumulative and hierarchical. This notion is based on a supposed increasing degree of complexity and sophistication of the cognitive processes. But, Marzano and Hutchins (1985) state that current research and theory does not support the distinction between a set of *lower* and *higher* levels of thinking skills. The lower–higher level distinction is made by the varied levels of complexity in the *content* teachers present to students and the problems they have to solve.

Teachers should distance themselves from equating thinking with a long list of discrete mental operations and concentrate more on a smaller number of critical attributes in their students that distinguishes between good thinking and poor thinking. Figure 5.3 developed by Glatthorn and Baron (1985) contrasts the attributes of good thinkers with poor thinkers. This is not meant to categorise an individual student as a person because a student may be a good thinker in mapping but a poor thinker in physical geography.

The previous section expressed the need for a reconceptualisation of the curriculum and instruction to foster thinking skills. Students can learn to think better if the teacher concentrates on teaching them how to do so.

Figure 5.2 Skills and attributes that constitute high-order thinking skills (Kearney 1986)

• Comparing and contrasting	• Understanding verbal analogies
• Making inferences	• Selection of a solution process
• Analysing events	• Selection of a way of representing a solution
• Synthesising information	
• Drawing conclusions	• Selection of a problem-solving strategy
• Identifying the problem	• Allocation of processing time
• Analysing the problem	• Sensitivity to feedback
• Suggesting possible solutions to the problem	• Translation of feedback into an action plan
• Testing consequences of possible solutions	• Implementation of an action plan
• Assessing the reliability, relevance, sufficiency, validity and meaning of data	• Testing hypotheses
	• Linear reasoning
• Analysing arguments	• Data gathering
• Judging credibility of sources	• Decision making
• Observing and judging observations and reports	• Classifying
	• Organising
• Induction	• Identifying alternative points of view
• Deduction	• Recalling
• Assumption identification	• Grouping/labelling
• Prediction	• Classifying/categorising
• Identification of fallacies	• Ordering
• Definition of problem	• Patterning
• Distinguishing between differences of kind and differences of degree	• Prioritising

The good thinker:	The poor thinker:
• Welcomes problematic situations and is tolerant of ambiguity.	• Searches for certainty and is intolerant of ambiguity.
• Is sufficiently self-critical; looks for alternative possibilities and goals; seeks evidence on both sides.	• Is not self-critical and is satisfied with first attempts.
• Is reflective and deliberative; searches extensively when appropriate.	• Is impulsive, gives up prematurely, and is overconfident of the correctness of initial ideas.
• Believes in the value of rationality and that thinking can be effective.	• Overvalues intuition, denigrates rationality; believes that thinking won't help.
• Is deliberative in discovering goals.	• Is impulsive in discovering goals.
• Revises goals when necessary.	• Does not revise goals.
• Is open to multiple possibilities and considers alternatives.	• Prefers to deal with limited possibilities; does not seek alternatives to an initial possibility.
• Is deliberative in analysing possibilities.	• Is impulsive in choosing possibilities.
• Uses evidence that challenges favoured possibilities.	• Ignores evidence that challenges favoured possibilities.
• Consciously searches for evidence against possibilities that are initially strong, or in favour of those that are weak.	• Consciously searches only for evidence that favours strong possibilities.

Figure 5.3 Good thinking vs. poor thinking (after Glatthorn and Baron 1985)

A Sequence for Teaching Thinking Skills

The following sequence has been adapted from the excellent work of Whitehead (1975, 1978) where he investigated thinking processes based on the factors underlying cognitive growth. He proved conclusively that students who are given training in specific thinking skills including an overview of the problem-solving process improve markedly in their ability to think. However, his original sequence was used only in training the teachers and students in the control classrooms.

The sequence has been adapted and developed over recent years and used in a wide range of inservice workshop with geography teachers. The sequence may be used:

1. in any school;
2. in any area of the school curricula;
3. at any year level, for example some teachers use the sequence when students first enter high school while others use it when students move into the final years of high school for their tertiary accreditation subjects;
4. with any content area;
5. as an intense course over a few weeks or spread evenly over a semester or year;
6. with increasing complexity over several years;
7. in a system directed towards internal assessment or external examinations;
8. where teaching is via a set text, class sets, school-based resources or combination of these;
9. in fieldwork;
10. with any forms of information, such as text, pictures, slides, movies, graphs, diagrams, cartoons.

This final point is important as the examples in the sequence that follows are based only on three black and white pictures. In the reality of the classroom, a large range of resources could be used and this is suggested in the sequence.

Phase One: Gathering Information—List, Group, Label, Sentence

This first phase in the thinking process gives the students an overview of the purpose of the sequence, why it is important, and outlines the type of activities in which the class will be involved. Also, the students are introduced to the procedure of collecting information from photographs, using the skills of listing, grouping (categorising), labelling each group and writing a sentence about each picture using the lists and groups.

1. Discuss with the students the way a geographer goes about solving problems. Point out that, unlike mathematics, there may be several possible answers to a problem and they may have to decide on the best answer. Explain that each problem for the rest of the sequence will relate to a photograph.

2. Show a photograph such as the one in Figure 5.4, but without the caption, and have students individually list all the things they can see in the photograph.

3. Students now sit in groups to compare each others' lists and add where necessary.

4. Develop a composite list on the chalk board or large sheets of paper for display and future use.

5. Students individually group (categorise) the list of words by common attributes and then give each group a label. Ask students to list the words in columns with the label as a heading.

Figure 5.4 Selling cooked seafood in Istanbul, Turkey (photography: K. Boyle)

6. Students compare their categories with other members of their group.

7. Conduct a class discussion about the range of groups and give praise to general or abstract labels. Record some group's labels for display.

8. Students individually write a sentence about the photograph using the words in the lists and the labels for the groups.

9. Students then read and discuss their sentences in groups in order to develop an agreed sentence for discussion.

10. Students hear each group sentence and praise the more creative ones. Record the group sentences on a wall chart for display and future use. A classroom display consisting of the photograph, word lists, groups and sentences could be developed and maintained over a few days or weeks.

11. Repeat this sequence with a second photograph to ensure that the listing, grouping, labelling and sentence procedure is well developed and understood. Variations on this procedure can be used at any time in the sequence.

Phase Two: Gathering Information—Observed and Inferred

The second phase in the thinking process develops the distinction between information that is directly observed in data and information that is inferred. Inferencing helps students to generate a range of possible solutions by going beyond perceived information. This is important in problem solving. Inferencing may be related to guessing and this latter term may be used with students.

1. Rewrite or show a word list from the first photograph and introduce the idea of observed and inferred information.

2. Students individually put an 'O' against things observed and an 'I' for things inferred, and then discuss their findings in groups.

3. Have a class discussion and record the answers on the wall chart or board.

4. The number of inferred words will usually be less so ask each student to list five extra inferences, Then compare these in groups. During class discussion, add all the new inferences to the list and praise the more abstract ideas.

5. Introduce a new photograph (or refer to one previously used) and ask students to draw up two columns in their pads as shown:

Things observed in photograph	Things inferred in photograph

6. Students individually list things observed and inferred and then discuss their findings in groups. Develop a composite list from class discussion on a wall chart or chalk board.

7. Students individually and/or in groups categorise the lists and give them a label. The grouping will spread across both the observed and inferred categories of information. Have a class discussion to find the types of categories developed and record these.

8. Ask students individually and/or in groups to write a sentence about the picture. Discuss these in class and record on the wall chart for classroom display. This could be a separate display or added to the previous display.

Phase Three: Generating Alternative Answers to Geographical Problems

This third phase encourages students to see problems from different points of view and to suggest possible alternative answers. Three sources of information are developed: information in the photographs, information from previous experience and guessing. The traditional approach of one correct answer is removed as all answers are valued.

1. Revise the section in the first phase on 'the way a geographer solves problems (Step 1)'.
2. Outline and discuss the three sources of information:
   (a) information from the photograph both observed and inferred;
   (b) information from books, television, teachers, other people and previous experience;
   (c) inferencing (guessing).
3. Introduce a new photograph and use the 'list, group, label and sentence' method, if needed, or refer to a previous photograph and its information on display in the classroom.
4. Using a photograph such as the one in Figure 5.4 introduce the problem: for example, 'Why do these people buy their food this way?'.

Keep the question simple. Have students individually try to write down *five* possible answers and then discuss their answers in groups.

5. During class discussion, record all answers on a wall chart (for display) or the chalk board. Encourage all answers, even the 'stupid' ones, but have each student outline the information source. (If the photograph relates to a particular area of content, move on to the 'answer' although this should not be done on all occasions so that some 'answers' remain uncertain.)

Phase Four: Problem Clarification

Problem clarification allows students to pause and reflect on the nature of the problem and to ask such questions as:

- What does the question mean?
- What does each word in the question mean?
- What information do I need to find?

Thus, clarification of the 'meaning' of the problem aids in establishing what evidence is needed to accept or reject possible solutions. The task of finding evidence to support possible answers now becomes more difficult as the search is limited by the particular problem statement.

1. Revise 'the way a geographer solves a problem' (Phase one, Step 1) and the 'three sources of information' from Phase three, Step 2.
2. Introduce a new photograph such as the one in Figure 5.5 and use the technique of 'list, group, label and write a sentence' if necessary.

Figure 5.5 Farmer in Novgorod, USSR (photography: P. Wilson)

3. Introduce the problem in the photograph. For example, in Figure 5.5 the question might be 'Is there poverty in this area?'.
4. Ask the students to discuss and establish the meaning of key words, for example 'poverty'. A dictionary is a valuable aid in this procedure.
5. Ask the students to draw two columns in their pads as shown:

Evidence for poverty		Evidence against poverty	
Observed	Inferred	Observed	Inferred

6. Students individually fill in the table and try for at least five pieces of information in each column. This is followed by a discussion to develop a group table. Have a class discussion to develop a composite table which is recorded on the chalk board or wall chart for display or future use.
7. Nominate half of the students in each group to write individually a sentence about evidence for poverty and the other half to write a sentence about evidence against poverty. Have the students discuss their answers with their neighbours.
8. Conduct a class discussion by calling for volunteered sentences or nominate students to read theirs. Occasionally, challenge students to justify their sentences. Ask for some of the better answers to be re-read, and ask other students to suggest why they are 'good' answers.
9. Collect some of the sentences and mount a new display in the room. This should consist of the photograph, composite tables and selected student answers.

Phase Five: Weighing Up Evidence

This phase introduces the skill of weighing up evidence in order to reach a solution to a given problem. The focus is on exploration rather than description and students are encouraged to find and use evidence in a careful, complete, deliberate and systematic manner. The idea of 'quantity' of evidence vs. 'quality' of evidence is introduced and that 'speed' on reaching a decision is not really valued in the classroom. Also, students are encouraged to:

- state the 'best' answer;
- agree and disagree with each other;
- explain why they will accept or reject alternative views;
- recognise other people's points of view.

1. Carry out a class discussion which centres on the ideas of 'quantity' and 'quality'. Again, a dictionary might be a valuable aid. Have individual students give personal examples of how they have been affected by 'quantity' and 'quality'.
2. Refer the students to the photograph and their composite lists of evidence for and against poverty from Phase four. Ask students to re-read some of their sentences.
3. Conduct a class discussion on the idea of 'weighing up' one thing against another. Now, refer the students to the composite list and rewrite it so that the evidence in one column is systematically matched or weighed up against the evidence in the other column. Note areas where evidence is missing and ask the students to see if this can be filled in. Point out that this may be impossible sometimes.
4. Ask students to write *one* sentence about the problem in the picture which weighs up evidence for and against and chooses which argument is preferable. Some students may even weigh up the evidence and reject both arguments. Have a class discussion about the sentences which are written.
5. Record some of the sentences and add those to the classroom display.
6. Introduce a new photograph with a problem and have students carry out the previous steps as a way of consolidating the skills.

Phase Six: Weighing Up Evidence— A Refined Approach

This phase continues the skill of weighing up evidence and introduces the 'tentative' nature of answers. Students should be reminded that solutions to geographical problems are difficult as many, sometimes unknown, variables are involved. Hence, conclusions are often drawn in the light of available evidence which is often sketchy. So any conclusions must be 'tentative'. Tentativeness implies that more evidence must be found for a satisfactory answer.

Stress that speed in reaching a decision is not valued, that quality of thinking is important, good answers are reached only after careful consideration of all evidence and impulsive answers often cause poor-quality decisions. Students should be encouraged to:

- use as much information as possible;
- draw upon the three sources of information (Phase three, Step 2);
- argue why alternative answers are not acceptable;
- be tentative in drawing conclusions, especially the final conclusion.

1. Show a new photo such as the one in Figure 5.6 and pose a question such as 'Where is this person going?'. If you pose a more complex question, make sure the students are clear about the meaning and the words in the question.

2. Ask students individually to list at least three possible answers. Students should discuss these in a group before a composite class list is developed on the chalk board or wall chart. Accept all answers even those that may be outrageous or silly.

3. Conduct a class discussion, asking students to justify their suggestions about each one. Carefully lead the class to *three* logical but dissimilar answers. All students to draw up a table similar to the one shown for Figure 5.7.

4. Students work individually to fill in some information for each column. Stress observed and inferred information as well as weighing up evidence. (This could be a lengthy classroom activity and could be used as homework.)

5. Discuss the table in groups and then conduct a class discussion to develop a composite class table.

6. Ask each student to write one sentence for each possible solution. Ask the students *not* to weigh up evidence about possible solutions. Then hear a range of responses from the students, asking them to comment on each others' answers. Try to highlight some of the better answers and explain why.

7. Display the picture, question, composite table and several selected sentences for each of the three possible answers.

8. Encourage the students to study the display and re-submit their revised answers. This allows for on-going individual reflection and group discussion.

Figure 5.6 Sheepherder in Greece (photography: K. Boyle)

Figure 5.7 Page format for weighing up evidence—a refined approach

Taking the sheep to the market		Taking the sheep to pasture		Taking the sheep to the slaughterhouse	
Evidence for	Evidence against	Evidence for	Evidence against	Evidence for	Evidence against

Phase Seven: The Model for Thinking

Phase seven introduces the student to a total, integrated, structured process for problem solving in geography. Duplicate the full page, structured, abstract model as a record sheet in this phase shown in Figure 5.8. At the conclusion of this phase, the students should know all the steps in problem solving and recognise the need always to make rough notes.

1. Revise the nature and difficulty of geographical problems introduced in Phases one and six, and the three sources of information from Phase three, Step 2.

2. Compare the solution of geographical problems with the role of a detective solving a crime. Individual and group work may be used to supplement a discussion which establishes the following problem-solving sequence:
   (a) Looking around the scene of the crime.
   (b) Establishing the nature of the crime.
   (c) Determining possible suspects.
   (d) Collecting evidence for and against each suspect.
   (e) Making an arrest based on the evidence.

3. Distribute a full page copy of Figure 5.8 and explain that the top box is for writing the problem, the next three boxes are for possible answers; then evidence for and against can be recorded and finally a sentence about the decision can be written.

4. Show a new photograph to the class and ask them individually to list, group, label and write a sentence about it. Introduce a problem such as the one for Figure 5.6 in Phase six which was 'Where is this person going?'.
   At this point still keep the question as simple as possible. Later a more complex question can be introduced perhaps using some of the higher-order key questions of geography such as how and why?; what impact?; and how ought?.

5. Refer the students to the model (Figure 5.8) and outline how it may be used to answer the question by working through the five steps:
   (a) Look carefully at all parts of the photo, including left and right, foreground, middle ground and background.
   (b) Identify and understand the stated problem.
   (c) Suggest possible answers to the problem.
   (d) Gather information for and against each possible answer, remembering observation and inferencing.
   (e) Make a final decision on the answer.

6. Ask students to write the question—Where is this person going?—on the top column of the model and then look at the photograph.

7. Tell students to turn the sheet over, and list as many possible answers to the question. Then they must choose the three best and write these in the next columns.

8. For each of the three possible answers ask students to find evidence for and against, remembering to use observation and inferencing.

9. Finally write a sentence in the bottom box which weighs up all the evidence and reaches a decision.

10. The teacher should move around the class as the previous four steps are introduced slowly. Then, conduct a class discussion which hears and criticises a range of answers.

11. Finally, mount a display in the classroom of the photo and a range of completed models.

Figure 5.8 Abstract model to record the thinking involved in solving a problem

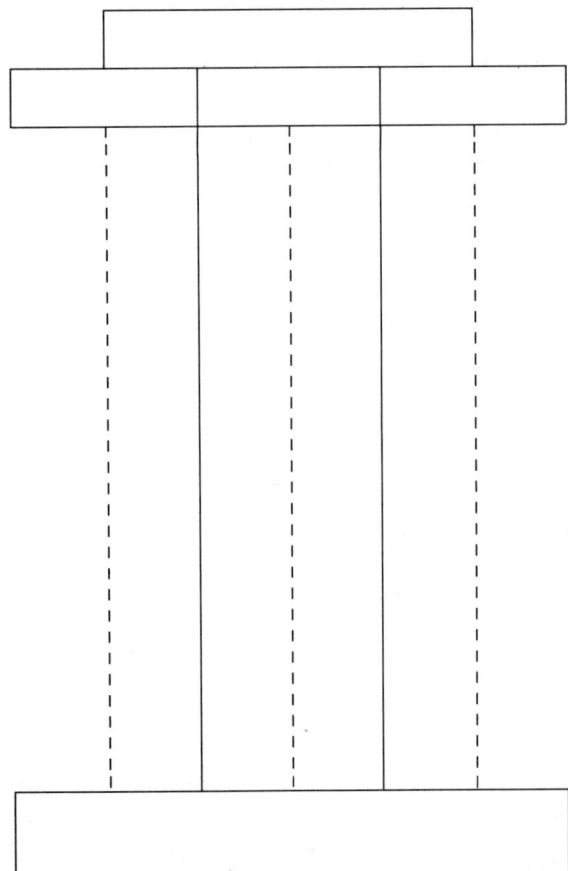

12. Repeat this procedure over a number of weeks, in a variety of contexts, with a variety of resources to ensure that all students can use the model according to their capabilities.

Phase Eight: Simplified Model for Problem Solving

Phase eight removes the special model sheet introduced in Phase seven and encourages students to follow through the same procedures, using notes jotted in their pads. Students often have difficulty in problem-solving due to their inability to coordinate the many subcomponents of the process. This method gives them an easy procedure for working through the subcomponents required to solve a geographical problem.

1. Review the nature and difficulty of geographical problems (Phase one), the five steps for problem solving (Phase seven, Step 5) and the three sources of information (Phase three, Step 2).

2. Show a new photo and explain that no model will be used but instead notes will just be jotted in their pads. Have the students write in note form:
    (a) the problem;
    (b) possible answers to the problem;
    (c) some evidence for and against each possible answer;
    (d) a concise sentence which weighs up the evidence.

3. Have a class discussion which hears and criticises a range of answers. Comment on the responses, offering special praise for ones that use complex or abstract words and ideas.

4. Develop a chart for the classroom (or distribute a copy to each student) of the following 16 steps. Explain that from now on, the classroom or home procedures for solving a geographical problem are:
    (a) Use the full sequence for solving geographical problems.
    (b) Search for information by both observation and inference.
    (c) Clarify the nature and purpose of the problem.
    (d) Do not centre on the one best or first answer.
    (e) Suggest possible alternative answers.
    (f) Always collect information from the three possible sources.
    (g) Weigh up evidence for and against.
    (h) Have a degree of tentativeness.
    (i) Speed is no longer important.
    (j) Strive for quality rather than quantity.
    (k) Explain any alternative answers and their evidence.
    (l) Jot down rough notes.
    (m) Address all sections of the question.
    (n) Use explanation as well as description.
    (o) Try to use complex words and ideas.
    (p) Write a clear, concise, quality answer.

Techniques for Evaluating Student Thinking

Teachers who have used this sequence for promoting thinking skills in the geography classroom usually ask two important questions:

1. What techniques can be used to evaluate students' thinking?
2. What method can be used to evaluate the quality of students' thinking?

Some examples of instruments for evaluating student thinking follow. Present the material and pose a suitable question related to the content you are about to cover, are studying or have just finished.

1. A horizontal, oblique or vertical coloured or black and white photograph with a suitable question attached, such as those in Figures 5.4, 5.5 and 5.6.

2. An historical photograph with a question related to a time period in the future, such as, 'What will this shopping centre be like in 20 years time?'.

3. A series of 10 photographs, for example, with five about cities and five not about cities (Whitehead 1975), and the students are asked to sort these into two groups. Then students are asked 'Which features in the photographs caused you to put them into a city pile?'.

4. A black and white sketch such as the one in Figure 5.9 where the students have to describe the position of the trees (after Haemon 1973).

5. A map of an area with a suitable question posed. An example is shown in Figure 5.1.

6. A range of graphic information (Figure 5.10) which includes such examples as maps, diagrams, tables and graphs followed by a question.

Question: Describe the position of the trees.

Figure 5.9 The Olgas in central Australia (after Haemon 1973)

Figure 5.10 Croft farming at Park on the Island of Lewis in Scotland (after Rhys 1966)

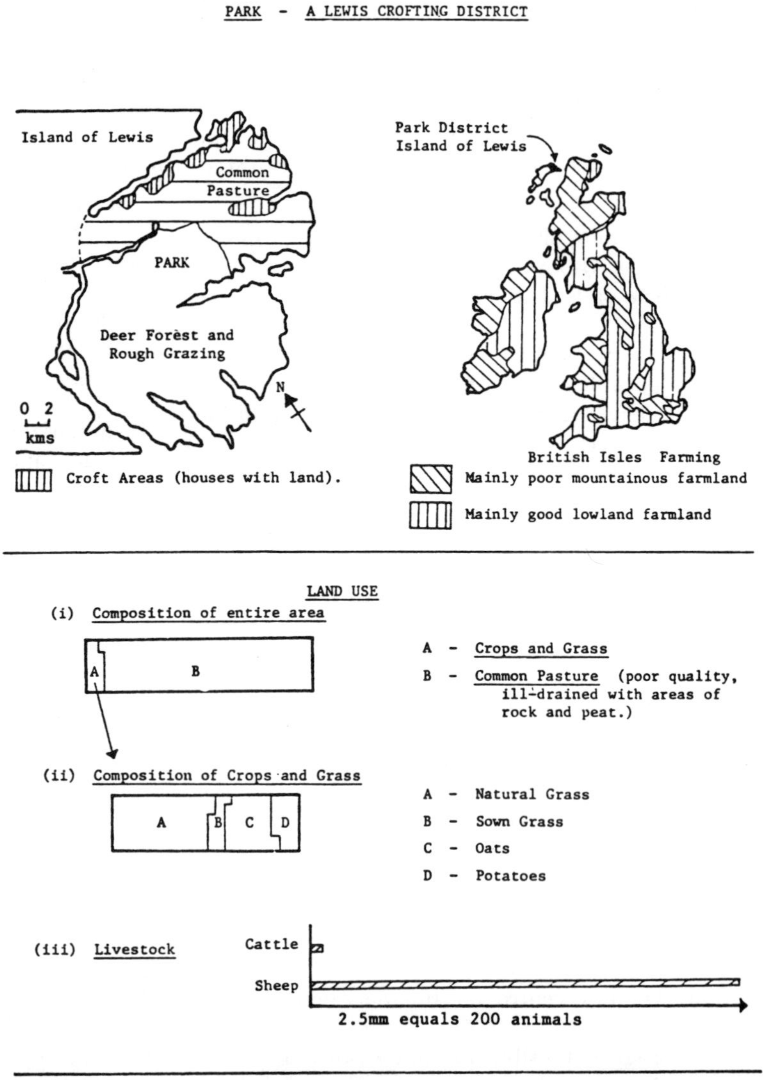

PARK — A LEWIS CROFTING DISTRICT

Island of Lewis — Common Pasture — PARK — Deer Forest and Rough Grazing

Park District Island of Lewis

0 2 kms

⊞ Croft Areas (houses with land).

British Isles Farming
▨ Mainly poor mountainous farmland
⊞ Mainly good lowland farmland

LAND USE

(i) Composition of entire area

| A | B |

A — Crops and Grass
B — Common Pasture (poor quality, ill-drained with areas of rock and peat.)

(ii) Composition of Crops and Grass

| A | B | C | D |

A — Natural Grass
B — Sown Grass
C — Oats
D — Potatoes

(iii) Livestock

Cattle
Sheep

2.5mm equals 200 animals

Question: What is a croft farmer?

7. A small section of text followed by a question, such as the example in Figure 5.11.

8. Identify any geographic term and ask 'What does . . . mean to you?' (Milburn 1972). For example, for the term 'alp' the question would be 'What does alp mean to you?'.

9. A piece of text with a key conceptual word replaced by an artificial nonsense word.

Students are asked to name the word and explain its meaning:

Horseshoe Farm is in south-west Shropshire near the Welsh border. The general climatic figures for this area show us that the farm may expect 1000 mm of rainfall per year. The maximum monthly temperature is 21°C (for July) whilst the minimum monthly temperature is 3°C (for January). Mr

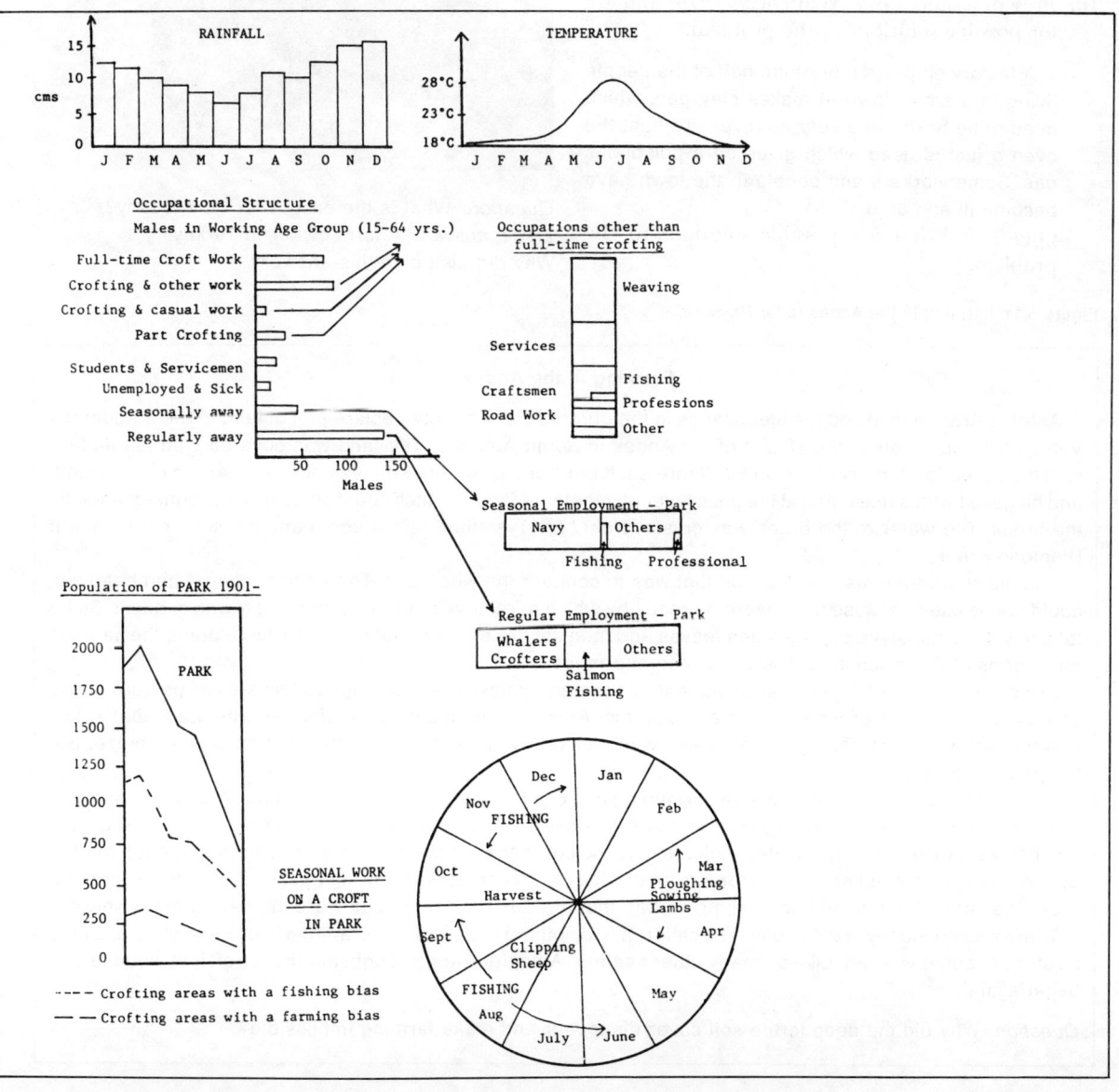

Williams is not likely to rely only on this information, for climate figures are always compiled from the *average* weather over thirty-five years. Average figures always conceal *extremes*. It is the extremes like gales, droughts, floods and late frosts that really upset the farmer. So Mr Williams has to take account of the extremes of weather that he may expect in his valley and so limit the risks that he takes accordingly. He will also be interested in the *zykam* of his farm, for conditions will be substantially different in the sheltered valley to those on the exposed moorland some 400 metres higher up. (After Warn 1984)

Question: What does *zykam* mean to you?

10. Present a short story (Whitehead 1975) and ask for possible solutions to the problem.

A factory employs more than half of the people living in a small town. It makes clay pots which need to be baked in a very hot oven. To heat the oven a fuel is used which gives off a poisonous gas. Some workers and people in the town have become ill and died.

Question: What are possible solutions to the problem?

Figure 5.11 Farming in the Andes (after Rhys 1966)

11. A puzzle (Figure 5.12) which involves a picture, a question and a suggested answer (after Whitehead 1975).

Figure 5.12 A picture puzzle (after Whitehead 1975; photography: K. Boyle)

Question: What is the occupation of this lady?
Lisa's answer: A farm owner in Turkey.
Why did Lisa give this answer?

Farming in the Andes

Antonio Arango invested his life-savings in the purchase of a twenty-hectare plot of land in the Magdalena Valley, high up amongst the slopes of the Andes in South America. The land was covered with tall timber, which showed the fertility of the soil. "Where such big trees grow, the soil is good," said Antonio to himself, and he gazed with satisfaction at the thick layer of vegetable topsoil which could be seen in the cutting made by the brook. The water of the brook was crystal clear, so crystalline that Antonio and his sons christened it Diamond Brook.

Antonio sharpened his axe, the iron that was to conquer the woodland. The timber, when it had been cut, could not be used because there were no roads by which to remove it nor neighbouring towns to buy it. So he let a few days' sunshine dry the fallen leaves and then set fire to them. Other settlers were doing the same in other parts of the mountains. The blaze was enormous.

At long last the ground was cleared and Antonio sowed it with maize, keeping two hectares as pasture for his cow. And, on the high ground he built a house, too. Antonio named the house after his wife, *La Isabella*. The first harvest was encouraging. Antonio was well pleased with his efforts, so he went on sowing maize. But things were changing.

Diamond Brook, which had once been brimming and crystal clear through the year, had shrunk to a thread of water in summer-time. In the rainy season it was a yellow flood tearing loose rocks and mud and lumps of soil. The harvests were growing smaller. Antonio sold the cow because the pasture was no longer enough for it to live on. Things were not going right for him at all. One day, when he was sowing maize, his spade struck solid rock. The vegetable topsoil had become so thin that already rocky outcrops were appearing everywhere.

There was no money in the house. The children had nothing to eat. Lastly, Diamond Brook vanished and only its stony channel was left. Like so many other settlers, Antonio Arango sought another stretch of woodland to begin again.

Question: Why did the deep fertile soil cover disappear and make farming impossible?

Figure 5.13 Problem-solving story (Whitehead 1975)

The land surrounding a small town is to be divided into housing blocks. Many new houses are to be built. It has been decided that a transport service must be provided to link the town with the city. Three suggestions have been made—trains, buses or trams.

Question: What transport service should be provided?

Trains	Buses	Trams
• Require a large amount of land for tracks and station • Provide a fast service • Can transport large numbers of people • Very expensive to build	• Fumes from buses pollute the environment • Roads already go between the town and the city • Some buses already travel between the town and the city • They only run on week days. The service is poor.	• Can carry more people than buses • They are a hazard for cars on the road. • The roads have to be ripped up to lay the track • Fares on trams are cheapest

12. Problem-solving stories, such as Figure 5.13, where possible answers (two in simpler versions) are given to a problem with information to support them and the student must decide which is the best answer and explain why (Whitehead 1975).

13. A diagram, such as a food chain with some written descriptions followed by a question about possible change.

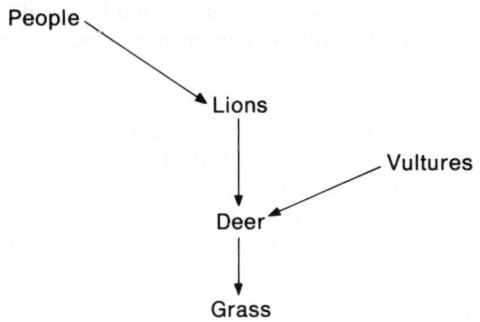

A food chain in a part of Africa

This diagram shows what lions, deer and vultures eat in a grassland area of Africa. The deer eat the grass which grows in the area. The lions eat the deer when they can catch them. The vultures eat the remains (leftover carcass) of any deer that has been killed and eaten by the lions. This way the vultures keep the area clean of dead meat.

For hundreds of years nature has been in balance in this area. Deers eat the grass, lions eat deer and vultures eat any remains of the killed deer that lions might leave. In this way, the numbers of the different animals are kept in balance.

Then, big game hunters came into the area and shot most of the lions.

Question: What will happen in the area after a few years
— to the number of deer?
— to the amount of grass?
— to the number of vultures?

14. A PMI (de Bono 1970) which asks students to analyse information in the Plus direction (good points), the Minus direction (bad points) and finally in the Interesting direction (interesting things that might cause or are worth noting, even if they are neither good or bad). Bailey (1978, p. 55) provides an example.

Example idea: Britain should export more tractors to India. Do a PMI on this. Here is mine as an example:

P [a] Mechanised agriculture would improve food production and help prevent malnutrition.
  [b] It would be good for Britain's exports.

M [a] Mechanisation causes rural underemployment and makes worse the shift of population to the towns.
  [b] There are insufficient spare parts and trained mechanics in India to maintain tractors in good condition.
  [c] High fuel costs increase India's dependence on oil-producing nations.
I [a] Export of tractor parts to India would help them to help themselves by developing agriculture and training mechanics.
  [b] Britain should export either tractors or parts to India or the Russians will probably do so instead.

15. Ask students to draw a free recall sketch map of an area they are familiar with, such as their bedroom, home or schoolgrounds, and mark the map using the scoring scheme provided in Figure 23.6, Chapter 23.

Evaluating the Quality of Student Thinking

Teachers already evaluate the quality of thinking but in a subjective way. For example, teachers read a 'good' or 'bad' piece of written work to other colleagues in their staff rooms and make favourable comments or have a good laugh. They write 6/10 and often comments such as 'a good try, but you missed the point so try to develop your argument more'. This usually aids the *quantitative* evaluation of thinking which shows how much information a student has learned. Teachers are familiar with the vast array of assessment items for this, such as field trip reports, objective tests, short answers, essays and oral responses in the classroom. But, the *quality* of thinking is equally as important and probably in this era, more important. The teacher and student also need to know how to improve this.

Biggs and Collis (1980, 1982) have provided a simple method for evaluating the quality of thinking which helps overcome the problem that each teacher's idea of quality is often so different. Biggs and Collis reviewed the various types of evaluation in areas such as English, geography, history and mathematics to evolve the SOLO Taxonomy, which is the Structure of Observed Learning Outcomes. The clue to this is the *structural organisation* of the answer which discriminates well learned from poorly learned material in a way not unlike mature thought is distinguishable from immature thought. This taxonony is based on the work of Piaget, but whereas Piaget's purpose was to classify the learner into a particular stage, for example a concrete thinker, this taxonomy is concerned with classifying particular responses. Hence, the stress is on the wide variety of responses that a student may make during class or on a piece of assessment.

Despite the daunting name, the SOLO Taxonomy is easy to use and will allow teachers to assess quality retrospectively in an objective and systematic way that is also easily understandable for both teacher and student. For this reason, the taxonomy may be used as an instructional as well as an evaluative tool. The basic features of the SOLO Taxonomy are shown in Figure 5.14.

The SOLO Taxonomy allows teachers to evaluate the quality of students' thinking according to four main variables which are:

1. The degree of abstractness—ranging from personal and subjective prestructural thinking through the next three levels, which are particularistic and concrete, to the final general or abstract thinking level.

2. The number of organising dimensions—ranging from the prestructural, which lacks any intrinsic organisation, through the uni-structural, which uses several, and finally to the abstract responses which involve general principles.

3. The amount of consistency—ranging from the inconsistent prestructural, through the uni-structural involving one internal element to the multi-structural with several unrelated elements, and finally to abstract responses which are internally and externally consistent.

4. The openness of conclusions—ranging from the lower levels which are either closed or indecisive to the abstract which is often qualified and leave are room for different interpretations.

The SOLO Taxonomy is a simple yet powerful technique to evaluate the quality of student thinking, especially when linked with the range of instruments outlined in the previous section or used during oral questioning in class. Geography teachers who consistently use this taxonomy will be able to:

1. develop the quality of students' thinking in their classroom;

2. evaluate the level of thinking of each student in their class on a particular topic;

3. determine the types of questions to use either orally or in written form;

SOLO level	SOLO response description	Response structure
Prestructural	Irrelevant response, including incidental information that has struck the student's fancy.	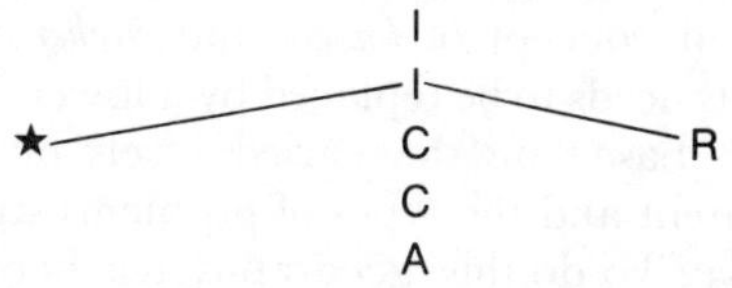 Cue, Data, Response — irrelevant data (I) leading to response (R)
Uni-structural	The use of one obvious piece of given concrete data that the student has seized upon.	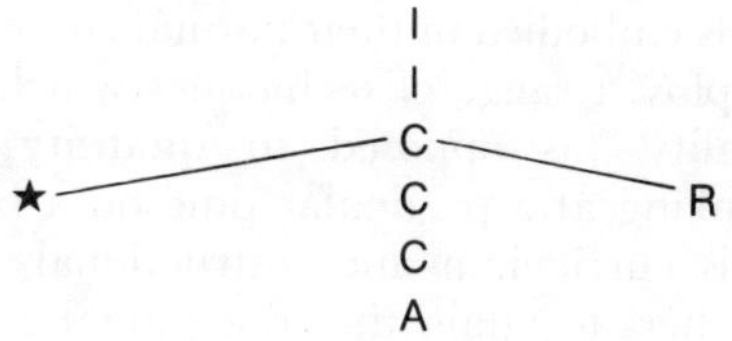 One concrete datum (C) used to form response (R)
Multistructural	The sequential use of two or more simple concrete facts based directly on the given information to form a unique conclusion, but without thinking about how they interrelate.	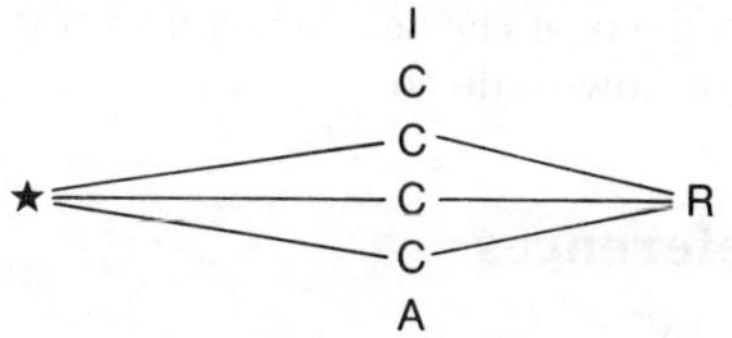 Several concrete facts (C) used sequentially to form response (R)
Relational	The integration of the given data to form a unique conclusion or generalisation, but sticks to the given context.	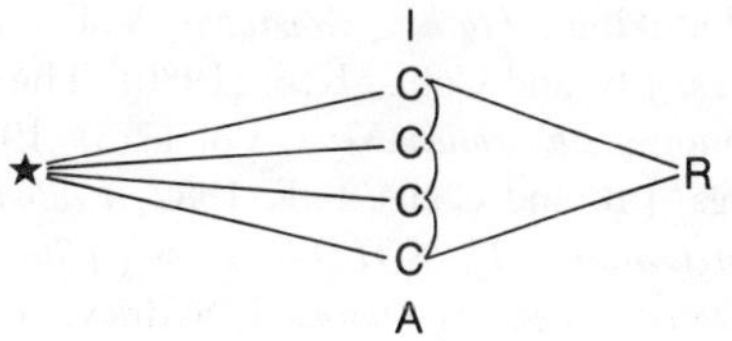 Integrated concrete facts (C) forming a unique response (R)
Extended abstract	Abstract principles are used to interpret concrete facts. This may include forming a general hypothesis, assessing the quality of models and accepting open-ended answers. Several conclusions are possible. In all cases, the response involves the use of 'information' not given in the stem.	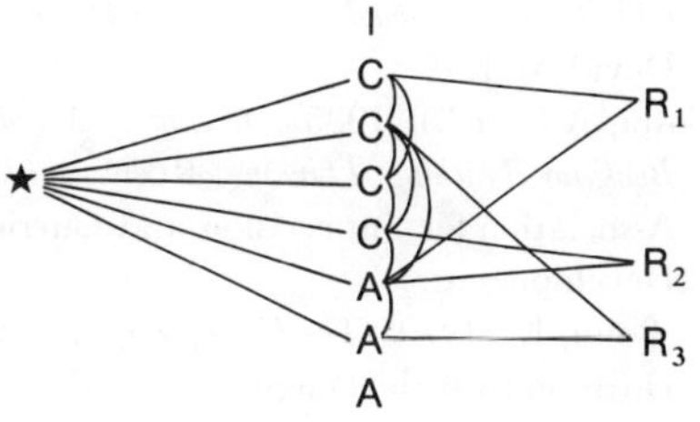 Concrete facts (C) and abstract information (A) producing multiple responses R_1, R_2, R_3
Kinds of data used in response: C = Concrete fact as given		I = irrelevant or inappropriate A = relevant, hypothetical, abstract not given

Figure 5.14 The SOLO Taxonomy (after Collis and Davey 1984; Biggs and Collis 1980)

4. receive some form of feedback into their teaching strategies and content selection;
5. determine the amount of background knowledge(content) that is required to lead students to higher levels of responses;
6. determine the teaching strategies which are appropriate to handling content at a given target SOLO level;
7. develop guidelines for what may be expected in terms of learning quality.

Conclusion

Teaching for thinking is regarded as one of the most important educational goals as many societies move from an industrial age to an information age, where the information half-life (the time period during which information in a field becomes outdated) of some fields is as short as six years — and reducing.

However, to attain this goal there must be a reconceptualisation of curriculum and instruction with the emphasis on learner-centred classrooms concentrating on teaching for thinking. The outdated concept of *lower-* and *higher*-order thinking skills needs to be replaced by a lower–higher distinction based on the varied levels of complexity of content and the types of problems students have to solve. To do this, geography teachers need to use a carefully planned sequence for teaching thinking skills embodied in their normal content lessons, and employ a range of techniques which evaluates the quality (as opposed to quantity) of students' thinking at a particular time on a particular task. This curriculum and instructional style will help teachers to attain the educational goal of teaching students to think. Students can learn to think more effectively if the teacher concentrates on teaching them how to do so.

References

Bailey, A. (1978) 'Geography and the Teaching of Thinking', *Teaching Geography*, Vol. 4(2), 54–56.

Biggs, J.B. and Collis, K.F. (1980) 'The SOLO Taxonomy', *Education News*, Vol.17(5), 19–23.

Biggs, J.B. and Collis, K.F.(1982) *Evaluating the Quality of Learning: The SOLO Taxonomy (Structure of the Observed Learning Outcome)*, Sydney: Academic Press.

Bloom, B.S. et al. (eds) (1956) *Taxonomy of Educational Objectives: Handbook 1: Cognitive Domain*, New York: David McKay.

Costa, A.L. (ed.) (1985) *Developing Minds—A Resource Book for Teaching Thinking*, Roseville, California: Association for Supervision and Curriculum Development.

de Bono, E. (1970) *The Use of Lateral Thinking*, Harmondsworth: Penguin.

de Bono, E. (1971) *Practical Thinking*, CAPE.

Educational Leadership (1984) 'Thinking Skills in the Curriculum' *Educational Leadership*, Vol. 42(1).

Ennis, R.H.(1985) 'Goals for a Critical Thinking Curriculum', in A.L. Costa (ed.) *Developing Minds—A Resource Book for Teaching Thinking*, Roseville, California: Association for Supervision and Curriculum Development.

Fraenkel, J.R. (1980) *Helping Students Think and Value: Strategies for Teaching Social Studies*, second edition, Englewood Cliffs, New Jersey: Prentice Hall.

Glatthorn, A.A. and Baron, J.(1985) 'The Good Thinker', in A.L. Costa (ed.) *Developing Minds—A Resource Book for Teaching Thinking*, Roseville, California: Association for Supervision and Curriculum Development.

Haemon, A.J. (1973) 'The Maturation of Spatial Ability in Geography', *Educational Research*, Vol. 16(1), 63–66.

Kean, (1986) 'Assessing High Order Thinking Skills: An Awareness of the Issues', in G.P. Kearney *Assessing Higher Order Thinking Skills*, TME Report No. 90, Princeton, New Jersey: ERIC Clearinghouse on Tests, Measurement and Evaluation.

Kearney, G.P. (1986) *Assessing Higher Order Thinking Skills*, TME Report No. 90, Princeton, New Jersey: ERIC Clearinghouse on Tests, Measurement and Evaluation.

Marzano, R. and Hutchins, C. (1985) *Thinking Skills: A Conceptual Framework*, Aurora, Colorado: Mid-Continent Regional Educational Laboratory.

Mayer, (1977) *Thinking and Problem Solving: An Introduction to Human Cognition and Learning*, Glenview, Illinois: Scott, Foresman and Company.

Milburn, D. (1972) 'Children's Vocabulary', in N. Graves, (ed.) *New Movements in the Study and Teaching of Geography*, London: Temple Smith.

McTighe, J. and Schollenberger, J. (1985) 'Why Teach Thinking: A Statement of Rationale', in A.L. Costa (ed.) *Developing Minds — A Resource Book for Teaching Thinking*, Roseville, California: Association for Supervision and Curriculum Development.

Perkins, D.H. (1985) 'What Creative Thinking Is', in A.L. Costa (ed.) *Developing Minds—A Resource Book for Teaching Thinking*, Roseville, California: Association for Supervision and Curriculum Development.

Presseisen, B.Z. (1985) 'Thinking Skills: Meanings, Models, Materials', in A.L. Costa (ed.) *Developing Minds—A Resource Book for Teaching Thinking*, Roseville, California: Association for Supervision and Curriculum Development.

Presseisen, B.Z. (1986) *Critical Thinking and Thinking Skills: State of the Art Definitions and Practice in Public Schools*, San Francisco, California: American Education Research Association.

Rhys, W.T. (1966) *The Development of Logical Thought in The Adolescent with Reference to the Teaching of Geography in Secondary School*, Unpublished MEd thesis, University of Birmingham.

Rhys, W.T. (1972) 'The Development of Logical Thinking', in N. Graves (ed.) *New Movements in the Study and Teaching of Geography*, London: Temple Smith.

Stiggins, R.J., Rubel, E. and Quellmalz, E. (1986) *Measuring Thinking Skills in the Classroom*, Washington DC: National Education Association.

The National Commission of Excellence in Education (1983) *A Nation at Risk: The Imperative for Educational Reform*, Washington DC: The National Commission of Excellence in Education.

The National Science Board Commission on Pre-

College Education in Mathematics, Science and Technology (1983) *Educating Americans for the 21st Century*, Washington DC: The National Science Board Commission on Pre-College Education in Mathematics, Science and Technology.

Warn, C. (1984) *The Geography of Production*, London: Macmillan.

Whitehead, G.J. (1975) *Cognitive Performance and Social Studies: The Impact of Two Training Programs on Grade Six Students*, Unpublished PhD thesis, La Trobe University, Melbourne.

Whitehead, G.J. (1978) *Enquiry in Social Studies*, ACER Research Series No. 101, Hawthorn, Victoria: ACER.

6

Making Inquiry Learning Work in the Geography Classroom

Bernard Cox

If teacher exposition is the most commonly used method of teaching in geography, student inquiry is the method most commonly advocated. Student inquiry may take many forms depending upon the amount of responsibility teachers are prepared to give to students in setting inquiry goals, selecting and gathering data, analysing them and determining conclusions to the inquiry. Bernard Cox suggests that if any form of inquiry is to work in the geography classroom, teachers need to be able to stimulate student interest and curiosity, know when and where not to intervene in the process of student inquiry, and provide them with opportunities to develop critical thinking skills. This chapter attends to each of these points and provides numerous examples of ideas for successful classroom inquiry.

What is Inquiry?

The topic had been discussed in class for some time. There had been a substantial amount of exposition by the teacher, but they had progressed with the subject to a point where pupils were able to offer informed opinions. Then, unexpectedly, 'Oh, Miss, I'd really like to find out what happens when . . .' from Charlie Jones who was not noted as one of the most serious scholars in the class.

Generally, such an episode would be music to a teacher's ears. Yet, it is one that raises several interesting questions, including:

1. Why is it likely to be a source of satisfaction for a teacher to receive such a reaction in class?
2. What preconditions provoked such a reaction from a student in the class?
3. What should the teacher (who may be following a carefully devised lesson plan) do about Charlie's remark?

Most teachers are delighted to receive such an expression of interest in the work of the class because it indicates intrinsic motivation, because the pupil has shown a desire to go on learning, because it almost certainly indicates that the topic students are studying is meaningful and is seen by them to be relevant and, finally, because 'keeping-the-lid-on' management problems disappear from a class where such an attitude prevails.

A spirit of inquiry is alive in any student who makes a remark about wanting to find out. An inquiry involves finding an answer to a question or the resolution of a problem. Inquiring is a state of mind which may be satisfied by numerous learning experiences. It may, for example, be satisfied by exposition from a teacher; or alternatively it may be satisfied by any of the range of learning experiences commonly called discovery learning or problem solving, or creative activity, or laboratory practicals.

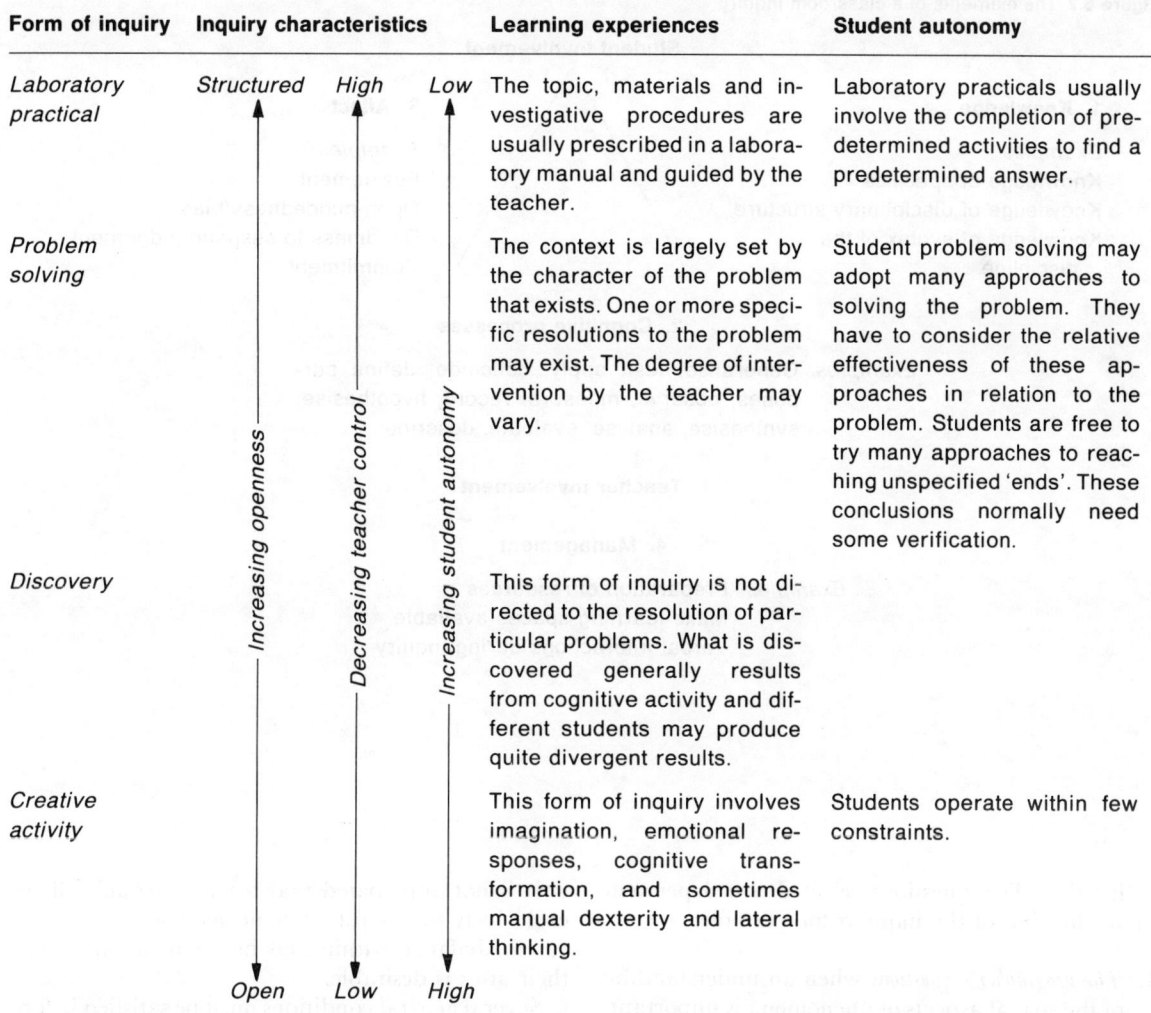

Figure 6.1 Four forms of classroom inquiry

Inquiry permeates all of these learning experiences. It is useful to distinguish them in terms of their relative openness or structuredness which is reflected in the degree of autonomy assumed by the student. Figure 6.1 illustrates four forms of a classroom inquiry in terms of these characteristics.

The descriptive statements about various forms of classroom inquiry in Figure 6.1 make it clear that all inquiries have some creative quality: that students engaged in each of these learning situations are likely to possess some degree of intrinsic motivation toward the task in hand, that the learning being undertaken is meaningful, and that students following any one of these activities are willing to do something to learn by themselves rather than wait to be taught. Many teachers value these qualities in their students.

Inquiry means many things to many people and there are many ways of undertaking an inquiry. However, it would appear that a successful classroom inquiry involves at least four elements:

1. Knowledge
2. Cognitive processes
3. Affect
4. Management

The close interrelatedness of these four elements is shown in Figure 6.2.

Some indication of the variety of inquiries that may be undertaken by students of geography is provided by the examples in Figure 6.3. Students attempting such inquiries may well guide their search for answers by using questions as a framework

Figure 6.2 The elements of a classroom inquiry

Student Involvement

1. Knowledge

Examples:
Knowledge of specifics
Knowledge of disciplinary structure
Knowledge of syntax of the discipline

3. Affect

Examples:
Puzzlement
Open-mindedness/bias
Readiness to suspend judgement
Commitment

2. Cognitive processes

Examples: Generalise, test, apply, conclude, define purposes, observe, measure, record, hypothesise, synthesise, analyse, evaluate, describe

Teacher involvement

4. Management

Examples: Preparation of resources
Make learning spaces available
Verbal interactions during inquiry

of inquiry. The questions selected are dependent upon the aims of the inquiry, for example:

1. *The geographer's questions* when an understanding of the spatial aspects of phenomena is important:
   (a) Where is (the thing concerned) located?
   (b) Why is it there?
   (c) With what is it associated?
   (d) What are the consequences of its location and associations?
   (e) What spatial alternatives should be considered in decision making?
   (f) Who decides, for whom?
2. *The geographer's questions* where the personal development of the student is important:
   (a) What is the place or thing we are concerned with?
   (b) What are my own perceptions of this place?
   (c) What are the perceptions of other people?
   (d) What is the language used to describe this place?
   (e) What does this place mean to people as evidenced by their reactions to it?
   (f) What are the causes and consequences of the perceptions of this place?

It is not anticipated that teachers would adhere exclusively to one list while neglecting the other. A careful balance within and between inquiries and their aims is desirable.

Several general conditions must be satisfied before successful classroom inquiries may be undertaken. These include:

1. Students require a question or a problem to pursue.
2. Teachers who commonly adopt didactic stances in classrooms must be prepared to adopt new roles, both to initiate and sustain inquiry. A student meeting a problem normally turns to the teacher and seeks the answer. Faced with this, the teacher in the inquiry classroom must make it clear that it is the student's responsibility to find out, though at the same time, there will be support and encouragement from the teacher.
3. The students will need to gather data. Teachers can facilitate this by making learning resources reasonably accessible to searching students.
4. Students need substantial blocks of time in which to undertake many inquiries.

Making Inquiry Learning Work in the Geography Classroom   67

Inquiry	Comment
1. What are the major rivers of Canada?	A closed question with right answers; although the word 'major' may need to be defined in terms of several alternative criteria, e.g. length, discharge, economic use.
2. 'Surprisingly, lakes exist in the world's hot deserts.' Where and Why?	This is a largely cognitive inquiry insofar as there is a problem to be solved by rational manipulation of data.'
3. 'A resourceful people without resources.' To what extent is it right to describe the Dutch in this way?	The slogan points to an incongruity—always a potential source for an inquiry. The question asks for an evaluative approach to answering.
4. Here is a drawing of an imaginary animal called a girapotabok. What would you deduce to be its habitat? Show its likely place in a food chain. 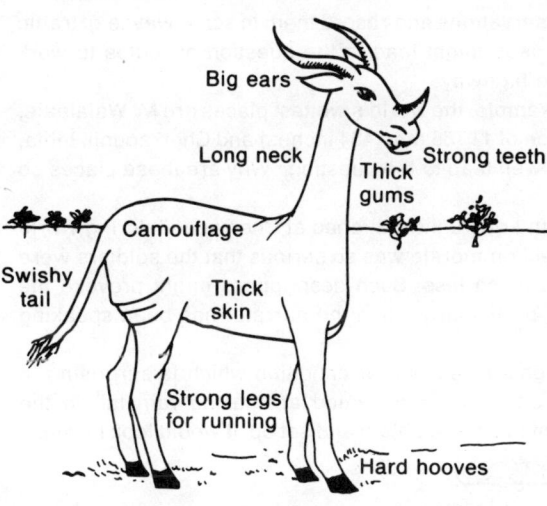	This question calls for: (a) analysis of the animal, for example, what is the purpose of its long neck, hard hooves, etc? (b) recognition of the animal's characteristics as adaptations to environment; (c) synthesis in devising a composite statement describing its habitat. One quality of this type of question is that the hypothetical component minimises the amount of simple factual recall.
5. Should there be a new state in northern Australia?	This is an evaluative question requiring judgement of choice. The answer requires reference to knowledge about the implications of a new state viewed from the values stance of the respondent.
6. Do mountain landscapes have more appeal than desert landscapes to you?	This question requires an effective response. It may lead to a subjective review of what appeals aesthetically to the student answering the question, or it may lead to a more cognitively oriented view of the characteristics of desert landscapes versus mountains.
7. What images come to mind when you hear of: • freeways? • famine? • mangroves?	Such a question emphasises personal response and individual perception. It has strong affective connotation.

Figure 6.3 Some examples of the variety of inquiries possible in geography

Stimulating Inquiry

The second question posed at the beginning of this chapter relates to the conditions that might provoke students to make pleasantly unexpected statements of interest in their work . . . or are they so unexpected?

The classroom conditions which provoked Charlie Jones into saying that he 'would like to find out' have almost certainly involved the teacher in several preliminary activities including the following:

1. *The teacher has stimulated interest by picking topics which are familiar to students or perceived by them to be*

relevant to their present needs and interests. For example, the topic may be drawn from the local area, or it may be currently newsworthy and receiving attention daily in the mass media. Student interest may also be kindled by a variety of teaching methods, including:
(a) providing opportunities for students to discuss issues in small groups;
(b) motivating learning experiences, for example field studies, graphic descriptions and remarkable examples;
(c) providing opportunities for students to offer opinions and make evaluative statements.

2. *The teacher has checked on the availability of learning resources for inquiring students.* This could have involved a visit to the school library and discussions with the librarian, the preparation of annotated reading lists and displays of newspaper cuttings, student work, photographs, maps and the like on the class notice board. In practice, very much of the success of an inquiry depends on the teacher's initial access to the learning resources. The customary sources of materials for students making inquiries include: accounts from current media—newspaper, radio, magazines and television—field studies and studies of the local area, opinionnaires sought from students, parents and the community, textbooks, reference works such as atlases, gazetteers, statistical handbooks, and company and government reports.

A summary of six ways that have been used to stimulate student inquiry is provided in Figure 6.4.

Figure 6.4 Six ways of stimulating inquiry

1. **Fieldwork** in which students make direct, personal observations and record them in some way; e.g. traffic counts during peak hour on suburban sub-arterial roads might lead to the question of routes to work motorists regard to be practicable alternatives to the highway.
2. **Quotation** of statistics or examples of extremes. For example, the world's wettest places are Mt Waialeale, Kauai Island, Hawaii, which has an annual precipitation of 11 786 mm (464 inches) and Cherrapunji, India, which experiences 11 608 mm (457 inches). This may well lead to the question, 'Why are these places so wet?'.
3. **A graphic story:** e.g. describing the fact that the British army units stationed at Cherrapunji during World War II were almost driven 'troppo' by the rain. Its effect on morale was so serious that the soldiers were moved out on to the Assam plain where the rain is much less. Such descriptions might provoke the question, 'What does it really feel like there, Miss?', particularly when the narrator has been speaking from personal experience.
4. **Laboratory practicals** in which an experiment highlights a reaction or condition which is surprising in some way. For example, the teacher may set up an experiment aimed at 'making rain fall in the classroom'. The equipment needed for this experiment is easy to obtain and set up. It would look like this.

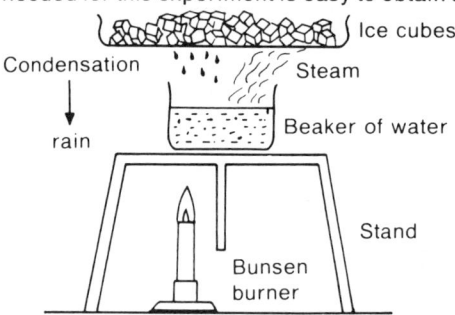

Such an experiment may well provoke the question, 'Why did it work, sir?'.
5. **Real world problems** (i.e. not hypothetical questions) needing solution can be discussed in current affairs sessions with the help of mass media reports as sources. For example, descriptions of planning activities undertaken when a new airport is being built near a big city might lead to such questions as 'Where is the best place to put it?' and 'Who should make the decisions?'.
6. **Conflict of values**
Students may be told a story in which some conflict of values is apparent. For example, traditional Eskimos used to leave aged members of the group to die in severe winters when they broke camp and moved to new hunting grounds. Making this statement in class might provoke such questions as 'Why did they do that?', 'Was it right for them to leave the old people to die?'.

Roles for Teachers During Student Inquiry

Teachers may adopt quite a broad range of roles during the course of inquiries undertaken by their students. These roles vary largely in terms of the dominance of the teacher and are influenced by such factors as the age and ability of students, the level of achievement in geography that they have attained, the nature of the question or of the problem being investigated, and the availability and character of the learning resources. Some teachers are quite obtrusive during the course of the inquiry. They select the problem, prescribe the steps in the method of solving the problem, and closely monitor the students' work at all stages. They justify their approach in terms of the lack of sophistication of students undertaking the inquiry ('They have little training in inquiry procedures'; 'Their span of attention is short'; 'They are not very bright and are unable to decide what to do next'). Other teachers feel insecure if there are too many surprises in their classrooms. They feel most comfortable within a carefully planned objectives model of curriculum and teaching in which they specify the intended learning outcomes for their students in advance of the occasions when learning occurs. Teachers who adopt dominant roles do not necessarily undermine the value of inquiry learning for their students, but they do need to bear in mind that their inquiries are likely to be found near the top of Figure 6.1, in the area of teacher-structured activities.

However, one of the principal values of inquiry learning is that it promotes the capacity of students to learn how to learn, and how to go on finding out for themselves. The teacher seeks to become dispensible to students while such an objective prevails, and consequently adopts less directive roles during the course of the inquiry. There are still numerous classroom tasks for such unobtrusive teachers during the course of student inquiry. These include:

1. Helping students choose a problem that is within their capabilities.

2. Helping students identify the goals of their inquiry and thus reduce random behaviour involving guessing and irrelevancies.

3. Acting as a sounding board on which the students may try out their ideas, hunches, guesses, hypotheses.

4. Showing students how to analyse their selected task, perhaps by making a topic analysis that points to the relatedness of aspects of the inquiry.

5. Helping students consider appropriate ways of recording the inquiry as it progresses.

6. Assisting students with the discovery and use of learning resources. Many teachers feel quite justified in making these readily available to students as they work.

7. Advising about alternative ways of recording and presenting the results of an inquiry.

Some teachers sometimes adopt a *laissez-faire* approach to their inquiring students. They provide little comment on the practicability of the student-selected questions. Students are not stopped from moving into what are (to the teacher) obviously unproductive lines of inquiry. Students are required to find their own data sources. Teachers are tolerant of time wasting. Teachers adopting this approach often justify it in terms of the importance of students to be free while conducting an inquiry of their own. The freedom contributes to the students' sense of independence and to their continuing motivation.

Promoting Inquiry

The many roles teachers can play in promoting inquiry in the classroom can be illustrated by examples of statements that teachers are likely to make in carrying them out. For example:

1. *Clarifying the problem for students:* 'John, when you say—what is it like there?—are you referring mainly to the weather?'

2. *Encouraging*: 'You're going well, Anthea. Have you thought about . . . yet?'

3. *Helping with the process of inquiry*: 'If you can't find it in the table of contents at the front of the book, Ali, try the index at the back.'

4. *Silence*: Listening to students. Saying nothing can encourage them to expand their thinking.

5. *Expediting classroom management*: 'Group 3 should be the next to have access to the newspaper cuttings.'

6. *Stimulating students to probe more deeply into their inquiry*: 'I like the way in which you presented your experimental results on a graph, Mary. Have you tried to explain why they are like that?'

7. *Finding alternative 'audiences' for the results of the inquiry, thereby providing students with interested parties to whom to present their results*: 'I think we should ask the shopping centre manager if you can display

your findings on litter around shops on one of his notice boards, don't you Gareth?'

It should be remembered though that all classroom teaching has an existential quality. The impact of these types of statements and behaviours by teaching is likely to vary according to various classroom circumstances. So, the roles and statements are proposed very much as guidelines rather than as definitive advice on how best to promote student inquiry in specific classrooms.

Inhibiting Inquiry

Teachers may inhibit inquiry through various inadvertent or intentional actions and statements. The following behaviours fall into this category, though the preceding cautionary note about the existential nature of classrooms applies here also. They include:

1. A heavy emphasis on written work.
2. Undue emphasis on evaluation.
3. Persistently requiring a tight definition of terms.
4. Proclaiming the teacher's own viewpoint firmly.
5. Rejecting material because it is irrelevant to geography.
6. Insisting on specific sequences of activity in the process of inquiry.
7. Strongly praising one student's work over others.
8. Repressing leadership shown by students.
9. Interrupting the inquiry often, and attempting to have agreement on progressive results.

Inquiry Activities for Students

Students engage in many roles during the course of pursuing an inquiry because there is a multiplicity of activities associated with inquiry and many sequences in which the activities may be undertaken. Many analysts of inquiry have provided sequenced lists of the steps that should be undertaken by inquiring pupils. Nearly all refer to groups of activities which include the following:

1. Recognising the problem or inquiry question and thinking about ways of resolving it.
2. Observing, measuring, recording, describing, classifying and representing data.
3. Hypothesising/guessing: engaging in various forms of logical and critical thinking, e.g. interpreting, analysing, synthesising, generalising, applying, testing, drawing conclusions; discussion and other group approaches to inquiry.

Classical models of scientific inquiry appear to suggest a given order in inquiry activities, but many sequences of these activities may be adopted in pursuing school inquiries. There is no universally appropriate approach to learning and it is rare for there to be only one effective way of working on any given problem or question. This viewpoint is argued on two grounds:

1. The spontaneity of students is stultified if they are required to make lock-step progress through an invariant set of activities.
2. Several major forms of inquiry seem to imply alternative sequences of activities. For example, there are inductive and deductive forms of logical inquiry, controlled physical science experiments, correlative studies (often in the social sciences), case studies, intuitive or random approaches to trying possible solutions to a problem, and theoretically grounded hypotheses.

The experience of many inquiry episodes in geography lessons suggest that the classic scientific model of inquiry is seldom followed in its ideal form by students investigating an issue problem. Many students rapidly scan the available data and often jump to premature conclusions. Some students give up quickly if they are unable to come to a ready conclusion. It is desirable therefore that teachers be prepared to intervene at least during the initial phases of an inquiry, or until students have developed competence in inquiry activities following an initial training period.

In practical terms, it seems undesirably limiting for teachers to feel that they must advise rigid adherence by their students to any given set of procedures. At the same time, there may well be limitations on the quality of students' thinking if too many steps in the inquiry process are omitted.

Developing Thinking Skills for Inquiry

Inquiry learning frequently requires students to engage in various forms of critical and logical thinking. The likelihood of students succeeding with inquiry will be enhanced if they are given opportunities to engage in various skills and forms of thinking using data of interest to geographers.

Some skills and forms of thinking that need to be developed in students include:

1. Finding and using basic sources of information.
2. Evaluating sources of information.
3. Distinguishing fact from opinions.
4. Recognising statements that are difficult to prove.
5. Classifying and ordering information.

Here are some examples of various exercises and problems that can be used to develop these five thinking skills.

1. Finding and Using Basic Sources of Information

Locating sources of information, be they in the form of books, tables, photographs or maps, is an important inquiry skill for students to develop. Chapter 8 surveys many ways of giving students practice in it. An additional exercise could be to present students with a list of sources and topics to 'match up'.

Exercise A
Imagine you have the following sources in your school library or home:

1. *The Canadian Arctic—A Geography*, published by a Canadian Government Resource Survey Branch.
2. World Meteorological Survey (statistical data).
3. A large map of the Arctic area centred on the North Pole.
4. A concise *History of Arctic Exploration*.
5. The log book kept by the Captain of the USS *Nautilus*, first submarine under the ice to the North Pole in 1958.
6. *Under Northern Lights*, short stories about life in the north of Canada, by Alan Sullivan.
7. 'A Look at Alaska's Tundra', an article in the *National Geographic Magazine*, in March 1972.

Which one of the sources would you use to find the following information or answers? Place the letter code of the source in the box.

1. The air distance between Irkutsk and Fairbanks. ☐
2. The pattern of ocean currents in the Arctic Ocean. ☐
3. Strategic locations for a radar warning system against missiles in northern Canada. ☐
4. Migratory habits of Arctic birds and animals. ☐
5. The formation of pingos. ☐
6. Difficulties of traditional Eskimo life. ☐
7. The agony of frostbite. ☐
8. The effects of very low temperatures on human capacity to work hard in the outdoors during tundra winters. ☐
9. The annual range of temperature at several tundra towns. ☐
10. Russian expansion into Siberia in the nineteenth century. ☐

2. Evaluating Sources of Information

Students frequently hear and read different accounts of the same subject during their studies in geography. It will help if they are able to evaluate which accounts are the most useful. Try this exercise now.

Exercise B
Three topics are presented. After each one, a list of four possible sources is included. Your task is to rank the list of sources, putting the most accurate and useful one first.

Topic A: Seasonal changes in the landscape on the Barkly Tableland in Australia.

Rank
☐ A The Commonwealth Bureau of Meteorology.
☐ B An Aboriginal stockman working on one of the pastoral stations in the Barkly.
☐ C The CSIRO Land Survey team that investigated the Barkly area several years ago.
☐ D The Postmaster at Camooweal.

Topic B: The best way of marketing cattle from the Barkly Tableland.

☐ A A scientist who has studied the problem of cattle marketing in Australia.

- ☐ B The manager of one of the cattle stations on the Tableland.
- ☐ C The owner-driver of a 'road train'.
- ☐ D The manager of a big meatworks in a city in coastal Queensland, for example Rockhampton.

Topic C: Life in the Nigeria savannas of Africa.

- ☐ A The author of the historical novel about Nigerian exploration, *Fresh Tracks in an Ancient Land*
- ☐ B A safari leader who operates out of Ilorin.
- ☐ C The chief of tribal group living north of Kano.
- ☐ D An Australian geographer who spent a year studying Nigeria at first hand.

3. Distinguishing Between Fact and Opinion

A very useful skill is the ability to tell the difference between facts and opinions. Students will use this skill frequently in all walks of life and in relation to many topics, including their studies in geography. It is important to recognise facts because these give substance and credibility to opinions. Students need frequent practice in this skill. Here is a sample exercise:

Exercise C
Indicate with a capital F or O which of the following statements are facts (F) or opinions (O).

1. Cairns is an urban settlement in tropical Australia.
2. The weather in Cairns is occasionally influenced by tropical cyclones.
3. These cyclones have increased in their frequency and intensity in recent years because of the atom bombs which have exploded in the Pacific.
4. The humid heat in the tropics makes north Queensland an unsatisfactory place for permanent white settlement.
5. Queensland cane farmers believe they would benefit from wider export markets.
6. Foreign ownership of tropical beef pastureland in northern Queensland is producing a destructive use of Australian resources.
7. The largest selva in the world is found in the Amazon basin.
8. The Amazonian Indians are technologically backward because they live in a hot, wet climate which makes them lazy.
9. Soil scientists tell us that leaching tends to reduce the fertility of soils in many areas in the tropics.
10. The government of Brazil says that their road-building program in the Amazon will open new areas of settlement and development.

4. Recognising Statements Which are Difficult to Prove

This skill is similar to distinguishing fact from opinion, but has the added dimension of recognising if a statement can be proven to be true or not by checking sources. For example:

Exercise D
Pick which statement from the following two groups is most difficult to prove true or false.

Group A
1. India has expanded its capacity to produce steel during the course of the Five-Year Plans.
2. India's plans to industrialise will succeed because they are being carried out in a broadly democratic context.
3. The growth of population provided unexpected problems for economic planners in India in the 1950s.
4. A substantial part of the industrial development of India has taken place in the Bengal–Bihar area because of the favourable combinations of natural resources there.

The answer is ☐

Group B
1. India is not an important participant in international trade except in a few special items such as the products from jute and mica.
2. India has experienced balance of payments problems in several years since Independence.
3. India would not have experienced so many problems with international trade if she had retained colonial status.

4. It seems unlikely that India will export large quantities of manufactured goods in the next decade.

The answer is □

5. Classifying and Ordering Information

Classifying and ordering are ways of making masses of data manageable as they contribute to the clarity of a description or explanation. Ordering requires the adoption of criteria which enable the presentation to be systematic. It may be undertaken in many ways, some of which are exemplified in the following items:

Exercise E
1. *Arrangement of items in a spatial order.* Groups of items can be arranged in a spatial order. Arrange the following sets of items spatially, and state the criterion used to put them in order.

   (a) Moscow, London, Paris, Berlin, Rome, Athens.

   (b) Rain, playa, bajada, rock island, canyon, badlands, pediment, alluvial fan.

2. *Whole to part ordering.* Arrange the following groups of items from the most inclusive or general to the most specific or particular.

   (a) Weathering, disintegration, landforms, geomorphology, exfoliation, frost wedging, talus slope.

   (b) Region, house, municipality, universe, world, state, country, metropolitan area.

3. *Sequencing.* Devise a flow chart to present the order of activities involved in the production and marketing of textiles.

4. *Concrete to abstract.* Rank the following items putting the most concrete first.

   (a) Brisbane is the state capital of Queensland.
   (b) There are capital cities in all Australian states.
   (c) The internal pattern of an urban centre is its morphology or form.
   (d) Site refers to physical terrain on which an urban centre is built.
   (e) Cities provide functions, such as manufacturing, commerce or administration.
   (f) Smaller urban centres may be dominated by one or two important functions.
   (g) Great cities are multifunctional.
   (h) Internal patterns of large cities are determined by the axial arrangements of the transport routes focusing on the city core.

5. *Identifying and ordering* a range of factors that should be considered in relation to an issue or problem. Consider the following problem and task.

Problem: Should the mining and processing of uranium be banned in Australia? Order these factors in terms of their importance to the Australian uranium mining issue.

Order of importance	Range of factors
. . . .	(a) Impending world shortages of liquid fuels.
. . . .	(b) The cost of alternative sources of power.
. . . .	(c) The danger of radioactive waste.
. . . .	(d) The difficulty of disposal of this waste.
. . . .	(e) Political instability in areas to which the uranium might be exported.
. . . .	(f) Trade union attitudes towards nuclear power.
. . . .	(g) The contribution of uranium exports to Australia's national income.
. . . .	(h) The stimulus to the settlement and development of remote areas of Australia.
. . . .	(i) Conflict with Aborigines over land rights.
. . . .	(j) Materialism in Australian society.
. . . .	(k) The creation of new jobs in the economy.

Conclusion

It is arguable that inquiries undertaken by school students contribute well to the quality of their learning. Inquiry classrooms require that teachers and students adopt new roles and these require additional pedagogic skills on the part of teachers. Some of the approaches that teachers may adopt to foster the success of inquiries undertaken by their students are outlined in this chapter. No universally appropriate procedures are proposed because of the extent to which there is an existential character in school learning settings. Teachers who are well aware of and sensitive to the needs of their students possess the basic qualities needed to foster inquiry by their geography students.

References

Asmussen, D. and Buggey, J. (1977) 'Teaching Geography Through Inquiry', in G. Manson and M. Ridd (eds) *New Perspectives on Geographic Education: Putting Theory into Practice*, Dubuque: Kendall/Hunt.

Bartlett, V.L. and Cox, B. (1982) *Learning to Teach Geography*, Brisbane: John Wiley and Sons.

Blachford, K. (1975) *Inquiry in the Classroom*, The Teaching of Geography Unit III, Melbourne: Education Department of Victoria.

Cox, B. (1973) 'Find Out or be Told?', *Geographical Education*, Vol. 2(1), 55–61.

Fien, J. (1980) 'Am I an Enquiry Teacher', *Project News*, Schools Council Geography 16–19 Project, No. 11.

Fien, J. (1980) 'Teaching Geography Through Inquiry', *Queensland Geographer*, Vol. 15(2), 35–40.

7

Designing Worksheets to Promote Student Inquiry in Geography

Brian Hoepper

Inquiry learning and the development of thinking skills need to be regular features of the geography classroom. Most often teacher-developed worksheets are the chief way of implementing inquiry-based, thought-provoking learning experiences.

In this chapter, Brian Hoepper explores three features of effective worksheets: the principles that underlie worksheet design, the variety of resources that can be provided to stimulate inquiry, and the variety of cognitive, skill and affective responses that can be asked of students. These features of effective worksheets are illustrated with an example of a local environmental issue—in this case, the arguments for and against building a bridge to link an island community to the mainland. This sample worksheet is derived from the variety of resources and range of student activities developed in the early part of the chapter.

For many teachers from the 1970s, and for many students from that time as well, the word 'worksheet' probably conjures up bad memories. At that time, many classrooms became the settings for 'paper warfare'! Of course, it was all done with the best of intentions but, for many teachers, pedagogy seemed reduced to the handing out, collecting and marking of worksheets.

What worksheets they were! While many were little more than 'busywork' comprehension exercises, their presentation often left much to be desired. Indeed, the only uplifting aspect of most was the welcome whiff of duplicating fluid on the page fresh from the machine.

Happily, we seem to have learned a great deal from those experiences. Teachers have resumed the positions they often abdicated during the days of the paper war. The idea of the teacher as motivator, coordinator and even lecturer has regained currency. With it has come a more balanced and defensible appraisal of the role the worksheet might play in teaching and learning.

Why are Worksheets Worthwhile?

Worksheets have much to recommend them. Properly designed, they can provide students with a structured approach to an issue or question. Often, they can be more comprehensive in the questions they ask and in the responses they require than conventional texts.

Worksheets provide a way for a teacher to pay more than lip service to the notion of individualisation. A worksheet usually allows students to proceed at their own pace. If activities are graded, and if different media are introduced for both the stimulus and the response, then a worksheet can go some way to matching students' different abilities. If choice is offered, a worksheet can also acknowledge students' different interests.

Because worksheets are usually self-contained, their use can ease the teacher out of the focal role of being the constant centre of attention for the whole class. Instead, the teacher is free to move about the class, checking on progress in an informal way, pausing to help a student in difficulties, or to explore some tangential aspect with a small group. Indeed, this perceived freedom was probably a major reason why so many teachers took to worksheets with such gusto back in the 1970s.

What Principles Might Guide Worksheet Design?

Worksheets probably are many things to many people. Accordingly, the following principles should be seen as relating to the type of worksheet that I consider most worthwhile in classroom use. Certainly, there may be other types of worksheets, derived from different principles, that also offer students very worthwhile experiences.

These, then, are the principles:

1. The worksheet should *focus on an issue of significance* within the geography curriculum. Preferably, the issue should be one that is distinctly problematic, in that people should be arguing in a significant way about the issue.

2. To highlight the debate about the issue, the worksheet should *provide key items of evidence* reflecting the debate.

3. The worksheet should provide questions and activities that exploit the potential of the evidence in that they *require a comprehensive range of cognitive responses*.

4. The worksheet should also *go beyond the evidence*, to exploit the students' own experiences, knowledge and feelings, and to exploit aspects of the wider world in which the issue is alive.

5. The work sheet should *exploit the different abilities and interests of students*, by offering chances for students to respond in a range of media.

6. The questions asked, and the activities promoted, should be *clear, unambiguous and practicable*. They should not ask the unknowable, nor require the impossible.

7. The instructions for the worksheet should be sufficiently *comprehensive, clear and understandable* for the students to proceed without assistance.

8. The worksheet should be presented in such a format that students see it as a *credible and worthy alternative to a commercial text*.

Resources to Put on a Worksheet

1. *Text*. This could be a textbook extract, a newspaper article, a government report, an expert opinion, an eyewitness account.

2. *Graphic*. This could be a photograph, drawing, painting, cartoon, map, table, graph, cross-section, block diagram, flow chart, poster, etc.

3. *Research books*. These are books, relevant to the issue, referenced on the worksheet and available to the students, but not reproduced in any way on the worksheet itself.

4. *Print display*. This could be a chart, map or poster displayed in the classroom, but not reproduced in any way on the worksheet itself.

5. *'Artefact' display*. This could be a model, replica or other relevant object displayed in the classroom.

6. *Audio-visual presentation*. This could be a film, videotape, slides, audio-recording, filmstrip or computer display in the classroom.

7. *Students' prior knowledge*. This could be what the students already know, or believe, or have experienced.

8. *Other people*. This could be the recounted experiences, knowledge, attitudes or beliefs of other students, teachers and accessible members of the community.

9. *The local environment*. This could be the physical and social features of the school and its immediate surroundings.

10. *Other places*. These could be places related to the issue, and reached through an excursion or through students' travelling in their own time.

The effectiveness of a worksheet may be increased if a variety of the above stimuli is included on any worksheet. Obviously, some will appear on the worksheet itself, while others will be referred to, with instructions given on how to follow them up. Where items appear on the worksheet itself, it is important that details be given of the source of the item. Such information is essential for students to evaluate reliability of the source.

In this chapter one issue had been selected to illustrate these principles for designing effective worksheets. The selected issue is the proposal to build a road bridge to link North Stradbroke Island with the nearby mainland, close to Brisbane, Australia. The issue first emerged in the 1940s but really hit the news in 1977. Over 12 years later, the bridge is still not built, but the plans remain and the controversy continues.

Three items of evidence are provided on the worksheet: a letter, a speech and a cartoon. Extension activities on the worksheet also make use of other types of stimuli mentioned in the above list.

What Might Worksheets Require as a Response from Students?

Just as the stimulus resources can be varied, so too can the types of responses asked of students. Consider this list:

1. *Cognitive responses.* These include a range of responses, from comprehension, through analysis, interpretation, synthesis and evaluation. Comparative questions may be asked, related to more than one piece of evidence.
2. *Research.* These activities require students to use other relevant materials, and may require them to locate those materials themselves.
3. *Fieldwork.* This could include mapping, field sketching, survey, questionnaires, photographing, filming and recording.
4. *Discussion/debate.* This could include forming a group to discuss the issue raised by the evidence, or even organising a formal debate.
5. *Imaginative response.* This could be a creative response in writing, speech, art, music or film.
6. *Role play.* This involves devising and performing a role play based on an aspect of the issue raised in the worksheet.
7. *Making a judgement.* This involves making and expressing a carefully considered opinion, based on critical investigation of the evidence provided.
8. *Empathetic response.* This involves demonstrating an understanding of another's point of view, particularly if that point of view differs from one's own.
9. *Values response.* This involves the student making a decision about the issue, based on the student's own values.
10. *Making wider links.* This requires the student to take up issues that have arisen through the worksheet, and to demonstrate an understanding of the wider implications of the issue in society.

An Example: The Stradbroke Island Bridge Controversy

The rest of this chapter applies these ideas in relation to the specific issue of the Stradbroke Island bridge proposal. After selecting appropriate pieces of evidence to include on the worksheet, the next task is to brainstorm a range of possible activities for students. Figure 7.1 provides some sample questions and activities based directly on the three pieces of evidence while Figure 7.2 provides sample questions and activities that require the student to go beyond the evidence. Once all these possible questions and activities have been brainstormed, there are two more steps in worksheet design that need to be looked at: the selection, from the possibilities available, of those questions and activities that are to be included in any particular worksheet; and the relationship between worksheet activities and other teaching and learning activities.

The Stradbroke Island worksheet which follows reflects these two steps and provides students with opportunities for:

1. *Investigating the evidence:* making a sequenced set of responses to the evidence, including the making of a judgement.
2. *Researching the issue:* gaining more information about the issue.
3. *Interacting with other students:* discussing, and taking part in a role play.
4. *Expressing opinions:* expressing a personal opinion reflecting knowledge gained and values held.
5. *Being creative:* making creative, imaginative responses in various media.
6. *Making links:* identifying and exploring links between this issue and other social issues.

No suggestions are offered about whether any questions/activities should be mandatory. Teachers might determine this for their own classes, or consult with their students to ensure that each student might be set a realistic challenge by the worksheet.

One last, and probably unnecessary note, concerns the place of worksheets in the overall approach to teaching and learning. The 'paper warfare' of the 1970s was an unbalanced but long overdue reaction to curricula that eschewed inquiry and student activity in favour of teacher-dominated transmission of unproblematic information from textbooks. Most teachers would now probably see both extremes as undesirable. Consequently, worksheets need to be

Figure 7.1 Sample questions and activities based on the three sources of evidence on the Stradbroke Island Bridge issue

	Source A: The Letter	Source B: The Speech	Source C: The Cartoon
1. Comprehension Understanding the explicit meaning of the source.	According to this letter, who will benefit from the bridge if it is built?	What 'label' does Mr Hinze apply to those people who oppose the bridge?	What does this cartoonist predict will happen to the beach-front land if the bridge is built?
2. Analysis: The Parts Identifying the constituent parts of the message in the source.	What are the different problems Mr Peters predicts will develop if the bridge is built? List them under the headings: environmental, social, economic.	What benefits does Mr Hinze predict will follow if the bridge is built? List them under the headings: social, economic.	What problems does John Mason predict will follow if the bridge is built? List them under the headings: social, environmental.
3. Analysis: Relationships Identifying relationships within the message in the source.	According to Mr Peters, how will increased traffic cause a number of different environmental problems?	According to Mr Hinze, why will the bridge benefit both tourists and residents?	How has John Mason shown that the development of large tourist accommodation will actually cause problems for those tourists?
4. Interpretation Understanding the implicit meaning of the source.	What do you think Mr Peters means when he claims that 'A bridge . . . would be . . . a tribute to the greed of private developers'?	What do you think Mr Hinze means by the claim that 'many conservationists could not work in an iron lung'?	Who do you think are represented by the three characters in the foreground of the cartoon? Why do you think they have been depicted shaking hands?
5. Synthesis Identifying the overall message of the source.	What is the overall message of Mr Peters' letter about the Stradbroke bridge proposal?	What is the overall message of Mr Hinze's speech about the Stradbroke bridge proposal?	What is the overall message of John Mason's cartoon about the Stradbroke bridge proposal?

Designing Worksheets to Promote Student Inquiry in Geography

6. **Evaluation: The Author's Style** Understanding the way in which the authors has used a particular style to achieve a particular purpose.	Why might Mr Peters have begun by stating that he and his wife had been visiting Stradbroke Island for the previous thirty years?	Why might Mr Hinze have stressed the supposed laziness, dirtiness and vested interests of the conservationist critics of the bridge?	Why might the cartoonist John Mason have given the buildings such names as 'The Pits', 'Cylinders Cesspool', 'Deadman's Motel', 'Turd Tower' and 'Boring Tower'?
7. **Evaluation: The Author's Values** Identifying the author's values, as reflected in the message of the source.	Judging by the letter, what things does Mr Peters seem to value? How did you decide?	Judging by this speech, what does Mr Hinze seem to value? How did you decide?	Judging by this cartoon, what does John Mason seem to value? How did you decide?
8. **Evaluation: The Reliability of the Evidence** Making a decision about the possible reliability of the source in terms of its accuracy, completeness, bias, representativeness, honesty.	Do you necessarily accept what Mr Peters says as being an accurate description of the effects of a bridge on Stradbroke Island? Why could some of Mr Peters' claims be inaccurate? Could some inaccuracies be unintended, but others deliberate? Explain.	Do you necessarily accept what Mr Hinze says as being an accurate description of the conservationists who oppose the bridge? Why could some of Mr Hinze's claims be inaccurate? Could some inaccuracies be unintended, but others be deliberate? Explain.	Do you necessarily accept what John Mason has drawn as being an accurate picture of what might happen to Stradbroke Island if the bridge is built? Why could the picture be inaccurate? Could some inaccuracies be unintended, but others be deliberate? Explain.
9. **Comparison** Identifying and understanding features which are common in the messages of the different sources.	What is one major point that seems to be made by both Mr Peters' letter and John Mason's cartoon?	What is one aspect of the Stradbroke issue that is referred to by all three—Peters, Hinze and Mason—even though they may deal with it quite differently?	Mr Peters' letter refers to 'the greed of private developers'. How has John Mason illustrated this in his cartoon?
10. **Contrast** Identifying and understanding the differences between the messages of the different sources.	Mr Hinze claims that people will 'welcome the chance to drive to Straddie's sunny beaches, instead of having to endure the long barge trip'. What claim by Mr Peters seems to conflict with this statement by Mr Hinze?		Mr Hinze claims that 'all' the residents would share the rewards of the tourist boom. What aspect of John Mason's cartoon indicates that the cartoonist disagrees with this claim by Mr Hinze?

Learning Process	Sample Activity
1. Research	Using the library research file on Stradbroke Island, find out about the proposed routes for the bridge, estimated costs, suggested sources of funding, estimated usage. Present your findings in a table.
2. Media Search	Collect newspaper and magazine articles about the Stradbroke Island issue. Devise a system of headings for filing them.
3. Fieldwork	Design and conduct a survey to find out people's opinions about aspects of the Stradbroke Island bridge issue. You might investigate whether people ever visit the island, whether they favour a bridge, their reasons for their opinion, whether they are aware of the various arguments being put forward, whether they have ideas on who should pay for any Stradbroke bridge. Present your findings using coloured graphs.
4. Discussion/Debate	'Someone' will be making the decision about whether the Stradbroke bridge is built. In a group, have a discussion about who you think should make decisions about such issues: consider the government at Federal, state and local levels, 'experts', residents affected, visitors affected, potential bridge builders, the wider population of Queensland, of Australia.
5. Imaginative Response	Decide what you think about the Stradbroke bridge issue. Then plan a way of trying to convince other people to agree with you. You might consider producing a poster, bumper sticker, T-shirt design, radio jingle, television commercial. Try to produce the finished product if possible.
6. Role Play	Devise and present a role play based on the Stradbroke bridge issue. For example, you could base the role play on a family discussion, in a family where the mother is a Stradbroke Island businessperson who would benefit from any increase in tourism, the father works for the existing barge company, the daughter is an ardent conservationist, and the son lives on the island but works on the mainland, and would benefit if he could drive to work over a bridge.
7. Making a Judgement	Given the pieces of evidence you have studied, which case do you feel is more convincing—pro-bridge or anti-bridge? Give reasons. How happy are you to make this decision, based on just those three pieces of evidence? Explain.
8. Empathetic Response	Imagine that you are someone who holds the opposite views on the bridge proposal to the ones you really hold. Write a 'letter to the editor', in which you argue as persuasively as you can for that 'opposite' view.
9. Values Response	Do you think that a bridge should be built to link Stradbroke Island to the mainland? Why, or why not? Would you like more information, or more time, before you make that decision? Explain.
10. Making Wider Links	Collect a number of 'political' cartoons from newspapers and magazines. Decide how much exaggeration there seems to be in such cartoons. Think about whether cartoonists are 'fair' when they produce exaggerated cartoons. Perhaps you could make a poster display to highlight the use of exaggeration by cartoonists. Survey the 'Letters to the Editor' section of a newspaper for a week. See what types of issues lead people to write those letters. Classify the letters by category—for example: political, economic, social, religious, media criticism. Perhaps use a poster to display your categories. Do politicians (and others) use 'tricks' in their speeches to get people on side? Study reports of politicians' speeches in the newspapers, and see them speaking on television. Try to identify any such tricks, and decide whether you find them effective. Are they offensive?

Figure 7.2 Sample activities for going beyond the evidence provided on the Stradbroke Island bridge issue

seen as significant components of a curriculcum in which teachers foster student investigation of geographical issues, while still demonstrating the crucial motivational, expository and interactive roles they play.

Year 10 Geography: The Stradbroke Issue Worksheet

Stradbroke Island is located southeast of Brisbane, the capital city of the state of Queensland, in Australia. The island possesses features of great physical beauty—long sandy beaches, rocky headlands, freshwater lakes, forests and creeks.

The permanent population of 2500 lives mainly in the townships of Dunwich, Amity Point and Point Lookout. The island attracts thousands of visitors each year. Many people work on the mainland, and a number of high school students make the daily trip to the mainland as well. Access to the island is by a number of passenger and vehicular barges travelling from Cleveland and Redland Bay.

Since the 1940s, there have been periodic calls to link Stradbroke Island to the Queensland mainland by a road bridge. No proposal has come to fruition, but the issue intensified in the mid-1970s, with a commitment by the Queensland state government to bring about the building of a bridge within a few years. Over 12 years later, the bridge had still not been built, but the issue is still alive.

People are divided over the issue. On both the island itself and the mainland, there are supporters of the bridge proposal and opponents of the scheme. There has been much public comment and debate—at meetings and in the media.

This worksheet provides you with an insight into the issue. You will read a letter from a citizen and a speech by a politician. You will also see a cartoon about the possible effects of the bridge.

By the time you have finished the activities listed in this worksheet, you may be ready to make your own decision about the issue, and to act accordingly.

Read the letter and speech and look at the cartoon. Then move on to the questions and activities which follow.

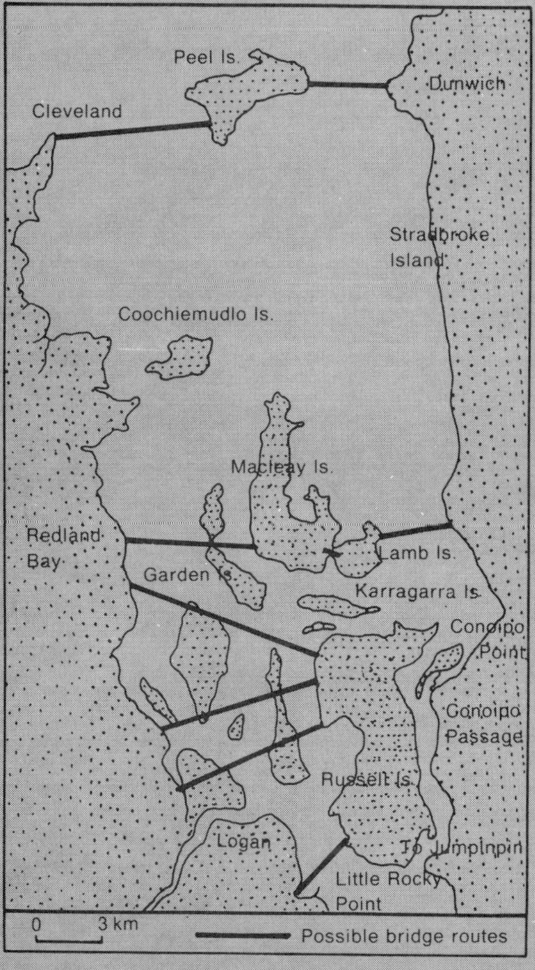

Figure 1 Proposed bridge links between the mainland and Stradbroke Island

Source A: The Letter

The *Courier-Mail*, Brisbane, 21 November 1977

Letters to the Editor

'SPARE ISLAND'

My wife and I have visited Stradbroke Island over the last 30 years and since 1973 have owned a seaside house at Amity Point.

On a recent trip to the island our fellow passengers on the vehicular ferry (which incidentally costs $16 a return trip) were canvassed as to their views on the proposed bridge.

Their replies to the questionnaire were strongly in the negative. . .

A bridge from the mainland would be an environmental disaster and a tribute to the greed of private developers. . .

There will inevitably follow the destruction of frontal dunes for high-rise development, upgrading of the road system to allow city motorists to go round and round in circles on an island approximately 20 miles long by 9 miles wide, pollution from motor vehicles, sewerage disposal (no doubt into the ocean), cutting down of vegetation, destruction of bird and animal life, interference with the Blue and Brown Lakes, residential allotments rising to $20,000 and more with uncontrolled speculation and progressive impoverishment and eventual removal of the pensioner population at Amity Point.

Already beach buggies, motor bikes and the like are speeding along beaches and destroying the fragile dune system. . .

No, we are not selfish. We welcome genuine development.

But a bonanza for the 'developers' and speculators and a shattered environment—not on your life!

G. Peters, barrister-at-law, Bonneville Street,
Holland Park West

Source B: The Speech

Mr Russell Hinze, Minister in the Queensland Government, at the opening of the Redland Shire public swimming pool, 11 March 1978. Extract.

. . . While I'm here with my friends from Redland Shire, I'd like to say a few words about our plan to build a bridge to Stradbroke. I'm sure all you fine people gathered here today will agree that a bridge to Straddie is long overdue. The benefits for the island's residents will be immense. They will have a safe and speedy connection with the mainland, and that's especially important in medical emergencies, and in bad weather.

The bridge will bring real progress to Stradbroke. Tourism will boom, and the rewards will be shared by all the residents. People from hundreds of miles around will welcome the chance to drive to Straddie's sunny beaches, instead of having to endure the long barge trip.

But what do we see every day in the papers? . . .the whinges and groans of the so-called 'greenies'! Now I reckon a lot of these conservationists couldn't work in an iron lung. And they look like they haven't showered or combed their hair for months. They don't know what real work is. All they do is sit in the bleachers and hurl abuse and peanuts. It's easier than getting into the ring and making positive points. These people who criticise the bridge are wearing blinkers. They've got preconceived ideas, and vested interests in seeing the project shelved.

Figure 2 The impact of building a bridge to Stradbroke Island

Source C: The Cartoon

After reading the letter and the speech, and studying the cartoon, answer the following questions and attempt the following activities. Where written answers are required, write them on a page of your notes headed 'The Stradbroke Issue: Worksheet'.

Investigating the evidence

1. According to Mr Peters' letter, who will benefit if the bridge is built?

2. What does the cartoonist John Mason predict will happen to the beachfront land if the bridge is built?

3. What are the different problems Mr Peters predicts will develop if the bridge is built?
   List them under the headings: (a) environmental, (b) social, and (c) economic.

4. What benefits does Mr Hinze predict will follow if the bridge is built?
   List them under the headings: (a) social, and (b) economic.

5. How has John Mason shown in his cartoon that the development of large tourist accommodation will actually cause problems for those tourists?

6. What do you think Mr Peters means when he claims that 'A bridge. . . would be. . .a tribute to the greed of private developers'?

7. What is the overall message of John Mason's cartoon about the Stradbroke Island proposal?

8. Why might Mr Hinze have stressed the supposed laziness, dirtiness and vested interests of the conservationist critics of the bridge?

9. Why might the cartoonist John Mason have given the buildings such names as 'The Pits', 'Cylinders Cesspool', 'Deadman's Motel', 'Turd Tower' and 'Boring Tower'?

10. Judging by his letter, what things does Mr Peters seem to value? How did you decide?

11. (a) Do you necessarily accept what Mr Peters says as being an accurate description of the effects of a bridge on Stradbroke Island?
    (b) Why could some of Mr Peters' claims be inaccurate? Could some inaccuracies be unintended, but others deliberate? Explain.

12. (a) Do you necessarily accept what Mr Hinze says as being an accurate description of the conservationists who oppose the bridge?
    (b) Why could some of Mr Hinze's claims be inaccurate? Could some inaccuracies be unintended, but others deliberate? Explain.

13. What is one major point that seems to be made by both Mr Peters' letter and John Mason's cartoon?

14. Mr Hinze claims that people will 'welcome the chance to drive to Straddie's sunny beaches, instead of having to endure the long barge trip'. What claim by Mr Peters seems to conflict with this statement by Mr Hinze?

15. (a) Given the three pieces of evidence you've studied, which case do you feel is more convincing—pro-bridge or anti-bridge? Give reasons.
    (b) How happy are you to make this decision, based on just those three pieces of evidence? Explain.

Researching the issue

1. Using the library research file on Stradbroke Island, find out about the proposed routes for the bridge, estimated costs, suggested sources of funding, estimated usage. Present your findings in a table.

2. Collect newspaper and magazine articles about the Stradbroke Island issue. Devise a system of headings for filing them.

3. Design and conduct a survey to find out people's opinions about aspects of the Stradbroke Island bridge issue. You might investigate whether people ever visit the island, whether they favour a bridge, their reasons for their opinion, whether they are aware of the various arguments being put forward, whether they have ideas on who should pay for any Stradbroke bridge. Present your findings using coloured graphs.

Interacting with other students

1. 'Someone' will be making the decision about whether the Stradbroke bridge is built. In a group, have a discussion about who you think should make decisions about such issues. Consider: the government at Federal, State and local levels, 'experts', residents affected, visitors affected, potential bridge builders, the wider population of Queensland, of Australia.

2. Devise and present a role play based on the Stradbroke bridge issue. For example, you could base the role play on a family discussion, in a family where the mother is a Stradbroke Island businessperson

who would benefit from any increase in tourism, the father works for the existing barge company, the daughter is an ardent conservationist, and the son lives on the island but works on the mainland, and would benefit if he could drive to work over a bridge.

Expressing opinions

1. (a) Do you think that a bridge should be built to link Stradbroke Island with the mainland? Why, or why not?
   (b) Would you like more information, or more time, before you make that decision? Explain.

2. Imagine that you are someone who holds the opposite views on the bridge proposal to the ones you really hold. Write a 'letter to the editor', in which you argue as persuasively as you can for that 'opposite view'.

Being creative

1. Decide what you think about the Stradbroke bridge issue. Then plan a way of trying to convince other people to agree with you. You might consider producing a poster, bumper sticker, tee shirt design, radio jingle, television commercial. Try to produce the finished product if possible.

Making wider links

1. Collect a number of 'political' cartoons from newspapers and magazines. Decide how much exaggeration there seems to be in such cartoons. Think about whether cartoonists are 'fair' when they produce exaggerated cartoons. Perhaps you could make a poster display to highlight the use of exaggeration by cartoonists.

2. Survey the 'Letters to the Editor' section of a newspaper for a week. See what types of issues lead people to write those letters. Classify the letters by category—for example: political, economic, social, religious, media criticism. Perhaps use a poster to display your categories.

3. Do politicians (and others) use 'tricks' in their speeches to get people on side? Study reports of politicians' speeches in the newspapers, and see them speaking on television. Try to identify any such tricks, and decide whether you find them effective. Are they offensive?

8

Learning Geography Through Classroom and Library Research

Louis Murray

One widely used form of inquiry in geography teaching is classroom and library research by students. Louis Murray explains five approaches to this form of inquiry with students using a variety of resources, including maps, atlases, aerial photographs, statistical reports, books and films. Detailed worksheets for these five activities are presented in the chapter. Lists of the basic resources needed in the geography classroom and school library and ways of using local or district libraries are provided also. This chapter illustrates how many of the ideas for inquiry presented in Chapters 5, 6 and 7 may be implemented but is also relevant to Chapter 4 as it shows several ways in which students may be actively involved in expository learning.

Terms such as 'research', 'inquiry', 'discovery', 'data-based learning', 'interpretation' and 'inference making' have become associated with many trends in geography classroom practices in recent years. Many of these terms involve quite complex chains of ideas and skills. Unfortunately, they are often too simplistically advocated as solutions to many learning problems without proper regard for the competence of students to handle them. This chapter on learning through classroom and library research suggests ways in which some modern approaches to inquiry learning may be utilised without leaving students too much alone, and without structure and direction in their inquiries. Five sample research activities are included in this chapter. They are expressed as instructions for teachers, rather than for students, so that the teacher's role in leading students in classroom and library research can be highlighted. Each is based upon resources and principles for teaching that are explained in the chapter.

Research Skills for Students

Classroom and library research is a practical search and recording strategy which incorporates skills such as library catalogue usage, structured note taking, and report writing. Together, these involve four major strands:

1. the location and collection of descriptive data;
2. the retrieval of data for use in the solution of predetermined problems;
3. the placement of the learner in an intimate interactive relationship with data, the problem under focus, and the implements associated with research techniques; and
4. the preparation and presentation of reports.

The perceptive reader will note that the term 'interpretation' is not included in this description of inquiry through classroom and library research. This is quite deliberate. The research process for school students, especially in the lower secondary school, should be primarily a guided, structured experience. It is not something that just happens or that students just do. Neither is it the mindless copying of material from encyclopaedias that is often described as 'project' work in primary schools. It is the systematic use of available data sources. 'Interpretation' in the sense that students are left to make their own conclusions from data needs to be carefully mediated by the teacher and controlled by the purposes and objectives of the individual lesson plan and units of work. Interpretive and imaginative use of data are important but it is unreasonable to expect them to occur randomly outside of a structured learning experience.

This fairly unequivocal assertion has significant implications for the teacher. The actions that constitute research (summarising content areas, using an index, tabulating distributions from current statistics, and so forth) need to be taught directly and practised routinely. For example, students in at least the first year of secondary school must be systematically inducted into the library skills of locating appropriate books in a shelf area and using the card catalogue to learn that:

1. a book is usually listed by subject, author and title;
2. all cards are arranged alphabetically either word by word or letter by letter;
3. cards have call numbers typically in the upper left-hand corner which indicate the location of the books on the shelves; and
4. some cards provide additional information such as an annotation of a book's subject matter.

We should not be too glib about the apparent simplicity of these skills. Neither should we assume that they are the prerogative of primary schools or English teachers. It is a rather sad fact that even many students in tertiary education cannot adequately use a library catalogue! Thus, we should be particularly careful in our expectations for school students and we must ensure that prerequisite research skills are, in fact, taught and learned. This requires a teacher to make judgements about the importance of certain research skills. The judgement should be guided by the centrality of the skills to geographical inquiry. For example, library research in geography is crucially dependent on a student's ability to read. It is inappropriate, therefore, to consign poor readers to the task of unsupervised research in the library.

Our thinking about the sequential logic of research skills within geographical learning is further enhanced by a consideration of *means* and *ends*. An essential 'end' of library and classroom research is to provide a set of tools which will endure as the student's learning capacity increases. We try to achieve this 'end' by structuring, as clearly and as practically as we can, research experiences that result in much shorter term gains in skill development. Adding these developments together over time and rehearsing them constantly with diverse content and functional problems facilitates the possibility that the research act will become an important part of a student's permanent learning behaviour.

The Research Venue

The Classroom

If the conventional classroom is to be used as a research venue, it may need to be transformed. Too often, geography rooms in secondary schools degenerate into dreary, sterile, austerely furnished closets, usually identifiable by a tatty and out-of-date wall map clinging wearily to the rear wall. Fitting a classroom out for research should take account of the following:

1. Furniture and fittings should be capable of varied configurations. At least one display table and cabinet should be available. The potential for physical movement of furniture and students should be enhanced, not hindered.
2. The walls should be well covered with (regularly changed) maps, charts, posters and pictures that have immediate and dramatic sensory appeal. Such things should also be capable of yielding succinct, summarised, up-to-date information.
3. Reference books should be visible, accessible and identified by content subheadings or subject. It is sometimes advantageous to bulk borrow a selection of books from the school library. This democratises borrowing and means that the teacher can ensure that library selections are used during the time set aside for research.
4. Information recording devices should not be limited to notebooks. Cassette recorders and polaroid cameras should be made available wherever possible, and particularly where verbal consultation and pictorial recording is required.

Graph paper is very useful for recording and ordering statistical information. It should be available in quantity as should various sizes of tracing and mapping paper and mapping pens.

5. Behavioural constraints can be relaxed and modified by reference to the ground rules of research. In this way, the exchange of information and movements to and from data sources can be encouraged. Students should be permitted to operate easily used technical aids such as the videocassette recorder, slide projector and tape recorder.

6. A range of reference aids should be available. These include a comprehensive world atlas, a globe, stereoscopes for use with aerial photographs, a concise geographical dictionary, and data sources such as national yearbooks, government and industrial reports, census summaries and reports on local and nearby areas. A good student atlas is one of the most important reference tools that a student can use in geography. Appropriate to the year levels of students, all geography classes should have a class set of at least one of the following: *Nelson Young Australia Atlas*, *The Macmillan Atlas* or *Atlas 2* or *3* from the *Jacaranda Atlas Program*. It is necessary to ensure that students understand the special properties of the atlas including its index, use of scales and symbols and the nature of its shading and colouring schemes.

The School Library

The most obvious advantage of the school library as a research venue is the presence of the reference collection of newspapers, magazines and books. Pupils should be introduced to the reference collection early in their school career and the full meaning of 'reference' broadened, as they progress through the grades. Newspapers should be a major feature of classroom and library research using local, metropolitan or national 'dailies' as well as weekly magazines and newspapers such as *The Guardian Weekly*, *The Christian Science Monitor* and *Time Magazine*. Of course, there are the many free local, state and national government publications as well as publications from overseas embassies and the United Nations that should be in the 'vertical file' of all school libraries.

Schools tend to make their own selections in regard to reference books. The following list of reference books, while not exhaustive, is indicative of the range of proven works that should be found in school libraries and that are suitable for use by geography students:

Australia
Australians: A Historical Library (1987) Sydney: Fairfax, Syme and Weldon.
The Australian Encyclopaedia (1983), 4th edition, Sydney: The Grolier Society Ltd.
The Australian Almanac (1985) Sydney: Angus and Robertson.
The Macquarie Book of Events (1984) Sydney: Macquarie Library Pty Ltd.
Year Book Australia (1988) Canberra: Australian Bureau of Statistics (or latest edition).

General
Everyman's Factfinder (1982) London: J.M. Dent and Co.
Far East and Australasia (1986) London: Europa Publications Ltd. (17th edition.)
The Global Resource Book (1987) New York: Global Perspectives in Education Inc.
Information Please Almanac (1986) Boston: Houghton-Mifflin. (39th edition.)
International Geographic Encyclopaedia and Atlas (1979) London: Macmillan.
Johnson, R.H. et al. (1983) *A Dictionary of Human Geography*, Oxford: Blackwell.
Macmillan Children's Encyclopedia (1985) Melbourne: Macmillan. (2nd edition.)
The New Global Yellow Pages (1987) New York: Global Resources in Education Inc.
Political Handbook of the World—1982/83 (1983) New York: McGraw-Hill.
Readers Digest Book of Facts (1985) Sydney: Readers Digest Services.
Statesman's Year Book 1985–86 (1985). London: Macmillan. (122nd edition.)
World Facts and Figures (1979) New York: Wiley Interscience.
The World— People and Places (1985) Melbourne: Macmillan.
World Resources 1986 (1986) New York: Basic Books for World Resources Institute.

Atlases
Australians: A Historical Atlas (1987) Sydney: Fairfax, Syme and Weldon.
Bay Books Concise Atlas of Australia and the World (1984) Sydney: Bay Books.

The Gaia Atlas of Planet Management (1984) London: Pan Books.
Jacaranda Atlas of Australia (1984) Brisbane: Jacaranda Press.
National Geographic Atlas of the World (1981) Washington DC: National Geographic Society.
The New State of the World Atlas (1985) London: Pluto Press.
Philips International Atlas (1981) London: George Philip Ltd.
Readers Digest Atlas of Australia (1977) Sydney: Readers Digest Services.

The Local and District Library

The local library is a useful supplementary resource venue and students should be encouraged to utilise its facilities. Given the very high cost of modern text and reference books, the feasibility of borrowing significant books should be fully explored. Note also that the local library is often the best starting point for advice and information on local and regional studies.

A major advantage of local libraries is that they usually stock current serials and periodicals. Most libraries usually offer most of the 'classic' geographical and environmental periodicals. These are: *National Geographic Magazine, Geographical Magazine, BBC Wildlife, New Scientist, Geo—Australasia*'s *Geographical Magazine* and *New Internationalist*. Geography teachers should seek to have their school libraries subscribe to these publications as well. They have the inestimable advantage of stimulating the reader by providing full-colour glossy photographs and 'conversational geography'. In so doing, they provide important geographical information but in the 'parlance' of the layperson. The latent, intrinsic motivation in these periodicals can be tapped not only to inform the research task but to give impetus to a lifelong interest in geographical matters.

Principles for the Conduct of Research

Classroom and library research in geography is dependent upon students being trained in appropriate research skills and upon accessible, well laid out and stocked research venues. It is also dependent upon several sound educational principles. Seven principles for the conduct of student research include the following:

1. *Success at research is in large part dependent upon a student's prior experiences with books.* Teachers should have a good knowledge of their students' reading capabilities before setting up research tasks, in particular those requiring information retrieval from complex reference texts. Also, teachers should have a good knowledge of the reference collections in both school and local libraries. This knowledge should include some notion of the technical complexity of the wording used in texts.

2. *Research implies a commitment to the individualising of instruction and the responsibilities that this entails.* Research need not be conducted solely by a single individual. Pairs and groups of students can and do work successfully together. However, research needs to be individualised in the sense that the learner is placed at the centre of the learning task. This demands trust on the part of the teacher and responsibility on the part of the student. Under no circumstances should the research venue (the school library in particular) be merely used as a 'car park' where students are deposited to while away their time under the care of a third party. Time in the library should be used maximally and productively.

3. *Research is a process that requires constant practice for its improvement.* It is unrealistic to expect pupils to produce good work when research exercises are set infrequently and without proper regard for their place in the total geography program. If library and classroom research is accepted as a valid strategy, then it should be incorporated into learning experiences as part of the normal class program. Remember also that research skills such as using a card catalogue and note taking must be taught directly and practised frequently if they are to become part of the repertoire of student skills.

4. *Success in research is often contingent upon the adequacy and specificity of the problem to be confronted.* Some years ago, I was being shown around a new resource centre at a high school in Western Australia. The students in the centre, a Year 9 geography class, seemed to be highly motivated and were busily engaged in individualised tasks. Their teacher proudly proclaimed that they were doing their 'projects'. During my circuit of the room, I glanced over the shoulder of one busy young lad. Sitting prettily on the top line of his worksheet were the words 'op. cit.' and 'ibid'. When I asked him what he was doing and what the words meant, he replied, 'I don't know; I'm only doing my project. I'm copying out this book just like the teacher wants'. He was, in fact, copying a Latin dictionary. Clearly, this student had no concept of research, was oblivious to any formal implications of the word 'project', and had only the crudest perception of his teacher's expectations. Tactful questioning revealed that the responsibility for this state of affairs rested squarely with the teacher.

It was suggested in the first part of this chapter that student research really meant one of four things: the location and collection of data; the retrieval of data for use in the solution of predetermined problems; the placement of the learner in a valid relationship with the objects and artefacts of his research; and the preparation and presentation of reports. *Teachers should use these four dimensions of the research act as a checklist for both the determination of research tasks and as a guide to the adequacy of the research problem.* If the research task cannot be seen to fulfil the requirements of at least one of these conditions, then it is doubtful if it is a valid task at all. Remember also that the determination of the research task is essentially the responsibility of the teacher, especially for younger students.

5. *Where research is problem oriented, the statement of the problem should itself suggest paths towards a solution.* Research problems are best phrased as informative questions. Consider the following tasks that could be set for students:

(a) Write a summary of the major characteristics of Western Australia's population distribution.

(b) Refer to page 120 of *Western Australia—An Atlas of Human Endeavour 1829–1979* (Perth: Government Printing Office, 1979). In what parts of the state has population increased since 1881? Now turn to pages 122–123. What relationships

Figure 8.1 Example of classroom researching, using a reference text

exist between population density, climate and land use?

The first problem as stated is far too general to be considered a satisfactory research exercise. It does not specify a research tool, rather, it relies heavily upon a student finding his own information. It also makes assumptions about a student's capacity to write. The problem is also heavily inferential. It specifies only a single population variable (i.e. distribution) and provides no guidance in the matter of the relationship of this variable to others (e.g. climate, soil). The term 'characteristics' is thus open to idiosyncratic interpretation by the student as researcher.

The second problem is much more satisfactorily stated. It specifies the research tool to be used (*Western Australia —An Atlas of Human Endeavour 1829–1979*). It structures in antecedent and successive relationships, both the stages in the research process (page 120 followed by pages 122–123) and the variables to be considered (i.e. population→ population increase→population density→climate→ land use). This second research problem maintains some inferential properties, whilst at the same time suggesting the criterion variables to be analysed and structuring the relationship between them. The problem statement assumes a relationship and directs the researcher to the information that helps in its elucidation.

6. *Research is often dependent upon a sound descriptive knowledge of the content area or substantive detail that surrounds a research problem.* Students undertaking research tasks (especially in the first few years of secondary school) must have an information base

---

Research activity 1

Research problem. How many independently governed countries are there in Africa?

Research purpose. Group use of a standard reference text in the retrieval of data; suitable for use with Year 8 or 9 geography classes concerned with broad issues in political and world regional geography.

Resources to be used. *The Statesman's Year Book 1985/86*, London: Macmillan, 1985, 122nd edition.

Stage 1. The teacher establishes criteria for the category of independent government. This takes account of broad related concepts such as 'Commonwealth', 'Dominion', 'Republic', 'Free State', etc.

Stage 2. The class is divided into five groups. At different times during a one-week period, each group uses *The Statesman's Year Book* to ascertain the number of countries in Africa that meet the criteria for independent government. Countries with idiosyncratic patterns of government are noted.

Stage 3. Each group reports its findings in a subsequent class session. Idiosyncrasies and variance between group findings are analysed under teacher direction, and by group cross-referencing against the original criteria established for the category of independent government.

Stage 4. Each group subsequently checks the ephemerality of independent government by reference to sources (pre-1960 atlas, historical geography text, etc.) that document different epochs in the social and political geography of Africa.

Research activity 2

Research problem. In what ways does the Avon River valley of Western Australia offer advantages of site for settlement?

Research purpose. A pairs-based observation and inference exercise using a stereo pair of aerial photographs. The exercise involves systematic observation, recall, verification and inference making. It is a particularly useful technique for it has the quality of being both an instructional technique (it structures a pupil's experience with aerial photographs) and a research activity. It is especially suitable when applied to rural landscape and settlement studies at the Year 10 level.

Resources to be used

1. A standard stereoscope per pair of pupils.
2. Aerial photographs WA 1410 Perth run 13 (5081–5120) 1 : 40 000 152.5 mm Nos 5107 + 5108—18/9/1972—available from Lands and Surveys Dept in Perth.

Stage 1. Student A views the aerial photographs through the stereoscope, and verbally identifies six examples of:
  (a) *relief features*, e.g. escarpments, plains, etc.
  (b) *hydrological features*, e.g. rivers, watercourses, etc.
  (c) *access features*, e.g. valleys, routeways, etc.

The identification is done as a concept-naming exercise. That is, student A gives a term or name which has public geographical meaning. Student B records on a card the information provided.

Stage 2. Student B now takes a turn at the stereoscope. Student A recounts (slowly) from the card the conceptual information generated in Stage 1. It is now student B's task to identify visually the concepts named; that is, to confirm visually A's original selection. Student B may seek clarification by demanding of the partner expanded meanings for the particular conceptual terms used, or question the relative position (relative to other features) of the objects in the photograph.

Stage 3. Both students now confer and the concepts visually identified are ranked according to their positive contribution to human settlement. That is, students make an inferential judgement about the concepts. The rank ordering is accomplished by using the following table.

	Relief features	Hydrological features	Access features
1			
2			
3			
4			
5			
6			

Stage 4. On the basis of the rankings, a report that directly addresses the original research problem posed is written. The veracity of the report is tested by comparing it with information contained in relevant books such as *Western Australian Year Book*, Number 23, and *Western Australia—An Atlas of Human Endeavour 1829–1979*.

Figure 8.2 Example of classroom research, using aerial photographs

> **Research activity 3**
>
> **Research problem.** What relationships exist between rural population density and distance from the city?
>
> **Research purpose.** Individualised, statistical-inference-based activity using conventional reference tools to demonstrate the generality of a relation between two geographical variables. Suitable for Year 11/12 pupils.
>
> **Resources to be used.** *Western Australia—An Atlas of Human Endeavour 1829–1979.*
>
> *Stage 1.* Relevant places (Perth and country towns) are nominated or randomly selected from the atlas. Population figures are derived for each place and a general estimate of density (number of persons per square kilometre) is derived from a population density map (page 122 in the Western Australian atlas).
>
> *Stage 2.* Using the following table, students plot rural population densities in relation to distance from the city by marking with an asterisk. The information is used to demonstrate the general relationship: *Rural population density decreases with increasing distance from the city.*
>
>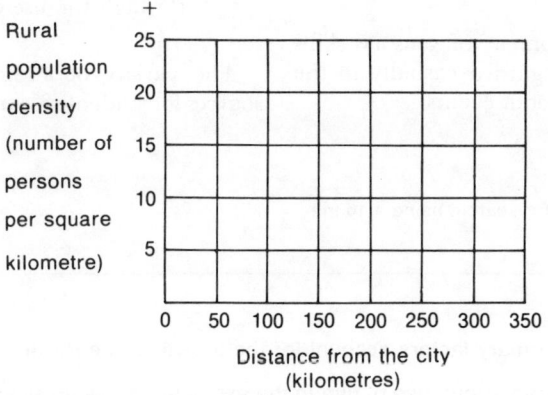
>
> *Stage 3.* The reliability and applicability of the inverse relationship is tested by repeating the exercise with places and distances selected from contrasting regions and cultures.

Figure 8.3 Examples of classroom research, using statistics

from which to work. For example, in order for students to examine the significance of the Panama Canal in world trade, they must first know where it is located, the names of the oceans it gives access to, its approximate latitude, and that it is a desirable alternative to the long, stormy route around Cape Horn.

It is easy to overestimate the knowledge possessed by students. If a teacher is committed to the centrality of research and inferential methods in geographical learning, she must ensure that students enter inferential experiences with good, prior descriptive knowledge *or* that they gain such knowledge during the research process.

7. *Any research task must take account of the ability and development experiences of students.* As a general rule, the younger the student, the more structured the research task should be. Careful control in the structuring of research problems can prevent research becoming an elitist practice for very able students or a time-filling exercise for the non-academically inclined. Remember also that good research exercises are incremental and cumulative in nature. What is done at the lower year levels has consequences (both in terms of the development of geographical knowledge and familiarity with the tools and routines of research) for formal study in upper school geography.

Research activities 1, 2 and 3 illustrate these principles, using books, an atlas and aerial photographs as data resources (see Figures 8.1, 8.2 and 8.3).

Audio-Visual Equipment and Non-Book Resources in Research

An interesting change in the naming of school buildings has occurred in recent years. In some school, 'libraries' are out and 'resource materials centres' are in. This indicates a shift in emphasis in the way in which information can be represented to the senses. Print materials remain as a primary information source and reference tool, but, increasingly, audio-visual equipment and non-book data sources such as charts and aerial photographs are being incorporated into the learning experience.

The use of electronic audio-visual equipment, its accompanying software, and other non-book data sources brings the following qualities to the research act:

1. It facilitates the development of sensory skills (vision, hearing) often regarded casually in the course of everyday classroom events.
2. It motivates and allows for a more complete address of the data under review in the sense that students can be placed in an interactive, reciprocal relationship with both data and media device.
3. Information not readily available in, or amendable to, printed form can be stored, analysed and transmitted through various types of audio-visual instruments.
4. The learning experience is enriched by the use of media, and non-book resources for students' options are widened in terms of:
   (a) their ability to gather information;
   (b) the possibility of variation in the reporting of information;
   (c) the emphasis they can place on details of content which are more readily revealed through the use of media.

The variety of audio-visual and non-book resources for student research in geography is immense

Figure 8.4 Example of classroom research, using a 16 mm colour film

---

Research activity 4

Research problem. What primary factors account for the growth of the Perth metropolitan region?

Research purpose. Structured, group use of film in the explanation of a problem in settlement geography; suitable for Year 11 or 12 geography classes.

Resources to be used. *Fremantle—First Port in Australia*, 16-mm colour film.

Stage 1. The class is divided into four groups for viewing the film. During the viewing, each group selectively focuses on one of the following organising concepts: *trade routes*, *primary produce*, *European migration*, *centralised manufacturing*, deliberately selected for their significance to the problem and their central emphasis in the film.

Stage 2. Under the general heading 'main ideas' each group member *independently* records descriptive information about the allocated concept on a small file card.

Stage 3. At the end of the film, members of a particular group confer to check similarities and variations in their note taking. This provides a group check on the reliability and adequacy of the data gathered.

Stage 4. Ambiguities or information that is unclear is checked for meaning:

(a) against standard reference texts; and
(b) with the teacher.

Each group now addresses the research problem posed by formally recording its *aggregated* information on one side of another file card. The cards are now circulated amongst the other groups for examination and comment.

Stage 5. The entire class again watches the film. At the end of the second showing the cards are considered *thematically* by the whole class under the direction of the teacher. That is, *the cards are considered according to historical and chronological continuity*. On the blank side of the group file card, each group is now able to summarise the independence, interdependence, inclusiveness and exclusiveness of their allocated concept in the perspective of the three other concepts.

and includes both hardware and software, for example:

Hardware	Software
Tape cassette recorders	Prerecorded cassettes, tapes and records
8-mm and 16-mm movie projectors	Moving films
Slide and filmstrip projectors	Slides and filmstrips
Television and videotape recorders	National television programs and video cassettes
Stereoscopes and geoscopes	Maps of various scales and types
Computers	Pictures and aerial photographs
	Computer programs in print, tape and disc form

Figure 8.5 Example of classroom research, using a range of print and non-print resources

---

Research activity 5

Research problem. What is 'shifting cultivation' and where does it occur?

Research purpose. Individual conceptual analysis exercise involving structured use of general reference texts and generating, by implication, a further focus on location. Suitable for Year 11 or 12 geography classes.

Resources to be used

South East Asia—Family, a 16-mm colour film
An atlas
A globe
A dictionary of geography
Asia 1980 Yearbook
Larousse Encyclopaedia of World Geography

Stage 1. Show the film. Through the eyes of the Sen family in Thailand, students gain a descriptive understanding of subsistence farming in hot humid climates. Individually, students now proceed to:

Stage 2. Define primary regions in which shifting cultivation occurs. The globe and the atlas are used to locate the countries 20 degrees north and south of the equator in which shifting cultivation is likely to be a dominant agricultural practice.

Stage 3. Define substantive and implied terms. The dictionary and encyclopaedia are used to derive the meaning of 'subsistence economy', 'humid tropics', 'humus', 'leaching', etc.

Stage 4. Establish a broad definition. The encyclopaedia is used to derive a definition of shifting agriculture.

Stage 5. Discover local and regional variations. The Asia 1980 Yearbook is used to derive information about: *ladang* in Indonesia, *bewar* in central India and *swidden* in Thailand.

Stage 6. Establish broad generalisations. The information so far gleaned from all sources is now pooled to help in the formulation of the following generalisations:

(a) Tradition and local variations in climate influence the farming calendar and the crops grown in any given locality.
(b) There are definite upper limits to productivity inherent in the shifting cultivation system and the additional use of existing labour will not significantly alter this fact.

Stage 7. Confront broad questions of a values kind. The whole class reconvenes to watch a second showing of the film. Discussion ensues to consider:

(a) Value questions—for example, is shifting cultivation the only possible response of non-technological societies to soils of restricted fertility? As soil management and improvement techniques develop, is shifting agriculture likely to be replaced by more contemporary forms of husbandry? In the long term, can it be said that the system is inefficient and deficient?
(b) Tentative broad hypotheses—for example, was the decay in Mayan civilisation causally related to the progressive over-use of significant agricultural areas close to centres of large populations?

These learning resources, used either individually or in combination, can greatly enhance the quality of the research done by students and their enjoyment of it. The scope of this chapter does not permit a review of the use and applications of these resources. Readers should consult one of the standard audio-visual or teaching manuals for details of the selection and operation of these resources in the classroom. Good examples include: Ticehaust, G. W., Irwin, H. et al. (1982), *A Competency-Based Guide to the Operation of Projectors, Sound systems, Videorecorders, Microcomputers*, Sydney, Kuring-Gai CAE and Wittich, W.A. and Schuller, C.F. (1979), *Instructional Technology— its Nature and Use*, New York, Harper and Row, 6th edition. The use of such resources in student research should be governed by the same principles for the conduct of research described earlier. However, the special nature of non-print resources involves the following extra guidelines for teaching through classroom and library research:

1. The *passive* watching of film or listening to records should be avoided. Students should be in an active, reciprocal relationship with their selection of audio-visual and non-book resources.

2. The range of *purposes* to which media resources and non-print information can be put should be made obvious to the student. Visual information, for example, can be used both in the motivational mode (at the commencement of the research task) and in the reporting mode where the problem (or part of it) is actually reported in visual form.

3. Non-book resources work effectively when a co-operative classroom atmosphere has been created. That is, an expectation is clear to students that they will operate media sensibly, understand the purpose of media in the context of the research task, and use resources selectively for the analysis of a particular problem.

4. In the case of audio-visual equipment, all users should be adept in its operation and care.

5. The choice and use of non-book resources should reflect the nature of the geographical problem to be confronted. For example, visual material is very appropriate when students have no first-hand experience of an issue or place under study.

6. Where possible, combinations of non-book and book resources should be used as this allows for variety in the way in which topics can be studied.

7. Remember that for the current generation of school students, non-print resources and technological devices are, in a sense, natural and commonplace. This is not the case for many teachers. Consequently non-book resources should be regarded as part and parcel of the process of instruction, and not subjected to mystification processes, such as once yearly use, confinement in locked storerooms, etc.

Research activities 4 and 5 in Figures 8.4 and 8.5 illustrate these principles of teaching using a variety of book and non-book data sources.

Conclusion

Research activities are an important, but not isolated, part of geographical learning. I have emphasised the centrality of *structure* in any research task: the need for a statement of purpose and a sequential action format that directs the mental operations that students must use in elucidating the research problem. I have also tried to emphasise the reciprocity of responsibility in the research act: the teacher must plan in accordance with her predetermined goals which the student must implement in line with the plan. Independently motivated and conducted research is commendable but it is dependent for its success (at least in part) on the quality of prior, developmental research experiences. In establishing an action format for research in school geography, one is essentially setting up a guide for cognition and for more effective participation in learning. In the long term, this results in improved abilities to gain insights into the geographical conditions of society.

9

Models and Reality: Integrating Practical Work and Fieldwork in Geography[1]

Barrie McElroy

Although we aim to teach students about the real world, we generally do so by presenting them with models of it. Barrie McElroy suggests that a solution to this problem lies in integrating practical work in the classroom on models with fieldwork. He suggests that students can develop more perceptive understandings of models and how they portray aspects of reality if they first learn the practical handling of models in close association with fieldwork. This principle is explained by brief descriptions of a range of integrated practical and field activities in the local area of a school. The many practical ideas in this chapter should be considered in relation to Chapter 8 on classroom and library research, Chapter 10 on fieldwork and Chapters 13 and 14 on qualitative inquiry and environmental sensitivity. It thus provides a good link between classroom and field inquiry.

The college in which I teach and its campus high school sit on the lower slope of a fault line escarpment. Geography students in the school generally learn to explain faults, block movements and scarps using simple, line diagrams from a popular text and get good marks by reproducing them. Although they cycle up and down the scarp face going to and from school, many of them do not recognise the real feature as the one they are studying. They do not connect the model with reality.

In fact they and we, their teachers, often prefer to handle the simple model which is unconfused by the complexities of reality. A simple diagram can be readily learnt, reproduced, explained and marked correct. The real thing is not so easily comprehended nor explained. Neither is it so easily taught.

It seems a whole lot safer to avoid the complexity of the real world. However, it can be quite unsatisfactory. It is a bit like falling in love with a photograph which is only an *image* of a much more lovable, real, whole, living person! Images of people and of the world have their valuable place in our lives by helping us to know and enjoy reality better. However, it is a different thing to let our students develop a habit of not seeking the reality beyond the model.

As an experienced geography teacher, you already have a mental image of the site and location of my college. You will have an enhanced realisation when you find out that the college is found at GR483 723 and the school at GR488 715 in Figure 9.1.

A major part of our task as geography teachers is to help children develop this facility in using models to increase their understanding and appreciation of their world.

The Basic Ideas

This chapter contends that a better use of geographical models for the purpose of greater aware-

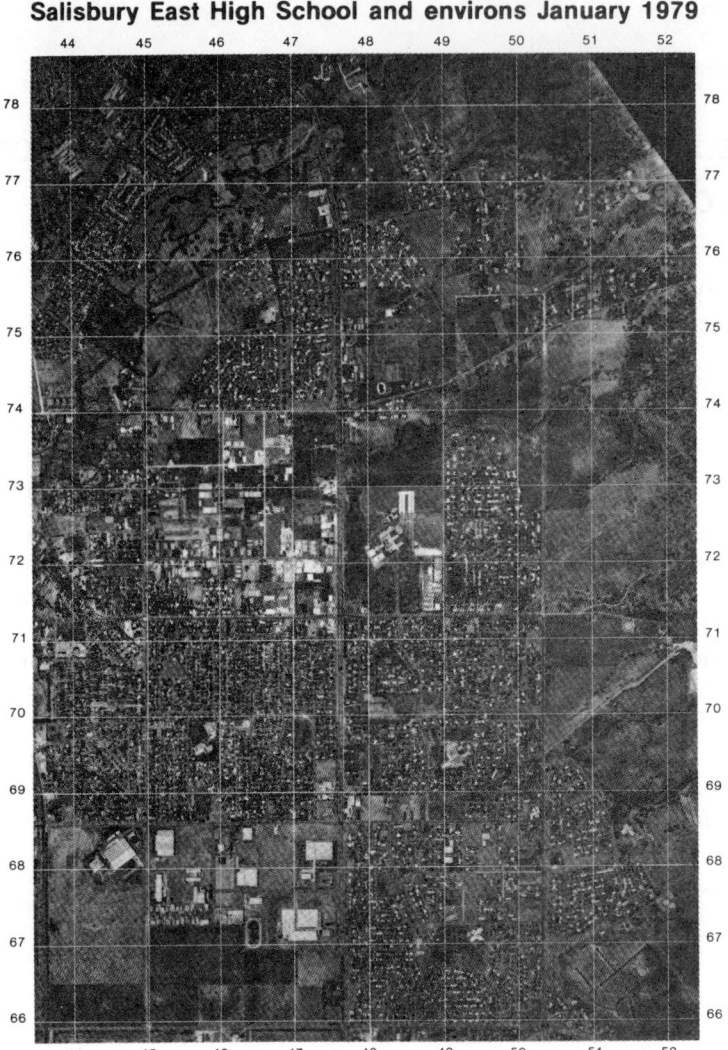

Figure 9.1 Aerial photographs and the local area

ness and understanding of the real world can be made by habitually integrating practical and fieldwork.

So that the simple thrust of this argument can be clearly grasped some definitions and premises must be stated.

Models and Reality

Models are images of reality that simplify and order it for easier study. They deliberately present one view, thereby providing an emphasis which highlights and exposes some aspects of reality while ignoring or diminishing others. Models are made necessary by the complexity of reality and, often, the remoteness or inaccessibility of the real subject of study. They assist conceptual understanding. However, they convey not the whole truth, but a useful and apparently comprehensible part of it. They are idealised representations that show something useful about the real world.

Figure 9.2 indicates that the range of models used in geography teaching is broad and the list long. The list includes Christaller settlement patterns, Loschian landscapes, road maps, ground and aerial photographs, written descriptions, retail gravity formulas, Hadley cells, topographic maps, climatic graphs and many, many more. Figure 9.2 is a useful planning matrix for organising the teaching of models in the geography curriculum.

Commonly used models vary considerably in their levels of abstraction from reality. Three classes may be recognised: iconic, analogue and symbolic.

1. *Iconic models* are easily recognisable images of some part of the real world. Photographs, especially coloured ones, videotapes, and 3D models are good examples.

2. *Analogue models* are those where the features of reality can be indirectly recognised through the use of stylised forms. Maps with obvious, conventional signs and where lines, points and areas retain the shape and proportions of reality are common examples.

3. *Symbolic models* are major abstractions where the easily recognisable features of the surface of the earth are obscured within mathematical formulas or graphs for instance.

As well as varying between models (e.g. photographs and maps) the level of abstraction can vary within a model (e.g. between the representation of street layout and the shape of the land shown by contour lines).

Practical Work and Fieldwork

For the sake of the argument presented here, *practical work* is defined as those activities of handling and using geographical models. This definition is useful in distinguishing such activities from those of *fieldwork*, which are concerned solely with the handling of the real world. By these definitions geography students study models by practical work and the real world by fieldwork.

It will be seen in the examples given later, of course, that they also can come to understand the real world through the practical study of models and similarly better understand models through the fieldwork study of reality.

Thinking and Study Activities

If this distinction between practical work and fieldwork is accepted as a way of discussing geographical study, there appear to be several excellent thinking and study activities that integrate the examination of models and reality. These have great potential for improving geographical understanding and awareness. Possible thinking and study activities include:

1. *Identification* of features of reality from clues in the model.

2. *Translation* of information about the real world from one form and/or level of model to another form or model.

3. *Analysis* of a model into appropriate components to expose elements of reality.

4. *Synthesis* of elements derived from other models or reality to create a new model.

Figure 9.2 Classes of models: a curriculum planning matrix

Examples (May be expanded by specifying types)	Class			Practical activity with model	Appropriate field activity
	Iconic	Analogue	Symbolic		
Verbal Written Oral					
3D—Solid Volcanoes Farms, etc.					
2D—Flat Pictures Maps Charts Aerial Photos, etc.					
Mathematical Gravity model Beta index, etc.					

5. *Application* of concepts and skills learnt to probe and solve puzzles raised by models and reality.
6. *Generalisation* from a model or reality to identify and comprehend the desired aspects and to ignore or remove distracting elements.
7. *Prediction* as a means of suggesting possible identifications, associations or explanations that can be tested later by gathering further evidence in the field or from other models.

Each of these activities may be used to integrate practical work and fieldwork. Examples may be included in the final two columns of Figure 9.2 to improve its usefulness as a curriculum planning matrix in this area.

The Crucial Idea

The pivotal, self-evident, but oft-neglected, aim of our geography teaching is that our students will develop an awareness and understanding of their world through the use of geographical approaches, ideas and skills. We generally seek to do this in the classroom by the use of models. This is forced on us by the demands of schooling and the complexity and remoteness of much of the real world. However, the range of models need not be as narrow as is common (maps and notes!). Nor need the treatment of those models be so divorced from reality.

Students have the best chance of conceiving reality through the models they handle in the classroom if they learn the strengths and limitations of those types of models by close association with real-world examples in the field. Too often, the ways we train our students to handle models, perhaps through 'data response' questions, map interpretation questions, or drawing cross-sections, rarely encourage them to imagine the reality represented. Teachers and textbooks sometimes present models as though they are what is important for the children to learn rather than the reality they represent. As most teaching time usually has to be spent in the classroom, geography teachers must learn to capitalise on the great variety of images of reality available, especially those most valuable ones held in our students' minds from their varied experience of the world outside the classroom.

The intention of integrating practical work and fieldwork is for students to learn to handle models via fieldwork so that they can learn about the real world via models.

Integrated Practical Work and Fieldwork: Some Examples and Strategies

The explanation of these activities relies on specific examples, all of which come from the local area of my campus high school (see Figure 9.1). I do not believe this to be parochial as the principles and methods can be applied anywhere. Besides, the local area of the school is by far the best place for regular fieldwork activities to be conducted.

Provided here is a series of illustrations of integrated practical and fieldwork activities that is used in geography courses in the campus high school. Although not generally made explicit, these examples are related to units of work where the children's increased familiarity with certain models can lead to a greater understanding of reality, especially where the latter cases are foreign or hitherto unknown. We believe that the use of local models is so critical that the school has invested in a reasonable range of them, especially models in the form of maps, pictures and aerial photographs.

It will be seen from the following examples that much of the integration comes from a comparative study of models and reality in various combinations such as:

$$\text{model} \rightarrow \text{field}$$
$$\text{field} \rightarrow \text{model}$$
$$\text{model} \rightarrow \text{field} \rightarrow \text{model}$$

Also implicit in these examples are many rationales for adopting particular procedures. The worth of the approaches and activities in the cases cited can be evaluated by readers in the context of their own settings.

To make but one rationale of my planning explicit it should be noted that the practical handling of models is generally designed to increase greatly the value of the always too limited time in the field. Consequently, I usually try to adhere to the following three principles:

1. During pre-excursion activities, students are encouraged to anticipate the reality they will encounter.
2. Post-field activities relate various models and model building to the observed reality.
3. Both before and after, analogies and examples are drawn from the students' collections of images formed from their wide experience of the world.

The treatment of the examples here is both cursory and fragmentary. They are *examples* to reinforce the messages in this chapter, and are intended to prompt planning appropriate to each individual teacher's needs rather than stand as total exercises to be taken and applied.

Aerial Photographs and the Local Area

This exercise is in line with the principle that students learn to handle models best in their local area so that they can later use such models to learn about other places. An aerial photograph with grid (Figure 9.1) of the local area is the main model. This is supported by a larger-scale version, about 100 cm × 60 cm, which can be used by students (Figure 9.3).

The aim is for the students to learn aerial photograph interpretation and mapping skills while learning concepts of land use in the rural/urban fringe.

A range of activities using the model before, after and during a local excursion includes:

1. *Pre-excursion*. The use of six-figure grid references. Bearing, distance and scale exercises. Identification of features, using a provided statement of clues. Land-use hypotheses, using a provided statement of clues. Preparation of a base map by tracing from the aerial photograph.

2. *Excursion*. Verify identification of features and land use suggested prior to the excursion through a process of 'ground-truthing'.

3. *Post-excursion*. Complete a land-use map with specified major features.

Towards Greater Abstraction

This exercise has Year 8 students identifying local land use from a large-scale aerial photograph, generalising and classifying it, and then translating it onto a more abstract base map. Students learn a good deal about the way that models generalise by removing less significant detail. The students in Figure 9.3 are working on this exercise.

The correctness of identification and classification can be tested through casual fieldwork. One does not have to take the class out into the local area for this as the children's aggregate experience is sufficient. There may, however, be good reasons for a class excursion for ground-truthing.

Students learn to appreciate the value of both models in locating land-use areas more easily than by field study. However, they chafe at the perplexing decisions relating to generalisation (e.g. removing houses from a predominantly rural zone).

Developing Concepts Through Practical Work and Fieldwork

Students are given a series of large-scale aerial photographs (1936, 1949, 1956, 1959, 1965, 1967, 1968, 1971, 1974, 1979, 1985) which illustrate aerial change through time in the local region centred on the school. They are asked to identify, generalise and

Figure 9.3 Translation: sketching land-use maps from a large-scale aerial photograph

Figure 9.4 Areal change through time: translating a series of aerial photographs onto base maps

translate land-use information in appropriate categories (rural, residential, industrial, public and recreational) from the aerial photographs to a series of simple base maps (Figure 9.4).

A final land-use map is drawn from data co-operatively collected in the field by the students. The problems of generalisation are heightened by the field visit and group decision making about boundaries and classification. They realise what is lost in such maps and are therefore cautious about the limitations and use of such maps in future.

Subsequently students are asked to write a description of the spatial changes over the period providing evidence from the sequences of aerial photograph and land-use map models.

From Fieldwork to Models

During the course of a Year 8 unit, 'Me, Us and Them', which includes the previous two activities, students prepare a range of models of aspects of the local area from data they collect in the field. This proves to be an excellent way of learning the benefits and limitations of particular models.

As individuals or in small groups they select and prepare what they believe to be appropriate models from the data gathered. This choice can be challenged by their peers. It is to them that the authors must justify the models and explain the limitations.

A wide range of models come from this activity including: maps (e.g. time/distance, location, paths), graphs (e.g. frequency, mode, distance), written descriptions, sketches, mental maps, flow charts, systems diagrams, and many more. Some examples from a class display of their models are provided in Figure 9.5.

We do not get the class to survey the whole community by questionnaire and interview, but only their families, other classes and teachers. This not only avoids the dangers and embarrassments of open survey, but in later analysis leads the children to understand the problems of a biased sample.

A Battery of Models for Fieldwork

Matriculation students in the campus high school make a study of the township of Gawler, just north of

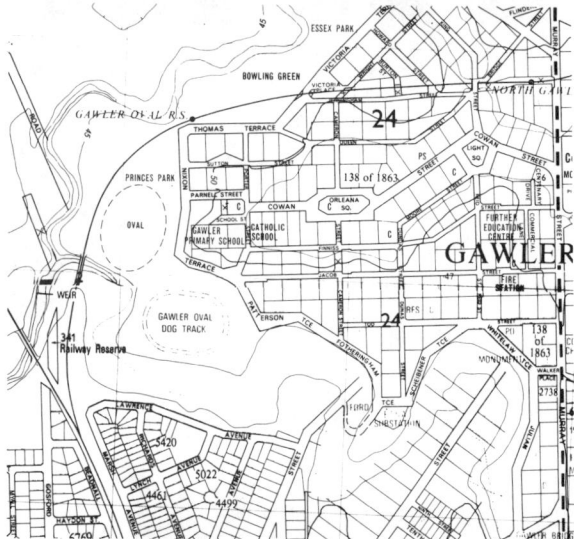

Figure 9.5 Models from fieldwork: aspects of the local area expressed as map and graph

Figure 9.6 Different models of the same area: contrasting levels of abstraction and generalisation aid learning

metropolitan Adelaide, as an introduction to a variety of geographical concepts related to the topics of 'Town and City', 'Manufacturing Industry', and 'The Changing Rural Landscape of South Australia'.

The town offers an historic and contemporary microcosm of the concepts needed for these topics as well as some of the more analytical models such as concentric zone, radial sector and multiple nuclei models of cities. Similarly its setting can also be tested within central place models.

A battery of models and practical activities is used to prepare for, and after, a one-day field excursion. Figures 9.6 and 9.7 show some examples. Models used include:

1. *Maps*—1:50 000 topographic; 1:10 000 topographic/cadastral; 1:2500 orthophoto/topographic/cadastral; street directory; 100-year-old town map.
2. *Documents*—histories of the town, recent planning reports, telephone directory, newspapers.
3. *Aerial photographs*—1:10 000; a sequence of 1:16 000 for 1949 and 1974.
4. *Statistical data*—population and socioeconomic data from census returns.
5. *Generalised models*—models of land use and the morphology of towns and cities.

Pre-excursion practical activities include:

1. An *identification* of features, including evidence of change over time of land use and settlement.
2. The *translation* of the models from one form to another (e.g. by drawing a cross-section across the scarp with pre-excursion prediction of land use relative to slope).
3. The *generalisation* of land use into zones for later verification. Base maps that suggest historical and current land usage are prepared for use in the field. These include walks near the railway station, through the main industrial region and the main shopping area. The cross-section passes through the original village.

The *fieldwork* begins with a circuitous tour of the town to check and plot features and patterns on a small-scale base map. This familiarises students with many ideas that they have previously imagined from the models. Also importantly, to students in a motor-car-dominated age, they have time to become accustomed to the place from the sanctuary of the coach

Figure 9.7 Ground photograph: iconic models can extend field experiences back in the classroom

before being exposed as researchers working in the field.

Most of the fieldwork is collecting and recording information in answer to questions raised by the pre-excursion practical work. It is also largely recorded on models (e.g. base maps or tables) prepared during those practicals.

Post-excursion practical activities include preparation of a further variety of maps. Besides the now verified or corrected historical and contemporary models (e.g. maps), there are the inevitable written descriptions and explanations. Having noted the changes and trends in Gawler, students find one of the most rewarding activities to be small-group forward planning for suitable changes to the town. This is generally modelled by using whatever forms they think are most appropriate. The challenge of peers encourages them to a deeper understanding of the models chosen and the concepts being illustrated.

Even in this final year of schooling the students need to have refined their perceptions from models. For instance, in preparation they are asked to describe what they will see when looking down a street which is shown on the 1:10 000 map as crossing a valley. In the field they are usually stunned to discover that the valley is a very deep gorge and that that part of the street does not exist in reality! Thereafter they are usually much more careful and perceptive in interpreting contour maps.

The Principle in Many Contexts

In preparation for the later textbook studies the geography teacher can plan to improve practical work and fieldwork. Anything that can encourage the comparative analysis of data from varied models (e.g. Figure 9.8 which in fact is two models in one, thereby allowing for a comparative display), including those prepared from field sources, is likely to

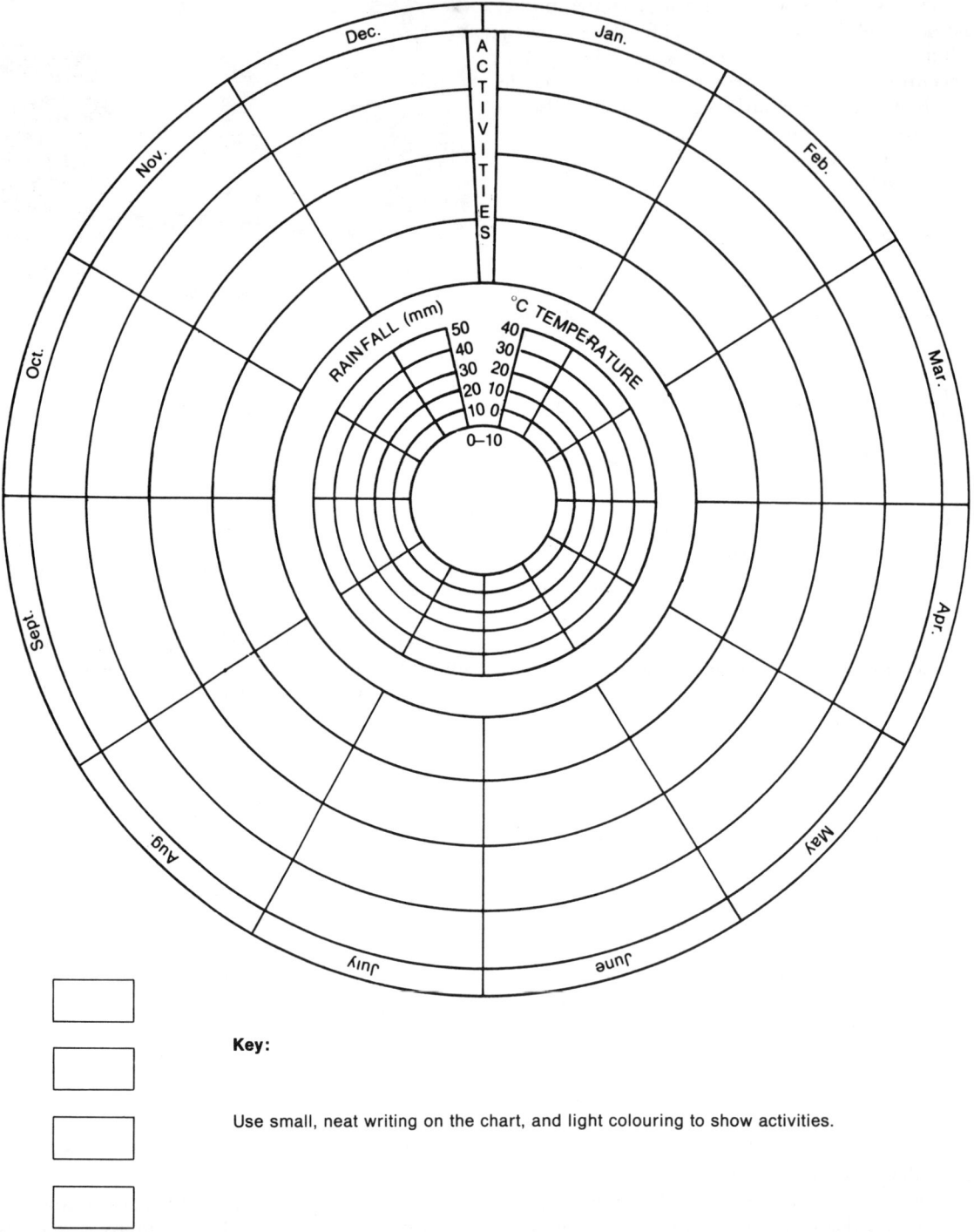

Figure 9.8 Two models in one: a comparative display

promote improved conceptual development.

The first curriculum-planning decisions for teachers therefore relate to the choice of a suitable range of models. These must be matched by selection of appropriate practical and field activities (see Figure 9.2). The potential of this approach is considerable and can be applied to most topics in geography classes. The following examples are but an additional few to supplement the predominantly urban orientation of the earlier examples.

1. *Beach profile studies* based upon tables of measurements (surveying exercises), section diagrams, photographs, wave tanks, . . .
2. *Farm studies* based upon aerial photographs, contour maps, field plans, taped interviews, systems diagrams, hardware models, . . . (see Figure 9.8).
3. *River flow and process studies* based upon aerial photographs, measurements of flow and stream form mapped and graphed.

Uses and Abuses of Models in the Study of Reality

The use of models may be seen as an aid to a more structured and purposeful approach to the study of the environment. Increased integration of practical work and fieldwork should enhance the students' ability to handle those models and improve their understanding of the real world.

Since models simplify and order a sometimes unruly reality, they can provide an entry to the understanding of that reality. Some *uses* of models in teaching include:

1. to analyse and illustrate patterns, processes, forms, etc.;
2. to explain patterns, processes, forms, etc.;
3. to allow an investigation of the abstract, remote or unobservable;
4. to assist communication, and the expression of one's understanding of concepts . . .

There are dangers in using models, however. Some *abuses* of models in teaching include:

1. Making the object of study the model, and not reality. It can be fatal to introduce the model too early, or to treat it in isolation from reality. That encourages rote learning about the model rather than the understanding of reality.
2. Failure to make students aware of the degrees of abstraction, generalisation and scale involved in a model
3. Neglect of the assumptions and intentions of the model.
4. The use of an inappropriate model, e.g. a map of the wrong scale, or models too abstract for young students to handle.

Nevertheless, there is much to be gained from using models, provided we understand their benefits and limitations and remember that reality is the object of study.

It is often the limitations of a model such as generalisation or abstractness that gives it its particular power to order, clarify and illuminate the desired attributes of reality. Students should become increasingly aware of the strengths and limits of the models they are using. The frequent, practical handling of models integrated with field studies provides students with their best chance of gaining the most value from both.

Note

1. I would like to acknowledge the help of Mr B. Locke of Salisbury College of Advanced Education for developing the local area aerial photograph exercise in this chapter, and the cooperation of Mr A. de Vries of the Salisbury East High School with whose classes I jointly taught many of the activities in this chapter.

10

Learning Geography Through Fieldwork

Kevin Laws

'Real geography' depends on good fieldwork. It is learned through the 'soles of one's boots', as the adage goes. But, just as most teachers of geography recognise this, they also are aware of the difficulties of time, distance, organisation and cost associated with fieldwork. We often resort to the use of models or secondary sources of data as a result. However, there is no substitute for well-organised fieldwork. Advice on field-based teaching is provided in three phases: a planning stage which actively involves students, a fieldwork phase based upon the skills of observation, data collecting and recording using tables, field sketches, sketch maps and transects, and a post-fieldwork phase in which the data collected in the field are interpreted and evaluated. Kevin Laws also provides guidelines for organising class fieldwork and fostering independent fieldwork by students in this chapter. Many examples are provided, including ideas for the investigation of stream morphology and the origin of street patterns in a city.

Scene: Geography staff room.

Teacher A: I really must organise some fieldwork for Year 11. We have been studying coasts for two weeks and they really need to get into the field and see the effects of the processes we have been talking about.

Teacher B: Where will you go?

Teacher C: I always take my group to Palm Beach. It's very close to the school and there are a good number of land forms to investigate.

Teacher B I hate fieldwork. It always takes so much time to prepare all the worksheets and organise the kids. I'd much rather go myself and take photographs of the important features. Then I can use them with my class and make sure they get all their notes complete.

Teacher A: I'm not going to have many question sheets for them to fill in. I want them to make accurate observations. I think we will spend most of the time measuring things like the wave interval. They can determine the direction of longshore drift, and from the headland you can see the pattern of wave refraction. Most of the work will involve the students. I think I will organise the class into groups and get them to draw a number of cross-sections from the parking area to the water line at the middle and each end of the beach.

Why do Fieldwork?

Most teachers of geography recognise the need to involve students in a range of activities which brings them into contact with the world outside the classroom. Sometimes slides, photographs and models can be used to help students develop their understanding of geographical concepts. However, there is no substitute for well-planned field activities. Group work, activities, worksheets and directed observations in an area relatively close to the school are key elements in planning a successful venture into the field. Through such experience, abstract concepts such as location, distribution, association, interaction, movement and change can come alive.

Quite frequently teachers find that students have no direct experience of many things which they themselves take for granted. It cannot be assumed that every student in Year 11 has visited the central business district of the nearest city. Not all students have been to a surfing beach despite the fact that the majority of Australia's population lives within 50 kilometres of the coast. Despite the rural image of Australia portrayed in many movies and television shows the majority of the population has not experienced rural living. Fieldwork can help provide these experiences. As well it can provide:

1. opportunities for developing skills in data collection;
2. a situation in which students can compare their personal perceptions of an area with the perceptions of others and with a map;
3. for the development of an appreciation of surface forms and an understanding of the processes which led to their configuration;
4. the concrete experiences which students need to grasp new ideas and incorporate these ideas into their cognitive structures.

Figure 10.1 Two approaches to fieldwork

Everson (1973) identified two approaches to the planning of fieldwork activities. The first, the *traditional approach*, is often referred to as field teaching. At its worst, this approach involves the teacher in directing the students' observations toward specific elements in a landscape at the expense of everything else. Often a mini-lecture is delivered from which students are expected to take notes. Little opportunity exists for student input and reaction. At its best, this approach involves students in the careful observation and description of an environment and in suggesting possible explanations based upon previously acquired information. The second approach, a *field research approach*, also involves the three elements of observation, description and explanation but adopts a problem-solving focus, utilising techniques similar to those used by a geographer involved in scientific explanation. This is the inductive approach to fieldwork that Clegg (1969) referred to as 'geographying'. These two approaches are illustrated in Figure 10.1.

Each of the approaches has relevance for geography classes, and the approach adopted for any particular field study will depend upon the purposes of the field activities. If students are inexperienced in making their own observations or lack confidence in their abilities to solve problems, field teaching can help, provided that opportunities for them to find their own examples of features and processes are included as an integral part of the experience. Field research requires a high level of planning on the part of students and teacher. Students must know very precisely what it is they are searching for and how they are to go about their searching. The teacher

Field teaching	**Field research**
Study of a geographic topic or theme in class—teacher talk, textbook study, note taking, slide viewing	Identification of a problem as the result of direct observations or from class work or from special interests of students
⇩	⇩
	Formulation of an hypothesis as a result of reading, discussion, thinking
	⇩
Observations, often teacher directed, and recording of information in the field—some field interpretation	Field activities involving data collection and recording
⇩	⇩
	Data analysis—processing information
	⇩
Further interpretation and explanation in class—writing up field experiences	Hypothesis testing—accept or reject

must ensure the students possess the necessary data collecting and recording skills and provide assistance to the students during the analysis phase. Students may work individually or in groups and explore various aspects of a single theme, sharing the results of their research with the whole class during culminating activities. What is important is that every opportunity is taken to get students involved in fieldwork as frequently as possible. It is not necessary for fieldwork to be exhausting, expensive and time consuming. Fieldwork can involve the preparation of a field sketch or sketch map during a 15-minute walk along local streets, the sketching of the skyline from the classroom window, interviews with local residents, the observation of physical processes such as erosion and run-off in the school playground or nearby, or testing hypotheses related to the effects of the end of the school day on local traffic flow.

The Purposes of Fieldwork

The range of objectives which can be achieved through fieldwork is great. Some objectives relate to the formation of attitudes and the development of an aesthetic awareness. Other objectives are concerned with the development of geographical understandings. Still other objectives relate to the development of skills, often those associated with the study and use of maps (Figure 10.2).

Although the teacher holds the ultimate responsibility for what happens during fieldwork, the experience can be used to help students develop a greater sense of their own responsibilities towards each other and the tasks on which they are working. When planning fieldwork it is necessary to match the activities selected with the objectives and purposes. The selection of objectives will depend to some extent upon the timing of fieldwork within the sequence of learning activities. For example, fieldwork can be used early in the learning sequence as a means of basic information gathering and increasing the motivation of students. Sometimes, fieldwork may be used towards the end of a unit of work as a means of drawing a number of themes together. At other times field activities may be integrated throughout a unit of work to develop students' understandings of concepts, generalisations and principles.

To be meaningful, fieldwork should be integrated with classroom activities. A sequence of activities for students can be identified involving pre-fieldwork, fieldwork and post-fieldwork activities. These steps are illustrated in Figure 10. 3.

Pre-fieldwork Phase

This phase can be of a variable time span. At the least it will involve a determination of the purposes of the fieldwork and an outline of the activities which students will be required to undertake. However, it should involve the search for documentary materials such as books, articles, maps and photographs for background information. Such a search can assist in the construction of a general frame of reference and the identification of the unknowns. It may also lead to the formulation of hypotheses which can be tested

---

Attitudinal and aesthetic objectives
To arouse students' curiosity.
To develop favourable attitudes towards learning.
To provoke students to ask questions and identify problems.
To sharpen students' perception and appreciation of changing landscapes.
To give students the experience of the pleasure of discovery.
To enjoy the study of geography and acquire a deeper interest in the subject.

Knowledge objectives
To develop better understandings of the nature of things discussed in the classroom and in books.
To enable students to observe and think, and acquire knowledge.
To understand the relationships between physical features and human activities.
To associate the different phenomena which together comprise the geography of an area.
To develop an awareness of problems relating to human occupance of the land.

Skill objectives
To develop an understanding of geographical modes of inquiry.
To distinguish between necessary and extraneous information.
To orient a map in the field.
To relate real features to map symbols.
To develop skills in data collection, recording and analysis.

Figure 10.2 Examples of the range of objectives possible for fieldwork

	Teacher	**Students**
PHASE 1: PRE-FIELDWORK	• Determine purposes of fieldwork • Revise essential prerequisite knowledge and skills • Fulfil all school/departmental requirements • Inform students and parents of purposes, costs, arrangements • Book site and transport • Visit site and plan activities • Brief guest speakers • Compile list of students' names and emergency contact numbers	• Be aware of the purposes of fieldwork (possibly contribute to their determination) • Master prerequisite knowledge and skills • Develop data-collection techniques • Know personal and group responsibilities • Be aware of arrangements, and necessary materials and equipment
PHASE 2: FIELDWORK	• General supervision • Provide assistance when required • Direct students by raising questions such as Why? How? • Provide additional information if essential at that time	• Make direct observations—identifying, naming, describing, ordering, constructing, measuring • Collect and record data • Make initial analysis and interpretations • Use specific field techniques—sketching, mapping, transect • Be aware of their own and other perceptions
PHASE 3: POST-FIELDWORK	• Evaluate the complete experience—including organisation and learning outcomes • Directing students to other resources to confirm their findings	• Organising information collected • Generalising on the basis of collected data • Checking findings with others • Discussing puzzling issues • Researching unanswered questions • Making presentations

Figure 10.3 A three-phase approach to integrate fieldwork and classroom activities

in the field. Involving students in discussion of the data-collecting techniques to be applied in the field can be a very useful exercise also. Data-recording devices can be generated and necessary skills developed before entering the field. A useful technique is to involve students in the development of any classification system which will be used in the field. This helps students identify the variables which are to be considered. Such a classification will be more readily and wisely used in the field because the students will know what is intended by the various categories. Any problems which may become evident when the classification is applied are also kept in perspective and students can then be involved in overcoming these difficulties. In a unit of work on recreation, the teacher may provide students with the following information:

The geography of recreation

Information for students:
This unit of work will compare the recreational resources and their location in our local area with the resources and location of those in the Port Macquarie area (the site of our three-day integrated field trip). You will work in groups of four. Each group will select a theme for study. Your selection can be made from those listed or you can discuss a theme among yourselves and then talk to me about it.

Possible themes

1. Plan a study of one type of recreational resource in our local area. Some suggestions are: registered

clubs, commercial amusement centres, boating facilities, parks, squash centres, cinemas.

2. Select an example in the local area where there is conflict among different groups over land use that involves recreation. Local newspapers may be a useful initial source for the identification of conflict areas.

3. How does the demography of our local area affect the recreational resources which are available? What role do government bodies and private organisations have in the provision of recreational facilities?

Once you have selected your theme, think of specific questions which will help focus your attention on the types of information you will need to collect. You should also identify the sources of information which may be useful, e.g. maps, street directories, local council directories, photographs, telephone directories, appropriate persons to interview.

Fieldwork Phase

Each educational authority has its special prerequisites which have to be met before fieldwork can be approved. However, a number of general principles can be enumerated concerning its organisation. These are:

1. Fieldwork should have direct relevance to a theme or topic in the course being studied and should make a worthwhile contribution to the needs of students.

2. Fieldwork should be integrated into specific units of work.

3. The effect of fieldwork upon the school routine and the remainder of the staff must be considered.

4. The precise study area should be selected and a reconnaissance carried out. The selected study area should be close to the school in most instances. Any site features which pose a potential danger to students should be noted during the reconnaissance and ways of avoiding problems worked out.

5. Armed with a provisional timetable, a statement of objectives and planned learning experiences, the teacher should consult the school principal and make formal application for approval to do the fieldwork.

6. At this stage, preliminary advice should be provided for students and their parents. It is important to inform parents of the purpose of any fieldwork, especially as it may involve them in added expense and inconvenience.

7. The dates, transport arrangements, accommodation (if necessary), and staff to be involved should be arranged. Letters should be sent to any persons or places to be visited.

8. Once the final itinerary has been arranged the students and their parents should be informed and money and permission notes collected.

9. Resource material should be consulted and students prepared for the activities which they will be required to undertake.

10. Any special notes or activity sheets should be prepared.

11. Immediately prior to the field trip students should be reminded of any special clothing and equipment required. Acceptable standards of conduct should be discussed.

12. The school administration, canteen and other members of staff should be advised of any impact the fieldwork may have on them. A list of students involved in the fieldwork should be left with the school administration. If overnight stops are involved relevant phone numbers should be provided to the school administration and to parents.

Actual fieldwork involves the collection and recording of information obtained through direct observation of phenomena, patterns and processes in the field. Good fieldwork involves making accurate observations and recording them precisely. In addition it is necessary for students to be able to analyse, interpret, draw conclusions and make generalisations from the data obtained through observation. The remaining part of this section looks at the types of general skills which can be developed through fieldwork. They include observational skills and the cognitive skills associated with the ability to analyse, interpret and generalise from the information obtained through observation. Such skills, developed through fieldwork activities, can provide students with a sense of fulfilment which often carries over into other work at school. Many of the skills are also applicable to leisure and recreational activities.

Observation

Fieldwork can provide many opportunities for all students to take an active part in the work and feel that they are making a worthwhile contribution to the activities. Unfortunately, students can become sceptical of their personal observations if they are not encouraged to discuss their perceptions during and after fieldwork. If the teacher conducts a field trip as a series of mini-lectures delivered when students are confronted by an object there are very few opportunities for students to discuss what they observe themselves. Often what they see raises more questions than answers, and so a continuing inter-

Figure 10.4 Developing information processing skills through fieldwork and classroom activities

active learning process m[...] to think of observations [...] Observations involve the c[...] the use of all senses. Fre[...] involve the use of sight with[...] sound, smell, taste and touch[...] awareness it is possible to [...] concentrate on the sounds and [...] to identify the direction from w[...] smells emanate. Obviously suc[...]vity can be followed by the use of sight to verify the source of sounds and smells. Another activity involves pairing the students, one being blindfolded, the other ensuring that no harm comes to the partner. Each pair is required to move through an area and the blindfolded students describe the sounds, smells and feelings around them. The use of the 'sense trails'

Unit theme: *How does our town help meet our needs? A study of functional zones.*

Activity	Process
1. **Fieldwork.** 30-minute bus trip along representative transects. Students required to think about what might be important. No other direction provided.	Unstructured observations
2. **Classroom.** Question/answer session *Teacher:* What did you see? hear? smell? feel? Students enumerate items and these are listed. *Teacher:* Which of these items belong together? Students group items by identifying what they see as common characteristics. *Teacher:* What could we call each group? Labels provided become concept names.	Concept formation strategy (after Taba 1967)
3. **Classroom.** Identified groups become a series of land-use/functional zones. Students formed into groups, each to investigate one specific zone. Discussion of subcategories within each zone and how information will be recorded.	Development of data collection and recording techniques
4. **Fieldwork.** Original transects or more detailed traverses can be studied to check ideas formulated in class.	Directed observations Hypothesis testing
5. **Classroom.** Each group records results on maps. Comparison of results. Identification of patterns. Further comparison with topographic and land-zoning maps. *Teacher:* How are the land uses in town grouped? What patterns can we see? Why do you think this has occurred? Students explain items, note interrelationships, make inferences and develop generalisations.	Interpreting data Making inferences Developing generalisations
6. **Classroom.** Generalisations developed are applied to other towns. Can involve further group work. Data supplied for planning and topographic maps. Testing of generalisations against further evidence.	Application of principles

...een developed in some national parks is a ...a in this regard. Similarly, students may be ...dfolded, led to a tree or other objects and ...lowed to touch it. When the blindfold is removed the students are required to identify the object which they were touching. Schoer (1979) has outlined a number of activities which involve a multisensory approach to field activities. It has often been suggested that observation skills can be developed through the use of searching questions posed by the teacher. However, if attention through the use of questions is concentrated on a small part of the field, little will be perceived in other parts. If attention is diffused over a large area no point will be clearly and accurately perceived (Vernon 1974). This suggests that planning for fieldwork should involve activities aimed at a holistic view as well as detailed study of specific elements. There is no certainty that students, through the detailed study of a number of discrete elements in an area, will be able to develop a complete view. Once students have made their observations it is necessary for them to be actively involved in the processing of the information which they have gathered. Taba (1967) outlined a number of cognitive processes which can be taught to students to enable them to become efficient processors of information. These processes are concept formation, the development of generalisations and the application of principles. Questioning plays a key role in these processes and the strategies are applicable in a wide range of field activities. Figure 10.4 outlines how the processes can be incorporated into a unit developed for students studying functional zones in their local town or suburb. Whether or not a teacher decides to use the specific strategies outlined it is necessary for fieldwork activities to provide opportunities for students to:

1. describe their observations systematically;
2. analyse or synthesise their observations;
3. separate and classify the different types of data;
4. evaluate the relative importance of the phenomena observed in light of the objectives of the fieldwork; and
5. explain the phenomena and attempt to develop general principles which can be applied elsewhere.

Data Collecting and Recording Techniques

The range of activities associated with the collection and recording of data during fieldwork is very wide. However, often activities are limited to taking notes when the teacher directs attention to specific phenomena, or the completion of question sheets prepared by the teacher. Unfortunately, the analysis of data has often involved little more than ensuring that students have the correct answers to all the

Figure 10.5 Student-developed form for collecting traffic flow data

Date and day of week .... Time .... to		
Mode of transport	**Number**	**No. of passengers**
Pedestrians male female		Total pedestrians ....
Car		
Truck		
Van		
Bus		
Motor cycle		
Bicycle		

questions on their worksheets. In this type of approach very little emphasis is placed on developing the students' own data collecting, recording and analysis techniques. This is not to say that worksheet exercises are always of little use. What is necessary is that a variety of data collecting, recording and analysis techniques be included to help students develop their own interpretations. A class discussion during the pre-fieldwork phase can be directed toward ways of recording information so that it can be analysed easily and relationships noted. One group of students devised the form in Figure 10.5 to collect information about the traffic flow in a busy intersection near the school. Data were collected during a series of ten-minute periods at different times of the day and on different days of the week. Because of the relative complexity of the data to be collected students spent considerable time discussing the most appropriate locations for each group to stand during the survey period. Each group then graphed and analysed the data they collected before the results for all groups were collated and further analysis proceeded.

Some data collecting and recording techniques are more appropriately devised by the teacher because the students may need more direction in the specific tasks of observing and measuring. Bendl (1981) provides some sample tasks for fieldwork in coastal locations and along streams (Figures 10.6 and 10.7). The data collected can be collated and analysed later in the classroom.

The survey technique is very useful when students are studying social phenomena. Each student can be assigned a particular area or theme to survey and the results of all findings are analysed and interpreted in the classroom. Graphs and maps of the data collected can be constructed and students can then begin to look for relationships and make generalisations. Lovett (1980) provides many useful examples of survey forms which can be used in the study of urban and industrial geography. A most interesting and useful task for students to undertake, especially when they initially encounter an unfamiliar area, is the recording of their impressions. This can be best done from a high vantage point. A variety of techniques can be used to record impressions.

1. *Field sketches* allow for easiest comparisons to be made, but students can write descriptive notes or record their impressions on a cassette for later analysis and comparison. Such activities emphasise the unique characteristics of individual perceptions when pairs and, later, groups of students discuss their observations. Field sketches are a valuable means of recording field observations. Advantages over photography are that it is possible to include only those features relevant to the purpose of the field trip. A useful introduction to field sketching can be the provision of a duplicated sketch for students to annotate and the use of a frame such as the one indicated in Figure 10.8.

Activities along a reach of stream

1. Sketch the pattern of the stream along this reach.

2. Indicate the location of pools and riffles.

3. Measure the width, depth and length of the pools and riffles.

4. Measure the distances from pool to pool and riffle to riffle.

5. Calculate the average pool to pool and riffle to riffle distance.

6. Measure the width of the stream at each pool and each riffle. Calculate the average width.

7. What is the relationship between the stream pattern and the location of each pool and riffle?

8. What is the relationship between the material which makes up the stream bed and banks and the location of pools and riffles?

Figure 10.7 Suggested activities in stream morphology

Standing on the headland overlooking the beach

1. Draw a sketch of the beach and headlands and indicate the wave refraction patterns.

2. Line up a point on the opposite headland and count the number of waves which pass during a five-minute period. Calculate the average wave period in seconds.

3. From which direction do the wave trains come? Label this on your sketch.

4. Draw a sketch to show how the beach deposition patterns are related to the direction of the wave trains.

Figure 10.6 Suggested activities in coastal geomorphology (after Bendl 1981)

(a)

	Centre **Background**	
Left margin	**Middle distance**	Right margin
	Foreground	

Prominent features are selected to mark left margin, centre and right margin. After these features are sketched those features of the foreground, middle distance and background are added.

Range of sight limited to 40° to 90°

2. *Sketch maps* are another useful means of recording data. These maps can emphasise those aspects which are relevant to the problem under study. Such maps should not be considered any more than a recording of observed information. The observer must feel free to annotate the outline map liberally, and maybe, unsystematically. A later classroom analysis of the map should lead to an understanding of how the observations fit together. It may then be necessary to prepare a more logically integrated sketch map. In the field a number of groups of students could each be required to map one or a limited number of landscape elements, say vegetation and topography, while another group maps human land use. Back in the classroom the work of each group can be brought together in an attempt to integrate the data collected.

Figure 10.8 (a) A framework for field sketching; (b) an example of a field sketch from Bungoon Lookout, Royal National Park

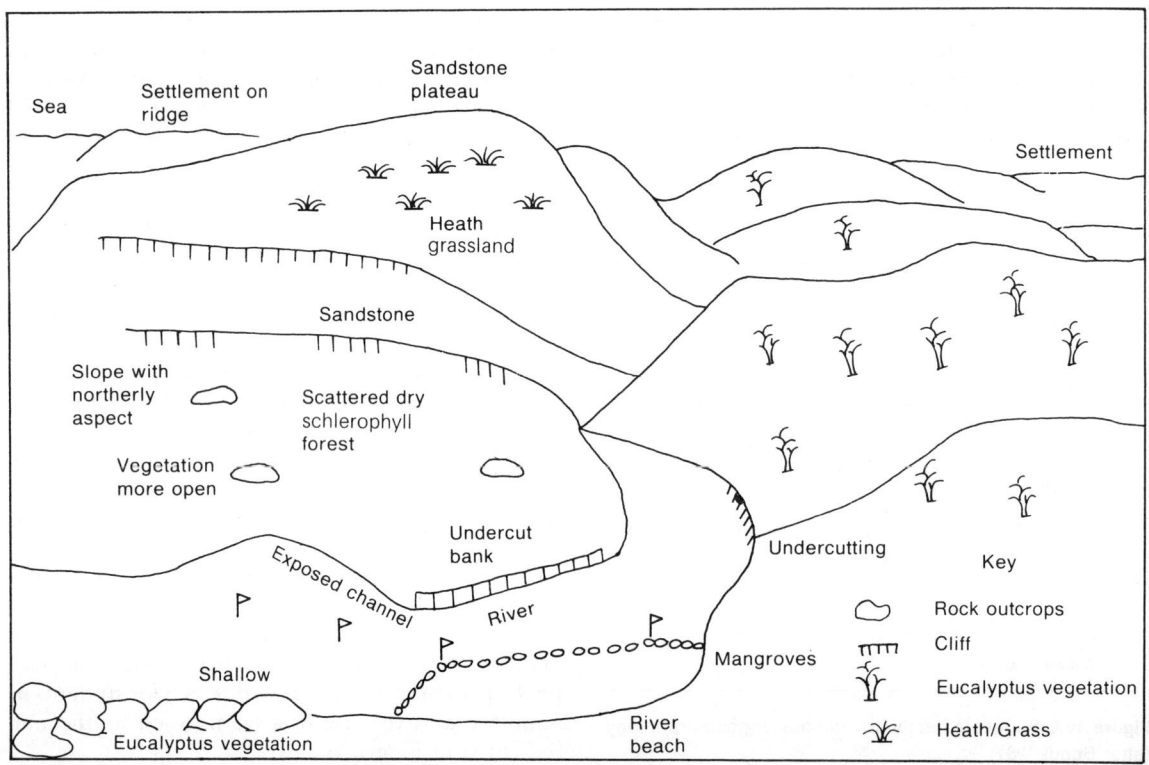

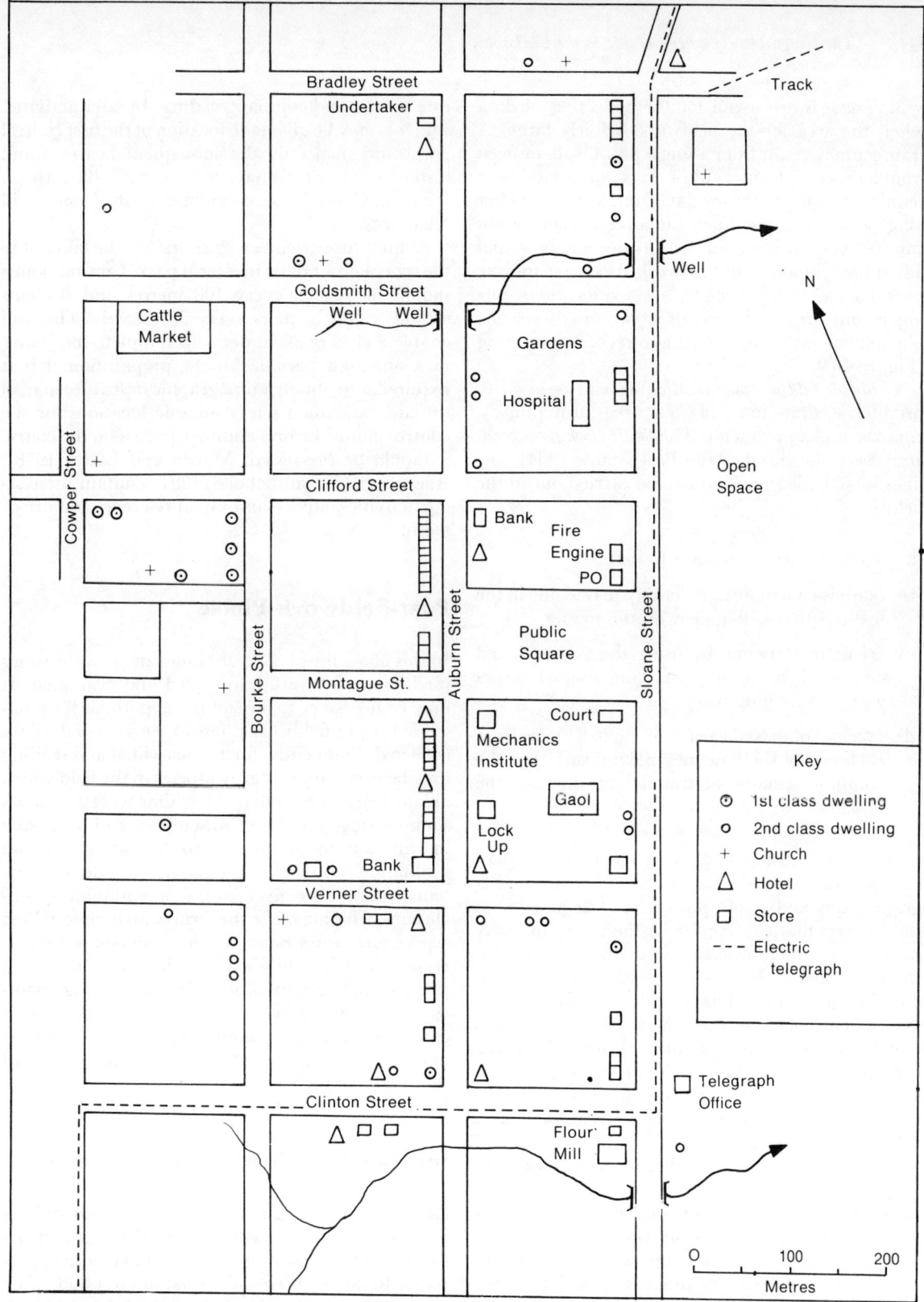

Students were organised to make transects of the area covered by the map to determine whether any of the buildings found today could also have been there in 1859. They were also asked to see how land use had changed.

Figure 10.9 A social map of Goulburn, NSW, based upon Jevon's 1859 map

3. *Transects* are useful for the collecting of data when the area for the fieldwork is fairly large. A transect may result from a single point walk or even from a coach or train window. It is not advisable to require students to record data from a transect over long periods of time. Such intensive concentration can be very tiring and defeat the motivational advantages inherent in fieldwork. It is best to note phenomena only in specially selected locations during a long trip and limit observations to specific phenomena rather than trying to record everything (Figure 10.9).

4. *Simple field mapping* techniques can be used to produce accurate maps of small areas, using only a compass and a protractor. The *Skills Book for Secondary Schools* (Jacaranda Atlas Programme 1984) outlines some basic steps that can be carried out in the field:

(a) select the area you want to map;

(b) establish a baseline, preferably to one side of the features that will appear on the map;

(c) orient the baseline by using the compass and measure it accurately, marking its extremities with two ranging poles;

(d) take a compass bearing of each feature from each end of the baseline, making sure that the compass remains accurately oriented to the baseline.

Once these bearings have been recorded the map can be accurately drawn in the classroom, using graph paper and a 360° protractor. The location of the features that will appear on the map can easily be found by the intersection of the bearings drawn from each end of the baseline. Students will need to establish a north point, a scale and symbols to identify each of the features. More advanced classes may be able to use triangulation methods of drawing a map by using a plane table, or even use simple surveying techniques with a dumpy level.

5. *Orienteering activities* can be of great value as a means of managing groups of students in the field and as a vehicle for instruction in many aspects of map work. Orienteering provides a valuable means of developing skills in the use of a map and compass while involving students in real-world settings. Students are able to develop skills in using an orienteering compass, and practise identifying scale, contour lines and conventional mapping symbols. Initial experiences may include a compass walk around a square or rectangle, along given bearings. Such an activity emphasises the importance of accurate pacing and compass reading. In later activities students may be given the location of the first control point and then only the subsequent bearings and distances. They then have to draw the full course on the map. Control points can be described using grid references.

A miniature orienteering course can be laid out in the school grounds or in a local park. Control points should be spaced every 100 metres and students should set off in pairs every 30 seconds. This will enable a class of 30 students to complete the course in a one-hour period. All the preparation that is required is to obtain a topographic or feature map of the area and then select suitable locations for the control points. Before sending the class on the course it should be pre-tested. Martin and Lotty (1978), Adams (1977) and Schoer (1981) contain interesting activities, advice and useful references on orienteering.

Post-fieldwork Phase

In this phase the ideas, techniques and classifications used in the field are interpreted and evaluated. It may be necessary to amend the hypothesis formulated in the pre-fieldwork phase in the light of the data collected. Frequently, further searching and reading may be necessary to clarify aspects in the field which are unclear or unfamiliar. Most time will be spent on summarising, recording, discussing and analysing information in an integrated format and in an evaluation of this information. Considerable attention should be given to the presentation of conclusions. This may take the form of an individualised report such as an essay, a field notebook or an oral report. A class exhibition may be set up involving project and group work. Displays using photographs, maps, tables, charts, graphs and models can be used as the culminating activity. Students may be encouraged to use a video camera to record the results of their work.

Independent Fieldwork by Students

Geography teachers involve students in fieldwork for many purposes. Frequently fieldwork is highly structured by the teacher and students have little opportunity to pursue their individual interest and relate geographical skills to problems which capture their attention. One way of allowing students to become more personally involved in their own studies is to encourage them to undertake independent field-based projects. The requirements of some final-year examinations have stimulated such activities. However, students in the early years of geographic study can gain much from independent fieldwork undertaken as individuals or in small groups. Obviously, such activities need to be quite limited in scope for younger students and greater teacher assistance will be required. The introduction of independent fieldwork which integrates geographical skills, methodology and content with students' interests provides teachers with opportunities to observe and evaluate teaching and learning in a situation not tested in written formal examinations.

When planning independent fieldwork it is useful to encourage students to use a model based upon the field research approach (Figure 10.10). If this is not done it is likely that students will collect masses of information without relating it specifically to their field activities and without ever defining the problem they are studying.

Figure 10.10 Stages in organising independent fieldwork by students. The example relates to urban patterns in Atlanta, Georgia (see Laws 1979)

Stages	Example
1. Topic selection Teacher provides list of themes • Local area problems, e.g. traffic congestion, beach destruction, vandalism, council planning, shopping habits. • Geographical theories, e.g. nearest-neighbour index, spheres of influence. Student selects own topic • Special interests and abilities. • Access to special information.	**Task:** To prepare a paper for presentation to rest of class. Topic to be selected from the area of historical geography. Reading about town planning in the United States. In nineteenth century, emphasis was on the use of a rectangular grid oriented to the cardinal directions.
2. Definition of a specific problem Expressed in clear and concise question form. Question may be convergent or divergent. Problem should be manageable and solvable. Identification of major elements in problem.	Student becomes hopelessly lost on first visit to Atlanta, Georgia. Wonders why street pattern is so complex when city developed at a time the rectangular grid was widely used. **Question:** Why did such a complex pattern develop?
3. Data collection and recording Background reading—variety of sources. Preliminary field observations. Search of published data. Consultation with experts and other teachers. Planning data-collection strategies. Development of field recording devices.	**Possible solutions:** Influence of topography, no overall planning for the area, the influence of the railroads. Discussions with other geographers, historians, librarians, archivists. Reading about the early days of settlement in western part of Georgia. Collecting early maps and photographs. Walking transects of the area concerned.
4. Data analysis Mapping and graphing of data. Use of simple statistical techniques. Identification of major elements.	Establishing chronology of the area. Drawing sketch maps of city's development.
5. Statement of conclusions and recommendations Conclusions: firm and precise or tentative? Planning format for presentation. Sharing experiences with others—other students, parents, the community. Recommendations forwarded to relevant authorities.	**Conclusions:** Street pattern found to be influenced by original Indian trails, original survey into land lots, the coming of the railroads, a land boom in 1846, and the role of the town council. Report written and presented.

A sequence of stages can be utilised when helping students plan independent field projects. By using these many of the problems which frequently occur can be overcome. The first stage involves the selection of a topic. This may be determined by the individual students, the teacher, or through collaboration between teacher and student. It is important to identify a specific problem for investigation. This should be expressed in question form and be able to be solved. Once the major elements of the problem have been identified students can begin the collection and recording of data. The data are then analysed. Mapping and graphing often helps analysis. Finally the student presents conclusions and recommendations. Figure 10.10 outlines the stages and shows how they were applied in solving a research problem. An additional useful reference has been written by Pilbeam (1980) in which ideas and examples for local projects for senior students can be found. Further ideas on individualising fieldwork are also provided in Chapter 24.

Problems and Constraints in Fieldwork

Despite the advantage of fieldwork as a learning experience, the problems and constraints have to be acknowledged. Many of the constraints are associated with organisational factors such as the difficulty of adequately supervising a large group of students and providing them with the assistance they may need, the lessons missed by the teachers conducting the fieldwork, the lessons missed by students and alterations which have to be made to the school timetable. The time needed to plan a worthwhile field trip and the costs of transport and accommodation, if required, also have to be considered. The argument that a teacher may lack the detailed knowledge of the geographical locality can be overcome by a reconnaissance, preferably with a colleague, and through reading. However, it must be acknowledged that the time factor is important. The safety of students is also something which must be kept in mind when planning activities. The problems and constraints emphasise the necessity to ensure that only meaningful field activities are undertaken. One way this can be achieved is through the specification of the anticipated outcomes of any field experiences. In this way it is possible to alert principals and parents to the importance of the work.

Some problems in fieldwork relate to the learning processes to be used by students. Observation, descriptive analysis and inferring are some of the skills required. However, there are many skills associated with data collection and the quantitative analysis of data which students must develop to get the most out of their fieldwork.

Despite concluding this chapter with a warning about the problems and constraints associated with fieldwork it should never be forgotten that perhaps the most meaningful and lasting learning takes place when students are actively participating in exploring the great variety of environments around them. In addition the fieldwork experience provides opportunities for teachers and students to get to know each other and interact outside the structures of the classroom and the school yard.

References

Adams, W.P. (1977) 'Geography and Orienteering', *Geography Bulletin*, Vol. 9(1), 16–22.

Bendl, J. (1981) 'Field Study of a Reach or Stream in the Royal National Park', in S.B. Codrington (ed.) *Senior Geography Case Studies*, Sydney: Geography Teachers' Association of NSW.

Clegg, A.A. (1969) 'Geographying or Doing Geography: An Inductive Approach to Teaching Geography', *Journal of Geography*, Vol. 68(5), 274–280.

Everson, J. (1969) 'Some Aspects of Teaching Geography Through Fieldwork', *Geography*, Vol. 54, 64–73.

Everson, J. (1973) 'Fieldwork in School Geography', in R. Walford (ed.) *New Directions in Geography Teaching*, London: Longman.

Hart, J. and Taggart, J. (eds) (1986) *Approaches to Fieldwork in Senior Geography*, Sydney: Geography Teachers' Association of NSW.

Jacaranda Atlas Programme (1984) *Skills Book for Secondary Schools*, Milton: Jacaranda.

Jones, P.A. (1968) *Fieldwork in Geography*, London: Longmans, Green and Co.

Laws, K.J. (1979) 'The Origin of the Street Grid in Atlanta's Urban Core', *Southeastern Geographer*, Vol. XIX(2), 69–79.

Lovett, P. (1980) *Local Studies in Towns: A Teacher's Handbook*, London: George Allen and Unwin.

Martin, T. and Lotty, D. (1978) *Map and Compass Fundamentals: Orienteering*, Sydney: A.H. & A.W. Reed.

Pilbeam, A (1980) *Local Projects in A-level Geography*, London: George Allen and Unwin.

Schoer, G. (1979) 'An Integrated Approach to Pupil-

Centred Field Studies in Geography Using Multi-sensory Observational Techniques', *Geography Bulletin*, Vol. 11(2), 42–51.

Schoer, G. (1981) 'An Orienteering Approach to Teaching Geographical Map Skills', *Geography Bulletin*, Vol. 13(4), 274–280.

Smith, D. et al. (1977) *A Handbook for Australian Geography Teachers*, Melbourne: Sorrett Publishing.

Taba, H. (1967) *Teachers Handbook for Elementary Social Studies*, Palo-Alto, California: Addison-Wesley.

Vernon, M.D. (1974) *The Psychology of Perception*, Harmondsworth: Penguin.

11

Developing Valuing and Decision-making Skills in the Geography Classroom

Brian Maye

Three essential goals of the geography classroom are knowledge, skills and values. Brian Maye presents the case for the importance of values education in this chapter. The concerns of geography, the demands of society and new emphases in educational aims make the development of valuing and decision-making skills essential goals of geography teaching. But what values and towards what ends?

This chapter answers the first question by explaining the contribution that geography can make to the three areas of social, environmental and citizenship education. Four broad approaches to the teaching of valuing and decision-making skills are then explained. They are: values analysis, moral reasoning, values clarification, and action learning. Strengths and problems in using each approach are explained. Examples of values-laden issues and teaching ideas are used as illustrations of each approach in the classroom. Further examples of values-based teaching approaches are provided in Chapter 12 on social and political literacy, Chapter 13 on qualitative inquiry in the geography classroom and Chapter 14 on developing sensory awareness and aesthetic appreciation.

Why Teach Geography?

Karen Harrison was being challenged.

'What good does geography do kids, anyway?' queried Allan Smythe, the science master. 'Teaching where places are and how people live around the world doesn't help them understand the real nature of the environment, or the technological changes which are taking place these days.'

'Well,' countered Karen, 'you should visit my classes sometime. You seem to have a pretty narrow view of what geography is. You need to consider what education aims to do in the first place, and then look at how geography, and every other school subject for that matter, contributes to this. What the subject teaches certainly is important, but how it is taught is important, too, if aims set for education are to be met.'

Geography and Aims of Education

Karen had a point. Education systems frequently state aims which refer to the need to 'assist the development of individuals as members of society'. Such systems often express intentions that they should, for example:

● develop perceptive understanding of our society and the societies of others;

- be flexible and adaptable in response to change;
- be able to make mature judgements;
- be able to make socially responsible contributions to the community.

Allan had a point, too. Embedded in his comment on what he saw as important content for students to learn about was the implication that rapid changes bring the need to re-examine the way we teach with a view to helping students cope with conditions these changes bring.

Statements of aims of the kind noted above emphasise our role as teachers of *people*, not simply of subject matter, and consider the needs of individuals in a societal context with the future in view. Such statements are value judgements, set to give guidance to teachers seeking to meet the needs of their students through curriculum planning and teaching. Geography teachers could well ask themselves: 'How adequately will the future needs of the students I teach be met by what I teach and the way I teach it?'

Teaching Geography in a Changing World

It is widely acknowledged that technological change is occurring at an accelerating rate, and that potential for impacts upon lifestyles and environments is enormous. Social change accompanies technological change, bringing increased needs for individuals, families, community groups and governments to respond to stresses brought about by competing demands. Some of these changes and tensions are alluded to in Figure 11.1, which suggests that stresses are placed upon value systems and bring a need for new competencies to cope.

All groups ultimately are made up of individuals whose decision making finally has an effect on the way in which people interact with each other and with their environments. The ability to make decisions which contribute constructively to social and environmental relationships comes in part from being well informed and developing the ability to think logically and rationally; but it also demands skill and ability in evaluating one's own values and those of others. How well geography teaching contributes to students developing this skill and ability will influence finally the relevance of the subject to students who are developing citizens in a world of rapid change (see Huckle 1983; Fien and Gerber 1988).

Figure 11.1 Forces of change place stresses on value systems (after Buckland 1979)

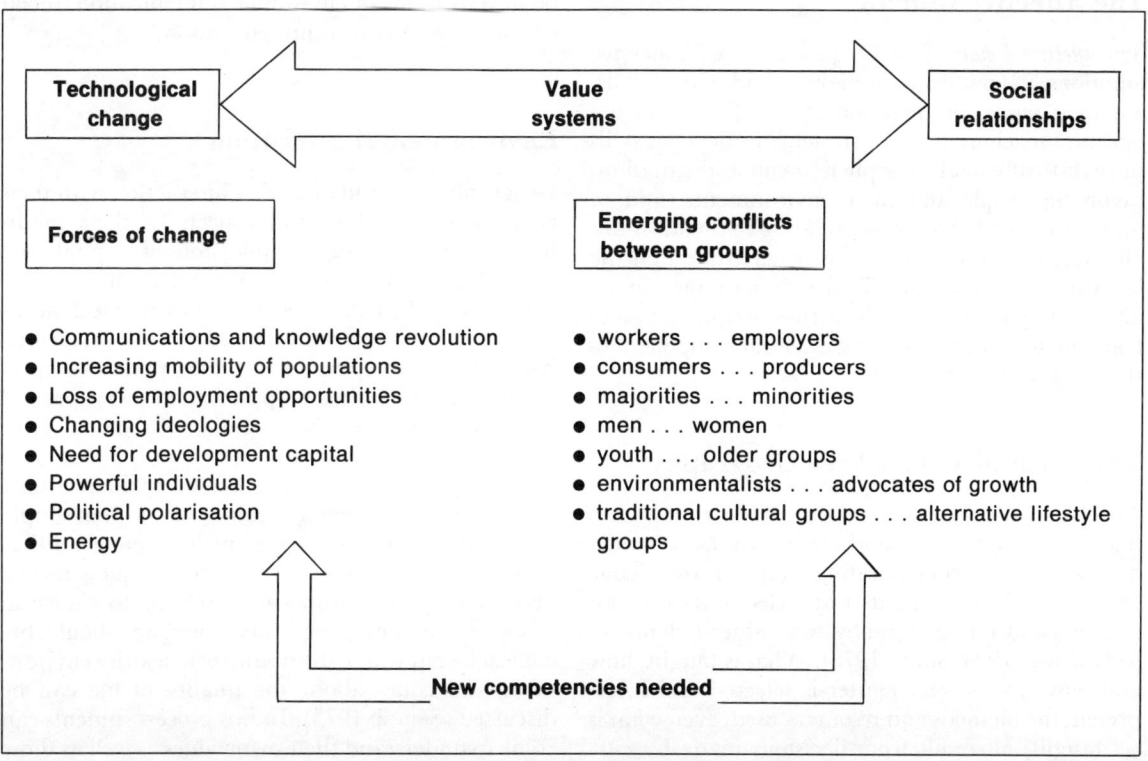

Geography Teaching and the Affective Domain

Geography teachers have traditionally regarded themselves as facilitators of objective learning, assisting students to draw reasoned inferences and reach conclusions based on analysis of factual data placed before them. They have, in other words, consciously concentrated on the *cognitive domain*, or knowledge aspects, of learning.

Where geography has been viewed as a study of people–environment relationships, teachers have concentrated on assisting students to understand the nature of the environment. In general, they have tended not to have examined people's opinions and feelings about the environment, or to have considered how and why people behave in their environment and change it in the ways in which they do (Lowenthal 1967). Where issues of community concern have been studied, they have similarly tended to concentrate on a 'facts of the case' approach, and usually have not probed the motivations, values and emotions of people involved in the situations studied, where in fact investigation of these aspects is essential to understanding why the 'facts of the case' are as they are.

The Affective Domain

The *affective domain* of learning includes self-concepts, emotions, values, decision making and action skills. There is scope for inclusion of all of these in geography teaching to assist students to develop skills in realistically analysing phenomena and situations involving people and their environments, and to make constructive responses relating to them. In this chapter, however, attention is concentrated on the last three aspects of the affective domain mentioned above. Treatment of each of these requires values teaching to be included as an essential component of the geography curriculum.

The Myth of Values-Free Geography

Teachers who protest that values education has no place in the classroom and claim to take a 'values neutral' stance, concentrating only on knowledge aspects, miss the point that knowledge itself is not neutral, and that geography is a values-laden subject (Cowie 1978; Smith 1978). What is taught, how and why that subject matter is selected and interpreted, the methods and resources used, even what is *not* taught, all result from decisions made by curriculum planners and teachers. All of these decisions are made on the basis of values held by such people. It is a matter of concern that teachers and others involved in deciding what and how to teach are not always aware of how their own values influence the way in which they do this.

Values Teaching in Geographical Education

Among the traditions which have been identified in geography, people–environment relationships and spatial relationships between phenomena are two which have endured (English and Mayfield 1972; Taafe 1974). Achieving aims of education which focus on future needs of students in a societal context requires that such traditions should be followed in a people-centred way in applying them to geographical education.

The study of people necessarily involves the study of values. People make decisions and act within their societies and environments according to their perceptions, values and emotions, so that balanced understanding of people–environment relationships and of spatial relationships must take account of this. In doing so, the purposes of teaching geography can be seen in terms of environmental education, social education and citizenship education.

Environmental Education

Geography can study people's interaction with their environment, and spatial aspects of their modification of it through application of capital and technology to meet their needs and wants.

Choices taken in using resources to meet needs and wants are heavily influenced by cultural factors, particularly perceptions and values, so that comprehensive understanding of spatial and other aspects of resource use requires investigation of these affective aspects.

Quality of life issues are becoming increasingly related to environmental quality, rather than simply to material well-being. This involves change, differences and even conflict in relation to judgements, decisions, opinions and values related to choice in resource use. The geography classroom should become a forum where the nature of a healthy environment and issues about the quality of life can be discussed (Senesh 1973). In this process students can come to understand their own values as well as those of others. Huckle (1986) extends this point of view by emphasising a necessary role for geographical education in promoting environmental citizenship.

Social Education

Geography can study interactions between people in terms of the distribution of status aspects such as employment, race and religion or in terms of movement, such as migration, or the transfer of goods and capital. Explanation of these distributions and movements must take account of people's values and behaviour, as 'groups with different value systems produce significantly different spatial distributions' (Fielding 1974). In studying these interactions, questions of quality of life, analogous to environmental quality, will arise as differences in social well-being are discerned. These differences sometimes become visible as social problems, often through media portrayal, and usually give rise to some kind of attempt to deal with either the immediate problem, or the wider issue of which it is a symptom. Differences in social well-being, and action to alleviate them, are values-laden topics, ultimately relating to concepts of social and spatial justice. There is great opportunity and need for geography students to learn about their own and other people's values in relation to social issues in order to develop balanced understanding which purely cognitive learning does not allow.

Citizenship Education

The notion of geography as citizenship education brings together both social and environmental education, and focuses on skills and processes which can be developed by the *way* in which these aspects are taught. As developing citizens, students will be expected ultimately to make positive contributions to their communities through their personal actions and conduct; and through more formal citizenship activities such as being members of groups, assisting community projects, attending meetings and voting. In exercising their roles as citizens, students will have direct and indirect influence on quality of life and environment within the communities of which they are part. To contribute constructively they will need to be well informed about both knowledge and affective aspects of social and environmental issues, to have developed processes which enable them to make soundly based decisions and judgements, and to be aware of the means by which these can be operationalised. Both Rob Gilbert (Chapter 12) and Huckle (1988) emphasise the role that geographical education has to play in developing political literacy, which contributes to citizenship education of this kind.

The reality of citizenship participation is that effective skills and/or the desire to participate are not possessed by everyone, nor will they be, but it is the responsibility of schools consciously to assist each student to develop his or her participation skills as fully as possible (Shaver 1981). The suggested values-teaching strategies which follow are intended to contribute to meeting this responsibility.

Approaches to Values Teaching

Values teaching is influenced both by the characteristics of values themselves, and by the reasons teachers have for implementing strategies designed to take account of values.

Some Characteristics of Values

It is always very difficult to define abstract notions such as a 'value', but several characteristics of values should be noted.

1. *Values are abstract concepts people hold about what is important in relation to aspects of life experience.* Values can be studied in terms of *aesthetics*, relating to what people consider beautiful; or in terms of *ethics*, relating to how people behave (Fraenkel 1977). In geography teaching most emphasis is given to the latter, although a case can be made for inclusion of the former. In practice, much attention is given to *moral values*, which are standards which guide decisions about what is right or wrong or just or unjust. The development of morality and the capacity to discriminate among values and beliefs are crucial aspects of the overall development of the rounded person. They are also a civic necessity (Curriculum Development Centre 1980).

2. *Values are closely related to actions and behaviour people engage in.* Kniker (1977) describes values in terms of their propensity to generate action or to make a deliberate choice to avoid action. Conversely, values can be *inferred* from people's actions, or failure to act, and from what they say. Values are held by individuals and by groups, both formal and informal. The relationship between values and behaviour is most significant in values teaching, as *analysing people's behaviour and actions* forms the starting point for investigating values.

3. *Values are often labelled by abstract terms.* Terms such as honesty, loyalty, justice are used to denote values, but it is not always easy to find a single word that is a satisfactory label. In practice, however, students are usually able to infer 'what is important to' a person or group in relation to particular actions. Where younger students have limited vocabulary and abstract reasoning skills this term can be used most effectively. It is not necessary to refer to the actual term 'values' in implementing teaching strategies with such students, although older students will cope readily with more abstract definitions.

4. *Values give rise to value judgements, value differences and value conflicts.* Value judgements are contained in statements about what is 'right—wrong': 'just—unjust'; 'fair—unfair'; in relation to particular issues. They are implied in statements of moral exhortation containing words such as 'should' or 'ought'. Value judgements have both direction and intensity. For example, good–bad judgements indicate differences in direction, while better–worse judgements also indicate intensity.

5. *Value differences occur between individuals and groups*, and can be inferred where different actions are taken in relation to similar aspects of life. Value conflict occurs where tension arises within an individual or group when a choice must be made between actions. Value conflict is also involved where a difference exists between the stated values of an individual or group, and actions taken in relation to a particular set of circumstances.

Established Values-Teaching Approaches

Several approaches to values teaching have been developed and can be seen as approximating a continuum in relation to the degree to which students are expected to be directly and deliberately involved in examining and acting upon their own values (Figure 11.2).

It should be noted that the differences in purpose of the approaches described in Figure 11.3 leads to differences in teaching styles. Values inculcation is teacher-centred, but the other four approaches are student-centred and teacher-guided, requiring an investigative approach to be taken. These four approaches share some common features, including:

1. The implicit intention that students be assisted to develop their own appropriate set of values and

Figure 11.2 Approaches to values teaching—degree of student involvement in examining own values and actions

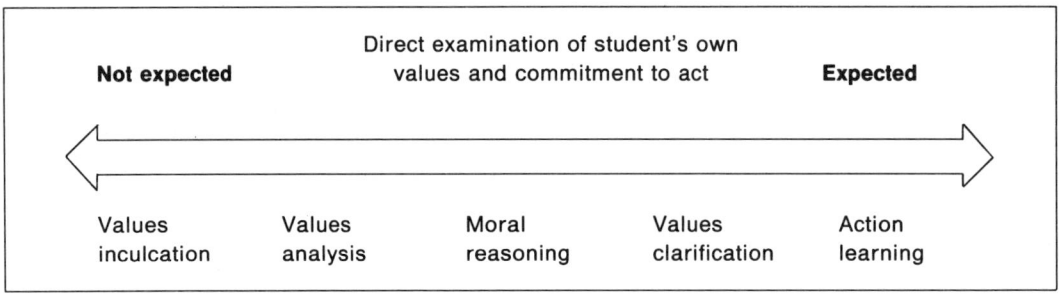

Values inculcation	has the objective that students will adopt a predetermined set of values.
Values analysis	uses structured discussion and logical analysis of evidence to investigate values issues.
Moral reasoning	provides opportunities to discuss reasons for value positions and choices with the aim of encouraging growth in moral reasoning ability.
Values clarification	has the objective of helping students become aware of their own values in relation to their behaviour and that of others.
Action learning	encourages students to see themselves as interacting members of social and environmental systems through having them analyse and clarify values with the intention of enabling them to act in relation to social and environmental issues according to their value choices.

Figure 11.3 Approaches to values education—a brief description

related behaviours to enable them to participate constructively in their community or society.

2. The use of stimulus materials to initiate teaching strategies; usually not immediately involving the students, but allowing scope for a progression towards having them specifically consider their own values and/or make their own decisions in relation to choice of appropriate behaviour.

3. The underlying notion that students face the need to make decisions about appropriate personal and social behaviour in their daily lives, and that as they move towards adulthood the requirement that they be capable of making such decisions independently will intensify. Emphasis is therefore placed on development of *skills* in valuing which can be transferred to situations other than those in which they were developed.

More of the theory behind these four approaches to values education in relation to geography teaching is provided by John Huckle (1981). He provides examples of useful texts and classroom activities, also.

Implementing Values-Teaching Approaches

The following suggestions relating to implementation of values-teaching strategies should be seen as designed to give guidance on applying the principles on which each approach is based. Within each approach, there are numerous particular strategies which can be used, and teachers are urged to seek and devise others which will supplement those outlined here.

Values Inculcation

The values inculcation approach has been used traditionally where the objective has been to have the students adopt a predetermined set of values. Methodologies used have included moralising, positive and negative reinforcement, guiding behaviour by set rules, modelling acceptable behaviour by teachers and others, and telling of 'inspirational' stories to provide worthy examples on which students might model their own behaviour.

Protagonists of the values inculcation approach claim that there are basic values in societies which can be identified and which should be transmitted to students. Opponents of the approach point to the difficulty of deciding which set of values to adopt and who shall prescribe them.

Shaver's (1981) enlightening comments on this dilemma note that individual freedom of choice can be maintained by identifying and inculcating basic ideals to which a society can be reasonably expected to be committed, but allowing for choice, differences and disagreement about the way in which these principles are translated into policy and applied.

It is worth noting that to do the latter, it is necessary to adopt investigative approaches to values teaching, as inculcation is unable to achieve these ends. There is some irony in the fact that adoption and use of such approaches affirms the necessity and worth of doing so, and that this in itself is a form of values inculcation.

Values Analysis

Figure 11.4 is a sample strategy for values analysis. It is a general one which could be applied to examine value positions of people in a wide range of situations. For example, a government housing corporation proposal to redevelop an area and re-house its inhabitants could provide opportunity to investigate values involved on both sides of the proposal, as could any other proposal to change land use. Controversy need not necessarily be involved, but concern almost certainly will be. People who become concerned about quality of community life and the environment often communicate this concern by what they say and do. This could include such things as organising clubs to provide youth recreational facilities, writing letters and taking up petitions to call for the improvement of community facilities for the aged or disabled, or publicising environmental degradation in an area used for public recreation.

The first step in analysing values related to these kinds of situations is to gather a variety of relevant resources, such as newspaper articles, letters to the editor, photographs, maps, and interviews with people who are involved. This then allows values-relevant behaviour to be identified, and values to be inferred from this.

Moral Reasoning

Moral reasoning strategies applied in classrooms have been most heavily influenced by the work of Lawrence Kohlberg, who developed methodologies based on moral growth as being both a cognitive and developmental process which can be assisted by involving people in using powers of reasoning to make moral choices (Scharf 1978).

Steps in the process	Student activity	Teacher activity
1. Select stimulus material		Preparation of stimulus materials containing values-relevant behaviour.
2. Initial analysis of stimulus materials to define values 'topic' or 'problem'.	• Read, view, listen to stimulus materials. • Describe situation. • Suggest a topic name.	• 'What is happening here?' • 'Where is it located?' • Encourage an accurate and factual description. • 'Can you think of a name for this situation?'
3. Identify the people involved in the behavioural situation.	• Identify and name people, roles and responsibilities verbally and/or in writing. • Assist teacher to list on chart.	• 'Who are the people in this situation?' • 'What are their names?' • 'What do they do?' (roles) • Build up an open-ended chart.
4. Describe behaviour of each person/group listed.	• Identify and describe behaviours.	• 'What did/does each of these people do in this situation?' • Encourage accurate description. • List behaviours.
5. Infer reasons for behaviour.	• State inferences verbally and/or in writing.	• 'Why does . . . do this?' • 'What reasons might these people have for . . . ?' • Seek evidence from materials to support inferences. • List reasons.
6. Determine value differences and conflicts.	• Make comparisons. • Identify groups. • Make up lists or charts to group people with similar values.	• 'Which of these people have similar things that are important to them? Which are different?' • Do any of these people act differently from what they say is important to them?' • Group people with similar values. Note value conflicts within persons and groups.
7. Hypothesise sources of values, reasons for values, differences and conflicts.	• Write individual lists of reasons, contribute to class discussion. • Form discussion groups to suggest hypotheses. • Draw a sketch or diagram to illustrate possible reasons for values, differences, conflicts. • Draw up background profiles of experience and training of main people involved. • Select from a list of possible value sources (e.g. belief, tradition, roles, economic stress) and add others.	• 'Why do you think . . . is important to . . . ?' • 'What reasons might there be for difference in what is important to . . . and . . . ?' • 'Why do you think . . . might say that is important, but do . . . ?' • 'What experience has . . . had which might influence what he sees as important?'
8. Consider behaviour alternatives and their consequences.	• Hypothesise consequences in response to teacher-posed problem; e.g. 'What would happen if . . . decided to . . . ? What might . . . do?' • Select most likely consequences from a jumbled list related to a given problem. • Select 'The Person Most Likely to . . . ' In relation to a set of given behaviours. • Respond to 'Not likely because . . . ' situations presented in cartoons. • Complete 'if . . . then . . . ' statements. • Role play behaviour alternatives and consequences.	
9. Seek evidence to support hypotheses.	Support choices by referring to analysis of evidence from materials.	Conduct follow-up questions for each activity to ensure rigour in reasons given by students.

Figure 11.4 A sample approach to values analysis

The most common stimulus material used for this approach is the unfinished moral dilemma story. These stories, which may be in written text, on tape recording, film strips or other electronic media, describe situations where the principal character in them is faced with making a choice about which course of action to take in response to some value problem. The stories stop at the point where the character must make the choice, and leave the possible solutions to be discussed and debated by the students engaged in the exercise.

Stories can be constructed on any geographical topic where choice is involved. For example, a scenario could be developed around a local government town planner charged with the responsibility of planning the location of an industrial area. The most advantageous site is close to the railway and highway, but this is in view of the high-class residential area in which he and the Mayor reside. The other possible site is close to a poor section of the community, where heavy traffic would have to be routed past their houses. The Mayor's real estate development company has an interest in this site. What should he do?

There are many activities which can be used to implement moral reasoning strategies. Some which can be effective include:

1. Review the 'facts of the case' as put in the story. Ask 'What should the town planner do? Why?' plus 'What would you do in this situation? What would be important to you?'. Then go on to predict consequences. 'What might happen if you did this? Why? What else could be done?' and so on. Values expressed by the main characters in the story could be analysed to hypothesise actions they might favour.

2. Draw up simulated maps and development proposals, then role play the story, complete with ending. Different groups of students could role play different solutions. All should explain why they have made their choice, and, in response to questioning, be able to say what would be important to them in the situation (i.e. describe values related to their choices).

3. Explain verbally or in writing to someone in the story who would be affected by the decision, e.g. simulate preparation of a public statement to justify the decision made, or in a role play, explain to a protest meeting why a particular choice has been made.

Values Clarification

The major aim of values clarification strategies is to enable the student to recognise his or her own values and to make values choices in a non-judgemental and non-threatening situation. Numerous strategies for values clarification activities have been developed. For detailed examples of each of these, see Raths et al. (1966), who initiated the term, Simon et al. (1972) and Volkmoor et al. (1977). Raths claims that valuing involves the three processes of *choosing*, *prizing*, and *acting*, where a value must be 'chosen freely, prized dearly, affirmed publicly, acted upon' and be 'part of a life pattern'.

Values clarification strategies use a variety of procedures which emphasise self-analysis techniques. Where students are new to the approach, techniques which minimise public affirmation can be used, with techniques requiring public disclosure introduced at a stage when class members feel more at ease with their peers and the teacher. Blachford (1979) makes a plea that in teaching geography, values clarification should be used to lead to examination of moral issues, and gives several examples of specific techniques which could be used, as well as outlining other approaches to values teaching.

Values clarification strategies often use techniques such as ranking, semantic differentials, rating scales, voting and activities which require introspection and making of values statements. Referring to the example used in the previous section, the techniques of semantic differentials, rating scales and list ranking could be used in relation to siting an industrial area in either location. Here are some examples:

1. Semantic Differential
How do you regard industrial areas? Mark 'x' on the appropriate line.

Clean	————————	Dirty
Beautiful	————————	Ugly
Quiet	————————	Noisy
Ordered	————————	Disorganised
Clean air	————————	Smelly

2. Rating Scale
Tick the point which shows how important you see the following qualities to be in industrial areas.

	Unimportant	Not very important	Don't know	Important	Very important
Cleanliness	—	—	—	—	—
Orderliness	—	—	—	—	—
Appearance	—	—	—	—	—
Landscaping	—	—	—	—	—

3. Ranking

Write down five qualities you feel should be considered in locating the industrial area. List them in order of importance.

Action Learning

Emphasis in the action-learning approach is on seeking solutions to values-related problems. Students are expected to become involved in issues, to analyse and clarify values in relation to them, to seek solutions and to take action to contribute to their implementation. This approach to values teaching is particularly designed to assist in the development of citizenship skills.

In this section two variants of this approach are outlined—*action strategies*, which emphasise early involvement leading to action, and *problem-solving strategies*, which emphasise extensive knowledge and values investigation as a prelude to canvassing solutions and anticipating consequences. In this latter strategy students may take action, but there is the option of stopping short of this step after having identified other persons and organisations who could take action.

Action Strategies

A general approach is outlined in Figure 11.5. This could be applied to a particular issue. For instance, students could seek to learn more about an environmental controversy or community issue by becoming involved.

Consider the situation of a small country town losing business through shoppers commuting to a nearby regional centre. Concern is voiced in the community about resulting decline in services and loss of job opportunities. Preliminary information could lead to a decision by students to become involved in overcoming the perceived difficulties. This could lead to active investigation, including fieldwork and surveys of shoppers, residents, businessmen to gather factual and values-related information. After deciding on possible actions, these could be implemented with the students becoming involved in, for example, public meetings, publicity campaigns, delegations to political representatives, petitions and letter writing campaigns.

Bartlett and Cox (1982) and Marsh (1987) draw distinctions between action learning and 'service learning', with the emphasis in the latter on learning in the course of providing a service needed by others. Action learning also contributes to outcomes which will be of immediate benefit to people and environments, but in addition emphasises longer-term goals. Involvement in strategies which require conscious use of knowledge and values-learning process is intended to assist students to become actively involved citizens.

Problem-Solving Strategies

These can follow several particular procedures, including those outlined by Senesh (1973), Shaver and Larkins (1973), Muessig (1975) and Banks and Clegg (1986). The problem-solving approach, also referred to as decision making, is particularly suited to geography teaching in that:

1. It provides the basis for the preparation of extended teaching units, where some other values-teaching techniques are short term in their application.

2. It seeks to balance cognitive investigation already included in geography units with values investigation necessary for full understanding of issues.

3. It allows informed decision making by students in a balanced way to counter polarisation which usually develops around issues.

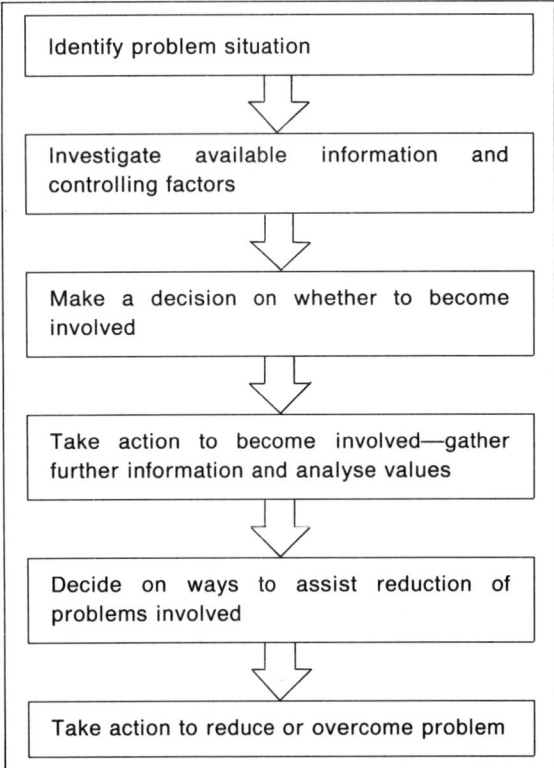

Figure 11.5 A general action strategy

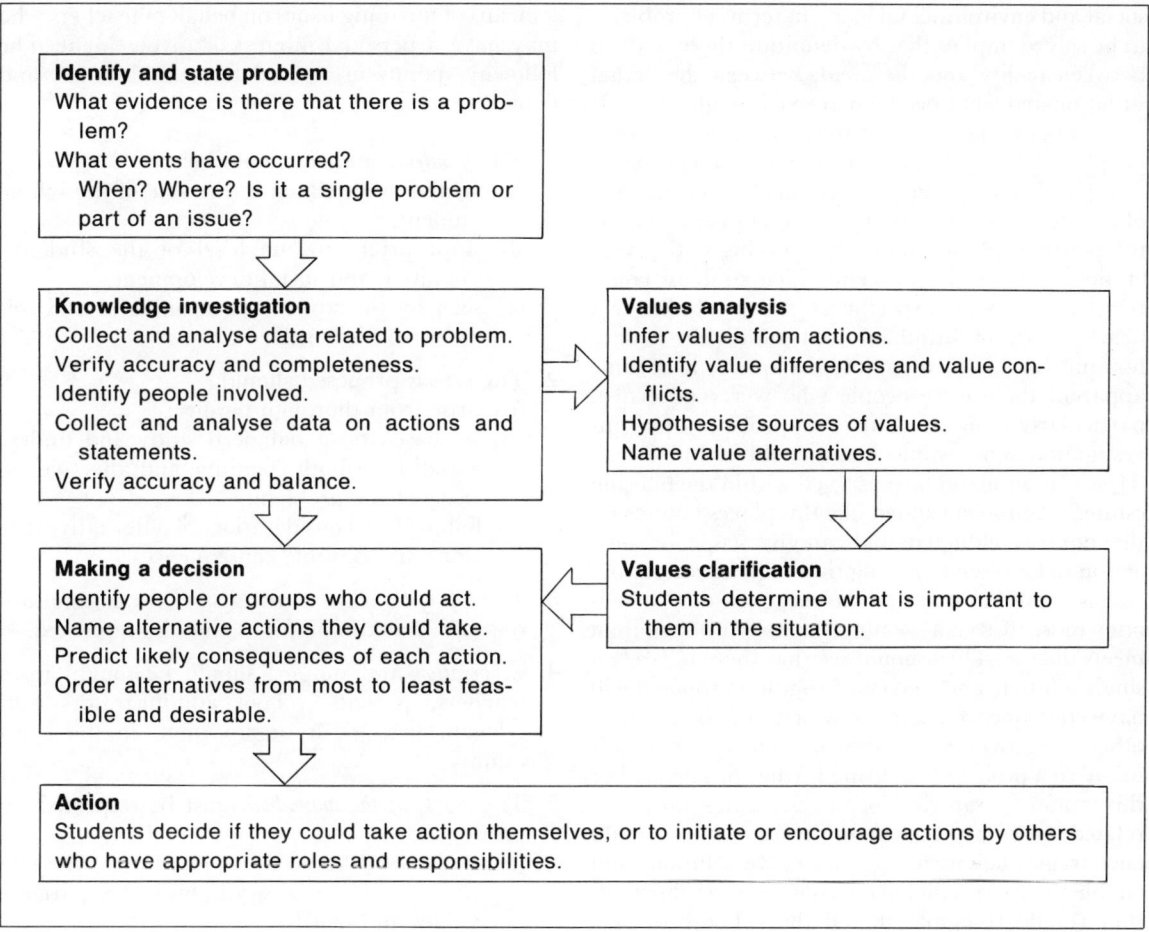

Figure 11.6 Problem-solving strategies—a generalised approach

4. It has high motivation and relevance dimensions, as issues and problems studied have a 'here and now' quality, and can draw upon the personal interests and geographies which all students have (Fien 1980).

Figure 11.6 outlines an approach to problem solving which is effectively a cognitive strategy integrated with a values-analysis strategy, which leads to personal clarification of values, decision making, and allows the option of taking personal action after the investigation has been completed. Appropriate practical exercises need to be selected at each stage, and associated teacher questioning is essential.

The approach to problem solving outlined has the potential to incorporate several major emphases which have developed over time in geographical education. It is particularly congruent with the *ecological perspective* outlined by David Hall in Chapter 2. Conceptually based knowledge investigation is balanced by analysis of values and clarification of the student's own values before proceeding to decision making and action. This process involves elements of empirical, rational and humanistic enquiry and is appropriately used in investigations emphasising people–environment approaches to geographical education.

Other strategies which emphasise similar processes can be found among those advocated internationally by the environmental education movement (Greenall 1986) and in recent geography curriculum projects. Most notable of these is the Geography 16–19 Project which includes a 'Route for Geographical Enquiry', where comprehensive analysis of factual and values related information precedes personal evaluation and judgement and a personal response to issues under investigation (Naish et al. 1987).

It is important for teachers and students to note the difference between problems and issues. Viewing social and environmental issues in terms of problems to be solved implies that by definition there is a gap between reality and the ideal; between the actual situation and what people perceive it ought to be. It also implies that value differences exist between people—if they did not, there would be no problem.

A problem is specific, and is usually symptomatic of a wider issue. For instance, a broad issue may be the perceived desirability of conserving native vegetation, but a local problem which students could relate to would be a specific proposal to bulldoze a nearby area of bushland and subdivide it into housing blocks. If, after initial inquiry, it became apparent there were people who were concerned, particularly the students, a problem for investigation and possible action could be stated: e.g. 'How can bushland be preserved within the housing estate?'. A problem stated like this places bounds on the inquiry conducted, and can give scope for some action to be taken following thorough investigation.

It is important that students recognise that there is often more to even a seemingly simple problem than meets the eye. They should see that there is rarely a single solution, and that every solution proposed will have consequences, some of which may be positive, others negative. Sometimes implementing a 'solution' to a problem can bring further problems. For this reason, strategies for investigating problems related to social and environmental issues should encourage students to see alternative solutions and enable them to predict likely consequences. Further, they should recognise that if they choose to take some action, they should not only be aware of possible consequences, but be prepared to accept them.

The Teacher's Responsibility in Implementing Action-Learning Strategies

Action-learning strategies, particularly problem-solving or decision-making strategies, are a particularly valuable approach to take to a geography teaching, emphasising as they do balanced and comprehensive understanding of social and environmental issues. This lays a basis for decision-making and citizenship skills, which can be developed through the ways in which strategies are implemented.

It should be recognised, however, that a high degree of professionalism is called for on the part of teachers. The use of action-learning approaches must *focus on the needs of the student*, and not be used as a means of pursuing issues on behalf of teachers who may have a personal interest or involvement. The following points are raised for consideration and debate:

1. The *problems* studied should be:
   (a) realistically related to the role of the school student;
   (b) appropriate to the level of the student's cognitive and social development;
   (c) seen by the student as meaningful and relevant to his/her needs.
2. The *actions* proposed should:
   (a) arise from thorough inquiry;
   (b) be based on a balanced study and understanding of both cognitive and affective aspects of the problem;
   (c) follow full consideration of alternative actions and possible consequences.
3. The *role of the school* as an agent of the community requires that community values be respected.
4. *Cooperation* and support should be sought from teachers, parents, school administrators and relevant persons or organisations in the community.
5. The *rights of the individual* must be respected, in particular:
   (a) no student should be expected to participate in any action contrary to his or her parent's values and beliefs;
   (b) where such differences occur they must be treated with sensitivity so that the student's self-esteem is not impaired in any way.
6. The *teacher* must be:
   (a) well informed and intellectually capable of understanding the problem or issue and of handling the classroom situation;
   (b) sensitive to the needs and values of the students, school and community.

Conclusion: A Final Word on Values Teaching

Successful application of values-teaching strategies in geographical education requires careful thought and choice of appropriate teaching styles. All the approaches outlined in this chapter, apart from inculcation, require use of an investigative approach, where students are actively involved in practical activities. There will necessarily be much questioning and discussion, so that accepting relationships between students and between teachers and students are vital.

When planning and implementing values-teaching strategies, teachers should give careful consideration to the following:

1. The teacher's own values will influence the topics and stimulus materials chosen and the way in which teaching strategies are implemented.

2. The teacher therefore needs to clarify his/her own values in relation to the topics chosen for study, and be clear in the purpose underlying the approach he/she chooses to follow. With very little change in emphasis and technique, a values-analysis strategy could become (intentionally or unintentionally) a values inculcation strategy.

3. An atmosphere of mutual trust and respect in the classroom is necessary for values teaching to be successfully undertaken. Teacher sensitivity is essential.

4. Skilful questioning based on a clear sense of purpose is essential to promoting investigative skills in students.

5. Students should not be pressured into answering or participating in activities about which they are self-conscious. Permit them not to answer or participate if they so choose.

6. Students' responses will be inhibited by teacher moralising, criticism, evaluation of student values, or overt display of the teacher's values.

7. Evaluation of students' learning can be based on their ability to *apply skills* related to valuing in practical activities and situations, rather than on values themselves.

References

Banks, J.A. and Clegg, A.A. (1986) *Teaching Strategies for the Social Studies*, 6th edition, New York: Longman.

Bartlett, L. and Cox, B. (1982) *Learning to Teach Geography*, Brisbane: Wiley.

Blachford, K. (1979) 'Morals and Values in Geographical Education. Towards a Metaphysic of the Environment', *Geographical Education*, Vol. 3(3), 423–457.

Buckland, J. (1979) 'Preparing Social Studies Teachers for Tomorrow's Classrooms', *First Australasian Conference of Social Studies Associations Report*. Sydney.

Cowie, P.M. (1978) 'Geography: A Value Laden Subject in Education', *Geographical Education*, Vol. 3(2), 133–146.

Curriculum Development Centre (1980) *Core Curriculum for Australian Schools*. Canberra.

English, P.W. and Mayfield, R.C. (1972) *Man, Space and Environment*, London: Oxford University Press.

Fielding, G.J. (1974) *Geography as Social Science*. New York: Harper and Row.

Fien, J.F. (1980) 'Operationalising the Humanistic Perspective in Geographical Education', *Geographical Education*, Vol. 3(4), 507–553.

Fien, J. and Gerber, R. (eds) (1988) *Teaching Geography for a Better World*, 2nd edition, Edinburgh: Oliver and Boyd.

Fraenkel, J.R. (1977) *How to Teach About Values*, Englewood Cliffs: Prentice Hall.

Greenall, A. (1986) 'Searching for a Meaning: What is Environmental Education', *Geographical Education*, Vol. 5(2), 9–12.

Huckle, J. (1981) 'Geography and Values Education', in R. Walford (ed.) *Signposts for Geography Teaching*, London: Longman.

Huckle, J. (ed.) (1983) *Geographical Education: Reflection and Action*, Oxford: Oxford University Press.

Huckle, J. (1986) 'Geographical Education for Environmental Citizenship', *Geographical Education*, Vol. 5(2), 13–20.

Huckle, J. (1988) 'Geography, Citizenship and Political Literacy', in J. Fien and R. Gerber (eds) *Teaching Geography for a Better World*, 2nd edition, Edinburgh: Oliver and Boyd.

Kniker, C.R. (1977) *You and Values Education*, Columbus: Merrill.

Lowenthal, D. (1967) *Environmental Perception and Behaviour*, Research Paper No. 104, University of Chicago.

Marsh, C. (ed.) (1987) *Teaching Social Studies*, Sydney: Prentice Hall.

Muessig, R.H. (1975) *Controversial Issues in Social Studies: A Contemporary Perspective*, Washington DC: National Council for the Social Studies.

Naish, M., Rawling, E. and Hart, C. (1987) *Geography 16–19. The Contribution of a Curriculum Project to 16–19 Education*, London: Longman.

Raths, L., Harmin, M. and Simon, S.B. (1966) *Values and Teaching*, Columbus: Merrill.

Scharf, P. (ed.) (1978) *Reading in Moral Education*, Minneapolis: Winston Press.

Senesh, L. (1973) *New Paths in Social Science Curriculum Design*, Chicago: SRA.

Shaver, J.P. (1981) 'Citizenship, Values, and Morality

in Social Studies', *The Social Studies*, National Society for the Study of Education 80th Yearbook, Chicago: University of Chicago Press.

Shaver, J.P. and Larkins, A.G. (1973) *Decision-Making in a Democracy*, Boston: Houghton Mifflin.

Simon, S.B., Howe, L.W. and Kirschenbaum, H. (1972) *Values Clarification: A Handbook of Practical Strategies for Teachers and Students*, New York: Hart.

Smith, D.L. (1978) 'Values and the Teaching of Geography', *Geographical Education*, Vol. 3(2), 147–161.

Taafe, E.J. (1974) 'The Spatial View in Context', *Annals of the Association of American Geographers*, Vol. 4(1), 1–16.

Volkmoor, C.B., Pasanella, A.L. and Raths, L.E. (1977) *Values in the Classroom*, Columbus: Merrill.

12

Promoting Social and Political Education in Geography Teaching

Rob Gilbert

In Chapter 1 it was explained that geography is a medium for the education of young people. As well as providing appropriate knowledge and a range of investigation and thinking skills for students, geography has a contribution to make to the wider education of young people as informed and active citizens. In this chapter, Rob Gilbert outlines the case for geography as a vehicle for social and political education.

The nature of social and political literacy, as developed by the British Programme for Political Education, is explained, as are the major concepts and skills needed for effective citizenship. The points are illustrated by a listing of potential topics for inquiry in geography as well as a detailed outline for a unit called 'Living in Isolation'.

Just what do we mean by a 'socially educated person'? Any answer to this question will be potentially controversial, for it must involve interpretations of the present nature of society and views about a desirable future. Take an example. Scrimshaw (1981) claims that socially educated people should:

1. be able to describe, explain, evaluate and justify their evaluations of the *significant* aspects of those *selected* individuals and groups that it is *important* to understand;
2. develop *desirable* attitudes and dispositions towards themselves and others that will incline them to act in *appropriate* ways in various social situations;
3. develop the social skills *necessary* to act *as they should*.

The interesting point about this definition is that Scrimshaw says it holds whatever our particular values might be. The disagreements will arise when we try to spell out the terms in italics. For while we might all agree that social education is about understanding the activities of groups and individuals, we still need to decide which groups should receive most emphasis: the local community or the global society? the Western industrialised nations or the less-developed world? the dominant culture or minority groups? And which aspects of these groups should be given priority: their problems or their achievements? those most relevant to our own concerns or to theirs? their economic, political or social activities?

We also need to decide what attitudes we want to develop with respect to these groups. A scholarly detachment or a commitment to justice? A concern for self-interest or altruism? A willingness to accept the world as it is or to try to improve it? We need to decide what skills are appropriate: the tools of academic inquiry or the skills of political rhetoric? the ability to act on social problems or to analyse and propose policies?

These are not black and white, 'either–or choices', but any attempt to fill out Scrimshaw's definition will require some selection, some choice of priorities, some value judgements with political implications. In geography, we cannot avoid these choices, for political considerations are inevitable parts of the subject, in a number of ways.

First, the geography curriculum itself is political, for the school curriculum is determined by decisions about what is thought to be important knowledge and this will reflect the views of those who make these decisions. These views will not necessarily reflect the priorities or interests of outside groups who do not have access to these decisions or the power to influence them. For instance, geography curricula have shown ethnocentric bias in treating other cultures (Wright 1985). Their explanations from particular economic and political perspectives have glossed over the knowledge needed for informed participation in the just solution of social problems (Gilbert 1984). Their assumptions about what is important in the study of society have ignored the perspectives of less powerful groups, as in the neglect of women as subjects of geographical study (Monk and Williamson-Fien 1988).

Second, there is increasing interest within the discipline itself in the geographical perspective on political issues, such that an understanding of contemporary geography would be incomplete if it did not include a study of political geography. The study of political boundaries, international conflict and competition, the spatial aspects of government decisions and administration, regional development and electoral systems are active areas of study for geographers, and provide useful material which demonstrates the potential of a distinctively geographical approach to political matters (Taylor and House 1984).

Third, and more generally, all the issues which geographers have studied are pervasively political, but their political aspects have often not been acknowledged. For instance, decisions about economic activity and settlement, world and regional and urban development, transport, trade, the natural environment are all subject to conflicts of interest, planning and policy decisions, and considerations of justice and human welfare which characterise political issues. If people are to be educated so that they can participate in decisions about these matters, then the value and political dimensions of these topics must be a constant part of the study.

Social and Political Literacy

Recently, a widely used approach to social and political education has been based on the idea of social and political literacy. People are said to be literate if they have a basic competence in a language so that they can communicate well enough to participate constructively in the normal activities of a group. And there is ready agreement that schools have a responsibility to promote literacy. In translating this idea to social education, we can say that to be socially and politically literate, people must be competent in the knowledge, skills and values required for constructive participation in society. If we can specify what these knowledge, skills and values are, then we have a strong base for justifying and designing social education.

The central elements are *competence* and *participation*, as illustrated in one definition that 'social literacy is an ability to understand the world, with a confidence in one's capacity to be an active, participating member of society' (Kalantzis et al. 1983). This emphasis on participation and action in society has meant that social literacy and political literacy have become very closely related, if not synonymous. The idea of political literacy was originally developed in a curriculum project sponsored by the UK Hansard Society, the Programme for Political Education (PPE) (Crick and Porter 1978). For the PPE, a politically literate person would be able to recognise and come to a position on major current issues, make realistic political judgements about how to deal with them, and take effective political action on them. These processes, the project claimed, are the basic requirements of political democracy.

The PPE argued that people need to have a range of knowledge, skills and attitudes which can be described as follows:

1. Knowing the system:
    (a) Knowing who stands for what, how to judge the value of their various claims, and what alternatives exist.

    (b) Knowing the major political institutions, how they can be used to settle disputes, how adequate they are, and what alternatives exist or might be created.

    (c) Knowing how to influence decisions on different kinds of issue, and what alternative forms of decision making exist in other groups and societies.

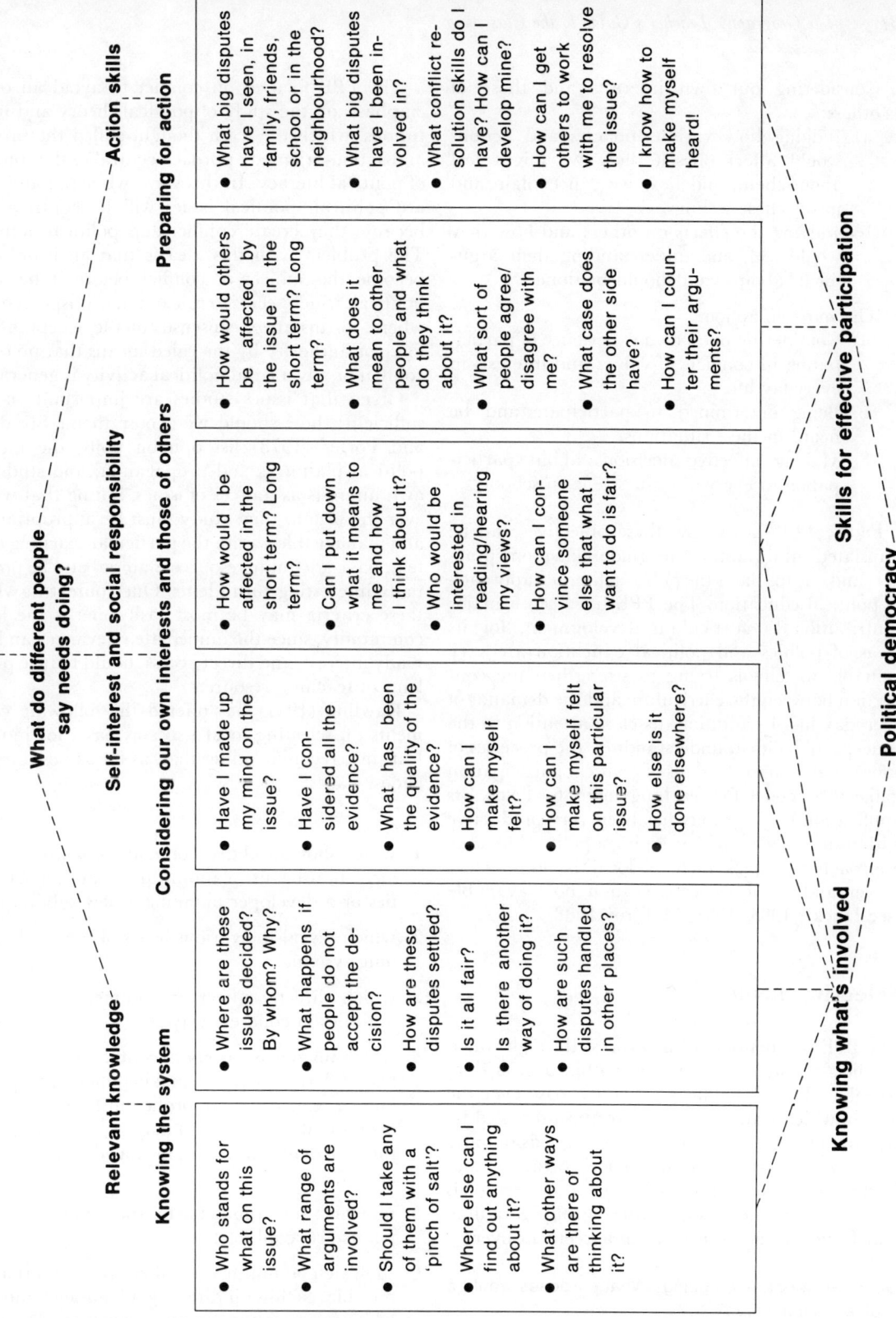

Figure 12.1 Political literacy: questions for classroom practice (after Crick and Porter 1978)

2. Considering our own interests and those of others:
   (a) Judging how various policies and actions would affect oneself, how we might feel about them, and how we can explain and justify these feelings.
   (b) Judging the effects on others and how they would feel, and understanding their arguments about what should be done.
3. The move to action:
   (a) Considering one's own experience of participating in conflicts of values and interests in everyday life.
   (b) Being determined to participate and be heard in these situations.
   (c) Making effective decisions about participation in everyday issues (e.g. in school).

Figure 12.1 shows how these objectives can be translated into questions to guide classroom practice, and encapsulates the political literacy approach to political education. The PPE has been a major contribution to curriculum development, for its ideas of politics and political education are very valuable to schools trying to strengthen the connection between the curriculum and the demands of everyday life. In addition, its close attention to the concepts of political understanding, the processes of issues analysis and the skills of political participation make it a model for teaching in many fields. Its application to geography has been very productive with practical ways of developing a political literacy approach to topics such as local issues, nuclear power and rainforest conservation now available (see Huckle 1983, 1988; McElroy 1988).

Selecting Issues

The PPE saw politics as 'the process through which conflicts of interest and values within a group are conciliated. Such conflicts generally arise over the way in which scarce resources are allocated, and by resources we mean not only material goods but also power, status, skills, time and space' (Stradling and Porter 1978). Conflict arises because some decision is a matter of dispute or issue within a group. Stradling and Porter define an issue as disagreement over:

Goals—where are we going? What purposes would a given action serve?
Values—in what way should we act or not act?
Methods—how should we do it?
Results—was it the right/the fairest/the best outcome?

The PPE focused on conflict to avoid an overemphasis on the study of political theory and institutions, which the team thought stifled the process and action-oriented approach central to the concept of political literacy. In this view, while not all issues are political, political issues will be controversial because they create conflict and political activity. The problem with this idea is that an issue must become the subject of conflict before it becomes political, which seems to rule out many aspects of life where, for instance, consensus or the acceptance of force or authority by the ruled means that no overt conflict or adversarial political activity is generated.

Given that issues studies are important (if not sufficient), how should we choose them? Stradling and Porter (1978) list opinion polls, the media, political platforms and propaganda, and students' own interests as sources of issues, noting that whatever the origin, their study must be appropriate to and manageable within the particular teaching context, have adequate resource material and be potentially interesting to students. One context in which these criteria may be most easily met is the local community, since the immediate relevance can lend ready interest, and direct access should reduce problems of teaching resources.

Rawling (1981) has offered the following comments on selecting local issues as bases for inquiry learning, recommending that, as far as possible, we choose issues:

1. which show local decision makers at work, e.g. a local industrialist planning an expansion of facilities or a developer planning a new subdivision;
2. which reveal the various levels of the local planning system;
3. which directly affect environments and communities in concrete ways;
4. which involve a variety of conflicting viewpoints and different ways of making these viewpoints known, e.g. different protagonists in a conflict over a building proposal may express their views through demonstrations, media advertising, legal action;
5. on which the arguments for the conflicting viewpoints are available;
6. to which a number of alternative solutions is possible, each with differing advantages and disadvantages for the various interests;
7. on which an official policy line has been taken and a survey or report exists.

A Broader View of Political Knowledge

Another definition of the political which goes beyond the study of issues is Leftwich's (1984) view that politics comprises all the activities of cooperation and conflict whereby people organise the production, distribution and use of human, natural and other resources. These activities influence and reflect distributions of power and patterns of decision making, the structure of social organisation, and the systems of culture and ideology in society or groups within it. This definition does not rely on overt conflict as a necessary aspect of politics, but recognises the important roles in the distribution of social goods, of power, social structure and ideology, all of which may be covert or unnoticed by those affected by them. This more comprehensive view seems more relevant to geography's contribution to political knowledge, for geography seeks to understand the spatial and environmental aspects of human activity whether this activity is generated by cooperation, conflict, force or authority.

But what does this broader definition mean for our teaching? Here we need a different set of criteria if we are to focus on political relationships which are not conspicuous issues at present, but which may have the potential (and perhaps deserve) to be, or which reflect cooperative forms of power and decision making. Guidelines for generating studies of this kind could be the following questions.

In the study of this place, social group or process, what are the resources (in the broad sense mentioned above) being distributed?

The term 'resources' here needs some elaboration, for in social and political education, the resources of greatest concern are those which determine the quality of life or social well-being. One list from welfare geography (Coates et al. 1977) includes:

Nutrition—the quality as well as the quantity of food
Shelter—basic provision and factors such as densities, amenities, related social problems
Health—including provision and accessibility of health care and services
Education—access, participation rates, literacy, vocational and adult education
Leisure—time free from work, cultural and recreational facilities
Security—personal safety, economic, social and political rights, social security
Physical environment—cleanliness, visual appearance, but also resource management and accessibility of amenities
Surplus income—surplus over basic needs which determines access to higher needs and aspirations

The geography of welfare will be a necessary part of political literacy, since only by considering ideas of social welfare and human needs, interests, and rights can we identify the resources which are organised through political activity. Any study of a place, group or process might use this list as a guide to the nature and levels of welfare and associated political activity.

What forms of cooperation and conflict are involved in the production, distribution and use of these resources?

To answer this question we would focus on the way social relations are bound up in the production and reproduction of social life. We might study the economic activities in a particular place, with an emphasis on the production of the resources listed above, but in particular we would ask about the forms of authority and power which organise this activity:

1. Are all the resources listed produced and distributed in ways that provide for the *welfare of all*?
2. If so, what forms of *social organisation* make this possible? If not, how are deficiencies or differentials sustained?
3. To what extent does the distribution reflect *consensus or conflict*?
4. How do levels of resources and welfare vary from place to place or group to group? What *social divisions* are created and sustained in the organisation of production, distribution and use?
5. What economic, political, and social *institutions* are involved in the process, and how do they operate?
6. How are decisions made, who has *power* to influence the decisions, and in what ways?
7. How do cultural traditions, especially *ideologies*, enter the system? Are they centres of consensus or conflict? How are they changing?

What alternative forms of political organisation might promote human welfare in this place, group or process, and how can they be instituted?

If we were wanting to emphasise an intellectual understanding of the politics of resource production, distribution and use, we might well stop at the second question above, but to develop political literacy we need to go further and look for the implications of the analysis for action. We would therefore need to ask what alternatives exist and how they might be brought about. This could involve identifying policies and platforms which address the problems, and political activity which would promote these policies. At this point we could return for guidance to the ideas in the political literacy curriculum framework (see Figure 12.1).

The contexts for political education in geography classrooms will generally arise in one of three ways:

1. The *teacher will intentionally raise the political aspects* of a topic which may have been chosen for other reasons, but which has clear political significance in its own right. For instance, any study of environmental geography would need to show how environmental issues have been the subject of political campaigns, policies and legislation. Any study of less-developed countries should include the politics of colonialism, aid and international capital. To do otherwise would hardly be an adequate study of these topics.

2. Inquiry will focus on *contemporary issues* or 'current events' either chosen for their own sakes or related to topics of study in the normal program. In May 1986 many geography classes may have turned their attention to the Chernobyl nuclear plant fire and radioactive emission and the links to the nuclear debate, superpower competition or the ideological aspects of media coverage.

3. Political geography will be *a scheduled part of the work program*, as when the politics of local government policy on town planning is studied for insights into environmental decision making, or when regional variations in voting in elections are related to other aspects of regional difference.

Government Who rules, how and for whom?	**Power** The ability to achieve an intended effect either by force or more usually by claims to authority. • Who makes things happen here? • Who could change the situation? • What things can be changed immediately and what not? Who says so? • Who might prevent change?
Relationships What kinds of relationship exist between rulers and ruled?	**Law** The body of rules made and recognised as binding by governments. • What are the laws, prohibitions, entitlements, regulations which apply to this situation? • Have they been properly interpreted and applied in this case? • What legal disputes have arisen? Are the laws legitimate? Are they fair (just)?
People What political 'goods' apply for the people in this situation?	**Natural human rights** What we claim as the minimum conditions for a proper human existence. What we can expect by virtue of being human. • What are the political, social, economic, legal rights recognised by rulers in this situation? • How do they accord with natural rights? What natural rights are not satisfied here? • Are there conflicting rights claims? How can they be resolved?

Figure 12.2 Basic concepts for political literacy. (A modified and expanded version of Crick's (1978a) tabular summary)

Force Physical pressure or use of weapons to achieve an intended effect—latent in all government, constant in none. • Has force been used in this situation? Could it be? What kind? By whom? • Are there threats of force, explicit or implied? • Is force or the threat of force an important factor in this situation?	**Authority** Respect and obedience given when an institution, group or person fulfils a function which is accepted as needed, and in which they have superior knowledge or skill. • Who claims authority here? On what grounds? Is it recognised by all? If not, why? • How does one achieve authority in this situation? What are its limits?	**Order** When expectations are fulfilled and calculations can be made without fear of all the circumstances and assumptions changing. • What are the main mechanisms by which order is maintained here? • Which aspects of order are necessary and which not? • How would changes to the situation affect order?
Justice What people accept as being done fairly. Hence justice is relevant to all forms of relationship between rulers and ruled. • Is this a fair way to decide the issue? • Is it consistent with other actions, or the way other people have been treated, or how we would want to be treated? • Are the rights of all parties being considered? • Will the action advance or inhibit the interests of the most needy?	**Representation** Claims of authority on grounds of being able to 'speak for' some group, for whatever reason. In modern democracies, representation is typically effected through some kind of elected parliament. • In this situation, who claims to represent/speak for whom? On what grounds? • Do the ruled accept these grounds? If not, why? • What competition is there among claimants to representativeness? How can it be resolved?	**Pressure** All the means by which government and people influence each other (excluding law or force) e.g. persuasion, economic, social or psychological influence, etc. • What forms of pressure are operating here? • What other forms of pressure could be applied? • Which are or are likely to be most effective? • Which are legal/illegal, fair/unfair? • How can these forms of pressure be most effectively used?
Individuality What we see as unique to each person—related to natural rights and respect for individual differences, but does not entail the ideology of individualism. • Does this policy/action/situation respect the rights of all affected by it? • Are any individuals not properly catered for? • Does it respect diversity in people's aspirations? • Does it provide for individual expression and achievement?	**Freedom** Making choices and doing things of public significance in a self-willed and uncoerced way. Freedom is to be practised, not just enjoyed. • Are people free to exercise their rights, both publicly and privately? • Are they able to engage in public life and citizenship, and to promote freedom, without coercive constraints?	**Welfare** The belief that the prosperity and happiness of communities and individuals beyond mere physical survival should be the concern of governments. • What aspects of human welfare are involved in this decision or issue? • How can the rulers improve levels of welfare? • Is the welfare of various groups in conflict in this situation? • How can the conflict be resolved?

Developing Concepts

The argument and questions above suggest problems, issues, principles, groups and social processes which might be studied in political education in geography, and give some directions for how we could study them. But we need to consider a conceptual framework through which these aspects of political study can be understood. Again the PPE is helpful here, for it suggests a set of basic concepts which a politically literate person should be able to use clearly, with some sense of their connotations

Figure 12.3 Outline of a teaching unit in social and political geography: 'Living in Isolation'

Topics and issues	Concepts and examples	Teaching strategies
1. Isolation and the quality of life • Assessing social well-being in a number of isolated communities • Differences within and between centres • Comparisons with metropolitan living • What do these comparisons show about how resources are distributed in society?	• Social well-being • Hierarchy of needs • Social and economic effects of isolation • Case studies of a mining settlement, a tourist resort, an Aboriginal community, and a regional service centre	• Rating scales • Community case studies • Mailed questionnaire (perhaps to isolated schools) • Group work on case studies
2. Cultural values and human occupance • How do culture and class lead to different views on needs, the environment and lifestyle? These can produce varying political priorities, allegiances and goals • Why do people live in these areas? What values are evident in their activities in various settlements? • What problems do people perceive? • Are the needs of all groups met fairly and equitably?	• Culture and tradition • Cultural diversity, conflict, dominance • Traditions of class and class conflict • Common needs and interests of communities • Tolerance, justice, equality • Case studies of Mapoon, Mudginberri, Tasmanian logging controversies	• Values analysis • Guest speakers • Cultural studies • Historical studies of class and cultural difference and conflict • Simulation games
3. Isolation and the political system • What demands do people in isolated places make for resources? How do they promote these demands? • Political histories of their access to influence and power, and their political allegiances • Relevant platforms of the political parties • Current policies on subsidies, special programs for isolated areas	• Costs and benefits of decentralisation • The electoral system, boundaries and weightings in sparsely populated areas • Principles of justice and fairness in representative government • Case studies of pressure groups and campaigns • Community cooperation and conflict in political action	• Debates • Guest speakers from political parties • Exercises in policy formulation • Simulations • Documentary and media analysis

and of how others use them. Figure 12.2 describes the PPE concepts and poses questions to show how they could be applied to particular studies. While not all the concepts will be central to every particular study, a comprehensive political understanding would require mastery of these concepts at least.

The concepts allow us to discuss political activity of various kinds by identifying the relationship between those who rule in any particular situation and those who are ruled (remembering that 'rulers' in some contexts will not necessarily be government, but those who have the predominant power, whether it is derived from economic power, status, charisma, etc.). The aim is to refine students' use of these terms to give meaning to political activity, not in any highly technical way, but by developing awareness, clarity and consistency in everyday speech in the belief that 'the ability to conceptualise and distinguish concepts is a real persuasive, moral and political skill' (Crick 1978a).

In addition, Crick (1978b) lists what he calls *procedural values* which must underpin any application of the concepts if it is to avoid indoctrination. He argues that 'it is proper and possible to nurture and strengthen these procedural values precisely because they are educational values, rational and public', and includes freedom, toleration, fairness, respect for truth and respect for reasoning. As a result, political education will presume these values as standards which will be applied not only by students in their judgements about political issues and situations being studied, but also by teachers in conducting classroom discussion and dealing with students themselves.

The Framework in Action: 'Living in Isolation'

Figure 12.3 is an outline of a teaching unit which tries to incorporate some of the ideas presented here. In studying the lives of people in isolated areas it recognises that such people have special problems and needs arising from their location, and that these problems and needs have implications for the way society's resources are distributed. To understand their situation adequately, we need to appreciate the historical legacies of their communities, their common needs and aspirations, but also the things which divide them. These needs and aspirations find expression in political activity and responses from the political system. Any society concerned for the welfare of its citizens need to find a just and equitable way of meeting these needs, and concepts of justice, equality and rights will consequently be a part of the study. Students are therefore introduced to the historical, cultural and social background of the isolated communities, to the way the political system deals with their needs and problems and the political activity of the communities themselves, and to the social values involved in any decision on policy or political action.

This chapter has focused on the 'what' more than the 'how' of teaching for political literacy in geography, for the strategies for political education are not unique, but involve inquiry, thinking and research skills, valuing and decision-making processes, and other more specific resources and strategies dealt with elsewhere in this book. Particularly important, however, are the following principles:

1. Thinking about politics involves *working out what is best for people*, but just what constitutes general human interest in any situation will not always be obvious. Any discussion of particular political policies or action must involve continuous reflection on how the welfare of all can be identified and promoted. Political education has an inevitable moral component.

2. Students should understand the various possible positions on issues, and the best way of doing this is to see, read or hear views presented by their real advocates, rather than second-hand or, worse, through hypothetical reports. *Actual policies, platforms, decisions and decision makers should be directly studied wherever possible.*

3. In treating issues, *it is not always wise to insist that students arrive at conclusions*, especially where the issues are particularly controversial, or where strong prejudices exist among students. In these cases it may be enough to show that there are plausible alternative positions. To ask students to express their own views can lead to premature closure of discussion or dogmatic polarisation of views.

4. Schools have no right to present views on particular issues as if they were the only acceptable ones, but it is equally indefensible to suggest that politics can be studied without a commitment to values. As the PPE concepts illustrate, political literacy in a democracy requires commitment to certain values, since literacy and competence presume participation, and the right to participate must therefore be upheld. Further, before we can say that a course of action is in the interests of all the members of some group, one requirement is that all the members have some effective way of expressing their interests as they see them.

These four points prove the need to see value concepts like justice, participation, representation, rights, freedom and welfare as *essential parts of political education*. They cannot be taught as abstract and objective concept whose definitions can be learned from a glossary, nor as value doctrines with finite and known content, but as *guiding principles whose meanings must be worked out in particular situations*. The process will be one of continuous testing, refinement and expansion, as students become practised in reflecting on these values and how they can be implemented in any situation, each time making the use of the values and concepts more discriminating, comprehensive and consistent.

References

Coates, B., Johnston, R. and Knox, P. (1977) *Geography and Equality*, Oxford: Oxford University Press.

Crick, B. (1978a) 'Basic Concepts for Political Education', in B. Crick and A. Porter, (eds) *Political Education and Political Literacy*, London: Longman.

Crick, B. (1978b) 'Procedural Values in Political Education', in B. Crick and A. Porter, (eds) *Political Education and Political Literacy*, London: Longman.

Crick, B. and Porter, A. (eds) (1978) *Political Education and Political Literacy*, London: Longman.

Gilbert, R. (1984) *The Impotent Image: Reflections on Ideology in the Secondary School Curriculum*, Lewes, Sussex: Falmer Press.

Huckle, J. (1983) 'Political Education', in J. Huckle (ed.) *Geographical Education: Reflection and Action*, Oxford: Oxford University Press.

Huckle, J. (1988) 'The Daintree Rainforest: Developing Political Literacy Through an Environmental Issue', in J. Fien and R. Gerber (eds) *Teaching Geography for a Better World*, 2nd edition, Edinburgh: Oliver and Boyd.

Kalantzis, M., Cope, W. and Hughes, C. (1983) *Making Curriculum in Australia: The Social Literacy 5–8 Project*, Stanmore, NSW: Social Literacy Project.

Leftwich, A. (1984) 'Politics: People, Resources and Power', in A. Leftwich (ed.) *What is Politics? The Activity and its Study*, Oxford: Blackwell.

McElroy, B. (1988) 'Learning Geography: A Route to Political Literacy', in J. Fien and R. Gerber (eds) *Teaching Geography for a Better World*, 2nd edition, Edinburgh: Oliver and Boyd.

Monk, J. and Williamson-Fien, J. (1988) 'Stereoscopic Visions: Perspectives on Gender—Challenges for the Classroom', in J. Fien and R. Gerber (eds) *Teaching Geography for a Better World*, 2nd edition, Edinburgh: Oliver and Boyd.

Rawling, E. (1981) *Local Issues and Enquiry Based Learning*, Schools Council Curriculum Development: Geography 16–19, Occasional Paper No. 2, London: Schools Council.

Stradling, R. and Porter, A. (1978) 'Issues and Political Problems', in B. Crick and A. Porter (eds) *Political Education and Political Literacy*, London: Longman.

Scrimshaw, P. (1981) *Community Service, Social Education, and the Curriculum*, London: Hodder and Stoughton.

Taylor, P. and House, J. (eds) (1984) *Political Geography: Recent Advances and Future Directions*, London: Croom Helm.

Wright, D. (1985) 'In Black and White: Racist Bias in Textbooks', *Geographical Education*, Vol. 5(1), 13–17.

13

Look into my Mind: Qualitative Inquiry in Teaching Geography

V. Leo Bartlett

The humanistic perspective on knowledge outlined by David Hall in Chapter 2 involves certain teaching styles and strategies. Many of these are explained in Chapters 11 and 14. Together, these imply a qualitative approach to teaching and learning which contrasts with a style that treats teaching as a craft, technology or a mechanical process and is based on what can be termed a positivist mode of inquiry. Two different ways of teaching the topic, 'What are the characteristics of black areas in Chicago? Why do such areas exist?', are explained to illustrate the different philosophical and practical aspects of using each mode of inquiry. Over twenty ideas for qualitative inquiry are suggested. These include both classroom and field activities. The chapter concludes with a list of teacher attitudes and behaviours which will foster the open classroom climate needed for qualitative inquiry.

Some years ago I was working on a research project with colleagues at Norwich in England. I recall setting off for home one evening after a particularly long day of discussions about the conceptual and ethical problems inherent in the project. My first action on arriving home was to settle into a comfortable chair. Such a setting might desirably stimulate discussion about the more important events of the family's day. But, as on many previous occasions, I succumbed to the 'heresy of good works' by silently sitting and rethinking some of the project's problems. My three-year-old son attempted to alert me to the inappropriateness of my behaviour by giving me a three-minute description about the sun and ending with a question about the origin of its heat. He received no reply at first; I barely heard him. So I responded by saying that I did not understand what he was talking about. He came over to me and pulling my head down to his, pressed his forehead to mine and said, 'Look into my mind'.

It brought me up sharply, not only to my inattentiveness to his needs but also to the profound significance of his comment. Here was a three year old telling me the problematic of knowledge. Can we ever know the world of people with a similar ability to know? And what is this similar ability to know? How *do* we know? Wouldn't it be an achievement if we could come to understand people's knowledge and experience of place, as they know it? How can we ever hope to accomplish this task in our teaching?

Six months later I was back to the reality of teaching a group of graduate student teachers at a university in Australia. One of my students in curriculum studies in geography was assigned the task of teaching an inquiry to a Year 12 class. The prescribed inquiry questions were 'What are the characteristics of black areas in Chicago?', 'Why do such areas exist?'. I provided my protégé with the only information I had at the time, an excellent publication from the Victorian Education Depart-

142 The Geography Teacher's Guide to the Classroom

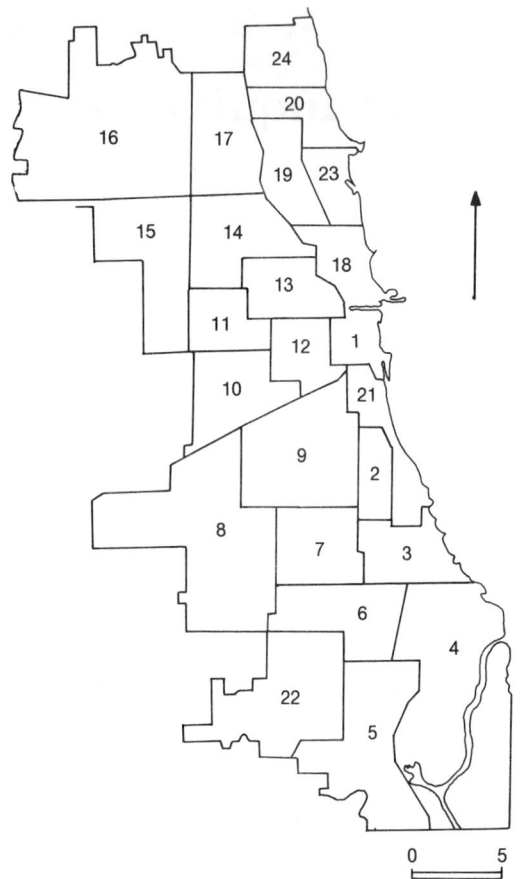

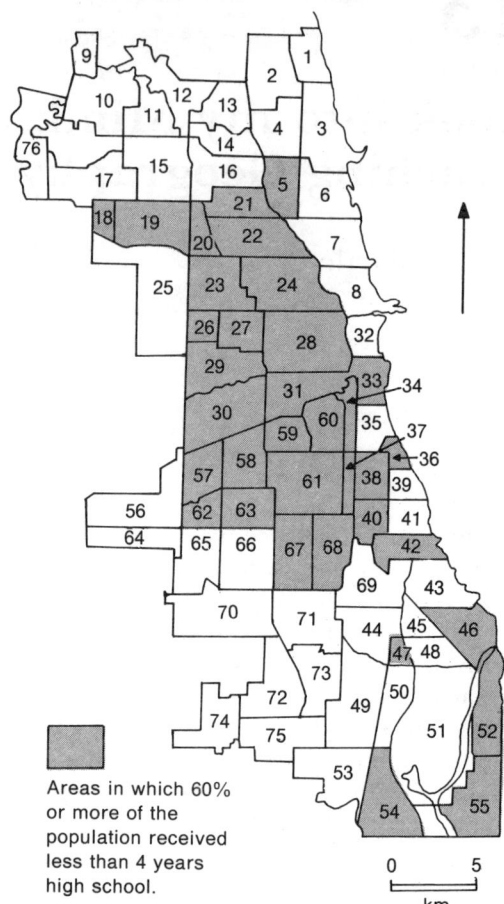

Robberies and homicides in Chicago by police district, 1978

Police district (see outline map of police districts for location)	Number of robberies (Critical level for mapping: 800 or more robberies per year)	Number of homicides (Critical level for mapping: 35 homicides per year)
1	322	14
2	1679	82
3	1197	46
4	602	26
5	467	43
6	692	22
7	942	56
8	271	17
9	391	31
10	857	50
11	1099	65
12	514	38
13	841	54
14	513	44
15	666	36
16	125	5
17	145	8
18	860	38
19	371	24
20	481	10
21	742	28
22	349	19
23	810	36

Figure 13.1 Sample data from the unit 'What are the Characteristics of Black Areas of Chicago? Why do they Exist?' (Curriculum Services Unit 1981)

ment (Figure 13.1). Statistical tables together with appropriate resources were included in the materials. The description in the teacher's guide said among other things:

The unit intends to examine the applicability of a stereotype in describing areas inhabited by black people. However the unit could be adapted for teaching about any group of people. . . the unit should reinforce previous learnings concerned with the structure of cities.

The objectives of the unit included the following kind of activities:

1. Describe the distribution of.. . .
2. Suggest reasons for the location of . . .
3. List several characteristics of . . .
4. Account for the characteristics of . . .
5. Detail difficulties in improving the lot of ghetto dwellers.
6. Express concern for other people . . .

Three weeks later I visited my student at his practice teaching school. I casually asked how his inquiry on Chicago had fared. He gave me a vivid description of how the unit had extended into ten lessons instead of six; how the students had responded with enthusiasm; how they insisted on dealing with the next topic in a similar way; and how their assignments were difficult to assess but made for absorbing reading. I looked at some of his assignments and to my astonishment and delight found they included quotations and interview data from Stud Terkel's book *Division Street*. The excerpts contained statements from Florence Scala who was born in Chicago but who 'hate[d] the fact that so much of it is inhuman in the way we don't pay attention to each other'; of John Rath aged 61 years who claimed that 'the only time they [Chicago people and its administration] find out if you're dead is when you don't go down and pay the rent'; and of Lois Arthur who spoke about Chicago as a 'lost town'. All the information was in the language of Chicago's inhabitants, providing a picture of their lived experiences. In addition to the above, there were maps based on census data and students' statements based on personal observations from pictorial and literary evidence.

I asked my student how he came to adopt what seemed to be such a different approach from the curriculum materials I had given him. He said he looked hard and long at the objectives. They left him dissatisfied; the first five seemed to be so analytic and searching for motives (of people) and explanations of a casual kind. The sixth objective was an appeal to understanding Chicago through the concerns of its people and the concerns of those who had had primary experience of the place. This suggested an entirely different approach so there seemed to be two ways he could 'jump'. He chose the latter.

My student was reinforcing for me the questions 'How can and ought we to know?','Can we have different perspectives of the same object?','Do students see what teachers see?'. The two approaches seemed to imply that it was possible. If this were so, how do we see or know? I had come the full circle to the questions that had been sparked by the experience with my son.

The purpose of this chapter is to discuss some aspects of the above questions and how they relate to the philosophy of phenomenology, firstly by briefly comparing ideas proposed by the two approaches and secondly by introducing teaching strategies and examples related to a phenomenological and qualitative methodology of inquiry.

Do You See What I See?

When we examine, investigate or 'look' into some phenomenon, for example a landscape, we are influenced by how we have come to know our world, that is our perspective. Take the following example from Asplund (1971). What could the following sketch show?

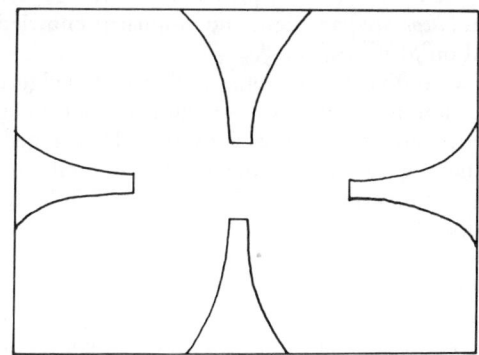

Figure 13.2

Perspective 1: A four leaf clover.
Perspective 2: Four elephants drinking out of a trough.
Perspective 3: A pair of Siamese twins joined back to back.

Each perspective is 'right', that is, valid in that the image seen is a 'true' image for the beholder. How do we explain this? Take an example from Hanson (1958):

Would Sir Lawrence Bragg and an Eskimo baby see the same thing when looking at an X-ray tube? Yes and no. Yes—they are visually aware of the same object. No—the *ways* in which they are visually aware are profoundly different. Seeing is not only the having of a visual experience; it is also the way in which the visual experience is had.

There is a personal dimension to our way of seeing things and to our knowledge. This dimension begins with perception and it shapes and determines the way we see our world. What we see is always interpreted within the framework of knowledge we possess. This personal knowledge is tacit in that 'we can know more than we can tell' (Polanyi 1966). I suspect my student teacher somehow had unleashed this tacit dimension of knowledge not by adopting the stance that all knowledge is relative for every individual student experience (epistemological relativism) but by assuming that there are many equally valid ways of knowing and that the scientific empirical way is only one of them (epistemological pluralism).

Two Perspectives on Knowledge

Let's pursue the notion of forms of knowledge for the moment by contrasting personal knowledge based on an *intepretive* paradigm and empirical knowledge based on *positivism*.

The positivistic view of scientific knowledge has been dominant in Western culture, particularly since the advent of industrialisation. It emerged as the 'new' geography in the 1960s. Its methodolgy can be summarised by the following principles from Phillips (1977):

1. Direct observation of perceived things and processes provides the ultimate link between scientific knowledge of the world and the world itself.
2. The reality of the world of which we have knowledge is independent of the observer.
3. There are observational terms which are themselves theory-dependent units of meaning.
4. The practice of science involving justification, testing and validation of theories and hypotheses is a rational process.
5. Scientific knowledge is cumulative and progressive.

The Chicago materials rejected by my student exemplify a positivist view of knowledge.

In contrast, the perspective that he did follow adopted a *qualitative* form of appraisal of Chicago based upon subjective interpretations of experience. The student used an *interpretive* view which could be labelled humanistic because it allowed for many ways of knowing; it brought to bear the personal knowledge of the students and the teachers, a knowledge embedded in a social context. The source of data was both direct (experiential) and indirect (census data). Cause–effect relationships leading to explanation of the type advocated by the empirical view were but a minor part of his perspective, if at all.

The two perspectives and their relationship to inquiry teaching in geography have been summarised in Figure 13.3.

The Empirical or 'Traditional' Method of Inquiry in Geography

So much (albeit sketchily proposed) for two perspectives on knowledge underlying the teaching of geography; how can we make them work in classroom practice? To answer this question, let us take a look at the syntax, the method of inquiry, based on an empirical or traditional scientific method oriented approach (Figure 13.4a). The activities in which many geographers are engaged may be related to one or more of the phases or steps in this 'scientific' syntax of geography. What steps in figure 13.4a do you think correspond to the activities of geographers listed in Figure 13.4b?

This empirical perspective and its syntax are well known to teachers of geography. It is represented in most texts and curriculum materials. See the descriptions of the lessons taught by Mark and Barbara in Chapter 2 for other examples.

Phenomenology: A Qualitative Method of Inquiry

The interpretive perspective in teaching geography requires a qualitative style of inquiry only recently recognised and described within the discipline as humanistic geography. There is a great differentiation of knowledge and 'world views' within this perspective. But all are related in some way to the ideas of phenomenology and it is to this philosophy we shall now turn.

Figure 13.3 Qualitative and scientific models of inquiry in geography (after Bartlett and Cox 1982)

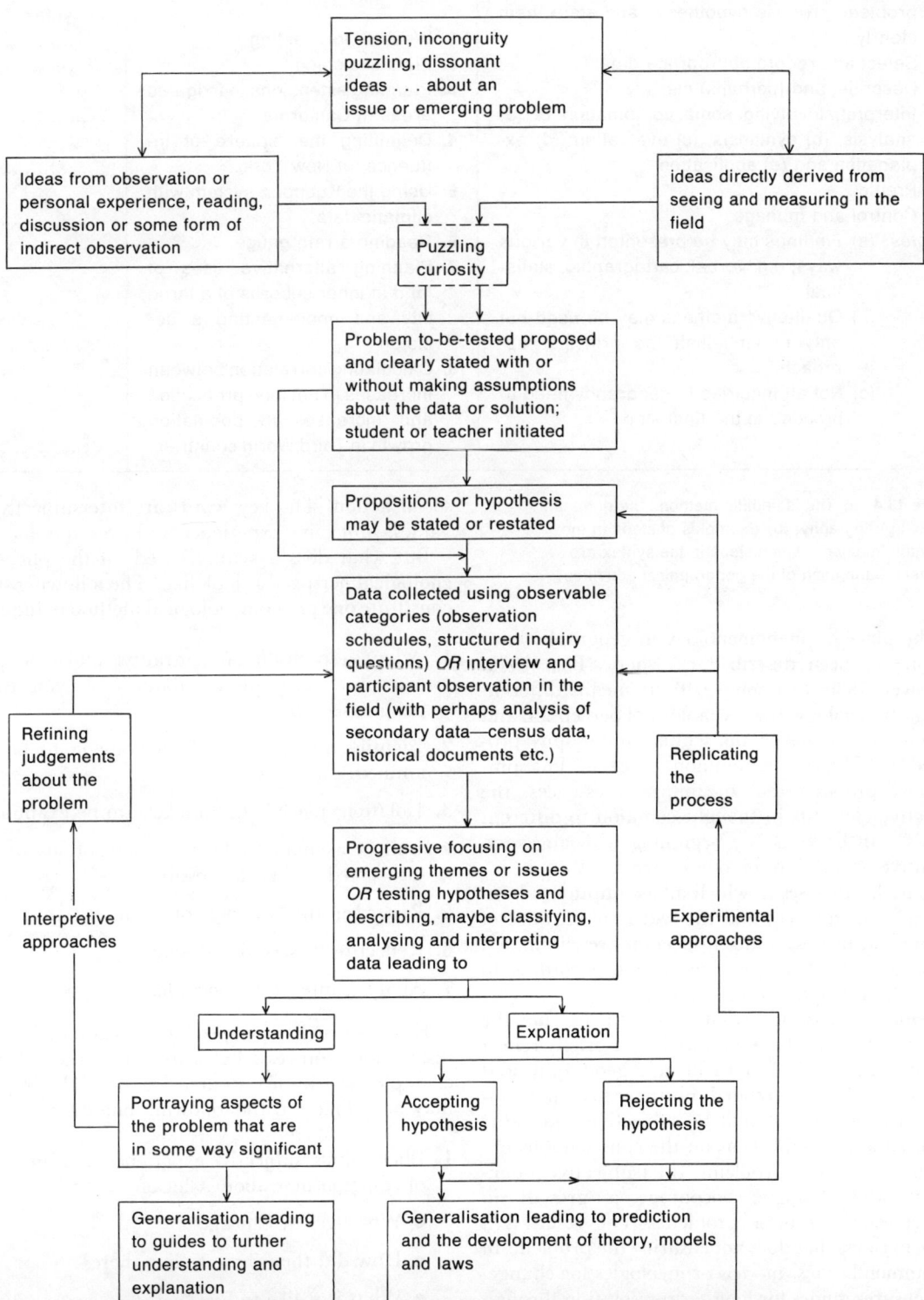

(a)

1. Observe the environment. Puzzle, identify problem, propose hypotheses and state them clearly.
2. Select and record appropriate data.
3. Describe, and (perhaps) classify.
4. Interpret, involving some combination of (a) analysis, (b) synthesis, (c) evaluation, (d) explanation and (e) application.
5. Predict.
6. Control and manage.

Notes: (a) Findings may be presented in various ways, e.g. verbal, cartographic, statistical.
(b) Qualitative methods may be used but only to embellish the principal approach.
(c) Not all inquiries in geography need to proceed to the final step.

(b)

Geographical activity	Step or phase in syntax
1. Weather forecasting.	....
2. Field sketching.	....
3. Planning extensions to irrigated areas in California.	....
4. Delimiting the 'sphere of influence' of New York.	....
5. Using the Koeppen system with climatic data.	....
6. Reading a rain gauge.	....
7. Planning alternative uses of land in inner suburbs of a large city and implementing a decision.	....
8. Calculating correlation between increases in net food production and increases in population growth in Third World countries.	....

Figure 13.4 (a) The 'scientific method': the dominant syntax in geography; (b) Examples of steps in the 'scientific method'. Which steps in the syntax are represented in each of the geographical activities?

The place of phenomenology in geography education has been described elsewhere (Fien 1979; McEwen 1980; Seamon 1979). It means different things in sociology, the psychology of perception and philosophy. However, the seminal philosophy proposed by Husserl is common to each. Phenomenology proposes that meanings or essences, the objective elements in thought common to different minds, can be found by beginning with an exact, attentive inspection of one's mental, that is intellectual, processes in which all assumptions about causes, consequences and the wider significance of the mental process under inspection are eliminated ('bracketed'). Phenomenology is not, according to Husserl, an empirical technique.

Before discussing what this means in the practicum of the classroom, we might briefly reflect on Buttimer's (1974) idea of how geography and phenomenology are related. Buttimer says that content in geography should be subjective and value oriented and should focus on the conscientious engagement of social problems. Intersubjective judgements and dialogue encourage a growth in consciousness of these problems. The result is a consciousness that does not enshrine the problems in recommendations, models or ideologies for change, but ensures guides for long-term resolution through social action. The key words are intersubjectivity, values, subjective experience and everyday life.

But what does a syntax based on the phenomenological perspective look like? The following steps constitute one phenomenological method of inquiry:

1. Adopt a position of neutrality: make no preconceived assumptions about how you think people 'see' things.
2. Identify the segment of phenomena to be examined.
3. Isolate it: place it in 'brackets' to be examined.
4. Collect as many subjective perceptions of the 'bracketed reality' as possible.
5. Record in the language of the informants.
6. Reflect on descriptive data.
7. Analyse, interpret, conclude.

How can such a perspective be translated in the instructional process? Let us return to the Chicago example used by my student teacher. The inquiry may have followed the following phases:

1. Elicit pupils' subjective experiences to raise levels of consciousness about Chicago.
   - Who lives in Chicago?
   - How did they come to live there?
   - What is it like to live there?

- How does Chicago, the city, influence the lives of its citizens?

2. Consider the social, political and economic structures that influence the lives of Chicago people.
   - What does it mean to be educated?
   - What do we mean by attainment?
   - Is socioeconomic status a manufacture of man to prop up a philosophy, for example capitalism?
   - How are Chicago's police boundaries defined? Who defines them? On what criteria and to whose benefit? How do such constructs distort or assist our understandings of crime in Chicago?

3. Define social problems and issues in Chicago. This might best be done by referring to the realities of Chicago citizens.
   - What does 'neighbourhood' mean to Chicago citizens? How has it changed? ('You don't really feel part of Chicago today, 1965').
   - How does the increase in size influence Chicago citizens? ('There's too many of us now').
   - How can problems of planning be overcome? ('Everything is wrong in Chicago, everything').

4. Engage in intersubjective dialogue (teacher and students).

This may involve recording, further reflection on descriptive data and interpretation, leading to understandings based on the objective elements common to intellectual processes of individual pupils. Note there are no conclusive models or recommendations that require explanation, prediction or control as in the empirical method of inquiry. The city was seen through the 'multiple realities' of its citizens; the realities were not 'givens' but were treated as problematic and complex and were used to raise students' awareness or consciousness of the problems. My student used census data (Figure 13.1: but note that census research data is not survey analysis data), not to draw inferences but to establish how and how much social problems and issues are 'framed' by the social constructs we use (socioeconomic status, progress, educational attainment . . .). The location and allocation of resources, for example facilities for health and protection influencing people's lives, were viewed as problematic and critical rather than as a given commodity from or about which inferences could be drawn. The emphasis was not on questions like 'Area X is crime prone because . . . ' but on the questions which stated 'In what sense is area X crime prone and how is it so because of the powerful social and political structures we have set up . . . between individuals and in society generally?'.

A final observation about the Chicago example: it is not suggested or advocated that judgements about societal structures and social problems be consistently guided by or overlaid with a specific ideology about the nature of society such as a conflict ideology which seeks to eliminate societal structures. By raising students' consciousness about change within existing structures, the teacher makes it possible for future change and social action to occur.

Phenomenological and Historical Contexts

There is a sense(s) in which phenomenology and history are related. We shall not discuss this here but alert ourselves to the fact that too often the phenomena we 'approach' in the classroom are treated as timeless. Perhaps this is best expressed in a passage taken from the story of a country doctor, Sassall:

In the human imagination death and the passing of time are indissolubly linked: each moment that passes brings us nearer to our death: and our death, if it can be measured at all, is measured by that apparent eternity of existence which must continue after and without us . . . The anguished are trapped in a moment which is born of all that has happened to them. Faced with the rigid irreversibility of events . . . it is their experience which bends in a circle: unable to catch time by the tail, they chase their own, revolving in one moment blindly throughout all their life. How much can a moment contain? And how can one moment be compared to another person's experience of the same moment? After it seems almost incredible that Sassall putting out a hand to touch a patient finds the patient there, co-existing. (Berger and Mohr 1967)

This mind-blowing passage from the book, *A Fortunate Man*, provides insights into the phenomena of medical practice and might cause a serious cough in the allocation and locational predictive models of some medical geographers. For the teacher of humanistic geography, however, it may provide insights into the meaning of place within a framework of inquiry in medical geography.

More importantly, it alerts us to the historico-social context of phenomenon. As teachers we can fix coordinates of time and space to establish the presence of the phenomenon we experience with our students. The coordinates are stable. It is the subjective experience of time and space (for example, by sharing the personal knowledge of Sassall and subsequently making it public) which becomes distorted. In some way we have to 'correlate' the pupils' subjective experience with our own.

Ideas for Promoting Phenomenological Inquiry

The initiating activity in any phenomenological inquiry is to elicit the subjective experience of students (and our own). We shall begin with some 'zany' ideas adapted from the Environmental Studies Project (1975) and Romey (1980).

1. Send students outside to map something they cannot see.

2. Ask students to look for a change in environment that is predictable and describe how the change could be made unpredictable.

3. Have students search for the three most geographical things they can find. Describe them.

4. Have students write down in as many senses as they can how predators in the community relate to prey. (Alternatively, producers and consumers.)

5. Require students to find a million of something in the geography topic under investigation. Describe it.

6. Have students look for something about a place that is responsible for something else in that place.

7. Name a place familiar to all the students and ask them to write down as many questions as they can about that place.

8. Have students look for two places, one of which is responsible for the other.

9. Take a passage from a 'traditional' text placing a question mark after *every* sentence. Discuss the questions and answers introducing the overriding question 'How and therefore what is it really possible for us to know?'.

Further Ideas

Fien (1983) has suggested many ideas for phenomenological inquiry based upon some concepts of humanistic geography, including place memory, sense of place and place meaning. Some ideas adapted from his work include:

1. *Memory of place with 'props'*. Use slides taken of a local area landscape or of a place that all students would have seen at some time. (For example, a sunset, a river valley . . .) Then ask questions similar to the following:
   (a) Think of a similar occasion when you first became aware of such a place.
   (b) Where was it?
   (c) Describe it in some detail.
   (d) What made the place memorable?

2. *Memory of place without 'props'*. Ask students '*Where* do you remember:
   (a) The happiest experience of your life, and
   (b) The saddest experience of your life?
   Next ask them to describe the features of each place that influenced their experiences in and of it. Form small groups to compare experiences isolating the experiences of individual people or events evoked by the comparison.

3. *Sense of place*. Ask students to name a place they think they 'know' through senses other than by seeing. Ask them to consider which sense was most important in 'knowing' that place.

4. *Place meaning*. Understanding places is a very personal response to environments that have meaning for people. Understanding may better be achieved through empathy with places. Some tasks for students include:
   (a) Cut out scenes of places (from old journals, magazines, etc.) that have special meaning. What is idiosyncratic or unique about one or more of these scenes?
   (b) Write down the name of the neighbourhood in which you live. What is important about this place. Truth-test the meaning of this place to you by collecting data from several sources: historical documents, your observations, interview data.

Place meaning may well be illustrated by looking at the real names given to places. In the United States the meaning of place is often encapsulated in the name of the place: Arsenic Springs, Coffin Canyon, Thermopolis (the biggest thermal springs in the US), Four Corners (the only place where four states join), Half Dome and so on. In Australia there are many examples of how the name of a place expressed its meaning in the lives of early settlers: Mulga Valley, Camels Hump, Deadman Hill, Stonehenge and a host of Aboriginal place names. Yi-fu Tuan's *Landscapes of Fear* (1980) admirably demonstrates how perceptions and feelings for landscapes form personal geographies of the mind. This act may evoke empathic understanding for the environment. It is also important to note that Tuan draws upon several sources of landscape experiences including diaries, poetry, documents, etc. Finally, while the above examples demonstrate how people have viewed the world in the past, present perceptions colour our world view. This may be illustrated in the following vignette.

I walked into my class of student teachers recently and said 'Think of bridges—here is a photo of the Victoria Bridge spanning our river city. What do you see?' Many responses were given: cars, noise, freeways, traffic jams, inner city pollution . . . all tending to be the product of logical or familiar thought. One of the class suddenly jumped up and said, 'Drunks'. I immediately asked the class to write about some of the experiences that led to their responses. The response of the student who said, 'drunks', was a poem.

Brisbane Bridges

I remember at school they talked about London Bridge,
People lived on it so they said,
   Built houses on each edge.
   Marvellous stuff history.
   Any more grog? You godda bottle 'arry!
It stops the stuff that makes you care.
It wins. You have to run and hide.
You are on the edge.
Your thoughts turn to the bridge.
Bugga you Mate! You godda cheque Wednesday!
It's a poor man's sanctuary from Winter's lack of thought,
Not too wet under each end.
Cardboard cartons make your castle,
And its from this that you fight.
those that want your fief—
Fight for the privilege of living underneath.
Stuff you 'arry—you mean sod of a Bastard!
One day you'll walk again on top,
You'll slowly climb the ladder,
   Hit the caisson in your fall.
Buoyant now you move with the tides,
Another world's toadfish gnaw greedily at your sides.
Have you seen 'arry? Is he in the Nick?
   It's a Bridge—that's all.

Peter Oliver

In the discussion which followed the reading of this poem, the social problems that were revealed were made more explicit by other students expressing their perspectives. The social structures of societies within the state were subjected to scrutiny and critical judgement. The common elements in each student's thinking were analysed. Each had a perspective that could not be carbon copied. Certainly, the student whose experience was unfamiliar and who, as it transpired, used to work for a team serving the needs of the community of less fortunate, represented an experience that was partly able to be shared, mostly personal, and influential in the intersubjective dialogue of the group.

Mental Maps and Multiple Realities

Much activity with mental maps in geography classrooms tends to incline to estimates of discrepancy between the mental picture and the 'real' world picture. (Which is the more objective?) The information or explanation of discrepancy becomes a conclusion in itself. This type of activity reflects the influence of the positivist perspective which dominates behavioural geography. In a phenomenological approach the mental map is a representation in graphic form of the individual's reality and this provides the basis for intersubjective dialogue. Here are some activities adapted from Romey and Elberty (1980) and Stoltman (1980). All are the basis for discussion. Have students:

1. Draw a mental map of Australia. Lead a discussion as to how the students' cultural backgrounds influence the map?

2. Draw a conceptual map based on the question 'Where are you coming from?'.

3. Draw a 'smell' map of the school from memory.

4. Write a set of detailed instructions on how to get from A to B in the local area. Draw a map based on the instructions and write a report based on each one's lived experiences with the local area landscape (cf. Yi-fu Tuan's *Landscapes of Fear*).

5. Cast themselves in the role of visitor from outer space. Have them assume they have never seen the objects they now see. Then sketch to describe them and discuss how they came to see them the way they do.

Establishing the Conditions for Phenomenological Inquiry

Ultimately, decisions about what to include in a particular lesson, a unit of work or program are very practical commonsense decisions. To a large extent the students and teacher will determine what subject-matter content is to be included. The teacher has the more important role of establishing the conditions for inquiry. Firstly, it is important to accept that there is nothing mysterious about the 'multiple reality' concept in phenomenology. If the teacher gives his or her students the power to participate actively, they will reflect and evaluate their subjective judgements and experiences. Nevertheless, there still remain some requirements for the establishment of classroom conditions which might serve to facilitate the humanistic perspective in the geography classroom. Figure 13.5 contains a list of appropriate teacher behaviours in this task.

The Qualitative Approach in Teaching Geography

Does a qualitative approach in the classroom require new roles for teachers? The answer as to the kind of roles is not immediately answerable although we could work on the assumption that the teacher who uses a qualitative approach would need to facilitate relationships in the classroom. Two characteristics of teacher–student relationships appear necessary:

1. The learning task activity is able to be shared.

Figure 13.5 Establishing classroom conditions for successfully teaching humanistic geography

1. **Try to respect the complexity of the reality**	Good lessons rarely result from establishing a few objectives and selecting the appropriate learning activities. Establishing a lesson or a phenomenological inquiry necessitates complex political processes because this style of inquiry requires interaction between people and their perspectives: hence it is political reactive.
2. **Insist on multiple outcomes**	No event or phenomenon should be judged on the basis of one or two measures. Resist attempts by anyone to reduce the output of the lesson to simplicity or 'standardised' statements.
3. **Be aware of classroom transactions**	Regardless of outcomes, the atmosphere of the classroom—where students spend most of their waking hours—is important. Monitor and perhaps measure this 'climate' (there are many instruments for doing this).
4. **Collect many kinds of data**	In the light of the above the more kinds of data, the better. Testimonials, interviews, observations, documentary analysis, art, graphic displays, census data, . . . almost all are appropriate. They provide a richness about the event studied and mitigate against making decisions that are based on too few data, invalid data and abstract information.
5. **Collect data from different sources**	It is critical to find out what students are thinking about the event and about how the inquiry into the event is proceeding.
6. **Rely on intuition and professional judgement**	You may use analytic tools such as Q-sorts, likert scales, event categories to assist intuition. But there is no substitute for human judgement after these techniques have been used to elicit subjective experience.
7. **Promote diversity within the lesson**	The most difficult task for you will be to generate alternative ways of doing things. The methods selected in the lesson should promote the development of divergent ideas, not inhibit them.

2. The teacher will be *seen* to help the student to learn.

Experience and skill in group dynamics may facilitate the personalisation of the learning process. Implicit in this is the need for the teacher to admit error (in student perception and judgement when it occurs) and to reduce competition of all kind in the classroom.

What are some of the specific behaviours that establish teacher–student relationships and a personal classroom climate? Several (from Saltmarsh and Hubele 1975) are listed here:

1. Listen to students, wait rather than talk.
2. Look at students when they talk.
3. Pause after questions which invite answers and comments.
4. Practise increasing your toleration of periods of silence.
5. Sit down next to students; never sit directly in front of them.
6. Reduce the frequency of 'correcting behaviours'.
7. Use structured activities, e.g. group work, that facilitate interpersonal relationships.
8. Minimise the threat of assessment and grades.
9. Relocate furniture freely and frequently.
10. Change the type and composition (pairs, triads . . .) of groups frequently.
11. Permit students to design and execute their own learning.
12. Observe the maximal span of attention in expository segments of learning. This rarely exceeds ten minutes for most students.
13. Focus on antecedents and consequences of inappropriate student behaviour (rather than reinforcing them).
14. Relinquish territorial space at the front of the room.
15. Make it possible for all to seek the 'truth'.
16. Give productive feedback.
17. Make direct investments in the 'here and now'.
18. Maintain a personalised consistency with students.

If the students' experience of the world is the central perspective in the geography curriculum, then the teacher's instructional behaviour and the teaching–learning activities used ought to be consonant with the philosophical assumptions underlying the curriculum design. Some formal didactic activities such as narration or exposition may be appropriate on occasions, but interactive activities (group work, triads . . .) which promote humanistic student–teacher relationships ought to be dominant.

Conclusion

There are many styles of qualitative inquiry that could form part or the whole of curriculum design in geography: connoisseurship, ethnography, history, hermeneutics, fiction, literary criticism, portrayal. Their relevance for instruction in humanistic geography has yet to be exploited. Although there are varying epistemological bases for each, phenomenology is related to all in some way. Hence, this discussion has focused on the latter perspective in teaching geography by including, in part, the phenomenal reality of the author's own experiences. In doing this, the aim has been to demonstrate that the adoption of this form of inquiry requires more than the eliciting from individual students statements about their perceptions. Their subjective experiences once 'bracketed' in this way provide the data leading to understanding, possibly through some form of empathy. The method of inquiry outlined in this chapter sought to provide a vehicle whereby the teacher can raise students' level of consciousness about themselves and their world. They may never be able to answer the question, 'Look into my mind; do you see what I see?' with a 'Yes'; but they can reply with certainty, 'I see'.

References

Asplund, J. (1971) *Om Ündran Infor Samhallet*, Stockholm: Argos.

Berger, J. and Mohr, J. (1967) *A Fortunate Man*, London: Allen Lane.

Buttimer, A. (1974) *Values in Geography*, Association of American Geographers Commission on College Geography, Resource Paper No. 24.

Bartlett, V.L. and Cox, G.B. (1982) *Learning to Teach Geography*, Brisbane: Jacaranda Wiley.

Curriculum Services Unit (1981) 'What are the Characteristics of Black Areas of Chicago?' Why do such Areas Exist?', Melbourne: Curriculum Services Unit, Education Department, Victoria.

Environmental Studies Project (1975) *ES-SENSE I* and *ES-SENSE II*, Reading, Massachusetts: Wesley Publishing Co.

Fien, J. (1979) 'Towards a Humanistic Perspective in Geographical Education', *Geographical Education*, Vol. 3(3), 407–31.

Fien, J. (1983) 'Humanistic Geography', in J. Huckle (ed.) *Geographical Education: Reflection and Action*, Oxford: Oxford University Press.

Hanson, N.R. (1958) *Patterns of Discovery*, Cambridge: Cambridge University Press.

House, E.R. (1972) 'The Dominion of Economic Accountability', *The Educational Forum*, November, 13–24.

McEwen, N. (1980) 'Phenomenology and the Curriculum: the Case of Secondary-School Geography', *Journal of Curriculum Studies*, Vol. 12(4), 323–30.

Phillips, D.L. (1977) *Wittgenstein and Scientific Knowledge*, London: Macmillan.

Polanyi, M. (1966) *The Tacit Dimension*, New York: Doubleday.

Romey, W.D. (1980) *Teaching the Gifted and Talented in the Science Classroom*, Washington DC: National Education Association.

Romey, W. and Elberty, W. (1980) 'A "Person-Centered" Approach to Geography', *Journal of Geography in Higher Education*, Vol. 4(1), 61–71.

Saltmarsh, R. and Hubele, C. (1975) 'Facilitating Humanistic Relationships in the Classroom', *Journal of Teacher Education*, Vol. 26(3), 229–32.

Seamon, D. (1979) 'Phenomenology, Geography and Geographical Education', *Journal of Geography in Higher Education*, Vol. 3(2), 40–50.

Stoltman, J. (1980) *Mental Maps: Resources for Teaching and Learning*, Teaching Geography Occasional Paper No. 32, Sheffield: Geographical Association.

Terkel, S. (1967) *Division Street America*, New York: Pantheon Books.

Tuan Yi-fu (1979) *Landscapes of Fear*, New York: Pantheon Books.

14

Developing Environmental Awareness and Appreciation

Robin Hall

The focus of this chapter is environmental education, one of the goals of values teaching in geography outlined in Chapter 11. Robin Hall provides classroom and fieldwork ideas for helping students observe, respond to, record, interpret, evaluate and appreciate the environment. He presents these within a three-way matrix for developing sensory awareness and aesthetic appreciation of the environment. This matrix should be used to select objectives and strategies for environmental education through geography. Some of the activities explained include: the studying of architectural styles, sensory walks, the use of 'motivation', 'townscape notation' and 'streetometer' techniques as well as annotated line sketches for recording personal responses to the environment, serial vision, town trails, regional literature, landscape paintings and landscape evaluation. Many geography teachers will be familiar with some of these approaches through their leisure pursuits, if not through their teaching. All, however, will discover new dimensions to the purposes and approaches to teaching geography in this chapter.

We are constituted as a Society for the purpose of diffusing geographical knowledge, and I trust that in future we shall regard knowledge of the beauty of the Earth as the most important form of geographical knowledge that we can diffuse.

(Younghusband, Presidential Address to the Royal Geographical Society, 1920, p. 8)

Theoretical Backdrop

It's first necessary to look at some of the terminology used in this chapter. Awareness is a vague term which is used to describe a wide range of mental activities. Basically, to be aware is to be conscious of someone or something, but consciousness can involve *inter alia* observing, identifying, sensing, responding, interpreting, discriminating, appraising and imaging. Aesthetic awareness and appreciation is taken to mean the experience of contemplating any object deeply (Hall 1980a). It is to talk of experiencing and feeling, as well as thinking, as forms of knowing.

Feeling is itself an ambiguous term and may refer to sensations, emotions or intuitions. It can refer to incidents such as feeling itchy, feeling blue or a feeling of impending doom. If we are talking about *educating* feelings then it is necessary to restrict our interest to feelings of an emotive nature. It is not possible to educate an itch or an instinct. We *can* consider educating the emotions because emotions consist of rational as well as physiological features (Peters 1977). To feel joyful or morose is to feel that way as a result of seeing some event or object in a particular way. Emotions involve judgements or appraisals, and appraising involves having reasons or grounds for those appraisals. We might feel angry, for example, to find our favourite beach polluted by an oil slick and may feel that way because we have been denied a source of great pleasure. There is some rational element in emotive responses and it is this rational element which makes it possible to talk about educating the emotions, in this case the aesthetic emotions.

In geography the objects of awareness are the phenomena in the geographical environment and generally speaking these can be defined as landscapes and places. Geographical environments may be regarded as either 'natural' or 'cultural', depending upon the degree of human interaction required to maintain them. A natural environment could be defined as one which is perpetuated without human support or interference. Alternatively, a cultural-geographical environment can be defined as one which does require human intervention, so that a commercial pastoral landscape would be regarded as cultural. Our interactions with these two forms of environment might also be of two broad types. We may have direct, primary or 'field' contact with them, or our knowledge of them might be via indirect, secondary sources such as literature, photography, painting and the rest. There is I think quite a difference between first-hand and filtered contact, but both types may be valuable sources of information for teachers and students.

The kinds of classroom activities involved in teaching environmental appreciation can be divided into three general categories of *identifying*, *interpreting* and *expressing* (Figure 14.1). The division is a crude one but it can provide some guidance for organising teaching practice. *Interpreting* in itself is a very broad class of activities and embraces at least three other processes of (i) responding, (ii) ordering those responses by analysis, synthesis, reasoning, discriminating, and (iii) recording one's observations. In turn these three processes may be variously combined for critically appreciating the environment, for evaluating it or for using it. *Expressing* is also a somewhat ambiguous term but at the simplest level it can be divided into communicating for the benefit of others and expressing for one's own sake. All of the activities in Figure 14.1 overlap to some extent and the conceptual connections amongst them are quite complex, but I find that this schema does help in the organisation of an inquiry-based curriculum.

To recap so far, environmental appreciation can be conceptualised along three major dimensions. We can divide the geographical environment into natural or cultural types; we can distinguish between the sources of our interaction with the environment as direct field experiences or as indirect filtered contact; and we can distinguish between different types of classroom approaches. Figure 14.2 is a summary of the interdependence of these three separate dimensions. For example, cell A in this three-dimensional matrix might represent pupils *reading* a piece of *literature* on the *natural environment*, e.g. W.H. Auden's poem, 'Bucolics' (Auden 1958). Cell B might represent pupils conducting *streetwork* and *expressing* the results in the form of a *photocollage*.

Several other points ought to be made before moving on to discuss some specific classroom exercises. Firstly, the experimental nature of this sort of curriculum requires the adoption of a learner-centred approach. The activities will be unproductive without active pupil participation. This is less likely to be a problem when the work is fun based. Broadly speaking, the purpose is to encourage in students a feeling response to the environment and then rework or reflect on that reaction in some meaningful way.

Secondly, it is assumed that the ultimate aim of developing aesthetic awareness is the development of the skills of critical appreciation of the environ-

Figure 14.1 Activities for developing environmental awareness and appreciation

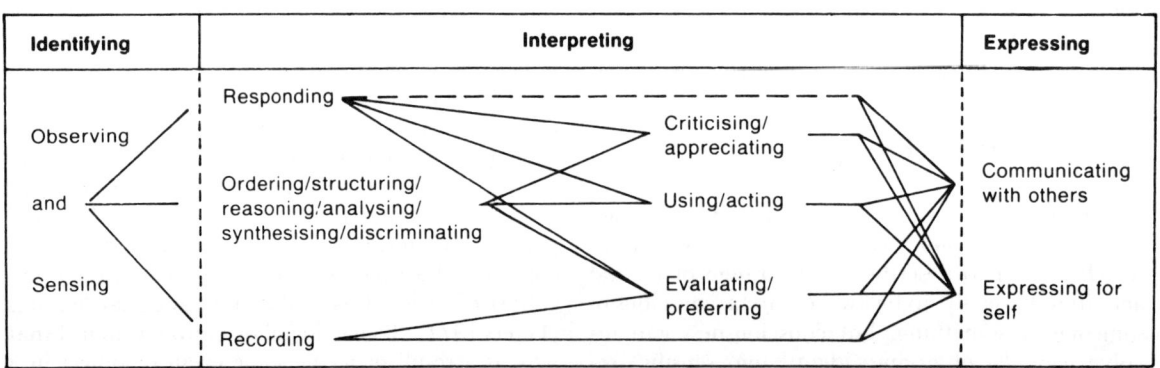

Figure 14.2 Three faces of teaching landscape aesthetics

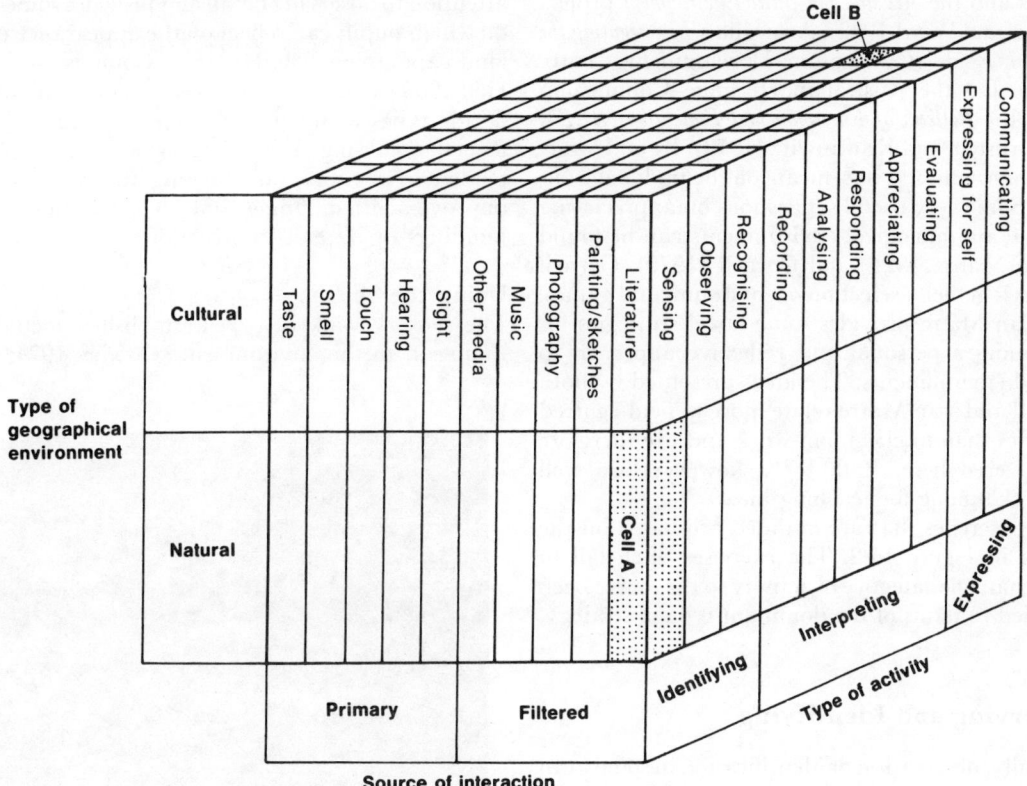

ment, and that this appreciation is primarily a function of one's perceptual abilities. Moreover, I am assuming that perception is best represented by a transactional or interactive approach between the perceiver and the object he or she perceives. That is, the perception of beauty and blight in the environment is not considered as lying wholly in the eye of the beholder, nor does it reside wholly in the object perceived. Aesthetic experience is considered as an interaction between the perceiver and the object perceived. Enriching one's appreciation then is largely a matter of improving one's perceptions.

Thirdly, there is a distinction between interaction of an *open* and a *closed* sort (Hall 1980b). Closed interaction means that the instructions a student receives are closely specified and the activity is highly directed. An open interaction is one permitting greater freedom of interpretation. There is some tension between the two but I think both approaches are useful.

Fourthly, there should be a willingness to study both the excellent and the prosaic in the environment. For example, we can learn both from grand and from vernacular architecture, and there is no necessity to restrict attention wholly to the excellent.

Any building or environment can be the object of critical appreciation, from the public toilets to the local Italianate courthouse.

Fifthly, the value judgements under discussion are aesthetic value judgements as distinct form moral judgements. The two are separate but often interconnected. For example, aesthetic judgement may be coloured by ecological righteousness, such as being appalled by the rainbow colours in an oil spill; and our moral position on an ecological issue may be influenced by aesthetic considerations, such as the recreational arguments for wilderness preservation.

Lastly, we should keep in mind that this is an area of study which until recently has fallen between the subject areas in the curriculum mesh. It is an area which benefits substantially from cooperation among history, social studies, geography, visual art and language teachers.

Classroom Strategies

Only a few of the exercises described below are original; nor is the list exhaustive. The principal sources of inspiration for these activities are Eileen Adams and the *Art and the Built Environment* project (Adams and Ward 1982), Australia's *Investigating the National Estate* (Curriculum Development Centre (CDC) and the Australian Heritage Commission 1980), the *Bulletin of Environmental Education (BEE)*, and Farbstein and Kantrowitz (1978). By and large, all of these authors concentrate on the built environment. Some very useful suggestions on appreciating physical and biological environments can be found in Van Matre (1974) and Cornell (1979). Cornell collects together a selection of environmental games, and Van Matre provides some novel strategies for developing a personal and reflective approach to natural environments. The ideas presented by both Cornell and Van Matre relate more to field-centred activities than to classroom work and are therefore not covered here. Both texts, however, are well worth skimming for teaching ideas.

The exercises that are outlined are based on the matrix in Figure 14.2. The exercises often fall in more than one category of activity so they have been classified in terms of the dominant type of strategy.

Observing and Identifying

Logically, observation or identification in some form is a minimum condition for higher-order activities such as interpreting and appreciating. Often, however, we take much of our environment for granted so that these are skills which are underdeveloped. Simple observation exercises can be quite literally eye-openers.

Observing Detail

Organise fieldwork on a thematic basis so that each trip focuses on one or a small number of specific landscape features, e.g. windows, chimneys, front doors, fences/boundary walls, street furniture, street vegetation. Alternatively choose one type of landscape feature (e.g. British pub, American gas station, Australian showground) and assemble a folio about it from fieldwork, magazines, etc. Or provide students with a silhouette of a familiar building and have them draw in the shapes and positions of windows and doors. Field check their design with the original.

Architectural Styles

Prepare and teach a unit of work with the intention of pupils being able to recognise major types of styles in the history of local domestic architecture. The advantages of this approach are that it directs attention to observing detail and provides some basis on which pupils can reflect on the appearance of the landscape (see CDC/Heritage Commission 1980; Hall 1980a). Some approaches include matching up façade types with verandah types, or chimneys, or fences; analysing a building which has been extended over time and labelling the date of each improvement; mapping and colour coding school buildings by date of construction.

Using the Senses

The sensory walk is a well-established method of tuning in to the environment (Goodey 1974; Hall

Figure 14.3 (a) The Breadknife, Warrumbungle National Park, NSW; (b) Shapes and lines

Developing Environmental Awareness and Appreciation 157

Figure 14.4 (a) The courthouse, Bathurst, NSW; (b) Shapes from the courthouse

1980b). It encourages students to concentrate on each of the senses separately (taste may be hazardous) and to break the habit of routine observation of the environment. Sensory walking works better where attention is partly directed, e.g. listening for hi-fi sounds, discriminating between natural and synthetic smells, relating tactile properties to visual form. The sensory information can be represented visually by drawing 'sound maps', or 'smell maps' (NSW Dept of Education 1986).

Studying Visual Form

The Australian *Investigating the National Estate* kit provides some interesting exercises in the visual analysis of place in terms of shapes, sizes, texture, line, direction, colour, light, space and mass. Examples of how this might be done include inviting students to:

1. Identify, draw and describe the dominant *shapes* in Figures 14.3a, 14.4a, 14.5a. (Figures 14.3b, 14.4b and 14.5b are sample answers to those activities.)

Figure 14.5 (a) Two war memorials, Bathurst, NSW;
(b) Spaces and mass in the war memorials

(a)   (b)

(a)   (b)

Figure 14.6 (a) Cottage: Pisé walls, Bathurst, NSW;
(b) Textures on and around the cottage

Figure 14.7 *Still Glides the Stream and Shall Forever Glide*, Sir Arthur Streeton, 1890

2. Redraw Figures 14.4a and 14.5a using shapes of different *sizes* and proportions.

3. Identify, describe, draw or make a photocollage of the *textures* in Figure 14.6a. What materials would you use to renovate the veranda and roof? Have students make rubbings of building materials used locally, or have them make up a feely-box and generate a word list to describe different types of textures.

4. Sketch the dominant *lines* discernible in Figures 14.3a and 14.7. Are the *directions* of the lines vertical, horizontal, oblique, curvilinear straight? Are they soft, heavy, thin, thick, sharp, blurred, dense, light, straight or jagged?

5. (a) Copy and paint a monochrome reproduction of a painting (e.g. Figure 14.8 by Australian Aborigine, E. Namatjira). Afterwards compare your effort with the *colour* composition of the original.
   (b) Discuss with pupils the reasons why the technique of Impressionism was so important to Streeton (Figure 14.7) and other members of the Heidelberg School in dealing with the intensity of *light* in subtropical environments (Smith 1947).
   (c) Have students take photographs of the built environment in different climatic conditions, e.g. fog, and note the differences in tone and sharpness.

6. The two war memorials in Figure 14.5a are quite different in terms of *space* and *mass*. Is the more massive more impressive? Figure 14.5b gives some indication of the operation of space in the two monuments. Perform the same exercise for the courthouse in Figure 14.4a.

Use of Contrasts

It is sometimes said that 'we don't know what we've got 'till it's gone'.

Figure 14.8 *Contrasting Colours After Rain*, E. Namatjira

Figure 14.9 Contrasts in time, William Street, Bathurst, 1905 and 1981

1. Contrasting what *is* with what *was* (or what *might be* as in sci-fi) can be a useful focusing device. Contrast, for example, the photographs of Bathurst in Figure 14.9 and discuss the aesthetic consequences of replacing verandahs and posts with cantilevered awnings.

2. Contrast the public and private faces of buildings, that is frontness and backness (see Tuan 1973).

3. Contrast the shapes we find in natural landscapes with those found in cultural landscapes (e.g. contrast Figures 14.3a and 14.4b) or search for bizarre juxtapositions of the natural and cultural, e.g. bird's nest in car wrecking yard. It is often the surprising and unexpected that holds our attention the longest.

4. Compare the appearance of specific botanical features from season to season, of streetscapes from day to night; or changes in a particular streetscape over shortish periods of time (e.g. changes in billboards, public notices, street vegetation); or contrast the visual messages found in shopping precincts/plazas and analyse them as examples of popular culture.

Observing Spaces

By exposition and field examples, teach about the different kinds of space described by Cullen (1975). Discuss notions of spaces in the built environment as being enclosed, divided, pierced or defined with reference to places like culs-de-sac, crescents, alleyways, back lanes, car parks and plazas.

Responding

Obviously the reactions and responses that people have to places is a major part of the feelings that places engender.

Egocentric Responses

There are two types of representations that might be appropriate here, i.e. the visual and the verbal. Visually, students could draw mental maps of their home or school and colour code their feelings about each part of the building or campus, e.g. strong colours for negative, weak colours for positive. In the field, a technique suggested by Adams is 'The Good, the Bad and the Ugly' (G/B/U) in which students draw thumb-nail sketches of selected places and describe them as G/B/U with reasons why. Good is defined as pleasant, comfortable; bad as unpleasant, unfriendly; and ugly as not good at the moment but with potential for improvement (Adams 1980).

Verbally we can ask students to make lists, or otherwise represent, places that stimulate particular responses and explain why they respond in that way, e.g. places that produce feelings of discomfort, nostalgia, claustrophobia, agoraphobia, anger, joy, fear, security, loneliness, awe, insignificance, or places where pupils like to hide, socialise, observe others, feel important. Alternatively, provide examples of places and have pupils react to them and explain their reactions (see Appleton 1975 and Hart 1978).

Empathetic Responses

1. Have students visit one or two distinctive places and imagine what those places might say if they could talk.

2. Have pupils experience a place through the experiences of a younger child, an adult or an elderly person.

3. Have pupils imagine they are a member of a culture with a lavish mythology about the creation and creator of the world. Ask them to relate how a particularly dominant feature in the landscape (e.g. Figure 14.3a) came into existence.

Recording

Most times we want students to record their observations and reactions. Recording may take many forms and only a few are mentioned here.

Diaries

Ask students to select one or a small number of places and to record critically, in a diary or journal, the changes which occur to the place over time. Try to establish whether the changes are cyclic, predictable, non-repetitive or whether they occur in some other sequence.

Scores

Have students experiment with different ways of recording their sensory-walk observations by making 'scores', as in music or dance. One of the principal proponents of sensory walking is designer Lawrence Halprin, who has also suggested a scheme for recording the experience of travelling through townscapes. This movement notation or 'motation' involves noting elements in the landscape in a symbolic form (e.g. Figure 14.10). It is accepted that music education benefits from children experimenting with their own notational schemes to represent musical sounds. The same strategy can be adapted to sensory walking by devising symbolic schemes to record visual elements in the landscape, or features in the soundscape/touchscape/smellscape. Examples of this type of symbolic representation can be found in Adams and Ward (1982).

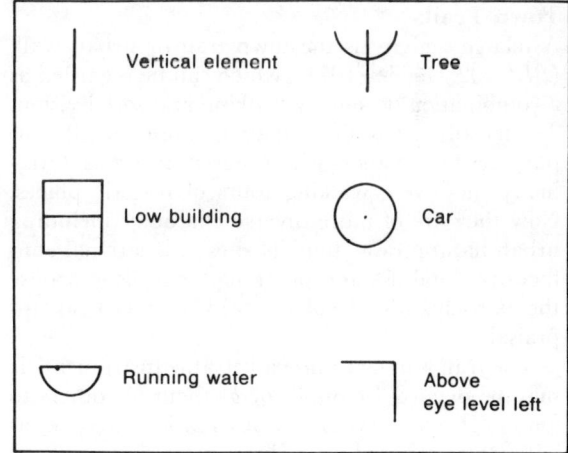

Figure 14.10 Examples of Halprin's landscape notation symbols

Sketching

Annotated sketching of individual features or of broader landscapes is a fundamental study technique. It is also useful to make labelled drawings of buildings, either from observation or photos, and to use the correct architectural language for roof types, window forms, roof-lines, etc.

Interpreting

To aim at educating aesthetic awareness it is necessary to expend considerable effort on developing the abilities to interpret, analyse, synthesise, order, structure and reason about aesthetic responses.

Serial Vision

One of the most interesting concepts defined and illustrated by Cullen (1975) is that of serial vision, involving the exploration of urban spaces as a continuous movement. The concept is a subtle one which is more easily taught in the older towns and villages of Europe than in the raw and regular townscapes of the colonial era. Interesting introductions to this sort of visual thinking are the 'steeplechasing' and 'homing-in' exercises (Wheeler 1976b). 'Homing in' involves viewing a landmark from a distance and then moving towards it, observing and recording changes in its appearance and relationships to surroundings. 'Steeplechasing' involves observing a landmark from a number of different points at different distances and directions from it. In both cases, sketches and/or photographs may be used and a rationale should be provided for the selection of perspectives.

Town Trails

A related activity is the town trail or urban walk (*BEE* 102; Goodey 1974), which can be regarded as a combination of sensory walking and serial vision. Town trails may be devised for a number of different purposes by designers with different interests. Originally they were walking tours of historic places. Now they are of more diverse character including urban nature trails, tours of sites of locational conflict over land use and planning, or trails to arouse the participant's visual curiosity and critical appraisal.

The trail is in fact quite a versatile instrument. It may be devised *for* pupils or *by* them for others to follow. It may serve as an observation exercise, as with memory trails in which a pair of students alternates with another pair to make observations and quiz each other about what they have observed and memorised along the route. Or the trail might be a vehicle for developing visual thinking (about chimney shapes, floorscapes, textures). Or again, it could be the means for analysing the invisible socio-political processes which bring townscapes into being. With all trails, it is advisable to think of them as being composed of focus points, links and outlooks. The route should be simple and clear, and the text or guide should do more than duplicate the participant's experience. It is more interesting to imply that the environment contains more than the participant is actually experiencing, perhaps by providing illustrations from a different physical perspective and information of a non-visual nature.

Local Issues

Some local information on these invisible socio-political processes can be gained from looking in the windows of real estate agents and cross-tabulating housing size, price and age/style. We can also regard landscapes as being, in part, a product of tension and conflict over land use. On many occasions the conflict is based on aesthetic considerations, so that it may be instructive to study local issues as a way of developing aesthetic awareness. The issues need not be real ones. They could be contrived and hypothetical, and involve games and sociodrama.

Biographical Insights

It can be useful to make a biographical study of people who have been deeply involved in creating and preserving the environment, or in refreshing our perceptions of it. Architects, painters and photographers all provide grist to the biographer's mill.

The goal is to come to understand the depth of concern and the passion which fires these devotees to care so intensely about their work. They offer us some of the best examples from which to learn. One of my favourites is Olegas Truchanas, photographer and conservationist of the Tasmanian wilderness, who was tragically drowned in the Gordon River in 1972 whilst attempting to reshoot the photographs he had lost in a domestic fire a few years before. 'He [had] perished in the river he sought to save' (Angus 1975).

Regional Consciousness

Geography has long been concerned with regional description, with the personality of regions and with regional consciousness. The progress made in this area has not been great but it seems to me that a more comprehensive way of developing a feel for a place is to study its culture in a broad rather than narrow, material sense. Regional cultural geography can embrace studies of the music (especially folk), painting and literature of and about a region (see Evans 1973; Zelinsky 1973).

Making and Using

There is an interesting tension between the aesthetics of a place and its utility, i.e. between form and function. Exercises can be designed around the ways in which pupils arrange their room or living space, how those arrangements might be changed and how they are constrained by functional considerations. Outdoors, there's much to be learned from designing, making and keeping a garden and linking that with a study of landscape garden design. Historically, it is in this context that concepts such as 'picturesque', 'scenery' and 'landscape' itself have their home. It's possible to construct an elaborate unit of work around the nineteenth-century designs of grand private gardens and to compare them with twentieth-century suburban gardens, oriental public gardens and contemporary public parks and gardens.

An exercise called 'personalisation' also has some possibilities (Punter n.d.). In this activity, a domestic streetscape is selected and each student adopts a house within it. She or he renovates the façade and garden and the class builds up a three-dimensional model of the revitalised street. In quite a different vein small groups of students could make a 'TV documentary' of local historic sites, using a cardboard box, rollers, a strip of paper for illustrations and a sound tape.

Evaluating

There is a fine, but important, distinction between evaluating and appreciating. Evaluating and assessing landscapes can have pedagogic value but should not be conflated with the deeper contemplation implied by appreciating.

Making Criteria Explicit

There is a need to be clear about the criteria on which evaluations are made. We might, for example, ask students to evaluate the cottage in Figure 14.6a and ask whether they would demolish, restore, leave or sell it if they were the owner. Ask them to make the reasons for their judgement explicit. The grounds for judgement might be historic, scientific,

Figure 14.11 Home improvements: personalised or trivialised?

functional, economic or aesthetic, and it is worth making the distinctions obvious.

Rural Landscape Evaluation

Evaluating landscapes has become quite an industry. The many methodologies for quantifying landscape quality take into account local relief, the character and nature of slope profiles, the degree of visual coherence and harmony, edges or zones of contrast between landscape features, visible water surfaces, textural and chromatic character, variety and sequence of views and the presence of highlights or singularly attractive features (see Penning-Rowsell 1981). Methodologies based on these parameters can be used in the same way as the streetometer outlined below.

Urban Landscape Evaluation

A distinction can be drawn between single-building and townscape evaluation.

1. Single building. Student exercises could take the form of:
   (a) A numerical assessment of the school in terms of site qualities, building layout and quality, vegetation and recreation facilities of the school grounds.
   (b) A study of a home that has been renovated or 'personalised', e.g. Figure 14.11. Make a list of the different renovations that the students can observe in the local neighbourhood and discuss whether they are appropriate or not.
2. Townscape evaluation. Exercises such as:
   (a) A numerical assessment of an individual building in the context of the surrounding townscape. Bishop (Adams and Ward 1982) has proposed the CRIG system based on studying:
      - *Context* (pattern of surrounding area, scale of development, form and site);
      - *Interfaces* (connecting interiors and exteriors of buildings);
      - *Routes* (and how they influence different types of vehicular and pedestrian traffic);
      - *Grouping* (relationships between subdivisions of a building).
   (b) Streetometer, designed by Midwinter, is a very valuable exercise for encouraging students to think about environmental quality (Adams and Ward 1982). The observer records environmental quality by rating the landscape on ten different criteria on a scale of 0–10. The parameters could include litter, floorscape, adverts, parked cars, wirescape, vegetation, building condition, road safety, air pollution, but the exercise works better if students generate the criteria themselves. The sums are totalled to produce a rating for the streetscape out of 100, and this value can then be compared with ratings for other streetscapes. The results are often surprising. With one class in northern England, a crescent of bungalows for elderly folk rates 'tops' because 'the houses are nice', 'it's safe', 'peaceful', 'modern', 'colourful', 'lots of vegetation', whilst a circular housing estate built around 1919 rates 'the pits' because 'it's a dump', 'colourless', there's 'pollution from chimneys', 'cracked pavements'. There may be stages of development in environmental appreciation that we should explore.
   (c) Infill exercises in which students appraise or design alternative ways of renewing a streetscape. The Geography 14–18 Project provides an example in which four different sketches portray the street before and after partial redevelopment. Students are invited to identify the original streetscape, and to select and justify the kinds of development to be avoided and the solution that a responsible planning authority might endorse (Tolley and Reynolds 1977).
   (d) The semantic differential is another technique that can be used to evaluate landscapes by rating them on a scale of bipolar descriptions or attribute pairs. Thus, a landscape might be rated from 1–5 points in terms of being artificial or natural, or harmonious as opposed to chaotic. The list in Figure 14.12 provides some examples of possible pairs that could be used. Having rated the landscapes on a number of pairs, an assessment profile might be constructed. Personally, I find this use of the semantic differential limited, except in so far as it introduces pupils to the vocabulary of describing landscapes.

Appreciating

The ultimate purpose of improving aesthetic awareness is to develop the faculties of appreciation, which can be considered as a form of expanded apprehension, deep engrossment, concentrated attention and disinterested contemplation (Osborne 1970). Appreciating is a complex process which involves the alternation of passive receptivity and active interrogation.

Developing Environmental Awareness and Appreciation

Attention:	Magnetic, attractive, boring, interesting, exciting, vital, stimulating, subtle, rich, poor, full, empty.	**Composition:**	Harmonious, contrasting, varied, balanced, unified, uniform, ordered, chaotic, quiet, noisy, sedate, symmetric, truncated, geometric, sympathetic, restrained, dissonant, entangled, complementary, static, dynamic.
Engrossment:	Awesome, wonderful, powerful, spectacular, exhilarating, depressing, bland, weak, ordinary, impressive.		
Style:	Elegant, pleasant, gauche, luxurious, spartan, ornate, graceful, refined, sophisticated.	**Personality:**	Threatening, angry, eerie, arrogant, menacing, dramatic, spiritual, ethereal, tranquil, tense, relaxed, friendly, comfortable, intimate, remote.
Proportions:	Grandiose, grand, stately, humble, majestic, weighty, massive, airy, spacious, pompous, bold, simple.		
		Impressions:	Novel, familiar, illusory, prosaic, eccentric, allusory.
		Texture:	Rough, smooth, fluid, soft, hard.
		Space:	Open, enclosed, exposed, mysterious, obvious, claustrophobic.
		Light:	Dark, sombre, vivid, raw, sparkling, dull, bright, dense, colourful.
		Virtue:	Natural, honest, virgin, degraded, idyllic.

Figure 14.12 Languaging landscape. Some examples of landscape qualities for developing word pairs for a semantic differential

Figure 14.13 Bathurst streetscape: critical appreciation

Appreciating Streetscapes

Appreciating streetscapes is essentially a process of paying attention to the ways in which the detailed features of each landscape resonate together. The experience is partly passive (in that one needs to absorb the scene contemplatively) and partly active (in that there are specific characteristics one can inquire about). In Figure 14.13, for example, one can direct attention to the variety and harmony of the roofscape, the ratios of house width to height and of solid walls to openings, the spaces between houses, the set-back, the verandahs and the fences.

More specifically, a frieze could be constructed around the classroom using roof silhouettes that students have observed and drawn. A second frieze of wallscapes could be made from observing different types of brick bonding.

Landscape Paintings

It is often the case that streetscapes just do not have the necessary qualities to sustain attention for long. Landscape paintings can overcome that limitation and can be put to good use. One should recognise though that the appreciation of landscapes is not quite the same as appreciating paintings of landscapes. The interrogation of paintings, such as those by Streeton and Namatjira, does require some familiarity with technique and composition so that the exercises on visual form are again useful.

Environmental Literature

Literature about the geographical environment can serve two purposes (Jeans 1979). It can provide us with a visual description and inform us vividly, or simply, or rudely, about the condition of the environment. Alternatively, literature can, like painting, provide us with special insights into the meaning and significance that places have for the author. We can consider the author, poet and painter as endowed with a special acuity of vision from which we can improve our own perceptions. For use in the classroom, we can think of literature as being generic or about different types of landscape (e.g. Leopold 1966; Paterson 1976; Salter 1971; Salter and Lloyd 1977); or as specific and about particular regions (e.g. Pocock 1979); or as composite and fictional geographies (e.g. Darby 1948). The study techniques applicable here are largely those of literary criticism and must be borrowed from that discipline.

My own particular favourite is W. H. Auden's (1958) group of poems on winds, woods, mountains, lakes, islands, plains and streams which is collectively entitled 'Bucolics'.

I can imagine quite easily ending up
   In a decaying port on a desolate coast
Cadging drinks from the unwary, a quarrelsome,
   Disreputable old man; I can picture
A second childhood in a valley, scribbling
   Reams of edifying and unreadable verse;
But I cannot see a plain without a shudder:
   'O God, please, please, don't ever make me live there!'

In much of his work, Auden delights not only in landscapes but in the poetic qualities of the geographical vocabulary which describes them.

Expressing

Expressing may refer to expressing for the benefit of others (communicating) and expressing for the benefits of one's self (self-expression). The distinction between the two is quite important for it represents the difference between the articulation of the critic's and the artist's response.

Self-expression

Features in the environment can become stimulus materials for the subjective expression of how one feels or how one might imagine one feels.

1. Students can be asked to imagine they are a tree, a house, a boulder, etc. and be asked to write poetically or expressively about their life history.

2. Draw in thought balloons for each of the buildings and street features in Figure 14.13 and write in what you think they might say about each other, or about life in the street at the end of the twentieth century.

3. Other art forms may provide completed representations of the landscape which may be used as stimulus material. For example, the use of expressive and evocative music, such as Milhaud's Rhodadienne symphony on the profile of the River Rhône or Mendelssohn's Scotch symphony, as stimuli for writing about the places represented by the music. Some help and other ideas may be gleaned from Wise's (1977) paper on music and geographical education.

4. Portray in mime, dance or theatre some conflict over land use, either real or hypothetical. In all this, the geography teacher is entering the domain of the creative arts teacher. Close cooperation between teachers of different disciplinary backgrounds may provide a gentle initiation into this unfamiliar territory.

Communicating

The geography teacher will probably feel happier about developing communication skills, especially those of an audio-visual and literary nature. The Art and the Built Environment (ABE) Project devotes a lot of attention to encouraging the learner to become a critic and to learn to interpret, discriminate and articulate critical responses through audio-visual and literary activities.

1. *Audio-visual*. It is here that the use of photography, annotated photographs and sketches can play a valuable role. Some tuition in the use of these media is essential and Day (1979) provides some guidance. The steps involved will usually include working out the purpose of the composition, the researching and investigation of the content, the selection of visuals, preparation of interpretation or commentary, and evaluation of whether the intended communication actually occurs.
2. *Language and literature*. Language permeates the whole learning process so it is equally important to develop literal thinking as well as visual thinking.
   (a) Learning an appropriate vocabulary is crucial, for language not only expresses thoughts but also brings them into being. Somehow, one ought to teach familiarity with the language that describes and appraises the sensible appearance of the geographical environment (Figure 14.12).
   (b) Language activites take a number of forms and again the issues approach has much to commend it. Students can be involved in discussing real or hypothetical landscape issues which involve aesthetic arguments. Both oral and written work have their place and the messages of the Language Across the Curriculum project are important here. Classroom interaction can be arranged in small groups so that students have opportunities for exploring ideas and talking to learn. And the teacher can present a variable, but specified, persona to the class and provide them with different types of audience reaction to their written work. The teacher could play the role of a philistine land developer, or an anodyne planner, or an overly sensitive citizen, or an ebullient newspaper editor, and stimulate correspondence among these factions. In so doing, there is initiated a dialogue between teacher and learner which goes beyond the customary assessment of performance. Necessarily this will also mean moving away from a dependence on textbooks and into providing access to a wide range of reading materials.
   (c) Designing a brochure for a town trail provides a good opportunity for students to use their own language to organise their thoughts, and avoid falling into the trap of guessing what language is acceptable to the teacher. Obviously, it works best if students devise the trail (as a small-group exercise) and then design the brochure with a particular audience in mind. By choosing their own audience, students are challenged to communicate their viewpoint to someone with whom they feel familiar and comfortable.

Conclusion

This anthology of exercises is a basis for providing students with opportunities to experience and develop the skills and insights involved in the sensory awareness and aesthetic appreciation of the environment. Readers may wish to develop similar ones for their classes or to select from the exercises described. It is necessary to work from a sound theoretical perspective, however, and to this end the matrix in Figure 14.2 is intended as a way of representing the context in which these exercises may be seen.

There are three broad observations I should like to make about the strategies outlined above:

1. There should be no surprise if students do not become *deeply* involved in exercises such as these. Mature appreciation is a fleeting and unpresaged experience which is likely to come over us when we least expect it. What one can hope to do in the classroom is to prepare the ground for the seeds of aesthetic awareness to grow.
2. There is much resistance from teachers against becoming involved in the affective domain, largely because of the nature of the outcomes from affective education. The results may be unpredictable and are often difficult and time consuming to assess. These are problems which simply have to be accepted and tolerated. In aesthetic education, the results are not known beforehand. The teacher is often learning with the learner and, of course, it is likely that it is the role model of the teacher-as-learner which the student remembers most.
3. It is necessary at this stage (as the ABE Project emphasises) for teachers to become involved in evaluating and appraising what goes on in the classroom when these strategies are tried out, and to share their experiences with other teachers and

curriculum developers. Developing, monitoring and reporting the progress of learning programs seem essential to improving educational practice. This is true in all areas of the curriculum and not only the teaching of environmental awareness and appreciation. Enjoy yourselves!!

References

Adams, E. (1980) *About Art and the Built Environment*.
Adams, E. and Ward, C. (1982) *Art and the Built Environment*, London: Longman.
Angus, M. (1975) *The World of Olegas Truchanas*, Hobart: OBM Pty Ltd.
Appleton, J. (1975) *The Experience of Landscape*, London: John Wiley and Sons.
Auden, W.H. (1958) *Selected Poetry of W. H. Auden*, New York: Random House.
Cullen, G. (1975) *The Concise Townscape*, London: The Architectural Press.
Curriculum Development Centre and the Australian Heritage Commission (1980) *Investigating the National Estate*, Canberra: Curriculum Development Centre.
Cornell, J.B. (1979) *Sharing Nature with Children*, Watford: Exley.
Darby, H.C. (1948) 'The Regional Geography of Thomas Hardy's Wessex', *Geographical Review*, Vol. 38, 426–443.
Day, D.H. (1979) *The Focal Guide to Photographing Places*, London: Focal Press.
Evans, E. (1973) *The Personality of Ireland*, London: Cambridge University Press.
Farbstein, J. and Kantrowitz, M. (1978) *People in Places*, Englewood Cliffs, New Jersey: Prentice Hall.
Goodey, B. (1974) *Urban Walks and Town Trails: Origins, Principles and Sources*, Birmingham: Centre for Urban and Regional Studies.
Hall, R. (1980a) 'Streetscape Appreciation', *Geographical Education*, Vol. 3(4), 489–506.
Hall, R. (1980b) 'Sensory Walking', *Classroom Geographer*, November, 3–8.
Halprin, L. (1965) 'Motation', *Progressive Architecture*, Vol. 46, 126–133.
Hart, R. (1978) *Children's Experience of Place*, New York: John Wiley and Sons.
Jeans, D.N. (1979) 'Some Literary Examples of Humanistic Descriptions of Place', *Australian Geographer*, Vol. 14, 207–214.
Leopold, A. (1966) *Sandy County Almanac*, London: Oxford University Press.
NSW Dept of Education (1986) *Curriculum Ideas: Geography and Environmental Education*, Sydney.
Osborne, H. (1970) *The Art of Appreciation*, London: Oxford University Press.
Paterson, J. (1976) 'The Poet and the Metropolis', in J.W. Watson and T. O'Riodan (eds) *American Environment: Perception and Policies*, New York: John Wiley and Sons.
Penning-Rowsell, E.C. (1981) 'Gauging Landscape Value', *Progress in Human Geography*, Vol. 5, 25–41.
Peters, R.S. (1977) *Psychology and Ethical Development*, London: Routledge and Kegan Paul.
Pocock, D.C.D. (1979) 'The Novelist's Image of the North', *Institute of British Geographers, Transactions*, Vol. 4, 62–76.
Punter, J. (n.d.) 'Personalisation', *ABE Working Party Newsletter*, No 2.
Salter, C.L. (1971) *The Cultural Landscape*, Belmont: Duxbury Press.
Salter, C.L. and Lloyd, W. J. (1977) *Landscape in Literature*, Washington DC: AAG Resource Papers for College Geography, No. 76–3.
Schools Council (1978–81) *Art and the Built Environment Project Newsletters*, London: Royal College of Art.
Smith, B. (1947) 'Art and Environment in Australia', *Geographical Magazine*, Vol. 19, 389–409.
Tolley, H. and Reynolds, J.B. (1977) *Geography 14–18*, London: Macmillan.
Tuan Yi-fu (1973) 'Ambiguity in Attitudes Toward Environment', *Annals of the Association of American Geographers*, Vol. 63, 411–423.
Van Matre, S. (1974) *Acclimatizing*, Martinsville: ACA.
Ward, C. and Fyson, A. (1973) *Streetwork*, London: Routledge and Kegan Paul.
Wheeler, K. (1976a) 'The Visual Approach to Environmental Geography', in K. Wheeler and B. Waites (eds) *Environmental Geography*, St Albans: Hart Davis.
Wheeler, K. (1976b) 'Experiencing Townscape', *Bulletin of Environmental Education*, No. 68, 1–28.
Wise, J. (1977) 'Music and Geographical Education', *Journal, Geography Teachers' Association of Queensland*, Vol. 12(1), 1–22.
Younghusband, F. (1920) 'Natural Beauty and Geographical Science', *Geographical Journal*, Vol. 56, 2–13.
Zelinsky, W. (1973) *The Cultural Geography of the US*, Englewood Cliffs, New Jersey: Prentice Hall.

15

Teaching Skills in Geography

Bernard Cox

Most geography teachers include the development of skills in their courses. By skills they refer to a wide range of competencies that they expect geography students to develop and to apply in their studies of different geographical issues. This chapter is built on the assumption that skills are important in geography teaching and that they should be developed in an integrated way rather than in isolation.

Before geography teachers can decide how to teach skills, they should become aware of the different types of skills that are relevant in geography teaching and how students learn skills. Then, they should become aware of the basic activities that are involved when a person learns a skill. Geography teachers are then ready to think of suitable strategies for teaching or developing skills in their geography classes. Several sequences for developing and applying skills in geography are incorporated in the chapter to demonstrate how to put the idea of skill development into practice.

The Joy of Being Skilful

'What superb skill!'

We've all exclaimed something like this in admiration as we've watched someone do something very well and with apparent ease. Perhaps it was a somersaulting dive from the high tower, or a driver taking evasive action to avoid a potential collision when travelling at speed on a freeway, or someone's fingers nimbly touching the keys on a word processor to turn out 150 words a minute without error. But we might also have expressed our admiration for a geography teacher who had just chalked a clear and proportionate outline of Australia on the board with the class watching closely, or perhaps the exclamation was made by the geography teacher more than satisfied with a hand-drawn map by one of the students.

The results may well be far less accomplished were you or I to undertake any of these activities. If this is true, it means that we have poorly developed skill in this activity. It is not only a joy to watch someone else's skilful performance. It is also a great source of personal satisfaction when we ourselves do something with consummate skill. However, it takes time and practice to achieve a high degree of proficiency in any skill. What then, are the benefits of possessing skill, particularly for students of geography?

The Benefits of Being Skilful

Skills are essential to people learning many things, from sports to music to geography; e.g. coaching in sports frequently takes the form of skill development, critics often refer to the technical skill of musicians playing instruments and syllabus committees in geography invariably include statements of skills which they believe should accompany the cognitive and affective learning being sought by students following that syllabus. These examples indicate that the importance of skills is widely recognised.

The term 'skills' is commonly applied both to psychomotor and to cognitive activity, which are, of course, closely related. Emphasis is placed in this chapter on psychomotor skills while other chapters in the book, for example Chapters 3, 4, 5, 6 and 7, emphasise cognitive skills. The importance of psychomotor skills is based heavily upon the benefits that learners derive from becoming skilful. The benefits include:

The joy derived from achieving an excellent performance. This was exemplified in the first paragraph. Enjoyment is powerful motivation to learners particularly where they cherish the skill. For example, most windsurfers are proud of being able to do a 360-degree aerial flip from a wave and will practise assiduously until they can do it well. Students of geography may derive similar satisfaction from having devised and drawn a graph representing statistical information. This is particularly so if the graph has been made relevant to the current topic of study. The link between cognitive and psychomotor skills is obvious in this case. Satisfaction derives both from how useful the graph is in enabling further study of the topic and from the sense of pride in having produced a fine-looking graph.

Practical consequences. The motoring skill referred to earlier may save the driver's life; the word-processing skill may well enable the possessor to earn a living. In short, there are often practical rewards for being skilful. These rewards are readily apparent to learners and are likely to provide the motivation needed to sustain their efforts to acquire the skill. The practical rewards of studying geography are rarely so direct. The skills used by geographers usually enhance their capacity to further their studies. So, a third benefit follows.

Becoming capable of finding out. Very many skills contribute to learning how to learn. All the following examples can help people learn geography: drawing a cross-section, reading a map, interpreting a map, visualising landscape from a graphic description, measuring distance, recording field observations, recognising different types of rocks, writing, using the catalogue in a library, counting and recognising the logic of an argument. None of these skills belong exclusively to geography but all can be turned to good effect by anyone learning geography because they contribute to one's capacity to find out. The motivation to acquire skills such as these relates quite directly to the extent to which the student perceives benefit from learning geography as a whole and from the particular topic with which the skills are associated.

Viewed together, these three benefits of acquiring skill are interesting because the major advantage lies not in the simple possession of skill but what one can get out of it. There are two important corollaries to this conclusion. Skills are taught and learned so that they may be used; and, skills are well learned at the time when they are needed. These two points may indeed be used by teachers planning programs of study, work units and lessons for their students. Illustrations of their application to geography in schools will be provided later in this chapter.

Thinking About Skills

The word 'skill' is often used to denote practical knowledge combined with ability. Practical knowledge refers to the level of proficiency in a specific activity or group of related activities. Four considerations arise from this. They are:

1. Many skills, such as driving a car or doing fieldwork in geography, are actually *clusters of related skills*. The fieldwork, for example, is likely to involve several different activities based on observation, other activities to record observations, and yet more activities are undertaken during the process of interpreting the data that have been recorded. All of these component activities require skill on the part of the person undertaking them.

2. Skill is related to *ability* in so far as many skills require the application of some cognitive ability. While ability refers to capacity, skill refers to the activities which the capacity enables. A person's abilities are acquired out of maturation and learning. Abilities once developed are quite enduring, while most highly skilled performances require frequent practice. Possessing an ability is an important basis of many skills because abilities are readily transferred; for example, ability in spatial reasoning underlies many approaches to interpretation in geography. While ability refers to capacity, skill refers to the activities which the capacity enables. Application, analysis, and evaluation are intellectual processes but they require studied thought and consideration, while motor skills can be practised to the point where they can be done with little thought. This chapter is

mainly concerned with the latter type of skill. Almost all 'skills' in geography require combinations of intellectual process and motor skill. The proportions vary, and are illustrated by Figure 15.1. The difference between map reading and map interpretation also illustrates the combination of intellectual process and motor skill. Map reading is literal translation of symbols, while map interpretation requires association of symbols, inference, deduction and synthesis.

3. Being skilful refers to the *level of proficiency* that has been attained by people undertaking the activities. The phrase 'a skilled worker' implies that the person can do the task and can do it well. Bearing this in mind, teachers concerned to foster the development of skills in their students should provide learning experiences aimed at improving performance. This will be elaborated later in the chapter.

4. Skills have been defined as involving proficiency in activities usually associated with practical knowledge. So, *manipulation and bodily dexterity* are often important aspects of skill development. A sense of timing, hand-to-eye coordination, a ball sense (in many sports) are often basic to skills. But, knowing and feeling can rarely be separated from the development of skills.

Several features of skills become apparent if we reflect on these four ways of thinking about them. These features can guide teachers helping their students learn skills.

1. We often use skills because they help us achieve something. For example, there really is no point in drawing a cross-section, neatly and accurately, if only to say, 'There it is, it's finished. Isn't it good?'. After the section has been drawn, it should be used to help someone find out about the relief of the area illustrated by the section, or to find out about intervisibility, or to help a decision about the practicability of building roads, farming or other human activity on the slopes in the area. Cross-sections have many applications.

2. We have to practise to improve our proficiency in a skill. Many skills are done well as reflex actions, without self-conscious thought about how to produce the action. Imagine the result if you were to change gears in traffic while thinking closely about the actions of both of your hands and legs. The purpose of practice is to perfect action. A geographer may learn how to make accurate readings from a prismatic compass while actually thinking more about the significance of direction for the study in hand.

3. It is possible to know about a skill without being proficient at it. All of us are armchair critics at times, for example when watching sports' telecasts, and it may be that our criticism is well founded on sound knowledge. Knowing about the processes underlying the skill often enhances performance. For example, the quality of a hand-drawn map is likely to be higher when the cartographer is well informed about principles of graphic design and about the uses to which the map will be put.

A Range of Skills for Students of Geography

Curriculum developers in geography, be they teachers or central committees, have to make decisions about the range of skills to be included. Some of the decisions can be made in consultation with the students following the course. But what principles

Motor skill		Cognitive process
Low		**High**
	• Problem solving in geography.	
	• Interpreting a map.	
	• Drawing a block diagram from data on a contour sheet.	
	• Drafting a map for a specific use.	
	• Drawing a cross-section.	
	• Reading a map.	
	• Laying out a statistical table.	
	• Calculating gradients.	
	• Walking through a forest.	
	• Reading a rain gauge.	
	• Ruling a straight line on a map.	
High		**Low**

Figure 15.1 This figure illustrates the varying proportions of motor skills and cognitive processes in selected 'geographical' activities. The actual placement of the items may well vary from one individual to another

Source	Acquiring information	Organising information	Interpretation	Presentation
Texts and documents	Reading Locating sources of data Note making	Classifying data • listing • grouping • categorising • labelling	Analyse themes or characteristics Interpret characteristics Explain characteristics Synthesise characteristics	Essay writing
Audio-visual materials	Listening to tapes Recognise features on a picture	Estimating scale in landscape photos Recognising viewpoint adopted in a cartoon	Recognising areal associations in a photograph Deductions from evidence in a photograph Interpret common symbols used in cartoons	Making poster displays Drawing a landscape sketch Making a cassette tape recording Lecturette
Maps	Reading symbols Calculating gradients Using linear scales Drawing cross-sections	Classifying features represented on the map Using map overlays	Drawing inferences from information represented on the map Recognising areal associations	Drawing maps for particular purposes
Statistics	Recognise unit of measurement Read quantities, times, directions, trends	Calculation of averages, means, median, etc. Sampling Making correlations	Using data obtained from quantitative methods of analysis to identify trends and sequences, etc.	Laying out statistical tables
Graphs and diagrams	Acquiring information by reading different types of graphs	Selecting and drawing the appropriate way of graphing the statistics	Drawing inferences from graphs and diagrams	Drawing a transect
Specimens	Collecting Describing	Recognising similarities and differences Labelling, identifying Classification	Analysing physical properties	Displaying collections
Field studies	Implementing questionnaires Measuring, observing and recording phenomena, e.g. estimating distance	Identification and classification Scoring questionnaires Using fractional notation	Comparing field records with relevant theory	Most of the above

Note: The matrix is not comprehensive

Figure 15.2 A selection of skills to be developed from different sources

can be used to guide such decisions? The following are suggested:

The Needs of the Learners

One of the reasons for including students in the process of making decisions about a course of studies is that they are often informed and articulate about their personal needs. But teachers also are able to use their experience and expertise as educators and as geographers to contribute to these decisions.

The Needs of the Discipline, Geography

One way of answering the question 'What range of skills?' is to identify those skills which are commonly used by people searching for answers to the series of questions distinguishing geographical study; for example:

1. What is being studied?
2. What is its location or distribution?
3. Why is it there?
4. What are the consequences of this location or distribution?
5. What spatial alternatives can be considered by people making decisions?
6. What actions should be taken after the decision?

Fairly obviously, the range of skills emerging from such a list of questions includes many associated with acquiring information (including observation), organising information, interpretation and presentation (including action). Figure 15.2 offers examples of skills organised into these categories and applied to sources commonly used in geography.

The Use that Students May Make of Their Geographical Studies in the Course of Living in Local, National and World Societies

This principle blends the two preceding ones and it may be focused through the question, 'What skills do geography students need for effective participation in Australian and world societies?'. Students should acquire knowledge and engage in reflection and action for effective participation. The range of contributing skills is shown in the following four-step approach which also shows how skills may be integrated with subject matter.

1. Students about to undertake the study of some topic, question or issue will do well to analyse the situation of the object of their study. Some aspects of the situation are internal to the topic itself while others consist of circumstances surrounding it. For example, if the topic was concerned with the effect of world prices of wheat on grain growing in the Darling Downs, the analysis of the situation may well involve studies of: (a) changes in the world price, seasonal weather conditions in other wheat-exporting countries, political decisions to sell surplus stocks of wheat in the USA and the EEC, the activities of the Australian Wheat Board, the construction of a new grain-exporting terminal on the banks of the Brisbane River; and (b) decisions made by farmers on the Darling Downs, for example about matters related to substitute crops, soil erosion, rainfall patterns on the Downs, and the traditional approach to farming adopted by many of the farmers.

   Students making a situational analysis will inevitably need many skills, including ways of observing people and events, constructive discussion of the topic with other students and conventional study skills.

2. The students must acquire knowledge on which to base their decision, conclusion or solution. Their ability to use information will be enhanced if they are aware of the value premises or the viewpointedness of information as they gather it. Again many skills are required, including:
   (a) locating information in libraries and references;
   (b) interpreting numbers and graphic displays such as maps and diagrams;
   (c) talking to resource people in the community;
   (d) analysing values.

3. Questions, problems, issues—that is, matters to be decided—require the establishment of priorities and a review of possible decisions. Again many skills are required. They include the skills that are used in association with problem solving, with logical and critical thinking, with probes of the values underlying the alternatives, and cooperation with other students. Throughout this chapter such activities as problem solving, logical and critical thinking, analysis, synthesis evaluation and the like are regarded as processes of thinking rather than as skills. Though these processes are often referred to as intellectual skills, they are distinguished from motor skills by

the fact that they require more self-conscious thought and judgement.

4. Making a decision, proposing a solution, coming to a conclusion and acting on it in some way. This aspect of the geographical study also requires many skills while it is being undertaken. This time the skills are used as the students take action based on the decision or conclusion and review the effects of the decision on themselves and on others.

The utility of this four-step approach to thinking about the use of skills in geography is that it emphasises the integration of skill development with the topic being studied and with the educational rationale for undertaking geographical study in secondary schools.

Considering the Sequence of Skill Development

Argument has been advanced for integrating the development of skills with other aspects of learning. Nevertheless the question, 'What should be considered when making decisions about the sequence of skills?' may still be asked. Two criteria are arguably important. They are (1) the readiness of students, and (2) the introduction of skills which enable increasingly elegant study by building on prior knowledge and skills. Gerber et al. (1986) have presented a detailed matrix of mapping and other skills that may be developed across the primary years. For example, in understanding the arrangement of features on maps, they suggest: discrete features at level 1, clusters of features at level 2, classes of features at level 3, large-scale area patterns at level 4, small-scale area patterns at level 5, and arrangements at differing scales at level 6. These suggestions apparently derive from application of the two criteria. Similarly, a Task Force of the National Council for the Social Studies (1984) proposed a detailed list of the essential skills for social studies and offered a general indication of grade levels when more or less instructional effort should be made in relation to each skill. The National Council's list places strong emphasis on reading from text as distinct from reading from graphic materials which are so important to geographers. Butler et al. (1983) have made good this deficiency in their statement about the sequential development of mapping elements. Their statement details aspects of map language such as signs, colour, size and shape, base data, representing relief, and lettering/numbers. Gerber and Wilson offer constructive advice based on primary research on the use of maps in the classroom in Chapter 17 of this book.

Overall, practical judgements by teachers are important when decisions are being made about the scope and sequence of skills in geography. A unit of work involving students in inquiry or problem solving will require that they use some different skills from those associated with expository learning.

How Do They Get That Way?

Reference was made early in this chapter to the admiration we feel when seeing some activity performed by a superbly skilful person. How do people develop such skill? Let's review the contributions of both teachers and learners to the development of skill.

Three Activities for Developing Skills

Characteristically, people wanting to acquire and develop skills need to engage in three kinds of activity. They are understanding, organising and perfecting.

Understanding
Some skills can be acquired by rote learning; for example, a seal can be trained to balance a ball on its nose. But it is generally agreed that a student should understand what the intended skill involves. It helps students become skilful when they know a lot about the skill—what it is, what it will help them do, who is good at it, what activities have to be coordinated. For example, students learning to draw maps would do well to know about the essentials of maps, such as scale, map symbolism, projection, orientation, and the steps involved in designing a map. Students can learn about map reading or about any other skill by watching someone else doing it well, by reading a step by step analysis of the skill, by watching a video clip showing the skill being performed, and by listening to expert commentary on the performance. But ultimately, it is hard to improve on direct personal observation at least as a first step in getting to know about the skill.

Organising
Organising involves learners in trying out the skill, finding out how to coordinate their actions, making mistakes and trying again. Practice sessions are aimed at the improvement of speed and coordination to the point where the skill can be performed

without deliberate thought. The emphasis is now on dexterity rather than on understanding. People learn how to use a clinometer so they can measure angles of elevation or depression when making field maps. Organising activities include holding the instrument to the eye, keeping one's eyes open, sighting on the object, and reading from the graduated scale. Most people are able to use a clinometer after they have organised these activities.

Perfecting

Refinement of skill depends on continuing opportunity to use the skill. At this stage motivation is enhanced when the practice is functional, that is when the skill is used to some purpose valued by the student. Coaching aimed at improvement in understanding and organising also contributes to perfecting the skill.

A Model for Teaching Skills

These three basic activities are needed by anyone learning a skill. The activities may be related in a model which guides teaching and learning. Principles guiding the learning are proposed, and more specific suggestions are made about relevant teaching procedures in Figure 15.3. In this figure, principles and procedures 1,2 and 3 relate mainly to understanding, 4 and 5 to organising, and 6 and 7 to perfecting skills.

A Sample of Skill Teaching

Drawing a cross-section is a complex skill and comprises several tasks. These are identified in Figure 15.4 with appropriate teaching procedures suggested for each task.

Principle	Procedure
1. Many skills used in geography are complex and can be broken into components, for example drawing a map involves drafting, lettering, shading and other operations.	• Make a task analysis of the skill to identify its components. • Find out about the readiness of students to undertake each component task. • Help students experiencing difficulty.
2. It helps students starting to learn a skill if they watch a good demonstration.	• Students should be able to see the demonstration clearly. • Demonstrate the whole, then the parts, then the whole again.
3. A commentary on the demonstration and/or a written plan of the sequence of actions helps people learn skills.	• Teachers should give a commentary as they demonstrate. • Provide guiding notes or require students to record their own notes.
4. Opportunities for supervised practice shortly after the demonstration help learners to minimise errors and refine their actions.	• Teachers should allow time for practice straight after the demonstration. • If the skill is complex, the demonstration plus practice should be applied in turn to the sub-skills.
5. Comments on their performance during practice help student organise the skill.	• Teachers should provide comments on the work of individual students.
6. Application and use of skills improve their meaningfulness and make them more easily transferred to other tasks.	• Devise several ways of practising the same skill, and build it into more complex operations such as problem solving.
7. The capacity to evaluate one's own performance can be used to improve the skill.	• Help students formulate criteria of a good performance and encourage self-analysis.

Figure 15.3 Teaching skills: principles and procedures

Task analysis	Teaching procedure
1. Note precisely the uses to which the section will be put as a guide to: (a) choosing the line of section, and (b) deciding the vertical exaggeration.	• Teacher and students discuss how the section will be useful in the study they are making. • Discuss the effects of using different vertical exaggerations.
2. Mark the line of section on the map or on an overlay to the map.	• Teacher demonstrates how to do this; students undertake the task individually. • Teacher checks students' work before going on.
3. Draw axes for the section.	• Teacher demonstrates on overhead projector or chalkboard. • Students undertake the task individually. Teacher checks.
4. Decide horizontal scale (usually same as map) and vertical scale. Calculate vertical exaggeration.	• Teacher explains why they should choose the map scale as the horizontal scale of the section. • Discuss choice of vertical scale. • Formula for calculating vertical exaggeration written on chalkboard. • Formula applied to this section and calculation made.
5. Gather data about altitude of points along the section by noting where the contours cross the line of section. This is usually done by putting a piece of paper along the section and marking the points on it. Where the same contour recurs successively, the symbols are used to indicate valley or hill between the contours.	• Teacher demonstrates on overhead projector or chalkboard. The demonstration is for the whole section and is accompanied by a commentary. • Where the same contour recurs the teacher explains how to recognise whether it represents valley or hill. • Students are then given an opportunity of making this decision for themselves. • Students mark points on the paper, noting altitude at each point. Teacher checks their work.
6. Plot the points which will be joined to make the section. Ensure accuracy on both scales.	• Teacher demonstration. • Student application. • Teacher supervision and checking.
7. Join the points to form a smooth curve. Estimate the depth of valleys and height of hills where the same contour recurs.	• Teacher demonstration. • Student application. • Teacher supervision and checking.
8. Note the annotations to the section, i.e. the scales, start and end points of the section, vertical exaggeration, title.	• Teacher demonstrates how and where these are noted on the section. • Student application. • Teacher supervision and checking.
9. Use the section for the purpose for which it was drawn; for example, to help a study of relief and slope in the area, to determine intervisibility, to provide the base diagram for a transect, to help make decisions about future land use.	• These applications may be undertaken in different ways, but usually require supervision by teachers in the first instance. • Teacher-led discussion, or inquiries structured by teachers are appropriate.
10. Students should be given the opportunity of judging the accuracy and usefulness of their own sections.	• Teachers and students devise criteria by which to judge accuracy. • Students apply these. • Revision questions by teacher focus on the uses of the section.
11. The teacher should provide opportunities in future learning situations for the students: (a) to practise drawing sections and making inferences from them, and (b) to decide for themselves that sections are an appropriate technique to adopt in some study being undertaken.	

Figure 15.4 Drawing a cross-section

Teaching Skills With a Purpose: An Example

The Year 11 geography class had been discussing the impact of mining activity on landscape. Several students had commented that the impact was very considerable, citing as evidence the effects of open-cut mining as in the central Queensland coalfields, and the waste heaps near some shaft and tunnel mines. It was also pointed out that the development of a mine in a remote area requires roads, water storage, housing for miners, plant and equipment. All of these contribute to the overall human impact on the landscape. Other students claimed that some human constructions such as artificial lakes in dry areas were aesthetically pleasing. Another defender of mining quoted figures showing the contribution of minerals to Australia's export earnings over the past decade while claiming that less than 1% of the total land area of Australia was affected in any way by mining activity. As the debate warmed up it was suggested that the students should make case studies of several mining sites with a view to examining their impact on landscape. Visits to mines were judged to be the ideal approach but the remoteness of many mines in Australia precluded this and it was decided to use the evidence available on maps as a substitute. The Swanbank 1:10 000 topographic sheet was chosen because (1) its large scale offered considerable detail in the landscape it represented, and (2) there were both mines and a coal-fired power station in the area.

The teacher pointed out that maps are specially useful to geographers because of the ways in which they show information. The layout of features is shown on the map and it is drawn to scale. This means that distances are accurate. Having the scale also makes it possible to calculate areas. The range of human and natural features shown on the Swanbank sheet was particularly useful for the impact study.

Map-reading skills were essential for the students to complete this study. The first task lay in reading as much as possible about the geography of the area from the map. Members of the class, individually and cooperatively, used eight steps to familiarise themselves with the Swanbank extract. These steps were:

1. Read the title of the map to find out its purpose.
2. Learn the symbols shown in the key to the map.
3. Read the scale and the contour interval.
4. Read the contours in order to describe the land forms. Steps in reading the contours are:
   (a) Trace over the rivers and coastlines to highlight them.
   (b) Find the highest and lowest points.
   (c) Identify rugged and nearly flat areas.
   (d) Look for particular topographic features, such as hills, valleys, ridges.
   (e) Form a conclusion about the overall type of land form in the area.
5. Note the different types of vegetation and their distributions.
6. Look for different types of settlement. Note their locations.
7. Find other uses made of the land. What are their locations or distributions?
8. Note the relations among various features shown on the map.

Having made this geographical reconnaissance of the area, the students were ready to consider the impact of mining. They did this through a series of structured activities demanding skills in map reading and interpretation prepared by the teacher. Her questions follow:

1. What evidence is there on the map that (a) mining and (b) power generation are major human activities in the area? *Skills used*: Map reading.
2. Comment on ways in which the local relief and slopes might have helped or hindered the construction of the power station. *Skills used*: Calculations of relief and gradients; map interpretation.
3. Note the main type of vegetation in the area. What proportion of the area shown by the map has been cleared of vegetation? What evidence is there that human activity has affected the vegetation? *Skills used*: Map reading; calculation of areas; map interpretation by inference.
4. There are no towns or cities in the area shown by the map extract. All of the buildings are associated with the power station or the coal mines. The nearest residential buildings (not shown on the map) are about two kilometres from the power station.

   Describe the pattern of land use in the area. Note the associations among specific features. What might be the main reasons for the separation of the power station/coal mines from the residential areas? What evidence is there of landscape design, regeneration, or other actions taken to reduce the visual impact of mining and power generation? *Skills used*: Map reading; measuring distances; map interpretation by inference.

5. The power station and the mines also affect places beyond those shown on the map sheet. Discuss possible effects of such things as the availability of electric power, the grid system of transmission lines, dust, and noise. *Skills used*. Social skills in discussion; inferential reasoning.

6. Discuss in class any actions that students might take as a consequence of the insights they have gained in working these activities. *Skills used*: Social skills; communication skills.

The use of skills in the study of the impact of mining on landscape focuses on two themes of this chapter: notably (1) skills are taught and learned so that they may be used to further one's study, and (2) teachers do well to help their students become skilful at the time when the skills are needed.

References

Butler, J., Clough, R., Gerber, R., Senior, C., Smith, S. and Wilson, W. (1983) *Jacaranda Atlas Programme, Resource Book 1*, Brisbane: Jacaranda.

Gerber, R., Tobin, S. and Wilson, J. (1986) *Jacaranda Active Social Studies. Book 1, Teachers' Book*, Brisbane: Jacaranda.

Task Force of the National Council for the Social Studies (1984) 'In Search of a Scope and Sequence for Social Studies', *Social Education*, Vol. 48(4), 249–263.

16

Teaching Graphics in Geography Lessons

Rod Gerber

Geography teachers often believe that by working with maps they are doing a good job in teaching their students how to understand and to use graphics in their geographical investigations. In fact, the graphics that students should use are numerous in type and function. This chapter aims to make geography teachers aware of the many motivating graphics that they can use in their lessons and it suggests a range of ideas for maximising the information represented on these graphics in geographical studies.

By focusing on the need to select or to design clear, effective graphics, this chapter introduces the importance of evaluating prepared graphics for their communication of information and the key guidelines that geography teachers should promote when they have their students design maps, graphs, cartoons or other types of graphics. This range of helpful guidelines and illustrations should convince geography teachers that a carefully selected graphic can promote learning in geography more easily than most other forms of information.

The students entered the classroom for their first geography lesson. Concerned to find out about their geographical skills and their understanding of their local environment, the teacher asked all of the students to 'draw how you came to school this morning'. The teacher's question was deliberately open ended and so the students responded as they saw fit. The results of the students' drawings were very individualistic. A few students designed very detailed route maps which were drawn to scale and which contained a comprehensive set of signs. Some students created mental maps which contained a range of personalised landmarks, their generalised route from home to school and inaccurate judgements of distance. For others, drawing meant non-map presentations. Some drew pictures of themselves riding on a bicycle or riding in a bus or a private car. The more creative students drew graphic diagrams such as a topological route diagram or a flow diagram. One student even tried to combine the times she took with the distances she travelled and to represent these on a form of bar graph. The class discussed their responses and expressed some amazement at the range of graphic responses they could make to a relatively straightforward task. The teacher collected the students' responses and analysed them that night for evidence of the students' abilities to draw graphics.

Scope of Graphics

The above scenario indicates that graphics are not only maps, but that they consist of a wide range of forms, many of which are used by students in different learning situations. The challenge for geography teachers is to recognise the range of legitimate graphics that can be used in geography lessons, become familiar with them, know their qualities and know when to use them in teaching geography.

Perhaps, you may wish to think for a moment about the items that may be classified as graphics in geography. A list of six kinds of graphics would include:

1. paintings, horizontally viewed photographs, oblique aerial photographs, vertical aerial photographs;
2. cartoons, comic strips;
3. quantitative data representations including: line graphs, bar graphs, pie graphs, pictographs, remotely sensed images, circular graphs, scatter graphs, Lorenz curves, tri-dimensional graphs, digital terrain models, cartograms, cross-sections;
4. diagrams including: flow diagrams, organisational charts, field sketches, wind roses, soil profiles, transects;
5. retrieval charts, word codes, wonderwords puzzles;
6. maps.

The obvious questions to ask is: 'How many of these graphics do I use in my geography lessons or do I encourage my students to use in their geographical studies?'.

Sharing Graphic Information

As the old saying goes, 'A picture is worth a thousand words'. However, it is not worth very much if its graphic message is not received effectively by the people who read it. After all, graphics are made up of a wide range of signs which go together to constitute a whole flow diagram, line graph or cartogram. Unlike words, these signs do not have fixed meanings, and therefore often need to be decoded by readers in order to understand the intended graphic message. In some cases these codes are explained in a legend (e.g. a legend to explain the components of a composite bar graph) and in other cases they are explained by labels on the graphic (e.g. the labels of the different elements of a diagram of a farm system).

The crucial aspect to this sharing of graphic messages between the people who design the graphics and the people who read them is that the graphics only receive their meaning from the people who read them. Therefore, the geography teacher or student who prepares a graphic representation of, for example, a coastal dune system must design it in such a way that the readers of this graphic can understand the intent. For example, was it designed to show the processes at work to form the dunes? Was it to show the ways that people can have an impact on dune structures? Or was it to show the location of the different parts of the dune system?

In addition, it is vital that each graphic is designed in such a way that it uses signs and labels which the readers can understand. This is achieved if the person designing the graphic knows something about the abilities of the people who are likely to read the graphic. Decisions about the abstractness of the signs to be used on graphics are important. For example, it may be easy to draw a bar and to shade it a solid colour. However, this abstract representation may be understood more easily if it consisted of a set of pictorial signs in a row with each sign having a particular value. These pictorial signs may take longer to reproduce but they will be understood much more easily. In this case, it may require the purchase of some commercially produced signs, e.g. Leteraset signs, and the use of these signs rather than the complete redesign of another set of signs. This is a small cost if it means that the resulting graphics are clear and useful.

Graphics–Text Link

It has already been inferred in this chapter that meanings to the signs in graphics are derived differently from the meanings to words. However, that is not to denigrate the importance of words in making sense of graphics. This vital link between graphics and text is evident in two ways:

1. the use of words as labels for key parts of a graphic; and
2. the use of words to explain the meaning of the information on a graphic.

In Figure 16.1, for example, the labels identify the major settlements in a part of Asia. However, the following description of the location of these places in China demonstrates the importance of words to interpret the graphic information:

Eastern China contains a large number of urban areas each of which contains more than one million people. These large urban areas have distinctive sites. Some of them, e.g. Wuhan, Guangzhou and Chengdu, are major river cities. Others such as Shanghai and Qingdao are large port cities, while cities such as Kum-

Figure 16.1 Settlement map in eastern Asia (source: *Atlas 2*, Jacaranda Press)

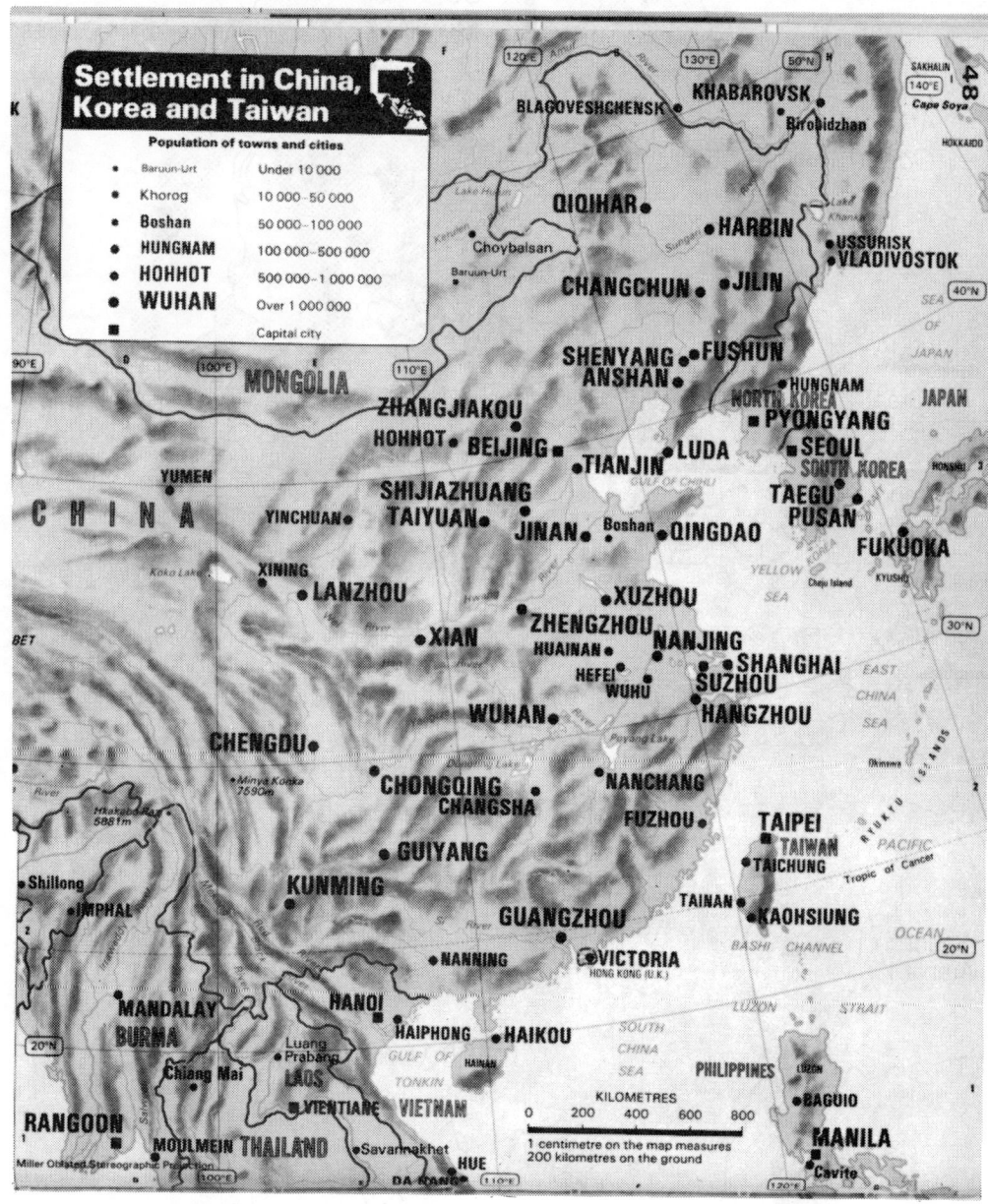

ming and Guiyang are cities located at strategic points in mountainous landscape.

The challenge for geography teachers and students is to link text and graphics efficiently so that the graphic message is received effectively. These aspects will be considered closely in the section on designing effective graphics.

Graphics as Propaganda

Just as bias can be detected in textual statements, it is also possible to detect obvious examples of bias in graphics. Sometimes, this bias is part of a deliberate political propaganda campaign to emphasise a point or to mislead readers (Burnett 1985). Figure 16.2 illustrates how graphics can be used for political

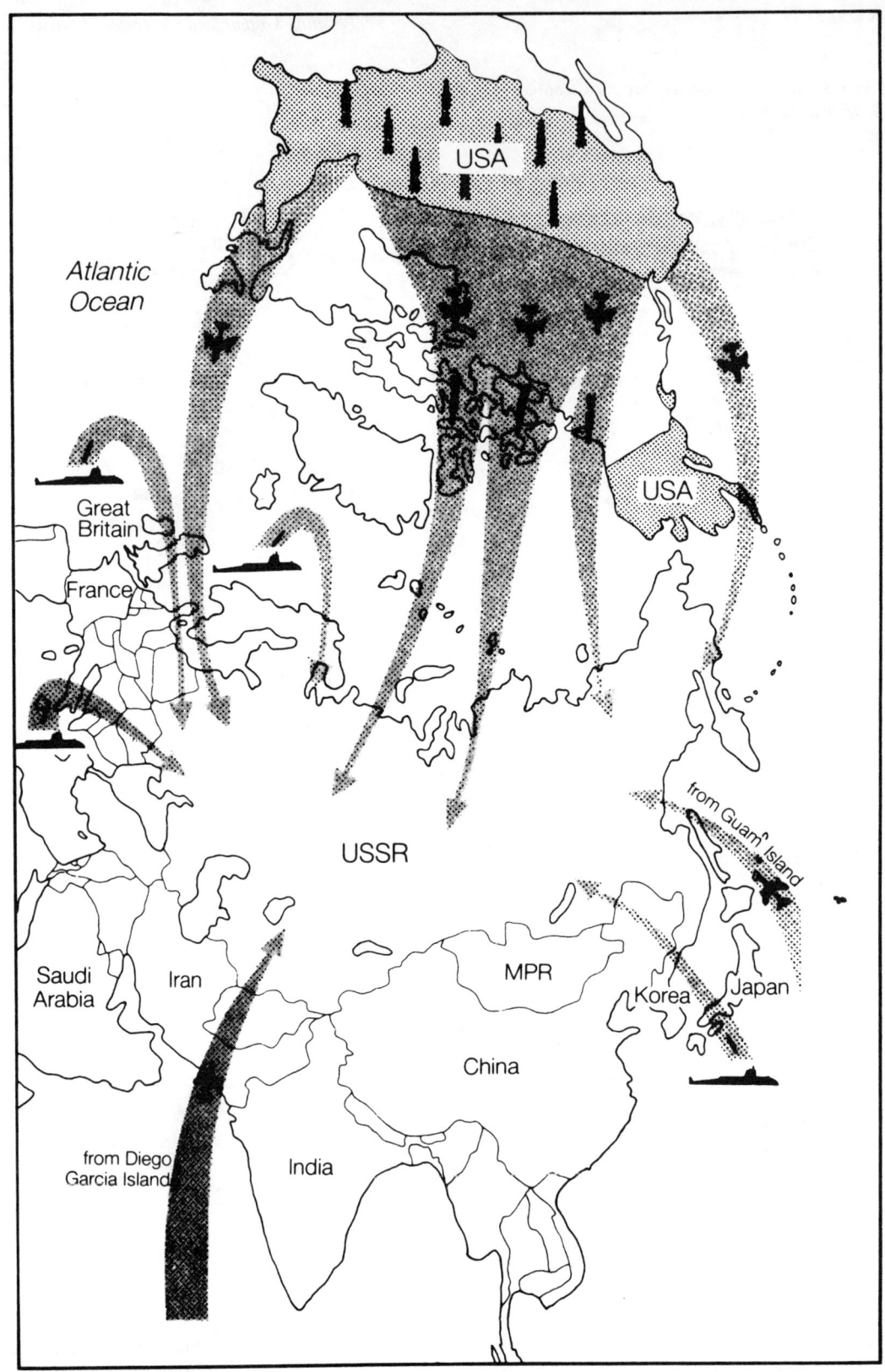

Figure 16.2 The USSR's view of the US military threat
(source: USSR Ministry of Defence 1982)

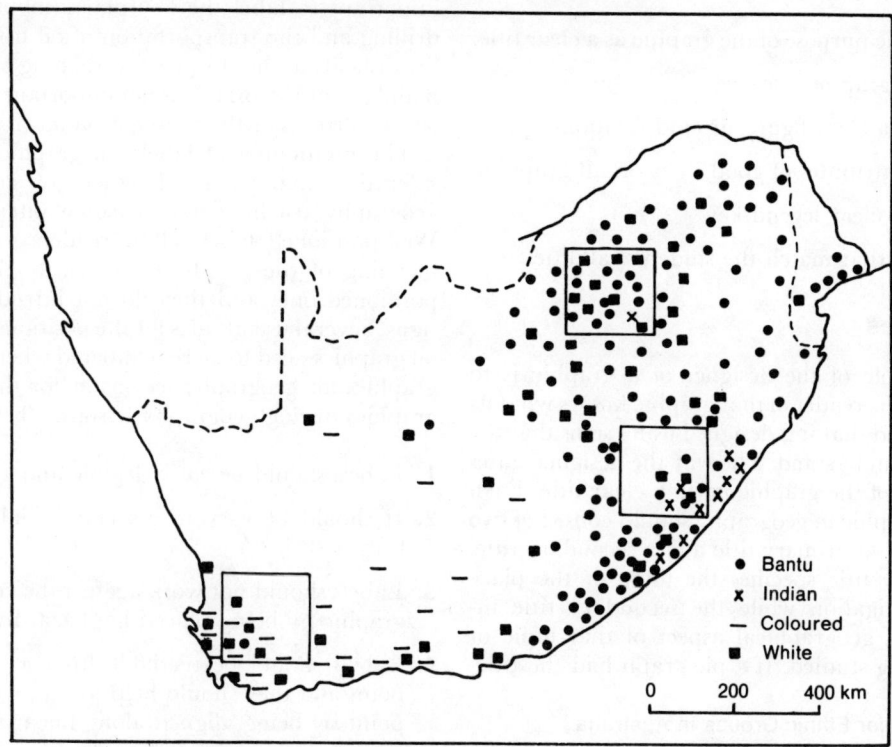

Figure 16.3 Bias in population distribution in South Africa (Wright and Pardey 1982)

purposes. In Figure 16.2 the USSR Ministry of Defence emphasised graphically the threat of the USA's military power to its people through the careful use of a stereographic map projection to reduce the size of the USSR and the use of heavy lines to represent the possible pathways of American weapons.

On other occasions the information on graphics can be biased through the inaccurate recording of the graphic information via the signs used on the graphics (Wright and Pardy 1982). In Figure 16.3 the inaccurate use of dot symbols on a map of South Africa gives a biased view of the numbers of blacks and whites in different areas of the country. Such a problem could easily have been overcome through a more careful selection of the information to be represented on graphics and by a more careful development of the signs for that graphic.

The aspect of bias has been mentioned here to emphasise the need for accuracy when preparing graphics and to alert readers to the fact that some designers of graphics can manipulate them for their own purposes.

Designing Effective Graphics

If geography teachers and students are to maximise graphics in their teaching and learning then they both must know how to design effective graphics. Six basic guidelines have been developed to make this task more efficient (Gerber 1985). They are:

1. Express the purpose of the graphic as a clear title.
2. Label all graphics clearly.
3. Establish a clear figure–ground relationship.
4. Use the attributes of good signs on all graphics.
5. Include a clear legend/key.
6. Use signs that match the students' abilities.

Clear Titles

The main role of the designer of a graphic is to ensure that the reader of that graphic knows what its purpose is and that it is designed using signs that the readers can understand. The way the designer states the purpose of the graphic is via a clear title. Each title for a graphic in geography should consist of two components—a primary title and a secondary title. The primary title specifies the topic or the place under investigation while the secondary title indicates what geographical aspect of that topic or place is being studied. If a pie graph had the title

Major Ethnic Groups in Australia

a reader could determine immediately that the primary focus of the graphic is Australia and that its specific focus or secondary title is major ethnic groups.

Clear, Well-Positioned Labels

Mention has been made earlier of the importance of the text–graphic link. It is important in labelling graphics to consider the range of labels to be placed on the graphic, their positioning on the graphic, their density and the need for any classification or hierarchy of labels. Labels are used on graphics to draw attention to key parts of each graphic. Therefore, it is important that only parts of the graphic which relate closely to its title should be included on the graphic itself. For example, on a graphic show-

ing the retrieval of oil from the North Sea it is important to label the key places involved in the drilling and the transportation of oil from the Bass Strait fields to the storage and refining areas on the mainland of Victoria. It is not important to label the ocean currents or the fishing grounds in Bass Strait.

The positioning of labels on graphics is a consideration that many designers of graphics for geography teaching and learning often overlook. Well-positioned labels allow readers to grasp the meaning of the graphic more easily than poorly positioned ones and they do not intrude on other signs. Several useful rules for the positioning of labels on graphics need to be remembered when preparing graphics for geography lessons or for the design of graphics during geography lessons. They are:

1. Labels should be easily legible and easy to find.
2. It should be easy to associate a label with its feature.
3. Labels should not work against the contents of a graphic by being placed haphazardly.
4. Labels should be readable from a single viewpoint and they should help to arrange this viewpoint by being aligned along linear features, by extending over areal features and by locating point features consistently.

Density of labels is important, for no graphic should contain an overabundance of words. Poorly constructed graphics usually consist of a mass of labels on a coloured or a black-and-white background. Graphics used in geography textbooks are usually reduced photographically. This means that many graphics which contain too many labels are almost unreadable. One important inference from this statement is that the density of labels should be greatly reduced for younger readers, especially those in primary and lower secondary levels. The basis for reducing the number of labels on a graphic should be one of retaining sufficient labels to highlight the key parts of the graphic.

The judicious use of colour and different type sizes and weights will enhance the effective use of labels in the design of geographical graphics by teachers,

Figure 16.4 Type hierarchies for labels on graphics

Population size	Map 1	Map 2	Map 3	Map 4
Under 100 000	• Victoria Downs	• Berlin	• Rainbow	• Annaton
100 000 to 500 000	• **Mt Enid**	• Joanspark	• Georgetown	• **Lindaville**
500 000 to 1 000 000	• **Brisbane**	• Port Alfred	• Julia Creek	• **Bonn**
Over 1 000 000	• **ROME**	• **KIMBERLY**	• FLORENCE	• **LAURA**

students or commercial publishers. For example, the use of blue labels for water features is helpful as is the use of type of different height and weight to show a hierarchy of settlements in a particular area. The hierarchies in Figure 16.4 demonstrate an effective range of type to show settlements of different populations.

Figure–Ground Relationships

All useful graphics should concentrate on the essential information that they contain to communicate their messages. To do this the designer should divide the information into that shown as the figure (i.e. the most easily noticed information) and the background or peripheral information. Since we read graphics as a whole it is important that the information that attracts the reader's eye is the essential information which is expressed in the title of the graphic. Therefore, in all graphics the most visible features must be closely related to the purpose of the graphic. For example, if the graphic focuses on the road network in a place then the figure on that graphic should emphasise the road network. Generally, an effective figure–ground relationship can be developed via the use of rich colours for the figural aspects and pale colours for the background information.

Attributes of Good Signs

The signs used on graphics vary from pictorial diagrams and representations to abstract symbols. Any of these signs can be effective if they adopt the characteristics of good signs. These are:

1. *Simplicity* of design so that readers can acquire their information with little distortion, e.g. blue winding lines to represent rivers.
2. *Elegance* to reflect the structure of signs, their symmetry and their links to the features which they represent, e.g. the symmetry of area symbols that represent forests and swamps.
3. *Ease of production, reproduction, repetition and transmission*, e.g. silhouettes of skiiers to represent the locations of snowfields or simplistic Christmas tree symbols to represent the extent of coniferous forests.
4. Use of *appropriately sized symbols* that occupy a substantial part of the field of attention, e.g. signs for railways should be of a suitable size to keep them in perspective with embankments, cuttings and adjacent roads on large-scale maps.
5. *Clear links between the sign and its referent*, e.g. the sign for a fishing ground could consist of pictorial representations of fish or that for a railway line could resemble a long, narrow, track.
6. *Clear links between the structures of the sign complexes and the structures of their referents*, e.g. linear signs such as highways should consist of narrow lines, point signs such as bridges should represent point-like features and areal signs such as deserts should focus on the areal extent of features.
7. *Easy discrimination amongst signs*, e.g. clear differences between different types of settlements, vegetation groups or commercial buildings.
8. *Suitability of operational synonyms*, e.g. the placement of labels to reflect the extent of a river or the area occupied by a desert.

The challenge is for designers of graphics to use as many of these qualities in the development of comprehensive legends for maps and other graphics.

Clear Legends

The use of clear legends on all graphics which contain encoded signs is fundamental to successful decoding of the meaning of these signs. A general rule to follow is to construct a complete legend before designing the actual graphic. This is an exercise in thinking through a graphic representation before actually drawing it. This enables the designer to visualise the information which will be represented on the graphic. It is also an exercise in thoroughness because all of the signs used on the graphic should be designed according to the above characteristics for good signs. All of the signs necessary for the reader to obtain the meaning of the graphic should be included in any legend for particular graphics.

Signs That Match the Abilities of the Users

It must be remembered that meaning is only given to graphics by the person or people who read each graphic. They can only do this effectively if their understanding of cartographic language matches that used by the designer of each graphic. Therefore, the signs used on graphics should be drawn at a level of abstractness which matches the reading abilities of the users. For younger readers who think in concrete terms the signs should be pictorial and uncomplicated on any graphics they use. Where possible, the challenge of effective signs can be overcome by emphasising making graphics rather than reading graphics. This enables the students to develop signs which they can understand and feel comfortable with. It minimises the risk of students being overcome by extremely abstract signs.

These general aspects give us some insight into the challenges of designing and using graphics for educational purposes. The real challenge for geography teachers is for them to know how and when to use graphics in their teaching and learning situations. The remainder of this chapter focuses on some strategies which can be used to improve graphic communication in geography lessons.

Strategies for Maximising Graphics in Geography Teaching

There are almost as many ways of using graphics in educational settings as there are graphics. The situation is no different for teaching and learning geography. As geography teachers, we are well aware of the need to use maps and graphs in our lessons, but how aware are we of the possibilities of using advertisements, graphic transformations and visual hierarchies in our lessons? This chapter concludes by suggesting a wide range of strategies for extending the use of graphics in geography lessons. The strategies mentioned here are:

1. Graphic transformations
2. Graphic searches
3. Graphics as stimuli for inquiry
4. Graphic codes
5. Photograph interpretation
6. Aesthetic awareness and appreciation
7. Visual hierarchies
8. Advertising
9. Using posters

Figure 16.5 Spending on health, drug consumption and pharmaceuticals in selected Commonwealth countries 1984–1985 (source: *New Internationalist*, November 1986)

Countries	GNP ($)	budget ($ mill)	(a) Pharma-cists	Pharma-acies	(b) Drug Consumption Total ($ mill)	Public ($ mill)	(c) Per cap. $	(d) % met by local prod'n
Low Income								
Bangladesh	140	98.4	197	N/A	102.50	N/A	1.10	90.0
Malawi	210	13.1	26	26	6.00	4.00	0.92	18.0
Uganda	230	11.6	28	28	2.16	2.16	0.16	–
India	260	536.0	180,000	182,000	1,874.00	400.00	2.61	91.5
Tanzania	280	36.8	200	170	40.00	32.00	2.02	10.0
Sri Lanka	320	53.1	460	5,800	20.00	6.00	1.35	10.0
Vanuatu	350	3.9	6	11	0.48	0.48	3.75	–
Ghana	360	117.0	579	N/A	36.00	36.00	3.00	N/A
Gambia	360	5.4	3	72	1.11	0.37	1.58	–
Kenya	390	114.3	403	213	27.00	7.00	1.50	20.0
Sierra Leone	390	11.6	16	88	13.00	1.30	4.06	15.0
Lower Middle Income								
Zambia	640	105.0	135	100	19.00	14.00	3.20	25.0
Solomon Islands	660	2.1	4	9	0.60	0.60	3.00	4.0
Guyana	670	10.7	32	29	6.00	4.00	7.50	40.0
Dominica	710	7.5	29	14	0.62	0.30	6.20	–
Papua New Guinea	820	75.5	33	17	3.50	3.00	1.06	10.0
Zimbabwe	850	118.2	265	96	26.00	N/A	3.50	40.0
Nigeria	860	136.0	2,780	N/A	324.00	N/A	3.56	30.0
Botswana	900	16.6	10	N/A	N/A	N/A	–	–
Belize	953	4.7	3	31	1.70	0.34	10.50	10.0
Upper Middle Income								
Jamaica	1,330	108.5	151	N/A	18.10	7.00	8.22	21.0
Malaysia	1,860	466.4	809	200	103.50	19.00	7.14	20.0
Fiji	1,950	30.3	3	48	5.16	1.72	7.37	–
Malta	3,800	48.0	307	158	9.80	2.50	24.50	5.0
Bahamas	3,830	57.0	46	21	N/A	N/A	N/A	N/A
Cyprus	3,296	39.0	359	251	16.80	5.60	25.85	5.6
Singapore	5,910	181.1	409	133	30.00	6.50	12.00	80.0
Trinidad and Tobago	6,840	235.6	415	411	34.80	34.80	31.63	9.0
Brunei	N/A	35.0	6	4	3.40	3.00	15.80	5.0
Industrialised								
Aotearoa/NZ	6,764	726.3	3,200	1,230	316.00	171.00	96.78	50.0
UK	9,660	27,140	31,955	11,597	3,128.00	2,507.00	55.36	75.0
Australia	11,140	4,184	10,000	5,600	1,052.50	675.00	69.24	70.0
Canada	12,983	4,596	18,356	6,120	1,400.00	700.00	57.00	70.0

The ideas for their use are intended to raise teachers' consciousness of the scope of graphics as valuable teaching resources rather than as fill-ins after the 'main' teaching has been done. Then, the intention is to demonstrate how graphics can promote effective learning in geography lessons. These strategies are illustrative of those which may be used in geography lessons.

Graphic Transformations

It is a favourite strategy of geography teachers to distribute a table of statistics like the one in Figure 16.5 to their students and to ask a variety of questions about the statistics, for example:

- Which countries are low-income areas?
- Which countries have most pharmacists?
- In which countries is drug consumption highest and lowest?
- How many countries produce most of their own drugs?

While such a question and answer session is valuable, it permits the students to make only one form of response — an oral one. A complementary or alternative strategy is to engage the students in making their own graphic transformations of these data to demonstrate their understanding of them. A graphic transformation is the process of converting data in other forms into clear, meaningful graphics. For example, the students using data from Figure 16.5 could construct a series of bar graphs for selected countries. Separate sets of bars could be drawn to show GNP, the number of pharmacists and total drug consumption. Visual comparisons can be made among the sets of bars to determine any relationships between these three sets of data. As well, the bars for drug consumption can be subdivided to indicate how much of this consumption is met from local production or the students can focus on single items in the table and present their own visual representations of each item.

These graphic transformations can be developed from other types of information than statistics. As detailed in their *Skills Book for Secondary Schools*, Butler et al. (1984) demonstrate how to develop graphics from words, numbers and from other graphics such as maps and photographs. Once students get started with these transformations they are limited only by their own creativity. Since these are exercises in graphic design, it is important to encourage students to develop clear graphics which convey a definite message.

Graphic Searches

Numerous existing graphics are available for geography teachers to use in their lessons. One very basic strategy for using the information contained in specific graphics is to encourage students to undertake a graphic search of the information, to make specific observations and to generate further questions which could lead to explanations. This strategy presupposes that students can read graphics efficiently. Here, it should be stressed that geography teachers should make sure that their students know how to examine the title of a graphic to obtain its message before they check for any legend or code of signs and for the meaning of the units along each axis. Then, the students can read off specific information from the graphic.

The graphic in Figure 16.6 focuses on the per capita research and development funding in OECD countries. Steps which geography teachers can undertake to conduct a graphic search are:

1. Ask the students to say or to write down the purpose of the graphic.
2. Have students discuss the meaning of the term 'per capita'.
3. Ask students to identify the four categories of research and development.
4. Ask students to specify the units along each axis.
5. Probe for students' responses to questions such as:

    - Which OECD countries have the highest and which ones have the lowest per capita research and development funding?
    - Where does Australia fit into this picture?
    - Which of these countries spend most research and development money on (a) defence and space, (b) advancement of knowledge and (c) economic development?
    - Does Australia's research and development spending match the OECD pattern?

Graphics as Stimuli for Inquiry

Many geographical studies take the form of inquiries to analyse the issues and to make decisions to resolve

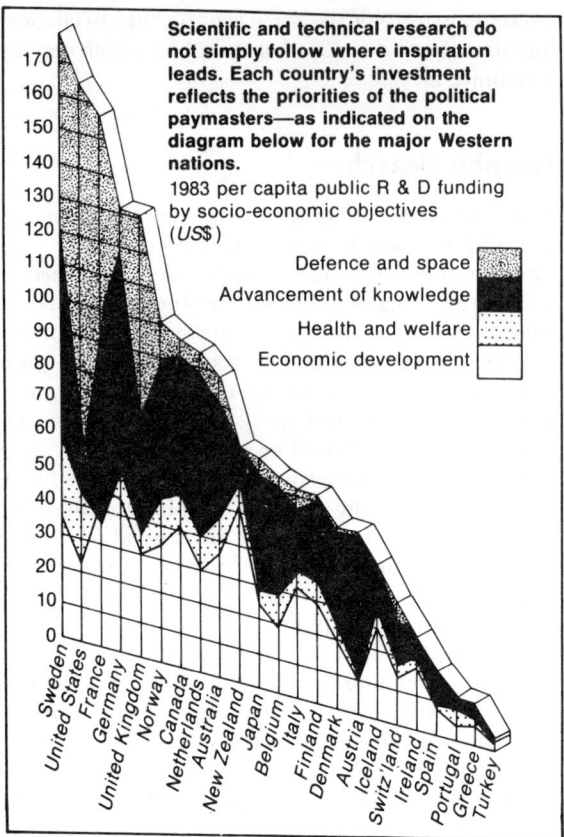

Figure 16.6 Research priorities in OECD countries in 1983 (source: *New Internationalist*, August 1986)

Graphic Codes

Since many graphics contain signs of different types which have different meanings, there is considerable scope for developing strategies associated with the development and use of graphic codes as part of geographical studies. While these signs should possess the qualities of good signs that were detailed earlier in this chapter they will vary in their degree of abstractness depending on who is designing them or for whom they are intended. For younger, more concrete-thinking students, these graphic codes will be rather pictorial whereas for older, more abstract-thinking students the codes will be rather abstract and bear little resemblance to the features represented in the graphic.

Strategies for developing and using graphic codes are evident in Figure 16.7. Here a graphic code has been overlaid on a map of the world. This code shows the average number of children for each woman in the different continents. These children are divided by sex using distinctive pictorial signs. The first strategy that could be used is that of completing a legend for the map. Here, students would reproduce the signs for male and female children in a legend and identify each sign. A second strategy is to interpret the graphic code on the map to analyse its information. One way of doing this is for the teacher to structure several questions which can start small discussion groups probing for understanding of the graphic code. Typical questions could be:

- How do average family sizes vary across the continents?
- What variations are there in the numbers of boys and girls?
- Australia is a part of Oceania. How typical is Oceania of the global pattern?

Once the students have answered these questions they might be ready to search for reasons for these patterns.

Photograph Interpretation

Photographs, along with maps, tend to be the most popular graphics used in geography lessons. Teachers should maximise the qualities of ground, oblique and vertical aerial photographs in making geographical studies. They should do so by using a range of strategies to interpret the patterns, associ-

each issue. The use of appropriate, motivational resources is integral to raising the students' consciousness of the fact that an issue exists and that it could form the basis of an excellent geographical study. Often graphics play a key role as an inquiry starter.

They do so because of their visual appeal. This appeal may not only raise the factual aspects of an issue, but often promotes the feelings of the students towards this issue. A very effective graphic for an inquiry starter is the cartoon, numerous versions of which abound in today's media. One strategy for using such a graphic as an inquiry starter could focus on the geography teacher raising a number of questions and inviting comments about the contents of the cartoon to try to clarify the issue involved. Upon answering these questions, students should then be keen and ready to investigate the issue as an interesting geographical inquiry.

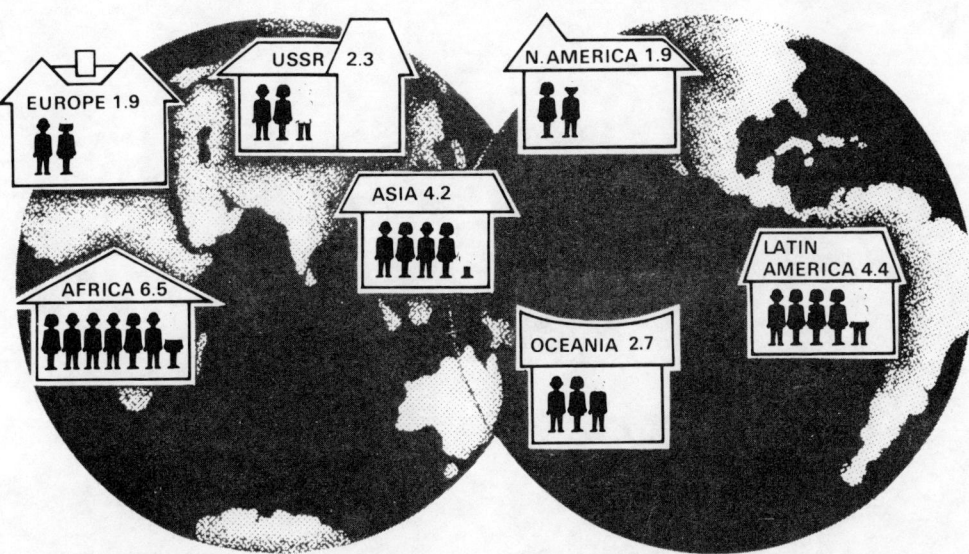

Figure 16.7 Using a graphic code to represent the distribution of children per woman throughout the world (source: *New Internationalist* Third World calendar, 1985)

ations and obvious impacts of people in photographs presented at different scales.

Geography teachers may use photographs in textbooks or from other sources and ask their students numerous interpretive questions and eventually probe for understanding. Beneficial though this approach, is, it does not encourage variety in photograph interpretation. The following is a strategy for enhancing this variety. For oblique aerial photographs such as the one in Figure 16.8 the construction of a photosketch is a useful interpretive activity. The photosketch enables students to identify specific geographical features and clusters of features from the photograph. These features are located on the sketch using a grid such as the one in Figure 16.10 which divides the area into foreground, middle ground and background, as well as subdividing each of these areas into left, right and central areas.

The strategy for constructing such a sketch is:

1. Select an oblique aerial photograph with a variety of information.

2. Decide on the size of the sketch and draw a border for its extent. Encourage students to enlarge or reduce the size of the sketch to suit their needs.

3. Draw the main boundaries or divides on the photograph, e.g. skyline or a river.

4. Draw in the outlines of the main features in the photograph.

5. Add detail to these features.

6. Name all of the main features.

7. Give the sketch a clear title, including direction and a date, if possible.

The result of the students' efforts for Figure 16.8 could look something like the photosketch in Figure 16.9. This sketch can now form the basis of a written or oral experience in which the students interpret the photograph by describing the main features in the scene and attempt to discern any relationships among these features.

Figure 16.8 Aerial view of Sydney looking south across the harbour

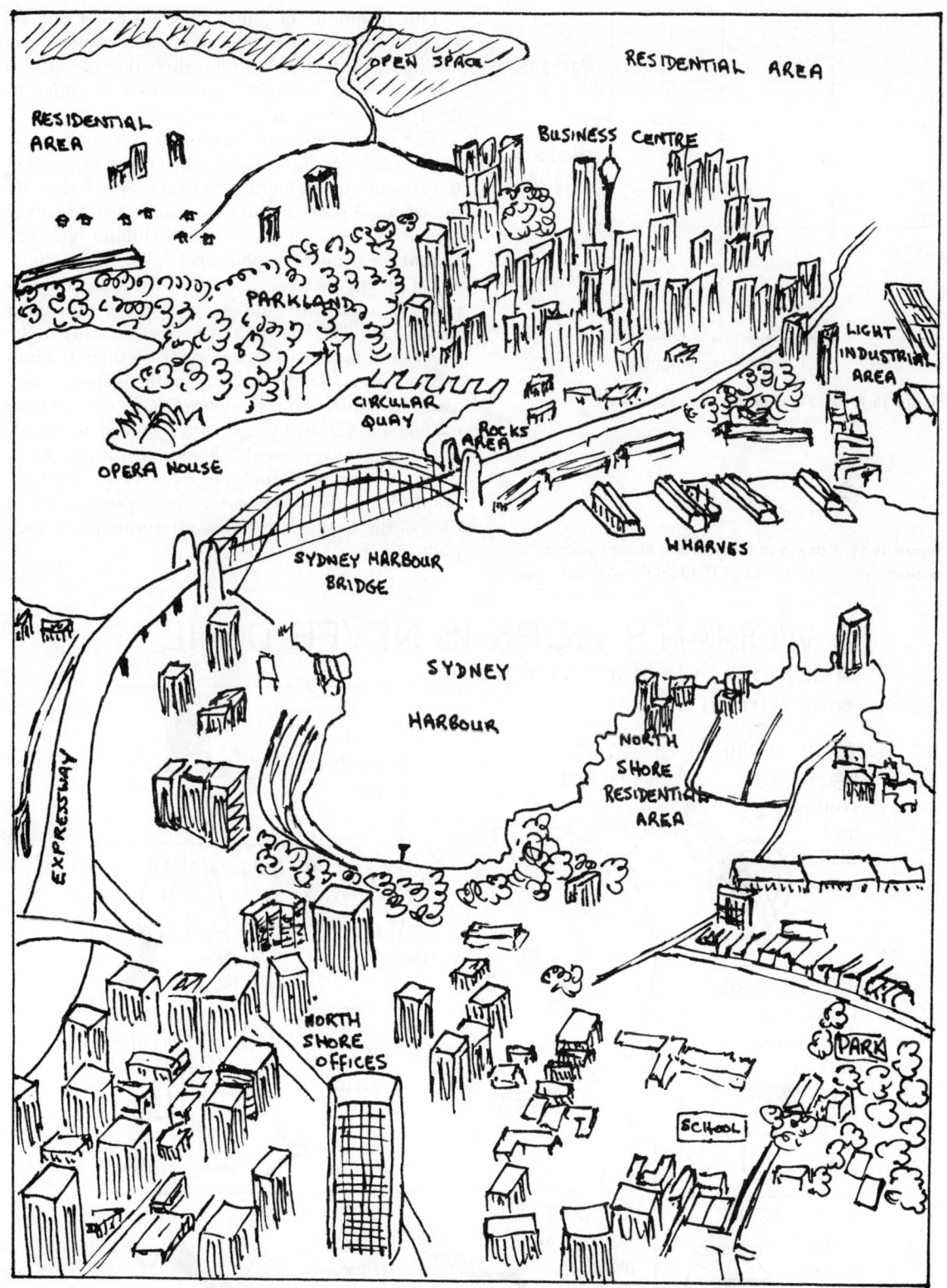

Figure 16.9 Photosketch of Sydney area

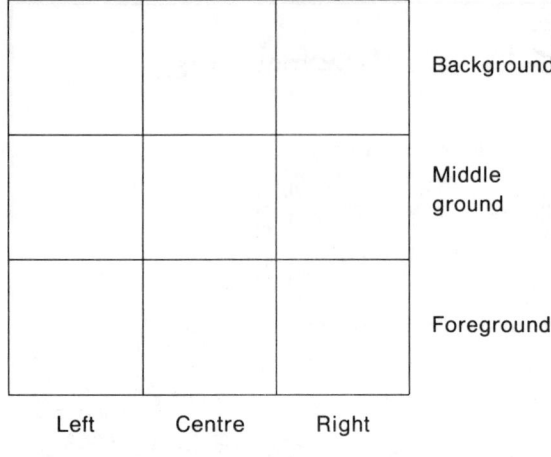

Figure 16.10 Grid for interpreting photographs

This technique of photosketching can easily be adapted for studies in the field when students are taken to a vantage point from where they are able to construct a sketch of the panorama in order to describe the visible pattern of land use.

The comparable strategy with a vertical aerial photograph is to have students construct a tracing of the area in the photograph so that they can simplify the land-use pattern and can analyse the impact of people on the environment. The technique is similar to that for drawing a photosketch from an oblique aerial photograph except that it is possible to interpret elevation only by the use of shadows on a vertical aerial photograph. A major difference in the way that interpretations are made on these tracings is that they make extensive use of the pictorial clues associated with aerial photographs. These clues are colour, shape, size, tone, texture, pattern and shadow. Geography students should be competent in using these clues for analysing vertical aerial photographs. They should be encouraged to verbalise how each of these clues is useful in interpreting any such photographs.

Figure 16.11 A day in the life of rural African women (source: *New Internationalist* Third World calendar, 1985)

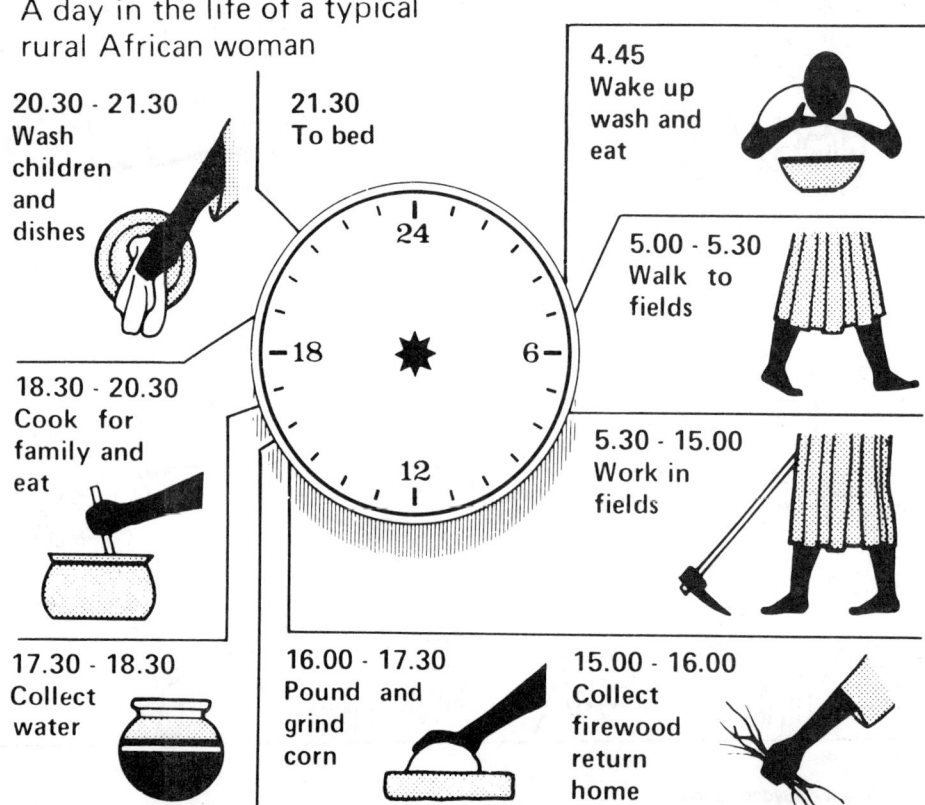

Aesthetic Awareness and Appreciation

Graphics often are able to generate emotive reactions from students as they search for meaning in their context. Since people's feelings and perceptions are now treated as valuable components in geographical studies then teachers should make greater efforts to enhance these affective areas. Clear strategies for using graphics to promote aesthetic awareness and appreciation are detailed in Hall's chapter (see Chapter 14). Use these especially in small groups so that the students can discuss the issue, clarify their own feelings and appreciate what other class members feel about the same issue.

Visual Hierarchies

Graphics which have been well drawn will contain a definite visual hierarchy, i.e. there will be one level of information which attracts the reader's attention and there will be some additional peripheral information. The trick is for the most visible information (the figure) to correspond with the title of the graphic.

Strategies associated with visual hierarchies can be related to students drawing their own graphics or to students interpreting visual hierarchies on prepared graphics. If the students are drawing their own graphics the appropriate strategy is for them to identify their graphic message and then by using the relevant guidelines for designing graphics they should attempt to encode the information on their graphic so that the most visible part of the graphic is the information that is essential to the intended message. For example, if the students are attempting to draw a graphic to represent a farmer's seasonal activities then the most visible part of it should illustrate these activities and link them together in a circular form to show that they are seasonal and that the seasons occur continuously.

For graphics which are going to be used for any kind of interpretation teachers should ensure that each one contains a clear visual hierarchy such as that in Figure 16.11 which deals with the tasks undertaken by many rural African women in developing countries. It is quite noticeable from the graphic that the emphasis is on physical work, much of which could be judged as menial labour. The dark colour on this graphic highlights the part of the body that women would use to do these tasks. This visual clue not only highlights the physical tasks, but it also relates closely to the title of the graphic and so is a very successful one for use in a geography lesson.

The appropriate strategy for interpreting this graphic should focus on maximising the most visible parts of it. Therefore, students could be asked to:

1. Find out the purpose of the graphic from its title.
2. Select those parts of the graphic which attracted their attention immediately they viewed it.
3. Explain how this part of the graphic helps to explain its purpose.
4. Explain how the circular shape of the graphic is important for describing the nature of the work done by many women in developing African countries.

This strategy should be used by the teacher to encourage students to become good judges of effective graphics. They should neglect those which do not have a clear link between the figure and its title.

Advertising

Many forms of media carry forms of advertising to offset their operating costs. These advertisements can be extremely useful graphic presentations which can be used in geographical studies. Like other graphics these advertisements carry clear graphic messages. Advertisements such as the one about the product Milo in Figure 16.12 can be used as important stimuli for a geographical study on diet and food consumption around the world. Other more specific advertisements such as ones for new real estate developments convey clear messages about the environmental qualities of the new subdivision and its links with adjacent places. It would be very useful in a geographical study of residential development in urban areas.

The strategy for dealing with advertisements in geography lessons could focus on the following aspects:

1. the purpose of the advertisement;
2. how the graphic relates to the particular geographical study;
3. the persuasive aspects of the graphic;
4. the impact of the graphic's message;
5. the method by which the designer attempts to convey this message;
6. the accuracy of the information;
7. the proposed audience for the graphic;
8. the expected results of the advertisement.

We finally reveal how you can make your own natural chocolatey Milo.

Step 1. **The Cereals**

Firstly you'll need to get out into the country.

Then you begin the search for the highest quality malt and barley, just the way we do.

Step 2. **The Milk**

Milo includes generous proportions of milk solids.

But you must be careful in selecting only the best full-cream milk, just like we are.

Step 3. **The Sugarcane**

It's best to go to Queensland for this delicious energy producing sweetener.

The sugarcane there is equal to the best in the world. That's why we get ours from there.

Step 4. **Cocoa Beans**

The unique chocolatey taste of Milo is derived from natural cocoa beans. Unfortunately, you'll probably have to go overseas for them.

Step 5. **Soya Beans**

Milo's source of lecithin is the soya bean.

Usually you can get them here at a good produce store.

But sometimes you may have to scout around the world's plantations to find the best.

It happens to us quite often.

Secret Recipe

These and other natural ingredients are the reasons why just three heaped teaspoons of Milo contain a quarter of the daily requirement of Vitamins A, D, B1 and Iron.

For generations we have been saying it's marvellous what a difference Milo makes.

Now you know why.

But, of course, knowing the ingredients is one thing, but nobody has ever cracked our secret recipe.

Build 'em up with Milo.

Figure 16.12 Geography in the Milo advertisement

Figure 16.13 Soils poster

Having analysed a number of such graphics the students could be asked to design their own advertisements to show how well they understand a particular geographical issue. This could well be used as a means of assessment in their geography course.

Using Posters

Somewhat like advertisements, posters have been developed to attract people's attention to an issue. Posters like the one in Figure 16.13 have a strong geographical connotation and should be maximised for visual purposes. The poster is intended for use as a stimulus for people to show greater concern for soil as the main site for the growth of plants. In this poster the relationship between the plant and the soil is quite explicit, as is the inference that human action is often involved in this process. A study on agriculture or forestry management would be enhanced by introducing this topic with this poster.

A most useful strategy with posters is to have students prepare their own posters to demonstrate their understanding of a contemporary issue. This can involve them in designing the contents for the poster, obtaining necessary newspaper clippings on the issue, drawing any special graphics to link the newspaper clippings, giving the poster an attention-grabbing title and demonstrating their ability to relate the content on their poster in a logical manner.

Conclusion

The above examples have illustrated the scope of graphics which can be used in geography lessons. The extent to which they are actually used really depends on the resourcefulness of the geography teacher and the information-gathering skills which the students possess. Associated with this observation is the plea that geography teachers consider carefully the qualities of any graphic for communicating an effective message. Any graphic which is found to be wanting on the qualities which have been enunciated throughout this chapter should not be used in geography lessons. Students should be encouraged to design their own graphics to match their own abilities. As teachers, do encourage them gradually to develop the desirable qualities in the graphics which they design. Also, encourage the students to become discerning evaluators of graphics so that they learn not to accept any graphic on its face value. Rather, the teacher and the students should only select graphics which have a clear message and which convey this message efficiently.

References

Burnett, A. (1985) 'Propaganda Geography', in D. Pepper and A. Jenkins (eds) *The Geography of Peace and War*, Oxford: Blackwell.

Butler, J., Clough, R., Gerber, R., Senior, C., Smith, S. and Wilson, W. (1984) *Skills Book For Secondary Schools*, Jacaranda Atlas Programme, Brisbane: Jacaranda.

Gerber, R. (1985) 'Designing Graphics for Effective Learning', *Geographical Education*, Vol. 5(1), 27–33.

Wright, D. and Pardey, D. (1982) 'Bias in Satistics and Statistical Maps', in A. Kent (ed.) *Bias in Geographical Education*, Department of Geography, University of London Institute of Education.

17

Using Maps Well in the Geography Classroom

Rod Gerber and Peter Wilson

Geography teachers use a large range of materials including maps, graphs, photographs, flow diagrams, paintings and drawings in their lessons. The essence of this chapter is that maps are critical to teaching geography. Rod Gerber and Peter Wilson highlight the fact that maps are often difficult to understand because they are part of a complicated graphic communication process which involves map creators and map users. Until now, suggestions for teaching mapping skills have not differentiated these roles, nor have they fully identified the essential properties of maps, or a sequence by which students can develop an understanding of these properties. This chapter presents such a sequence for understanding the components of a map, arising from research with children conducted by the authors. They suggest a sequence for teaching mapping skills based upon an understanding of four map properties: orthogonal or plan view, arrangement, proportion and map language.

Introduction

The class had been in high school for only two weeks when the geography teacher entered their room and distributed a 1:50 000 topographic map (called an ordinance survey or military survey map in some countries) of the local area to each student. She then proceeded to introduce the class to the intricacies of contour mapping. After 30 minutes, the chalk board was covered with examples of contour lines and patterns of steep slopes, gentle shops, V-shaped valleys, ridges and plateaux. She concluded the lesson by asking whether the students had any questions of their own. She was dismayed by Nicky who nervously raised his hand and asked: 'Please Miss, what are all those wiggly lines like spaghetti on the board all about again?'.

This is a true story and unfortunately we do not think it recounts an uncommon experience. Geography teachers seem imbued with a pioneering zeal to introduce their students to the pleasures and challenges of topographic maps. Few of us regard a book as a suitable text unless it contains some topographic map extracts and exercises. Yet, a topographic map is one of the most difficult, complex abstract types of map available for use by geography teachers with their students. Figure 17.1 shows that this type of map should be the last used in any sequence to introduce maps to geography students. This continuum of map types is discussed later in the chapter.

'If it is not mappable, it is not geography' so the saying goes; yet, mapping is only one form of communication available to geography teachers and

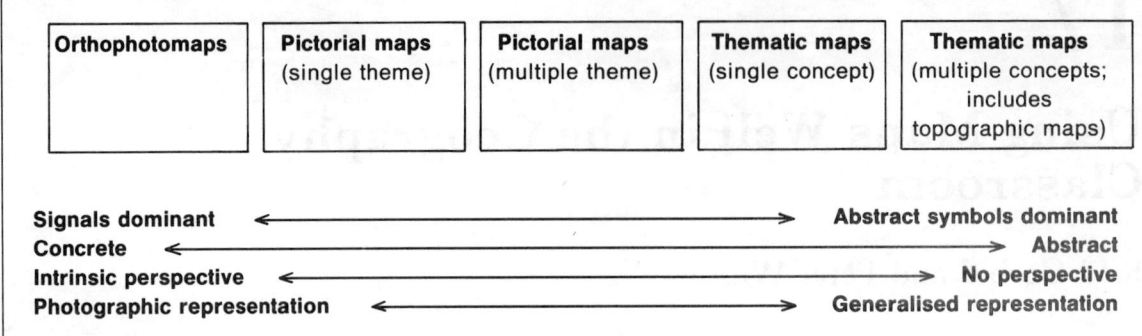

Figure 17.1 Continuum of map types based on the abstractness of their signs

their students. For example, in response to a simple map question such as, 'How can we describe the location of Hong Kong?' students have a variety of modes to communicate their answers. These modes include oral, verbal, numeric and graphic expressions.

Teacher: How can we describe the location of Hong Kong?
Debra: It's near China, Miss.
Teacher: Yes, that's right, Debra. But can anyone be more exact?
Frances: It's off the coast from Canton.
Teacher: You mean Kwangchow, don't you?
Frances: Oh, yes, I forgot.
Teacher: How far is it from Kwangchow?
Frances: I don't know, Miss. I'd need a map to work it out.
Angie: Wouldn't it have been better if we'd looked at a map to start with?
Teacher: You're right, Angie. What we are going to do now is draw a map to show the location of Hong Kong. Sean, will you help me give out these atlases, please?

Balchin (1972) described these responses as four complementary modes of communication in which all students need to develop competence. These are shown in Figure 17.2. Teachers have been bombarded in recent years with the need for literacy and numeracy and with demands that students be able to 'read, (w)rite and do (a)rithmetic'. Geography teachers should take great care that equal stress is placed on the visual and graphic aspects of learning.

This chapter first looks at the map as a form of communication and, building upon this, presents a sequence for introducing maps and mapping skills to students.

Cartographic Communication

Geography students use a wide range of visual materials in order to study the environment including landscape drawings, oblique and vertical aerial photographs, orthophotomaps, satellite images, graphs, plans, diagrams, models and maps. Each of

Graphic communication	Verbal communication
Oral communication	Numerical communication

Figure 17.2 Four complementary modes of communication (after Balchin 1972)

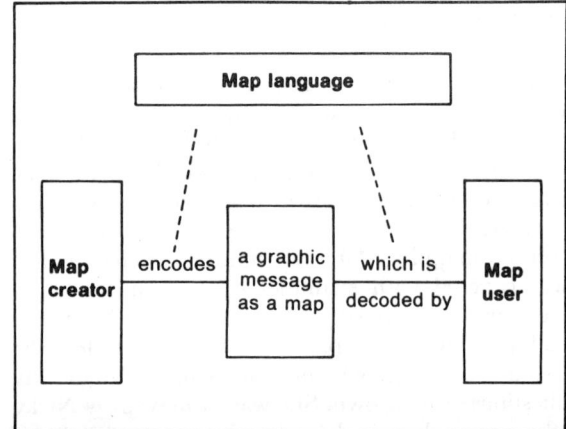

Figure 17.3 A simplified cartographic communicatiion system

these forms of visual materials is based upon a different process of communication, of which the map is the most difficult to understand. Cartographic communication via maps is a very complex task involving many learned skills. Students confronted with a map face a problem similar to being addressed in a foreign language complete with signs, grammar and expressions beyond their understanding. The geography teacher just cannot present a class with a map and expect every student to understand and learn merely by looking at it.

The process of communication via maps involves the student and the teacher of geography in either or both of two roles—the *creator* of maps and the *user* of maps. A simplified version of this process is shown in Figure 17.3. The map creator devises (encodes) a graphic message in the form of a map using a code of signs which are defined usually in a legend. The range of signs used in maps and their degree of abstractness are shown in Figures 17.1 and 17.4.

As map users, students have to decode the message via their understanding of the meaning of the signs in the legend. Great difficulty can occur when a legend is not presented. Consequently, students need to be familiar and competent with the processes of encoding and decoding in order to understand them effectively. This understanding and use of *map language* involves familiarity with all the properties of a map.

The Properties of Maps

The word 'map' is derived from the Latin *mappa* signifying the cloth on which maps were first printed. A map is a complex graphic document which is an abstracted and generalised, scaled representation of part of the curved surface of the earth. As such, a map may be seen to have eight interrelated properties.

Orthogonal or Plan View

A map demands an orthogonal, head-down, aerial view which is foreign to the experiences of most students. Maps are the only form of graphic communication which have no intrinsic perspective. Aerial photographs and orthophotomaps, for example, allow the viewer to appreciate the elevation of buildings and relief features over an area through a language system.

Spatial Relationships

The concept of absolute space means that the spatial relationship among any objects marked on a map is not affected by the position of a map user viewing the objects. Rather, the location of any object on a map is determined by three coordinates via triangulation. Students need to develop an understanding of the spatial relationships of proximity, separation, order, enclosure and continuity to be found on maps. A full description of each of these concepts and an explanation of their importance is outlined in Piaget and Inhelder (1956: 5–9). Some ideas for developing these concepts with students are found in the next section of this chapter.

Proportion

This property involves the two concepts of distance and scale. A map reduces a spatial unit of the earth's surface to a specific size of reproduction which is usually a small sheet of paper. Consequently, detail

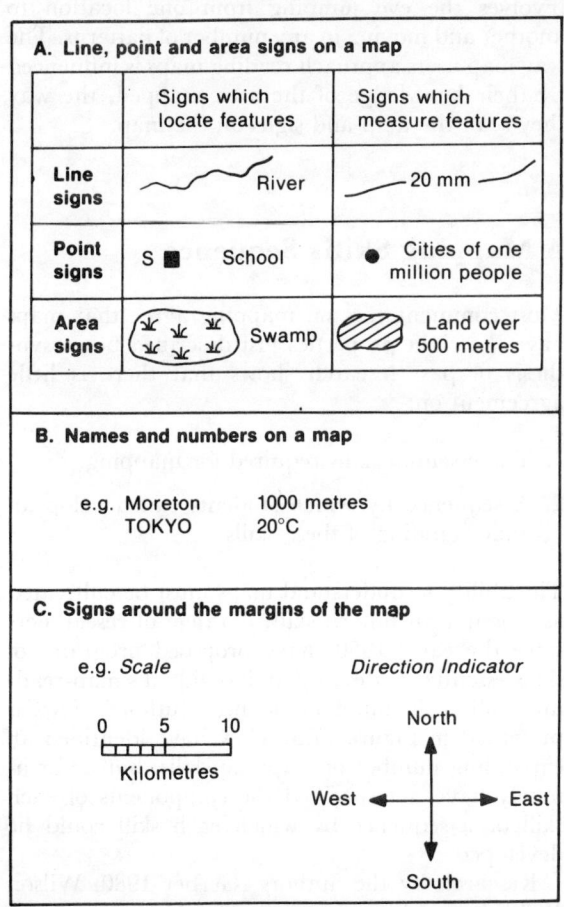

Figure 17.4 The range of signs used on maps

is lost and abstracted. The maintenance of the relative size of objects is very important and is closely linked to the degree of generalisation and abstraction. For example, rivers and coastlines are generalised as thin lines and towns as small dots through the process of reduction to scale.

Generalisation

Maps are not a photographic representation of the earth's surface. It is impossible to show all the features of a particular area on a map so a selection of features based upon the purpose of the map has to be made, which leads to a generalised view of that particular area.

Abstraction

The features selected for display on a map are represented through a system of signs. The nature of a sign system used on a map depends upon the desired degree of resemblance between the map signs and the real features they represent. For example, contour lines, which are highly abstract signs, bear no resemblance to a hilly area, whereas a small pictorial sign of a cow marked on a map does closely resemble cattle.

Isomorphic Properties

Maps always retain a structural relationship similar to the area they represent. For example, rivers, mountains and towns are drawn in the same pattern as they occur on the earth's surface. With this one-to-one correspondence between the signs used and the features they represent, a map is always isomorphic to the area it represents.

Map Language

The sophistication of maps is closely related to the degree of abstractness of the signs drawn on them. The signs used on a map range on a continuum from ones where *signals* are dominant to ones where abstract *symbols* are dominant. A continuum of map types based on the abstractions of their signs is shown in Figure 17.1. Orthophotomaps are examples of maps composed of signs which are signals. The student sees the roof lines of a house and usually there is an immediate behavioural response. Tourist maps with signs such as the silhouette of a skier to indicate the location of a ski resort are another example of maps which use pictorial symbols. Signs on topographic maps must be regarded as abstract symbols. Abstract symbols do not elicit an immediate behavioural response unless some form of recognition and understanding occurs. For example, a red line on a topographic map does not resemble a road. So, students must recognise that the red line signifies a road, and understand the concept of road, before reading or making a map. Also, the signs on a map consist of line, point and area signs which either locate or measure features. The range of signs used on maps is displayed in Figure 17.4.

A Way of Reading

There are dictionary meanings for word symbols, but there are no such definitions for the lines, dots and shadings used in map signs. Students can use the rules of grammar to understand the construction of a sentence, but no such grammar exists for maps. Map reading is made more complex by the individualistic manner in which people scan maps. Map scanning involves the eye jumping from one location to another and moving in any number of patterns. The way map users approach reading maps is influenced by their knowledge of the area mapped, the way they scan the map and signs on the map.

A Mapping Skills Sequence

Most commentators on mapping agree that maps have the eight properties just described but a synthesis of past research shows that there is little agreement on:

1. The essential skills required for mapping.
2. A sequence by which students can develop an understanding of these skills.

The ability to understand maps must be cultivated as it is not an inborn skill. A range of researchers since the early 1950s have proposed programs of skills essential to develop an individual's map-reading ability. A summary of these authors' views is presented in Figure 17.5. They have identified an expanding number of mapping skills, but unfortunately have not specified the components of each skill or a sequence by which each skill could be developed.

Research by the authors (Gerber 1980; Wilson 1980) has indicated that a suitable sequence to introduce the essential skills for mapping may be

Skill	Author									
	Kohn	Thralls	Warman	Sabaroff	Davies	Farrar	Kennammer	Hanna et al.	Stringer	Askov and Kamm
Orientation	*			*	*	*	*	*	*	
Direction	*		*	*	*	*	*	*	*	*
Scale	*		*		*	*	*	*	*	*
Distance	*				*	*	*			
Location	*			*	*	*	*	*	*	*
Relative location	*			*	*	*	*	*		
Distribution					*		*			
Symbolisation	*	*	*	*	*	*	*	*	*	*
Map comparison	*						*	*		
Inferencing	*	*					*	*	*	
Map language		*			*	*			*	
Map drawing		*			*	*		*	*	
Projection			*		*	*	*	*		
Legend			*			*	*			
Map titles			*							
Observe landscape					*		*	*		
Globe as model				*	*			*		

* = listed by author

Figure 17.5 Summary of essential skills needed for map reading (after Wilson 1980)

Figure 17.6 Map of an area in a schoolground

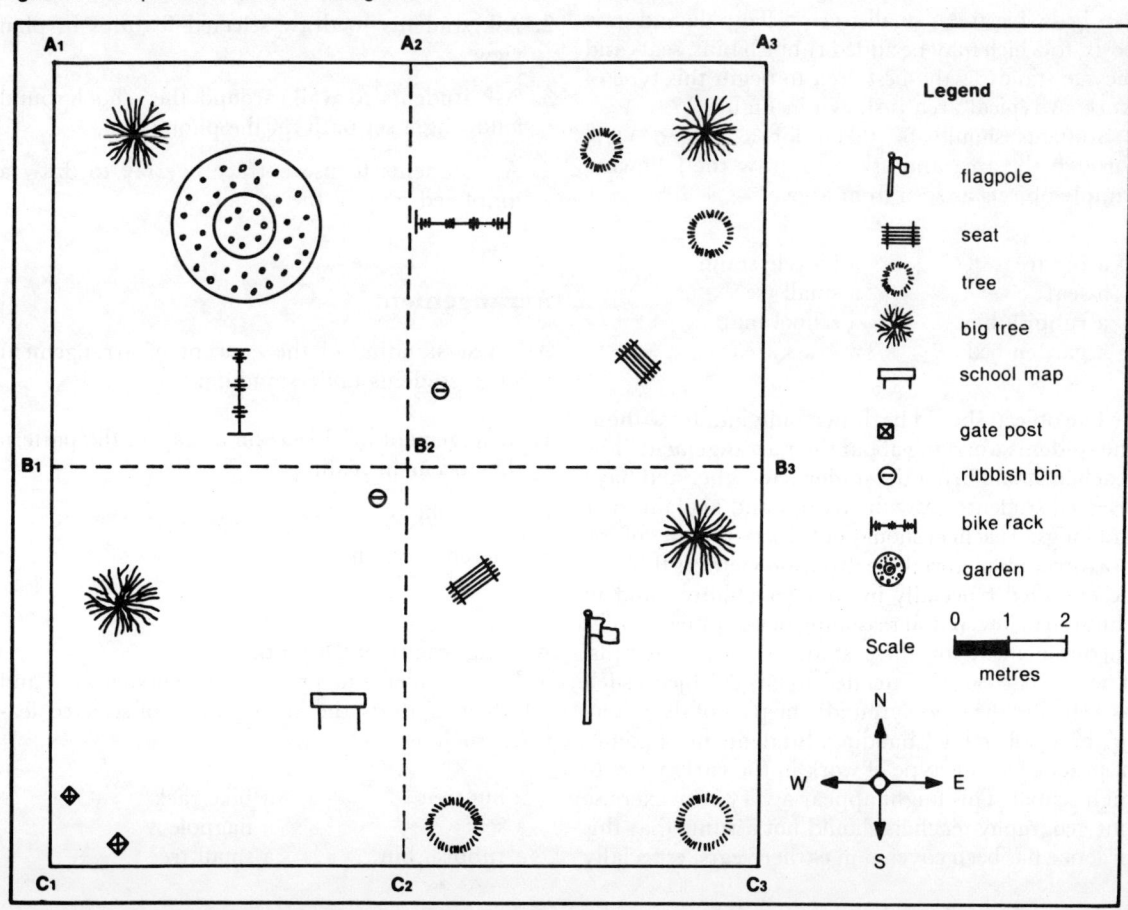

developed using four of the key properties of a map, namely:

1. Plan view
2. Arrangement
3. Proportion
4. Map language

These key properties can be developed in the above order, first individually and then integrated together, to form a logical sequence of cumulative map understanding. The sequence culminates in the students' construction of a map of their classroom (or any other suitable micro-sized area) which, paradoxically, is often the first mapping activity undertaken by students when they enter secondary school.

Plan View

Students should begin mapping by developing the concept of orthogonal or plan view. Any area of the schoolground with distinctive features such as garden beds, big trees, small trees, a flagpole and gate posts, to which may be added rubbish bins, seats and bicycle stands, is the best area to begin this type of work. A typical area is shown in Figure 17.6.

Students should be taken for a guided walk through the area and asked to draw the following sample objects as seen from above:

a big tree	a bicycle stand
a seat	a small tree
a rubbish bin	a school map
a garden bed	a grass area

The objects should be drawn individually without the students worrying about their arrangement. The teacher should bring the students together and have selected students show their work and explain their drawings. Teachers should not worry if many of the drawings are pictorial (i.e. front-on view) as this is to be expected especially in younger children and in children of low spatial reasoning development. Offer encouragement for these students' next attempts. The exercise can be extended to larger objects such as a shed in the schoolground, one part of the school or the whole school building. Students need plenty of practice at this type of work in the early years of high school. This might appear a very easy exercise but geography teachers should not assume that this practice has been covered in earlier years, especially given the generally poor teaching of social studies skills in many primary schools.

Vertical aerial photograıhs and/or orthophotomaps should form an integral part of early work in mapping. A vertical aerial photograph of the schoolgrounds is best for this purpose. Most government departments of mapping can enlarge the schoolgrounds (so that it covers the whole area of the photo) from the vertical aerial photograph of the local area. The local area photograph can be used at a later date with older students as a basis for drawing a map of the area or for interpreting its land use.

A class set of these photographs is useful as it allows students to work with their own photographs. Many schools now have the facilities to produce a class set of photographs at minimal cost by offset printing from one photograph where permission to reproduce it has been obtained.

A vertical aerial photograph of the schoolgrounds and the local area may be used in many ways, for example:

1. Ask students to identify selected features in the field, such as the large tree, the grass area, the back gate.
2. Ask students to draw selected features in plan view.
3. Ask students to walk around the schoolground following a set path on the photo.
4. Ask students to use a trace overlay to draw a simplified map of the school.

Arrangement

An understanding of the concept of arrangement involves students understanding:

1. arrangement of objects on a map in the pattern they occur in reality;
2. direction;
3. reference systems.

Arrangement of Objects

Take the students to a part of the schoolground and ask them to draw the arrangement of selected features, such as

a big tree	a bike rack
a seat	a flagpole
a rubbish bin	a small tree

The students have done a similar exercise on the element of 'plan view' and so will usually ask if the features should be marked that way. Leave this answer up to the students, but stress that they should concentrate more on the arrangement of the objects. Once students have finished, the teacher should walk the class round the objects, pointing out the location of each one, and ask the class to compare their drawings.

Most students will take a long time to complete the above exercise, often painstakingly drawing the plan view of a complex object. Now, have the students concentrate on the same six objects and tell them to:

1. draw the arrangement of the objects;
2. make each object distinct; and
3. complete the task in less than a minute.

This exercise concentrates on arrangement, but also introduces the student to the idea of map language (developed later in this chapter). So, a complex plan view of a bicycle rack may become an abstract X.

The vertical aerial photograph suggested in the section on plan view should be used also for work with arrangement. All the relative aspects of arrangement may be introduced through questions based on the photograph, for example:

1. What is to the left/right of . . .?
2. What is in front of/behind the . . .?
3. What is up/down from the . . .?
4. What is near to/far from the . . .?
5. What is between . . .?

Direction

All maps should have direction indicators, such as a north point or latitude/longitude lines, but few students who leave primary school have properly developed a concept of direction to use it in a functional sense. The relative arrangement terms used above should be fostered at first. Stand in the schoolground and ask such questions as:

1. Where is the rubbish bin?
2. Where is the big tree?
3. Where is the seat from the flagpole?
4. Where is the grass area in relation to the small tree and the rubbish bin?

Repeat these exercises at two different locations to show that relative terms are often not useful. Students will see that the rubbish bin has not moved but that the relative location description has changed. Then, introduce the need for a fixed form of reference via the compass.

Compass work should be an integral part of all mapping programs and needs to be carried out in all years at increasing levels of sophistication. Students must have 'hands on' experiences with compass work, so a class set of simple compasses is essential. The following steps are a guide to the development of a sequence for using the compass. The steps should be introduced slowly for students with little experience, but may be condensed for students who have had previous experience with compass work. The steps are:

1. Use the compass to introduce north. Have the students move around the schoolground with their compass and ask the question. 'What is north of us now?'. Relate north to the sun.

2. Use the compass to introduce the three points of east, south and west. Have the students move around the schoolground with their compass and ask the question, 'What is north/east/south/west of us now?'. Relate the four points to the sun.

3. Introduce the north, east, south, west compass rose and draw the rose on the vertical aerial photograph of the schoolground. Pose questions in class about the direction of one object from another. Repeat this exercise in the schoolground and introduce orientation of the map to north.

4. Introduce the four quadrants NE, SE, SW, NW and repeat exercises in (1), (2) and (3).

5. Introduce the 360° compass and relate this to the 8-point compass. Show the students how to take a bearing then move around the schoolground taking bearings to selected objects.

6. Carry out simple orienteering activities around the schoolground using compass bearings and distances to walk.

7. Use the compass and the vertical aerial photograph of the schoolground and have the

students carry out exercises of bearings from point to point.

8. Set up two markers in the schoolground about 50 metres apart on a north (0°) to south (180°) line. Ask the students to take bearings from the two markers to five selected objects, but only the 0° to 180° arc. In class, have the students use a 180° protractor to draw the rays from each point on a piece of paper. Tracing paper may be used then to plot the feature. (This activity has been carried out by the authors with 10-year-old students.)

9. Repeat the exercise in (8), but take the bearings in the full 360°. This time the students will need a 360° protractor to complete the exercise.

10. Repeat the exercise in (9), but have students stand at any two points. However, the students must record the bearing between the two markers, as well as obtain the bearings from the markers to the selected objects.

Reference System

Reference systems or grids enable the student to locate features accurately on a map. The following exercises should be used before work on street directories, latitude and longitude, and grids on topographic maps are attempted.

Select an area in the schoolground which has a number of clear features.

Take nine ranging poles and arrange them to form a visual grid of four squares. Number the poles A1, A2, A3; B1, B2, B3; and C1, C2, C3. You now have a visual grid about which you can pose a large variety of challenging questions concerning the relative and specific locations of discrete features.

Examples of the types of questions which may be posed by the teacher about an area of a schoolground shown in Figure 17.6 are:

1. In which square is the garden, flagpole, school map?

2. What is the grid reference for the seat, rubbish bin, gate post?

3. Stand at the following points and answer the questions:
   (a) At A1 where is the garden located in relation to other objects?
   (b) At A3 where is the seat located in relation to other objects?
   (c) At B2 where is the rubbish bin located in relation to the big tree?
   (d) At C1 where is the school map located in relation to the seat?
   (e) At C3 where is the flagpole located in relation to the big tree?
   (f) At B3 where is the flagpole located in relation to the small tree?
   (g) At B2 where is the flagpole located in relation to the seat?
   (h) At C2 where is the flagpole located in relation to the small tree?

Proportion

Students can develop an understanding of proportion using four increasingly sophisticated skills:

1. Judging distance.
2. Measuring objects.
3. Drawing objects to scale.
4. Arranging objects according to scale.

Judging Distance

Have students stand outside of their classroom and show them a metre length. Have students estimate the distance in metres between the following sample groups of objects.

1. Objects close together.
   e.g. How far is the big tree from the rubbish bin?
   How far is the seat from the garden bed?

2. Objects which are distant, but visible.
   e.g. How far is your classroom from the school office?
   How far is the toilet block from your classroom?

3. Distant objects which are not visible.
   e.g. How far is your home from the school?
   How far is Sydney from London?

Measuring Objects

Select a number of measurable objects in the schoolground. Arrange the class into small groups of three students. Ask each group of students to measure and to record the length and breadth of the selected objects, e.g. a seat, a bicycle rack, a parade ground, a set of stairs.

Drawing Objects to Scale

Provide students with sheets of graph paper which are marked off in centimetre intervals. Ask the students to draw the objects to scale so that the sizes of the objects drawn on the page are proportionate.

Students must first calculate a scale to 'fit' the objects on the page. This is done by establishing the maximum length or breadth measurements and then dividing this distance by the number of centimetres available on the page of graph paper, e.g. for a maximum length of 20 metres over a graph paper area of 20 cm, the scale is 1 cm to 1 metre. Students should then be asked to name each object represented on their sheets of graph paper.

Arranging Objects According to Scale

The most difficult aspect is to locate each object on a map in a distribution similar to that in the real area. This involves the use of direction, distance and scale in the following sequence:

1. Measure the size of each large object to be mapped.

2. Measure the distance between each pair of objects (from the centre of one object to the centre of another).

3. Use a compass to obtain bearings of each object (read to the centre of each object).

4. Establish a scale to fit the mapped area on a page of graph paper.

5. Plot the location of the centres of objects on the map using the compass bearings.

6. Include the scaled dimensions of the objects represented on the map.

Map Language

The following exercises introduce students to line, point and area signs (see Figure 17.4) as well as to the fact that they can be signals or symbols. Take the students to a selected area in the schoolground (see Figure 17.6) with a pencil, biro and colouring pencils. For each of the three groups of signs, students are asked to draw two types:

1. A simple sign for the object as they see it—a pictorial symbol.

2. A simple sign for the object which does not look like the object—an abstract symbol.

Students should use two columns on a sheet of paper to aid this activity. The three groups of signs are taken separately:

1. Ask the students to draw two line signs for:

road	concrete path
school driveway	creek
fence	powerlines

2. Ask the students to draw two point signs for:

garden bed	small tree
big tree	light
shed	flagpole

3. Ask the students to draw two area signs for:

grass area	oval
parade ground	forest area
garden bed	concrete area

Then, ask the students to devise a legend which contains the following range of signs.

a shrub	a stream	football goal posts
a big tree	an oval	an incinerator
a small tree	a forest	a line scale
a fence	a rubbish bin	a direction indicator

Assess the students' responses in terms of the relationship of the sign to the feature it represents. For example, is it pictorial? Is it abstract? Is it linear, point-like or areal?

The Whole Mapping Experience

Once students have completed a variety of the above individual activities at varying levels of sophistication, they should attempt to construct a whole map of a small area such as their classroom. Geography teachers should not dismiss this task as overly simplistic. The ability to draw a large-scale map of a small area such as a classroom involves all the key properties of a map and often indicates the level of map language of the students.

Students need the following equipment to construct a map of their classroom:

a measuring tape	an eraser
a compass	paper to record informa-
a ruler	tion and to draw the map
coloured pencils	on a backing board on
a sharp lead pencil	which to rest their paper

Students should use this equipment in the following sequence:

1. Measure the length and width of the classroom, using a measuring tape.
2. Measure the distance between large objects.
3. Use a compass to establish where north is. Draw the direction indicator on the floor or ceiling of the classroom.
4. Devise a scale for the map based on the size of the classroom and the size of the piece of paper.
5. Group objects in the classroom into a range of classes of objects.
6. Devise signs to represent each class of objects. This range of signs forms the legend for the map.
7. Draw the margins of the map based on the earlier measurements of the classroom.
8. Superimpose a simple grid over the map area.
9. Insert the key, line scale and direction indicator around the margins of the map.
10. Locate and represent the main features in the classroom on the map.
11. Give the map a clear title which indicates exactly the purpose of the map.

Range of Maps

Maps really consist of a number of layers of information. Normally, these layers involve a layer of base information which provides the structure of the map area together with a number of layers of specific geographic information which constitute the distinctive message on the map. This combination of layers of geographical information can be manipulated to produce a wide range of maps each of which has a distinctive use. Maps in this range may be grouped under a number of types. They may be (1) base maps which provide specific geographical data that is often used as a basis for other mapping; (2) thematic maps which focus on specific aspects of geographical information; (3) shaded or choropleth maps which represent thematic information on a map which is organised into a variety of definable areas; and (4) isoline maps which represent thematic information on a map in a series of isolines (i.e. lines of equal value for specific data such as air pressure or relief). Some examples of the first two types of maps are:

1. Base maps
   Cadastral maps
   Topographic maps
   Orthophotomaps
2. Thematic maps
   Political maps
   Tourist maps
   Geological maps
   Weather maps
   Road maps
   Social indicator maps
   Physical indicator maps
   Orienteering maps

Each of these four types of maps has distinctive qualities which lead it to be widely used for specific purposes, e.g. cadastral maps are used to show land tenure boundaries and orienteering maps are used to conduct the sport of orienteering.

Reading Maps

Each person can have a number of roles when it comes to mapping, e.g. a person can design a map or a person can use or read a map. While the role of a map designer is an important one when it comes to making maps, the role of map user is of paramount importance when it comes to giving the map its meaning for the person who reads the information from the map, who establishes if the map's message is effective. The range of signs which constitutes a map will be useless if the user cannot establish this meaning by decoding the meaning of each of the signs on the map and by injecting meaning into the geography expressed in that map.

The challenge is for the reader of any map to be able to maximise the reading process. This may be done by following a number of simple steps. Each of these steps is applied to the map representing the distribution of television and radio sets in Italy (see Figure 17.7).

1. *Locate the map area.* In Figure 17.7 students will have little difficulty recognising the distinctive shape of the country of Italy. In other cases of larger and smaller scales of maps this may require students to consult another reference, e.g. an atlas.
2. *Establish the cartographic message.* In Figure 17.7, as in all maps, the cartographic message is recorded in the title of the map. If the title contains two

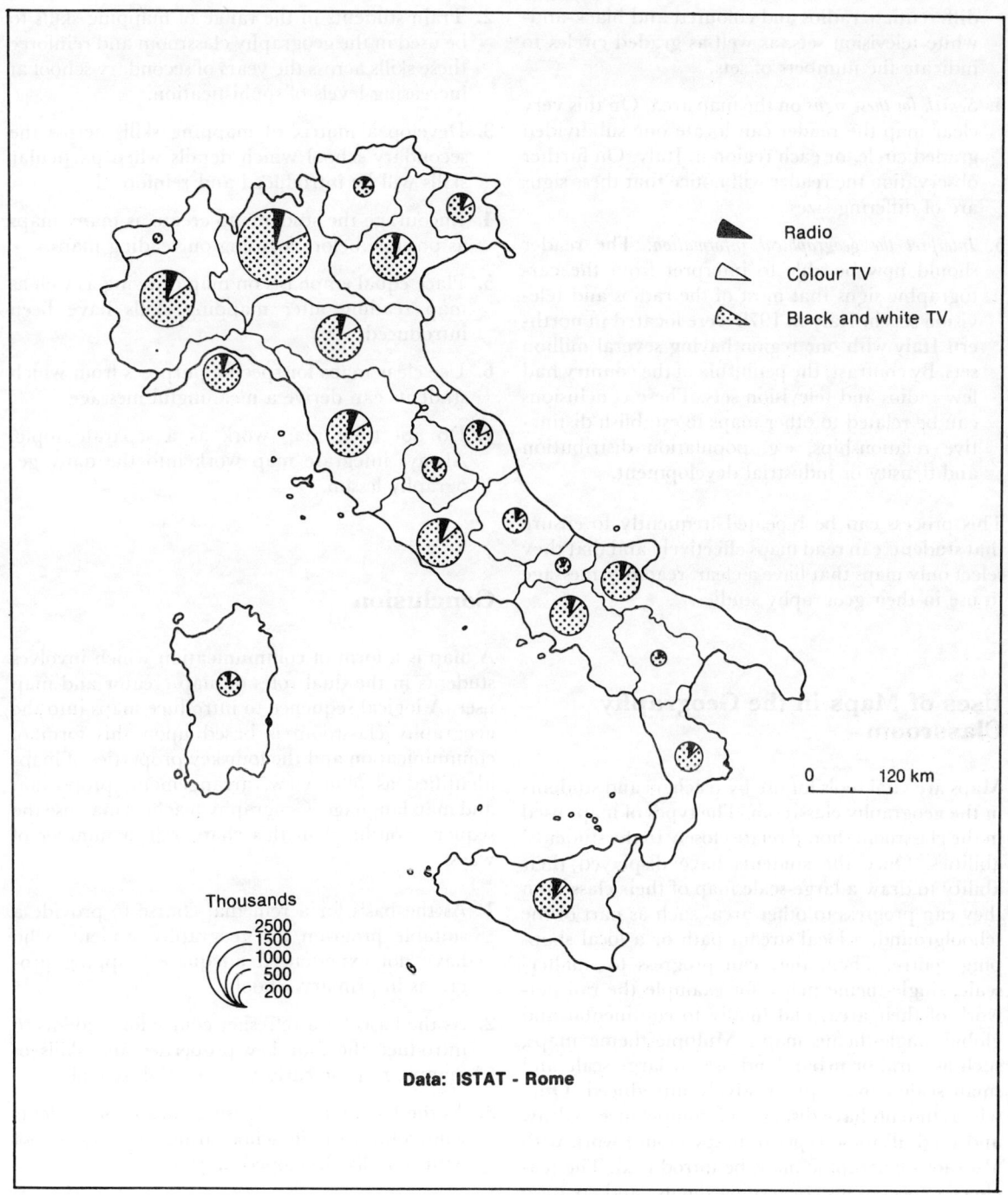

Figure 17.7 Distribution of television and radio sets in Italy in 1978

elements: (a) a clear statement of the area of the map, and (b) a clear statement of the type of geographical information shown on this area, the reader should have little difficulty in finding out the purpose of the map. In Figure 17.7 the purpose of the map is to show the distribution of television and radio sets in the different regions of Italy in 1978.

3. *Work out the code of signs* used on the map. The map of Italy, as a thematic map, contains few different signs. In fact, the signs consist of ones to

differentiate radios and coloured and black-and-white television sets, as well as graded circles to indicate the numbers of sets.

4. *Search for these signs* on the map area. On this very clear map the reader can locate one subdivided graded circle for each region in Italy. On further observation the reader will notice that these signs are of differing sizes.

5. *Interpret the geographical information.* The reader should now be able to interpret from the cartographic signs that most of the radios and television sets in Italy in 1978 were located in northern Italy with one region having several million sets. By contrast, the peninsula of the country had few radios and television sets. These conclusions can be related to other maps to establish distinctive relationships, e.g. population distribution and density or industrial development.

This process can be repeated frequently to ensure that students can read maps effectively and that they select only maps that have a clear, readable message to use in their geography studies.

Uses of Maps in the Geography Classroom

Maps are vital tools for use by teachers and students in the geography classroom. The types of maps used in the classroom should relate closely to the students' abilities. Once the students have displayed their ability to draw a large-scale map of their classroom they can progress to other areas such as part of the schoolground, a local stream path or a local shopping centre. Then, they can progress to smaller-scale, single-theme maps, for example the rail network of their area, and finally to continental and global single-theme maps. Multiple-theme maps, such as rural or urban land use, at large scale and small scale can be progressively introduced. Only when students have displayed a competence to draw and read all these types of maps should work with abstract topographic maps be introduced. The teachers' expectations concerning the use and creation of maps should reflect their understanding of their students' spatial abilities. Therefore, the geography teacher should:

1. Diagnose his or her students' mapping ability before commencing any map work or selecting the maps to be used in geography lessons.

2. Train students in the range of mapping skills to be used in the geography classroom and reinforce these skills across the years of secondary school at increasing levels of sophistication.

3. Develop a matrix of mapping skills across the secondary school which details when particular skills will be introduced and reinforced.

4. Encourage the students to create as many maps as possible before focusing on reading maps.

5. Place equal emphasis on map drawing as well as map reading after mapping skills have been introduced.

6. Use clear maps for specific purposes from which students can derive a meaningful message.

7. Do not treat map work as a separate topic. Always integrate map work into the daily geography lesson.

Conclusion

A map is a form of communication which involves students in the dual roles of map creator and map user. A logical sequence to introduce maps into the geography classroom is based upon this form of communication and the four key properties of maps identified as plan view, arrangement, proportion and map language. Geography teachers may use the sequence outlined in this chapter in a number of ways:

1. As the basis for a remedial course to provide a suitable program for geography students who have not experienced adequate mapping programs in primary school.

2. As the basis for a refresher course for students to introduce the four key properties and skills of mapping in the early years of high school.

3. As the basis for an extension course for students who come to high school from primary school with soundly developed map skills.

References

Balchin, W.G.V. (1972) 'Graphicacy', *Geography*, Vol. 57, 185–95.

Gerber, R.V. (1980) *Development of Competence and*

Performance in Cartographic Language by Children at the Concrete Level of Map-reasoning, PhD thesis, University of Queensland.

Piaget, J. and Inhelder, B. (1956) *The Child's Conception of Space*, London. Routledge and Kegan Paul.

Wilson, P.S. (1980) *The Map-reasoning Development of Pupils in Years Three, Five and Seven as Revealed in Free Recall Sketch Maps*, PhD thesis, Ohio State University.

18

Language in the Geography Classroom

Bryan Stephenson

Language is the medium through which most knowledge is acquired, processed and presented in geography classrooms. This makes the development of the verbal skills of listening, talking, reading and writing as essential a goal as the fostering of geographical knowledge and skills in mapping and fieldwork. The latter have been a particular concern of geography teachers by tradition. The former has been left to the language teacher by tradition. Various language development projects in Britain and Australia have put an end to this, however.

The teaching of any subject is dependent upon students' language skills and these can be developed to their fullest only within an across-the-curriculum policy on language within a school.

Bryan Stephenson suggests strategies through which the speech and writing of teachers can complement various approaches to developing the reading, writing, speaking and listening skills of students in the geography classroom. He draws upon recent research, language project recommendations and transcripts of student talk and writing to illustrate his advice to teachers on the use of language in the geography classroom. This chapter complements Chapter 19 which provides practical advice on improving students' reading and use of textual materials.

Chapters 16 and 17 are concerned with cartographic language as it is used in map and graphicacy skills. Chapter 22 deals with numeric languages in the geography teacher's use of quantitative methods in the classroom. This chapter focuses on verbal language and on the skills of listening, talking, reading and writing. Because much of Chapter 19 centres on students' reading, the primary focus of this chapter will be on ways of facilitating students' listening, speaking and writing skills.

Nicky: What are we supposed to do, then?
Debby: Decide if the film was fair. You know—I mean, did it just show things from one angle?
Nicky: I know that! But how do you do it?
Karen: He said we had to think who else should have been shown talking. You know, the other people, not just the poor people.
Debby: Yeah, like all the people who would get better water because of the new reservoir.
Nicky: Electricity. It was for making power, not just water. It said the towns needed electricity for industry. If they don't have that then they can't have industries. . . jobs and that.
Karen: So they should have shown people who didn't have jobs in the towns. Asked them if they liked the dam.
Debby: All right. That's one lot of people.
Karen: Anybody else?
Nicky: What about the electricity people.
Karen: Yeah, they got criticised, but nobody talked for them, did they?

This short extract from a discussion by a group of students in a Year 9 class about bias in a documentary film represents one kind of learning situation in the classroom. The same points could have been made, perhaps more elaborately, by the teacher. This alternative approach would have provided a different learning experience, however. It would certainly have limited the opportunity for the students to compare their own recollections and perceptions and to evaluate each other's views.

The nature of the learning experiences available to our students is obviously directly affected by the strategies we choose. Since language is the medium through which most learning occurs the choice of teaching methods should reflect its importance. Of course, in addition to verbal language geography teachers use a wide variety of other forms with their students: numeric, graphic and physical forms. In this chapter we consider features of oral language used in the presentation of information, in the teaching of concepts and generalisations, and in the development of reasoning skills and geographical perspectives. Some of the abuses of language in the classroom are also examined. The greater part of the chapter, however, is concerned with the skills of listening, talking, reading and writing. Practical guidelines are suggested for teachers who wish to develop these skills in their students.

The Uses and Abuses of Language

How and why is language used in the geography classroom? One dominant function is in the transmission of information, when the teacher uses skills of oral description and exposition, or when the student acquires knowledge through reading. Another use is in eliciting feedback, so that the teacher can monitor the quality of the understanding being achieved by the student. Often, as in the following example, the teacher will expand upon the student's response, hoping to reinforce the essential idea or fact for the wider classroom audience:

Teacher: Right then, so what have I been saying about the problems of farming in India? We've talked about the physical problems and you read about the human problems. Now, John, give me an example of one of the physical problems.
John: Drought.
Teacher: Yes! That's right. Good. They can't rely on the monsoon, can they?

Another teaching skill lies in the frequent use of clear examples, often in pictorial or cartographic form. Such illustrations enable students to classify and differentiate their own experiences and to encode information. This labelling is effected by means of language and represents an early stage in concept formation.

Teacher: Do you know what we call that? It's called a 'delta'.

Generalisations are formed when concepts are linked to develop principles which can then be applied in deeper understanding and reasoning. Language is the principal means through which such generalised constructs are expressed.

Teacher: So in a case like this, where you have poor methods of production, with large amounts of unskilled labour and little mechanisation, we can say that there is a low level of technology.

Training pupils to think like geographers means that we must train them to use appropriate vocabulary.

Teacher: Right. Two ways by which the water got back to the sea. Can you remember the two distinct ways by which the water got back into the sea once it's actually reached the ground? What's the first one?
Student: Ground flow.
Teacher: Ground flow and. . .
Student: By streams.
Teacher: What do you call that?
Student: Run-off.
Teacher: Good. What's that nice long word? That beautiful word? I also want you to give me what it means, that one long word.
Student: Evapotranspiration.
Teacher: What is it?
Student: It is when the sun draws moisture from plants and shrubs and trees.

The correct use of technical vocabulary and concepts is one means of demonstrating geographical learning. Another is the use of such concepts in solving geographical problems. The more advanced levels of reasoning needed to deal with such problems seem to depend upon the ability to formulate hypotheses. We use language to set out the possibilities, to reflect upon and reshape them.

In the following example, two 15 year olds are discussing a hydrograph. Both of them are prepared to speculate in their search for explanation and both understand the value of using appropriate technical terms.

Leo: I think—um—it's to do with the amount of rainfall. You know, if it rains more the river will have to carry more. There'll be more volume.
Barrie: Yeah, but then it's—it would always be the same. If you get the same, roughly the same, rain then the volume isn't the same in the river. I mean, look at the graph.
Leo: Well, perhaps it's how long it's been raining then. So—yeah—I know, so if the ground is very wet. . .
Barrie: Saturated.
Leo: So, if it's saturated, then, er, you know the run-off's bigger and you get more volume in the river.

The geography teacher presents a particular view of the world and its problems through his or her subject on the assumption that this subject perspective will be helpful to students. The teaching of geography can, therefore, be seen to provide an initiation with this perspective. The intention is that the students will be able to adopt this perspective in order to make better sense of their own experiences and perceptions. The danger is that the initiation can become a ritual, in which the correct use of academic geographical terms is more important than the essential process. Language can then become an impediment, closing down the access to understanding.

Teacher: OK! Now, let's settle down, fourth year. Now, remember last lesson we talked about central places and I told you about hierarchies. Well, if you look at that table on the worksheet. . .the one headed 'Range of functions'. . .what can you tell me about the differences between those settlements in terms of their functions? Come on! You can see that they're all ranked according to size, so what's the difference in the range of functions?

Sooner or later the teacher recognises that there are obstructions to learning and may make an accurate diagnosis and modify his or her own behaviour accordingly.

Teacher: We've already talked about the CBD haven't we? And whereabouts. Describe the location of the CBD, Janet.
Janet: I don't know what you mean.
Teacher: Well, you see those black blobs on the pink sheet there.
Janet: Yes.
Teacher: Whereabouts would you say that they were with regard to the rest of Stevenage when you look at the map? Whereabouts are they on the sheet?
Janet: In the centre.
Teacher: They're approximately—close to the centre. Fine. OK? That's what I wanted you to say.

To a greater or lesser degree all classroom strategies influence the language behaviour and therefore the nature of the learning achieved by the students. Some techniques have a very direct influence, as in the case of dictation. Others do so implicitly, by laying emphasis on the use of language which obscures information. Most of all, perhaps, the approach used by the teacher can either encourage or discourage the students in the development of their thinking by determining the opportunities available to practise cognitive skills through the constructive use of language.

Thus, our finest intentions are that our pupils should be enabled to address themselves thoughtfully to the problem, carrying out an internal review of information previously assimilated and making a reasoned selection of that which is relevant. This would then be articulated coherently. Instead, all too easily we become involved in guessing games, adopting strategies which are optimistically described as 'class discussion', when the students search for clues to the answer we want. Once that is achieved the lesson must roll on, the pace sustained by the demands of the syllabus content rather than determined by the competence of the learners.

Teacher: What is the shape that the rivers form?
Mike: A sort of 'V'.
Teacher: Yes. What do we call it? A 'V' shaped what?
Tony: A vapour.
Teacher: No.
Mike: A 'V'.
Teacher: You're right. A 'V'-shaped what? What would you call it? You would go down into what?
Nigel: A valley.
Teacher: A valley. That's right. Now how did the valley form? Joanne, how do you think the valley formed?

Some of these brief examples may seem to illustrate minor points. Cumulatively, however, they represent the kind of influence exerted through language on the nature of classroom learning. Thus, in its more negative effects, language can be used in a way which:

1. gives the subject matter (geography) an academic mystique, distinguishing strongly between 'school knowledge' and the everyday knowledge with which the learner deals outside the classroom;

2. endorses the knowledgeable authority of the teacher, who derives power from the possession of the special language code;

3. sets the learner in a position of inferiority, ignorant of the code and dependent upon the teacher as translator.

Specialist words are useful in making possible high levels of abstraction. But it is very important that students are encouraged first to deal with information in their own words, in order to 'make that knowledge their own' *(Learning through Talking 11–16* 1979). Otherwise, learning difficulties can be enlarged rather than eased.

Far more prevalent and pervasive, however, are those unintended and hidden influences which deny the importance of language in the learning process. These influences stem from the pedagogical view which associates language primarily with the product of learning, the oral answer or the written examination script. The work of educational psychologists such as Vygotsky and Bruner emphasises the central role of language as an instrument of thought.

The act of verbalising involves the sifting through of one's existing knowledge. At times it is impossible to distinguish the point at which information, brought to the surface as talk, becomes conscious understanding. When new information is made available it is accommodated by the reformulation of existing mental constructs which need to be tried out in a further verbalisation. This verbalising takes place through student–student and teacher–student interactions. These allow the interplay of words, ideas and generalisations as the students develop geographic understanding. It is necessary for teachers to recognise the essential importance of this and to distinguish between classroom approaches and strategies which facilitate this interplay and those which do not.

A second fundamental issue concerns the quality of the geographical thinking which is being promoted in the classroom. To begin to make some sense of our complex world, students need to exercise a range of cognitive skills, extending from the ability to understand information to the capacity to evaluate it. New forms of assessment now lay explicit emphasis upon the use of these skills, as well as on the ability to carry out investigations (Department of Education and Science 1986). Such inquiries require the student to formulate hypotheses, determine methods of data collection and evaluate results.

If at the age of fifteen or sixteen, our students are to demonstrate competence in the use of such cognitive skills then their preparation must of necessity provide adequate practice. The role of the teacher is one of the facilitator of inquiry, in which the students are encouraged to ask the questions, determine the means of finding answers and assessing the validity of the information available. This represents an active role for the student in generating and discussing ideas. It suggests also that the teacher will be concerned to prompt students into regarding issues as problematic, rather than offering definitive solutions.

This shift of emphasis to processes and skills which involve active student participation can be recognised in other changes in the aims of geographical education. Thus syllabus aims which propose, for example, that students should 'develop a sensitive awareness of the environment' or appreciate 'the significance of the attitudes and values of those who make decisions about the management of the environment' (Department of Education and Science 1986) suggest that students will be able to demonstrate and communicate an intelligence of feeling at the personal level and be capable of identifying and analysing the complex factors which underlie human motivation and behaviour. If such skills and levels of awareness are accepted as realistic and desirable outcomes of school geography, it seems necessary to ask what development program would make them attainable by the majority of students.

Are there basic skills which, practised through the years of schooling, will enable the students to speculate, to evaluate or to empathise? Whatever non-verbal skills are used in the geography curriculum important abilities such as these will depend upon the competent use of language.

Developing Student Language Skills

The first part of this chapter has concentrated on aspects of oral classroom language and on the implications of these language experiences for the learning opportunities of the students. In the balance of the chapter these implications are discussed more fully, with a consideration of the skills of listening, talking, reading and writing. Teaching attitudes and methods which affect the development of these basic skills are analysed and suggestions made for ways of improving language experiences in the geography classroom.

Students' Listening

Generally, teachers talk too much. In many cases they talk too much for the good of most of their students. In a study by Wilt, referred to in the report of the Schools Council Oracy Project (Wilkinson et al. 1974), primary school children were found to spend 57.7% of the school day listening, but most studies show that people are poor listeners.

To assist students who are subjected to great quantities of words teachers can use changes in intonation, volume, pitch and other means of emphasis. In this way critical points in a lesson can be signalled, facilitating the operation of selective hearing strategies. If students are to improve the quality of their listening to the talk of both teachers and peers, then it will be a part of their general language development. The authors of the Oracy Project report, having reviewed research on techniques to improve listening, argue that teachers should attend to the social context in which their subject is taught. Good classroom management, with an interesting presentation of material and pleasant interpersonal relationships, leads to an improvement in the range of language functions. This improvement will include an enhancement in the quality of listening.

An example of a strategy which can be used with younger pupils is based on pairs of students working together. For example, one student may be given a photograph which he or she then describes to the partner. When the oral description is completed the partner writes down, or repeats orally, the key features in the description. This exercise, which also promotes accuracy of observation and description, has many variants. It can, for example, be used to demonstrate how perceptions can distort information (e.g. as in the game of Chinese Whispers). With groups of pupils engaged in a problem-solving exercise or discussion one role can be that of rapporteur, responsible for a record of the various points made in the discussion. To avoid its becoming too onerous the role can be switched around the group. This method also has subsidiary objectives, not least that of encouraging some order in the discussions. This strategy can have wider applications, for example in the use of role-play exercises when a range of arguments is developed, leading to a summary of the evidence. Students can also appreciate the importance of accurate listening when conducting interviews as part of a field investigation. At an early age they can be encouraged, when listening to peers, to use reassuring behaviour such as eye contact and supportive body language.

Students' Talking

Classroom language interactions have been likened to a game in which there are four basic moves (Bellack et al. 1966):

1. the *structuring* move which establishes the context and purpose of talking;
2. the *soliciting* move through which responses are elicited;
3. the *responding* move which is an attempt to meet the expectations of the soliciting move; and
4. the *reacting* move which modifies or evaluates what has been said previously.

The initiative in the language game rests with the teacher in most geography classrooms and for most of the time. To illustrate:

Teacher (structuring): Today we're going to study coal mining.
Teacher (soliciting): Can anyone tell me why coal mining has been in the news recently?
Richard (responding): There's an argument about a new coal mine somewhere.
Teacher (reacting): Good. That's right. There's a big argument about whether a new coalfield should be opened up in Leicestershire.

The model is not of universal application. Teaching styles vary according to the pedagogic stance of the individual, personality factors and the circumstances in which a teacher is operating at a given time. Often, however, responding to the soliciting move is the only talk which teachers formally approve or organise for their students in class time. The lesson transcript in Figure 18.1 is typical of this situation.

This transcript has many of the regular features of strategies by which students are rehearsed in their recently acquired knowledge.

Stage 1:
Teacher →transmits information →Student receives

Stage 2:
Students →relay back the same →Teacher evaluates
    information

Curriculum context: High-ability 14 year olds studying the economic geography of British Columbia. This is the sequel to an earlier lesson and homework.		
Teacher:	Right . . . sawmills . . . Em . . . so that's a bit of a recap there. Now the homework you did on . . . em . . . on what area was it you were writing about? What was it you were writing about . . . ?	• Structuring move: switch of focus. • Soliciting move (closed recall question).
Paula:	Kitimat development.	• Responding (factual recall answer).
Teacher:	Right, Kitimat and . . . what was at Kitimat?	• Reacting, accepting the answer and eliciting more information.
Tony:	Dam, reservoir.	• Several students supply the wrong answer.
Teacher:	Not at Kitimat.	• Reacting, rejecting.
Nigel:	Aluminium.	• Responding.
Teacher:	Right, aluminium works. Where was the reservoir?	• Accepts, amplifies and returns to correct the earlier answer.
Nigel:	Hills.	• Responding.
Teacher:	Right. Reservoir in mountains (mumble, mumble). What was the name of the dam, can you remember?	• Reacting. • Soliciting.
Kevin:	Kenney Dam.	• Responding.
Teacher:	Yes, Kenney Dam. Now what was . . . the name for the em . . . of the raw materials? Now, this is where some people got muddled up. The raw material that is used in aluminium . . . Come on some more hands up. Jackie . . . Don't know, right, Colin?	• Teacher intent on rectifying a factual inaccuracy. • Social control.

Figure 18.1 Lesson transcript: teacher structures, solicits and reacts—students respond

The feedback process is a weak one, with 30 individuals 'competing' for a place. The interaction is dominated by closed factual recall questions; the students make short responses; the teacher takes a disproportionate share of the utterances and asks a large number of questions during the lesson.

Talking in Groups

One way of increasing the variety of language 'moves' by students is to work with students in smaller groups. Compare the varied talking experiences of the students in the lesson transcript in Figure 18.2 with the limited ones in the transcript in Figure 18.1.

Group work may be organised by teachers for many purposes during geography lessons. Chapter 24 provides many examples for using small-group work to individualise learning and practical advice for ensuring its success. The emphasis in this chapter's focus on group work is on the way language is used by students in groups, and how this can be monitored by teachers.

Providing small-group opportunities for students *to use talk to learn* necessitates the consideration of several practical questions:

1. How is the task to be presented?
2. Should it be closely structured with tight guidelines, or can it be more open?
3. What knowledge and skills can the student bring to the task?
4. Will inter-student relationships obstruct or facilitate cooperative learning?
5. How are student motivation and task focus to be maintained?
6. How can group conversation and the learning that occurs be monitored and evaluated?

The last two points constitute the major reservations held by some geography teachers for the use of

> **Curriculum context:** Three 14-year-old boys are presented with photographs of irrigation methods; they are given no preparation for this exercise, but are asked to 'describe, explain and compare'. This is a short extract from the middle of their discussion.

Richard:	So we've come to the conclusion that the ox is much better than the diesel pump, the shaduf and the Archimedes screw.	• Structuring move, summarising the thinking so far.
Colin:	I think so. Some people think different. The diesel pump is quick—I must agree there.	• Accepting, but reviving an earlier point.
Richard:	It can pump water further probably than oxen could.	• Reacting by introducing new idea.
Colin:	Yes.	• Accepts
Mike:	No, all the things that oxen do are lift water from one level to another and then it runs of its own free will so if you want to get it up to the top of a hill you can't do it with oxen.	• Responds to first statement, diverging and broadening the consideration.
Richard:	You can with a diesel pump, so a diesel pump is better on that score so if you had a diesel pump and an ox pumping the water.	• Reinforces and amplifies.
Mike:	A diesel would do more than the oxen would.	• Summarises.
Colin:	But the diesel also needs new bits and in a primitive society like that you could be miles away from any.	• Rejects and expands the argument, introducing three new ideas—maintenance, level of technology, inaccessibility.
Mike:	Yes, but you could have the ox as a back-up, something like that.	• Modifying to accommodate new understanding of possibly restating the previous point by student 1.
Richard:	Yes.	• Accepts.
Colin:	How could they afford to look after it while you're not using it?	• Challenging.
Richard:	Yes, but you could still be using it for lifting one thing to a level, all you need the diesel pump is for pumping up hill because the ox can't do that.	• Stressing the technical problem of raising water.
Colin:	It's not very hilly there.	• Challenging.
Richard:	How do you know?	• Challenging.
Colin:	I know but they have the water.	• Attempts to provide an alternative response.
Mike:	If you've got to pump up a hill how're you going to do it? You can't.	• Asserts the need for the diesel pump.
Colin:	You don't need to though because you're near the river.	• Uses available information to strengthen argument.
Mike:	It doesn't matter. The village might be situated near but the fields might be a mile away.	• Uses available information to strengthen argument. • Draws upon previous understanding.
Colin:	Normally the village is situated around the crops.	• Countering move.
Richard:	The diesel pump and the oxen are the best ones. You can't really define the difference.	• Structuring or conciliatory move which acknowledges the difficulty of establishing relative merits.

Figure 18.2 Lesson transcript: group work—students structure, solicit, respond and evaluate

student talk. What serious misunderstandings will be perpetuated? Is not the consideration of the problem likely to be superficial? The role of the teacher during such episodes needs to be changed as control over knowledge is transferred to the pupils.

This transfer of control over knowledge is the essential purpose of the strategy. The teacher, having access to the group discussions, is yet an interloper and should restrain the impulse to intervene impatiently whenever the pupils appear to be making slow progress. What we require, as teachers, is a record of the discussion for later analysis and follow up. In addition to the keeping of a record of the discussion, the value of group talk can be increased by students writing summaries of their deliberations. The exchange of reasoning and argument between groups is helped where summaries are presented on the overhead projector for class discussion. Groups using tape recorders can report directly to the teacher.

The recording of complete discussions provides the teacher with insights into what and how students have been thinking. The artificiality of such situations is lessened where tape recorders are used regularly for such purposes. Some teachers have extended the use of tape recorders to allow feedback and evaluation of lessons by students. They are able to communicate their own learning difficulties in this way, treating the teacher as a sympathetic and trusted adult.

The following questions have been suggested by Barnes and Todd (1977) as starting points for listening to recordings of group discussions:

1. What are the children doing in their talk? What are they using talk for?
2. What signs are there of learning going on?
3. Where does the presentation of well-formed ideas occur and where is the talk an exploration of our thinking?
4. Are the pupils working together in constructing lines of thought?
5. What devices do the pupils use to regulate their social relationships?
6. What seems to influence the relative success or failure of these discussions in promoting useful learning?

Questioning Strategies

Teachers can ensure variety in student speaking functions, even when working with whole classes rather than small groups, if attention is paid to questioning strategies. Questioning is a principal means of managing a learning environment and of controlling the level and quality of student talk in a learning situation. Important elements in questioning strategies include:

1. *Frequency:* Quick-fire questions give an impression of a lively interchange of information, sustaining a brisk momentum. In some classrooms a rate of one new question every 12 seconds has been observed. But, this is characteristic of low levels of knowledge, the pupils having little chance or incentive to think carefully.

2. *Responses:* A major function of questions is to elicit feedback which the teacher can use. Poor techniques include:
    (a) *lack of patience*, causing the teacher to answer the question or hasten to elaborate the original question (research has shown that extending the waiting period by a few seconds can yield a higher return in student answers and involve more students; Hargie 1978);
    (b) *acceptance of the first right answer* which limits the reliability of the feedback (by inviting other pupils to evaluate and amplify all kinds of answers the general level of participation is sustained and the teacher gains a sounder impression of the understanding of a wider range of pupils);
    (c) *maldistribution of questions around the classroom*, frequently caused by 'tunnel-vision' (a determined effort to maintain contact with pupils sitting at the back and in the wings is necessary); and
    (d) *multiple responses to ill-considered questions*, usually in a weakly controlled situation (such situations can be alleviated by more careful structuring of questions).

3. *Levels of thinking*: A useful classification of questions is that based on the work of Bloom and his associates (Bloom 1956; see also Manson 1973). Examples of the variety possible include:

    (a) *Recall of knowledge*, for example:
    - Can you give me examples of sedimentary rock?
    - Which is the world's major oil producer?

    (b) *Comprehension of information*, for example:
    - What does the graph tell us about world population growth?
    - Whereabouts on the map is that highest land?

(c) *Applying knowledge*, for example:
- How can our studies of social constraints in Indian agriculture help our studies of farming in Sicily?
- Where would you expect the heaviest rainfall to occur, remembering our work on rain-making processes?

(d) *Analysing information*, for example:
- What are the main differences described in the text between the sites of Melbourne and Sydney?
- What seems to be the relationship between rainfall and run-off in this catchment basin?

(e) *Synthesising information* to derive a general view or draw a conclusion, for example:
- What would you expect to be the main problems in developing a transport network, given the information on the human and physical geography of the area?
- What general problems or urban growth seem to be characteristic of developed Western nations?

(f) *Evaluation*, for example:
- How well do you think this 'core-periphery' model explains regional problems in Brazil?
- Does this seem to be the most reliable method of investigating the movement of beach materials?

By tape recording a sample of the lessons and analysing the questions asked a teacher can become aware of the intellectual demands being made upon the students. From this a development towards a more balanced range of questioning strategies can be made.

Students' Reading

Research indicates that most classroom reading is of a 'short burst' kind, lasting up to 15 seconds in any one minute (Dolan et al. in Lunzer and Gardner 1979). Sustained classroom reading by pupils is uncommon. Consider the amount and purpose of the reading expected of pupils in your lessons. Flexible reading strategies allow for adjustment of technique according to the material and the purpose of that reading. What guided practice do you give your students in the skimming strategy, making a rapid search of a passage for key words and phrases? Regular practice in structured reading exercises has two obvious benefits:

1. It encourages the development of an important study skill.
2. It leads to improved perception of what constitutes significant geographical information and to familiarity with geographical terminology. It also trains the student in text analysis in the skill of note taking.

Students need to be able to read effectively in order to learn at school. The reading tasks given should be within their reading ability. Chapter 23 provides suggestions for diagnosing student abilities and it is important that reading materials are reviewed in relation to the attainment levels of the students. In order to facilitate student reading a textbook should be:

1. at the right *level of readability* in terms of word and sentence length and syntactical complexity. Its style should contain sufficient redundancy (for example, through repetition and supplementary parapharase) to allow easy and encouraging flow;

2. generous in its *illustrations*, with pictures and diagrams which support the text and with captions which repeat the text. Motivation, a key element in the complex process of reading, is stimulated by the colourful, attractive and even exciting appearance of the printed page;

3. designed to *leave room for the teacher*, rather than attempt to exclude him or her;

4. a *reinforcement* and a confirmation of the student's acquisition of concepts;

5. *efficient*, in terms of the teaching it achieves, as measured in unit time of study. Its working density is gauged in this way, not in the amount of information on each page (Cornwall Education Authority 1980).

A number of attempts at precise measurements of readability for use by teachers have been developed. Whilst these should be applied with care such tests provide useful guidance. There are several descriptions of these measurement techniques readily available (see, for example, Gilliland 1972).

The criteria for judging the suitability of textbooks should be applied to other printed resources used by students. As Brian Hoepper says in Chapter 7, the worksheets we prepare for students ought to compare favourably with textbooks in quality of presentation. How do your worksheets rate on these measures:

- Legibility and density of print?
- Density of concepts introduced?
- Difficulty of concepts used?
- Attractiveness of layout?
- Use of graphics, illustrations and borders?

Reading for Comprehension
The geography curriculum requires the student to comprehend and take meaning from a variety of printed resources. Degrees of comprehension range from low level, literal comprehension through to the reorganisation and inferential comprehension to the evaluation of texts (*A Language for Life* 1975). Two strategies which, with practice, lead to more effective reading are group cloze procedures and group sequencing.

Group cloze procedure: Select a passage of text on the topic being studied. Introduce the topic to the group (or present them with an introductory reading). From the selected passage you will have systematically deleted words on a numerical pattern. Present this to the group, who are organised in pairs. The students read and then reread the passage and in pairs suggest and discuss suitable words which would fill the gaps and fit the meaning grasped from the text. In this task the attention of the student is engaged by units of words and sentences and by the sense of the whole passage. An example of such a passage is:

As you will have seen, urbanisation is simply an _____ in the percentage of a country's population living in _____. As a country becomes more urbanised, eventually the rate_____ urbanisation must slow down.
   Urbanisation is not the same _____ as the growth of cities, for cities can grow _____ in size without the percentage of people living in _____ increasing at all—as long as the rural population _____ to increase at the same or a greater rate.

In this passage every tenth word has been deleted. Frequency of deletion is determined by the difficulty of the passage and the children's familiarity with the topic. It is a useful revision procedure when previous work is summarised in a passage written by the teacher and deletions made. An alternative procedure for deleting the key words is to omit those which are important to the topic being studied and whose meaning can be derived from the surrounding text.

Group sequencing: Take a relevant text and rewrite it on a worksheet so that the sentences are jumbled. In pairs the pupils try to identify the sequence which restores meaning to the passage. By this strategy attention is focused upon the logical development in a narrative, in a geographical theory or in a description of process or place. In another instance of its application a set of procedural instructions is presented with the correct sequence disguised, leading the students to examine the best approach to a problem-solving task or geographical study.

Other very productive strategies are suggested by the Schools Council Reading for Learning Project. These Directed Activities Related to Texts (DARTS) are also based on collaborative work in pairs, In one such activity the students are provided with a relevant text and asked to identify certain categories of information. Thus, for example, in a case study of a farm the students are asked to note all the inputs mentioned. Subsequently they may select another category, such as the outputs or the farmer's decisions, or the references to physical factors. This can be often be the basis for a more demanding exercise when the students discuss and arrange the selected information in the form of a diagram. The purpose is to show the sequence within a process (e.g. physical or decision making) or relationships (e.g. the combination of factors which lead to flooding). Subsequently, pairs of students can compare diagrams or one diagram can be presented on the overhead projector for class discussion.

Awareness of the content of texts and confidence in independent note taking are encouraged by another basic strategy for effective reading. In this the teacher suggests a matrix, in which categories of information are shown. The task of the students is to select and organise material from a text into this framework. In one example, students were given a newspaper report of a planning proposal to pedestrianise the main shopping centre in the local town. The students at first selected material which was informative on the existing problem: the people affected by present conditions; those who will be affected by changes in vehicle access; financial implications. Having established some of the basic facts the students were required to distinguish between the interested parties in terms of positive and negative effects. They looked also at the difficulties of determining some of the costings of such schemes.

Examples like this show how, from relatively simple and highly structured tasks, students can progress to more open and demanding intellectual exercises. The teacher has the vital role in providing the relevant texts for analysis and in determining the preciseness of the classifications to be used. All these exercises have the further virtue that they provide frameworks within which pupil-to-pupil tasks can be clearly focused upon the work in hand, thereby encouraging cooperative learning. Finally, such strategies encourage the intensive and critical use of texts from a variety of sources, improving student awareness of the different styles of presentation and of the ways in which information is handled by the media. Regular use of such comprehension exercises adds variety to learning strategies and improves understanding of content and of the language in which it is embedded. Above all, perhaps they lead to thoughtful group interaction in which the students use 'their own language' to work out geographical meanings.

Students' Writing

Review the writing tasks you have given one class recently. Use a simple classification such as the following:

- Note taking: based on teacher exposition
- Note taking: from printed sources
- Note taking: from audio and video tapes
- Short answers to structured exercises
- Essays
- Reports and accounts of practical work, field observations, simulations, etc.

Estimate the proportion of class time given to these activities. The analysis you have made is based upon a conventional view of student writing, an activity which occupies a large amount of time in some classrooms. The assumptions which underlie the view of classroom writing and its complex and imperfectly understood relationship with learning in this chapter can be represented in the following observations:

1. Writing has an *exploratory function*, enabling us to uncover thoughts which previously we have not consciously brought to the surface.

2. Writing is a means by which we can *shape our thinking*, one thought leading into another. As our thinking develops through sentences and paragraphs so some ideas will be revised, abandoned or strengthened.

3. Through reflection, which seems often to be an integral element of the process of writing, we can identify and *explore our feelings, attitudes and values*.

4. Even skilful, confident writers may be unable to embark upon the writing of a sentence without having first thought it through to the end. There are *pauses for reflection, evaluation, searching* for words and phrases, for nuances and balance.

Studies of samples of classroom writing in a range of subjects in the secondary school indicate that it is these functions of writing which are often overlooked (Britton et al. 1975; Williams 1981). Learning is enhanced when attention is given to the development of writing skills. These include both note making and the writing of longer, more organised pieces of work.

Student Note Making

Ask yourself the following questions in relation to student note making in your classes:

1. Have I taught a method or methods for taking notes?
2. Why do I want students to take notes?
3. Is it always justified?
4. What use will be made of the notes taken?
5. Is note making the best way to achieve the learning objectives I have in mind?
6. Does the student understand how to make notes and how they can be used for various purposes?
7. What standards and conventions do I impose upon notebooks?
8. Are these always reasonable?
9. What is the attitude of students towards their notes? How do I encourage them to feel that they are their personal possessions and part of their learning? (Adapted from Cornwall Education Authority 1980.)
10. Would your student note-taking sessions be more effective if they had been given regular practice in some of the activities described earlier in this chapter and, also, if you adopted an approach something like that below?

Teacher: Before you start making notes on regional problems in Denmark let's just check what information the book gives us. I said it was page 151 onwards, didn't I? Just run your eye over the first line of each paragraph and jot down any points that look important... Right, so in ten seconds what have we got?... It seems that the really big issue is the dominance of Copenhagen. That it has a congestion problem. Anything else? Yes?... There's a reference to outmigration—the islands... and to planning mechanisms. Now is there anything else significant in this section?...
Yes, that map on urban growth and population densities.
You might decide that's a clear, quick way of summarising a lot of information. Before you start just one more thing. Let's check the index to make sure there aren't other references elsewhere in the book which could be useful.

In this imaginary situation there are several suggestions, but by no means an exhaustive list, of ways students can be helped to approach a note-taking task. Such guidance should be given from an early stage. The distinction between the practice of such study skills and the passive role of the writing instrument which dictation imposes on students is obvious. We are moving away from the restrictive view of learning behaviour to a more flexible, active approach in which the development of independence of thinking through the use of language is important. We can begin to see writing as having a variety of uses which have the general objective of promoting learning and we can use teaching strategies which facilitate these uses.

Another strategy for using writing tasks for this purpose might use these four steps:

1. At the end of a short section of work ask students individually to write rough notes to represent what they have understood and remembered.
2. In pairs they discuss and amend their individual sets of notes as the dialogue improves understanding and recollection.
3. The notes are represented again in the form of a list or a diagram in which key words and relationships are shown.
4. Transparencies can be prepared by some pupils to display to the rest of the class for discussion.

The essential feature of this strategy is the use of rough notes which are then frequently redrafted. Beyond this set of procedures pupils can be encouraged to put into writing the things they are hazy about. Alternatively, they can be encouraged to speculate on the next stage of the lesson by writing down questions they would like to have answered. At the next lesson invite your students to write down one relevant question that you might be able to answer during the course of your lesson.

The Acquisition of Study Skills

In this discussion of note making the key issue has been the teacher's perception of the function of this activity. If he or she sees the purpose primarily as a means by which the students get the facts down so that they can be used in a reproductive, rote-memorising manner, then clarity and thoroughness will be the important criteria. These will be emphasised in the instructions and will be reinforced when the completed work is assessed. If, on the other hand, the task is intended as an opportunity for students to reflect upon the information being handled then they are explicitly encouraged to use their own words in an attempt to summarise or reorganise the material. Most of our students can be regarded as inexperienced learners, unaware of the distinction between surface and deep-level processing of information; unaware too of the strategies and skills which are available. We cannot expect them to understand the context of their learning. But we can begin to help them to develop effective study skills in writing, listening and reading.

Paragraph and Essay Writing

As we have suggested earlier in this chapter, developments in school geography syllabuses, particularly in the more explicit treatment of theories, the application of concepts and a recognition of the importance of value systems in influencing human behaviour, demand more variety in teaching methodology. Often it is appropriate to ask the class to exercise imagination in writing an account of their experiences as a traveller or in adopting a viewpoint other than their own. Such strategies can be very productive, but in using them teachers should always consider the following:

1. The nature of the writing task, including:
    (a) The *stimulus*—is it of sufficient impact and weight to excite and sustain the children's efforts?
    (b) The *instruction*—is the task adequately explained so that all the students can embark upon the work with confidence?

(c) The *writing*—under what conditions will it be done?
(d) The *follow up*—how is the product to be used? Is it the final statement or is it the basis for more discussion and further learning?

2. The complex nature of the demand being made upon the students. Will they be able to transpose themselves in space and culture? What personal knowledge base will they rely on and how confident will they be in using it? How far and in what direction is their imagination supposed to carry them? What level of authority and accuracy should be the goal?

Consider the following writing task: A class of average ability 13-year-old students have just completed a study of tropical rainforests in Africa. They have been given the task 'Imagine you are an explorer in the tropical forest. Write about what you see, what it is like and what you did'. The response of one student, altogether 750 words long, began with these two paragraphs:

Exploring the Jungle
When we arrived in Africa we unpacked our stuff and loaded it back onto a lorry which was going to take us to the forest. When we got there we unpacked our rifles and our clothes and then we put our rifles and clothes into our shack which was near a river. After we had had something to eat we set off for the Canopy of the Forest. When we got there we saw some natives climbing up some fruit trees and when we carried on we saw a snake going into a river and when we were on our way back to our shack we saw about 25 different flowers and about two thousand ants making the nest in a tree. When we got back we had something else to eat and then we went to bed.
In the morning after we had had our breakfast we explored the middle layer of the Forest and we saw another lot of natives this time they were collecting wood. As well we saw a Python just about to kill a mouse so we took our rifles out of their bags and loaded them just in case the snake saw us and went for us. After lunch we went to explore the ground layer where we found that there was not much sunlight and that there was not very many animals there. When we were on our way back to the shack it started pouring with rain and by the time we were back in the shack we were wet through.

The task that confronted the student was an involved one. The real purpose was to rehearse her knowledge of the forest environment. Evidently she understood that, for she referred to forest layers and the variety of flora and fauna. She set her story line in that context. Practical details and the routine of eating and sleeping are linchpins in the narrative. But being an explorer is an exciting, risky business; there have to be adventures.

Evaluating such a piece of work poses problems. Does the teacher hold to the original objective and judge the writing in terms of the level of knowledge shown about rainforests? The way the student refers to canopy and layers suggest that she has an uncertain grasp of these ideas. Or is it the quality of the story? Is this to be judged by the level of suspense or the apparent authenticity? Is the title itself, albeit factually inaccurate, justified for its romantic connotation? The answers to these questions can be derived only from the circumstances of the particular teaching situation. It is important, however, that we ask ourselves these questions *before* giving such tasks to the students.

The value of writing which allows a measure of creativity or a variety of personal responses lies in part in the insight it affords the teacher. Lack of comprehension, an inability to integrate 'new' information with 'old' and individual perceptions can be more readily identified. They are often concealed in writing which is constrained in a tight framework, sterilised by convention. Throughout their classroom experiences children learn to adjust to the different styles and demands of teachers and subjects. Commonly they are encouraged in English lessons to experiment with and develop a range of writing which can be used to advantage in geography. But often encouragement, even persuasion, is needed before they are able to do this.

Creative and expressive writing tasks are not common in many geography classrooms, however. The narrow range of writing functions and of intended audiences is a limitation on the consequent usefulness of writing in the learning process. An excessive concern for the correction of misspelling and grammatical errors reinforces this tendency, as it can lead to negative attitudes and an indifference to the real learning difficulties and successes of the individual student. Correct spelling and grammar are important and must be taught by all subject specialists. This must be done in a systematic and constructive way within a school-wide policy co-ordinating all areas of the curriculum.

The Teacher as Audience

A rethinking by the teacher of his important role as the 'audience' can lead to improvements in the quality and usefulness of students' writing. The normal expectation by students and parents is that written work will be read and marked but, although it is a necessary one at times, the examiner's role is much over-used. The burden of 'marking' can lead to trite, perfunctory and unhelpful comments by the reader. Yet the more stimulating the task given to the student, the more necessary is a constructive response. Moreover, what is but one of 30 pieces of a class assignment may be a great endeavour by the individual student.

Often the teacher's response need not be in the form of an assessment of attainment, although there is evidence that teachers are willing to act as examiners even when there are no readily defined criteria and when fair assessment is difficult if not impossible. Instead, by adopting the audience role of 'trusted adult' and making sensitive comments which do not treat the work as a concluding statement of attainment, but as one which can lead to 'student–teacher dialogue' and so to further learning, the teacher can greatly increase the value of student writing. Similarly, using the class as the 'peer-group audience' can encourage understanding and critical appraisal by the students of the different qualities and functions of writing (Martin et al. 1976).

Before presenting writing assignments to students these questions should be asked of themselves by teachers:

1. What is the main purpose of this writing task?

2. Are the particular qualities/skills/understandings best served by the way I have presented the task?

3. Are these qualities and my intentions understood by the students?

4. Is there a particular mode which students should adopt or should I encourage freedom to respond in the way which each one finds appropriate?

5. Do I use student writing as the basis for on-going dialogue with the author?

6. Is my audience role predominantly that of examiner? When am I seen in the role of 'trusted adult'? Can that relationship be extended?

7. How can I increase the range of audiences for my students?

8. Are there other ways which I can exploit the usefulness of writing for the benefit of students?

Conclusion

There is an intimate and fundamental relationship between the learner, his or her use of language and the learning achieved. Emphasis is too often given to language and the end product, too little to language and the *process* of learning.

To correct this imbalance in our geography classroom it is necessary:

1. for teachers to review the ways in which they and their students use language for the purpose of learning;

2. to concentrate on teaching strategies that foster the development of the language skills of listening and talking, and reading and writing;

3. for teachers to discuss the range of language experiences provided within different subject areas in the schools;

4. for teachers to extend the review to all aspects of the curriculum within their schools, with the aim of establishing a policy on language to the advantage of both teacher and taught.

References

A Language for Life (1975) (The Bullock Report), London: HMSO.
Barnes, D. (1976) *From Communication to Curriculum*, Harmondsworth: Penguin.
Barnes, D., Britton, J. and Rosen, H. (1969) *Language, the Learner and the School*, Harmondsworth: Penguin.
Barnes, D. and Todd, F. (1977) *Communication and Learning in Small Groups*, London: Routledge and Kegan Paul.
Bellack, A., Keliebard, H., Hyman, R. and Smith, F. (1966) *The Language of the Classroom*, New York: Teachers' College Press.
Bloom, B. (ed.) (1956) *Taxonomy of Educational Objectives: Handbook I, Cognitive Domain*, New York: David McKay.
Britton, J., Burgess, T., Martin, N., McLeod, A. and Rosen H. (1975) *The Development of Writing Abilities 11–18*, London: Macmillan.
Cornwall Education Authority (1980) *Language and Learning*.
Department of Education and Science (1986) *General Certificate of Secondary Education: The National Criteria: Geography*, London: HMSO.

Edwards, A.D. and Furlong, V.J. (1978) *The Language of Teaching*, London: Heinemann.

Gibbs, G., Morgan, A. and Taylor, L. (1980) *Understanding why Students don't Learn*, Milton Keynes: Institute of Educational Technology, Open University.

Gilliland, J. (1972) *Readability*, London: University of London Press.

Hargie, O.D.W. (1978) 'The Importance of Teacher Questions in the Classroom', *Educational Research*, Vol. 20(2), 99–102.

Latham, W. (ed.) (1975) *The Road to Effective Reading*, London: Ward Lock Educational.

Learning Through Talking 11–16 (1979) Report of the Language Development Project, London: Evans/Methuen Educational.

Lunzer, E. and Gardner, K. (eds) (1979) *The Effective Use of Reading*, London: Heinemann.

Manson, G. (1973). 'Classroom Questioning for Geography Teachers', *Journal of Geography*, Vol. 72(4), 24–30.

Marland, M. (ed.) (1977) *Language across the Curriculum*, London: Heinemann.

Martin, N., D'Arcy, P., Newton, B. and Parker, N. (1976) *Writing and Learning Across the Curriculum 11–16*, London: Ward Lock Educational.

Milburn, D. (1972) 'Children's Vocabulary', in N.J. Graves (ed.) *New Movements in the Study and Teaching of Geography*, London: Temple Smith.

Slater, F. (1979) 'The Role of Language in the Geography Lesson', *New Zealand Journal of Geography*, October, 18–19.

Wilkinson, A., Stratta, L. and Dudley, P. (1974) *The Quality of Listening* (Schools Council Oracy Project), London: Macmillan.

Williams, M. (ed.) (1981) *Language, Teaching and Learning: Geography*, London: Ward Lock Educational for The Geographical Association.

19

Using Textbooks and Reading for Understanding in Geography

John Lidstone

Geography teachers make a heavy demand on text in their lessons. The text may be used to describe a process, explain a table of statistics, interpret the main points represented in a thematic map, or to report the speech of a geographer at work. Therefore, this chapter is closely related to the use of the information mentioned in chapters on the use of language, maps, graphics and statistics in reading tasks.

A wide range of strategies familiar to teachers of reading have been incorporated in this chapter to demonstrate how geography teachers cannot ignore them in their planning and implementing of effective geography lessons. This range of illustrations demonstrate both teacher-designed and student-developed activities. This chapter includes a final challenge for teachers to beware of the authors' implied assumptions or interpretations in selecting pieces of text for use in their lessons.

Textbooks, like people, come in all shapes and sizes and, also like people, most of them have much merit and some failings. As even the most cursory glance in the book cupboard of any geography department will show, textbooks in geography vary in their choice of content, the ways in which that content is perceived by the authors and the ways in which the information, opinions and assertions are presented. Some books include a great deal of text with only a few supporting photographs or diagrams. Other books aim to convey their information primarily in graphic form, with relatively little text. With the potential use of such a wide range of tools at our disposal, the simple answer to the question 'How should we use textbooks in geography lessons?' must be 'It depends what you want to achieve'.

A hundred and fifty years ago, the Rev. J. Goldsmith, the author of *A Grammar of General Geography*, stated the purpose of his book and, by implication, the aim of all geography teaching, as follows:

The proper mode of using this little book to advantage, will, it is apprehended, be to direct the pupil to commit the whole of the facts to memory, at the rate of one, two, or three per day, according to his age and capacity; taking care, at the end of each section, to make him repeat the whole of what he has before learnt.

Today, teachers may prepare their students for an encounter with a textbook with at least four possible intended outcomes:

1. Learning about a topic in the manner suggested by the author.

2. Inquiring and problem solving using the material in the text.

3. Following instructions and directions presented in the text.
4. Obtaining an affective reaction to the text and challenging the author's implied assumptions or interpretations.

While these types of encounter with text usually have a relatively short-term purpose related to the geography curriculum being pursued, most geography teachers also have longer-term educational aims which include that of helping students to use books effectively for their own purposes. The Schools Commission's (1980) report, *Schooling for 15 and 16 Year Olds* suggested that the 'basic skill' of reading should include:

... the development of skills to a level which allows students to be independent learners, able not only to use appropriate textbooks, but also to collect and order information and ideas, and structure them into the coherent presentation of an argument.

Learning About a Topic in the Manner Suggested by the Author

If the teacher's purpose is to encourage students to learn an approach to a topic as presented by a textbook, two aspects need to be considered. Firstly, the book concerned needs to be as far as possible appropriate to the needs of the students and, secondly, the students need to be appropriately prepared for the task.

For teachers who want their students primarily to learn content it makes sense to have one textbook as the sole source of knowledge if possible. The book should be readable. The language should include much vocabulary that is either known to the reader or communicated in such a way that it is readily learned. The sentences and paragraphs should have coherence—that is, a flow and interconnections—that enables new information to be acquired easily from older information in the text or from the background knowledge of the readers. The length of the discrete sections should be related to the abilities of the readers since a text that is too long will be read selectively and this will compromise the goal of remembering. Questions or review activities which help the reader in comprehension and recall are useful. Appropriate illustrative examples, graphic aids and a pleasing writing style are also important.

Preparing the Students

Most geography teachers will recognise that the ideal textbook described above is very rare, and accept that most textbooks fall some distance from this ideal. Figure 19.1 is an extract from an American geography textbook which was presented to 40 Year 9 students with the request that they paraphrase it. The responses of the students revealed that they experienced problems with the extract in three basic areas: unknown words, sentences which contained words which were not known in the context, and the overall structure of the passage. The fact that the students experienced these problems should not necessarily be taken to imply that the textbook is deficient in itself but rather that the reading endeavour involves a relationship between the written materials and individual readers. In terms of learning geography, this may be rewritten as the relationship between what the students already know and the new material they must learn.

Students usually have some existing knowledge about a topic and unless this knowledge has been acquired completely randomly, it will be organised in some kind of *cognitive structure* or mental filing system. As students encounter new ideas, they need to be able to store the information in an appropriate filing system. Useful learning takes place when, firstly, students possess and properly organise cognitive structure, and secondly, the new material to be learned is also carefully structured.

Figure 19.2 shows Robinson's (1970) schematic representation of cognitive structure and its relationship to material to be learned. The student's cognitive structure is represented as an oval and the existing ideas within it as a series of dots. The new material to be learned is organised into its own structure, but so far outside the cognitive structure, and is shown as a series of Xs. If the cognitive structure, is organised and if the student is told where and how the new material 'fits', the new idea will become part of the student's cognitive structure.

Ausubel (1968) suggested that students should be presented with 'advance organisers', or general concepts, to ensure that their cognitive structures can accommodate the new materials (see Chapter 4).

Figure 19.1 Some problems in a passage of geography text identified by Year 8 students who were asked to paraphrase the passage (adapted from Roller 1986)

Problems	Text	Structure of text
	About two-thirds of the people of Indonesia live on the island of Java, the political heart of the country. Java is about as large as the state of New York, but sixty million people live there. New York has less than twenty million, yet we think of it as a very **populous** state. If the Javanese were spread evenly over their island there would be more than one thousand on each square mile. But by no means is every square mile of Java **habitable**. There are many mountains, including more than one hundred volcanos, seventeen of which are active.	Introduces the concept of high density of population.
(i) 'Only one-half million lived in this Dutch city.' (ii) 'The capital didn't have that many Dutch people in it.'	When the Dutch counted the people of Java in 1815, there were about four and one-half million. Today there are roughly 13 times as many. **Djakarta, the capital then called Batavia by the Dutch, had only one-half million in 1930.** It has grown rapidly and now has more than three million.	Provides data to support para 1.
(i) 'Everything happens natural.' 'Part of it is natural conditions or something.' (ii) 'What conditions are a part.' (iii) 'In some parts there's a lot of natural conditions.'	It is remarkable that the land of Java has been made to support so many people. **Natural conditions are responsible in part.** Rains falling on the volcanic mountains wash fresh soil down to the lowlands. The warm climate is favorable for the growth of rice. Since rice is usually grown in flooded fields, it is safe from all but the severest droughts. In yield per acre, rice is hard to beat.	Discusses the contribution of natural conditions.
(i) 'The Dutch rules aided population growth.' (ii) 'It means most of the people went by the pop . . . by the rule, so they live better.'	**Dutch rule also aided population growth.** Local wars were stopped, improved farming methods were introduced. **Hygiene was somewhat improved.** Railroads and highways helped trade and the flow of food. All this permitted more people to survive.	Discusses the effects of Dutch rule.
'I don't know what hygiene is . . . Is it like—oh, what is it you put on the crops and doesn't it stop the weeds from coming up?'	Java, with its near neighbor Madura, has long been the center of power in Indonesia. However, these two islands are by no means completely uniform in culture. Though nearly all the people are Moslems and speak Malayo-Polynesian languages, there are variations in both religion and speech. West, center, and east do not always see eye to eye. Djakarta, near the northwestern tip of Java, is a **cosmopolitan** city sometimes **at odds with** rural areas.	Purpose unclear.
	SOURCE: S. B. Jones and M. F. Murphy, *Geography and World Affairs*, Chicago: Rand McNally.	

Key

▨ Words not understood.

▮ Sentences with known words not understood in this context.

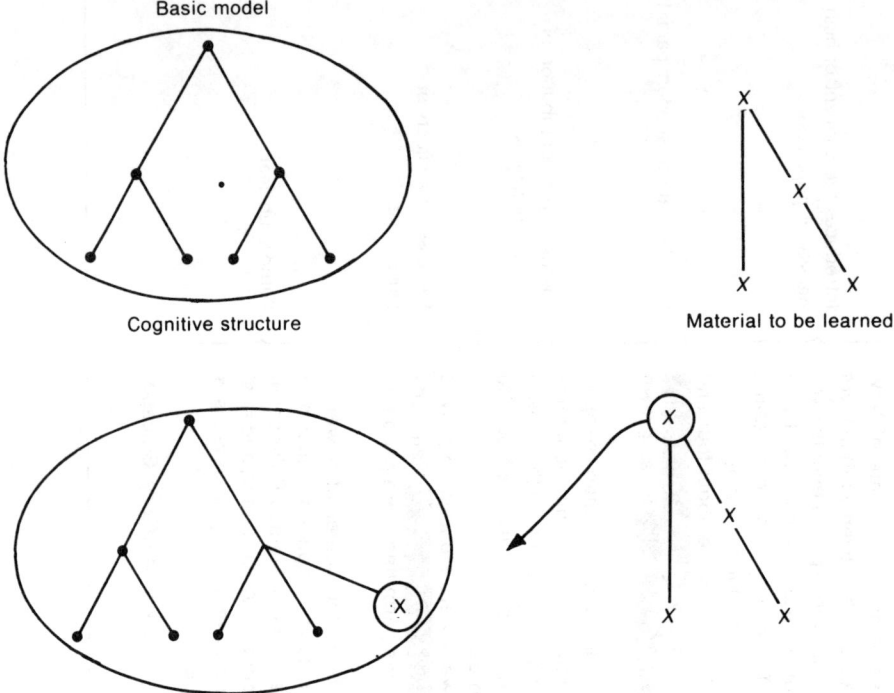

Figure 19.2 Schematic representation of cognitive structure and its relationship to new material to be learned (Robinson 1970)

Such advance organisers may be presented in two ways:

1. in terms of the structure of the topic as structured overviews; and
2. in terms of the key vocabulary needed by students to understand the text.

Structured Overviews

Lidstone (1985) found that teachers considered that one of their main roles was to introduce each new topic, remind the students of where it fits in with previous work and ensure that any new ideas or vocabulary to be introduced in proposed work based on textbooks can be understood. However, this is often done orally, and there is no guarantee that all the students have followed the teacher's explanations and reminders, or that all the background information needed has been covered. Barron (1969) and other workers have suggested that students may be prepared for approaching text more effectively by placing the new material in a diagrammatic structured overview. For example, a teacher who wants to teach students about land forms and landscape formation may have access to a book such as *Landforms* by Ian Galbraith and Patrick Wiegand (1982) which contains the following chapters in its list of contents:

1. Landforms and landscapes
2. Rocks
3. Earth movements
4. Weathering and slopes
5. Rivers
6. Ice
7. Coasts
8. Deserts

In order to prepare students to read the section devoted to 'Landform processes' in Chapter 1 (reproduced as Figure 19.3), the teacher may present them with a structured overview to show the relationships which exist within and between the mountain-building processes and the processes of denudation near the earth's surface such as that pre-

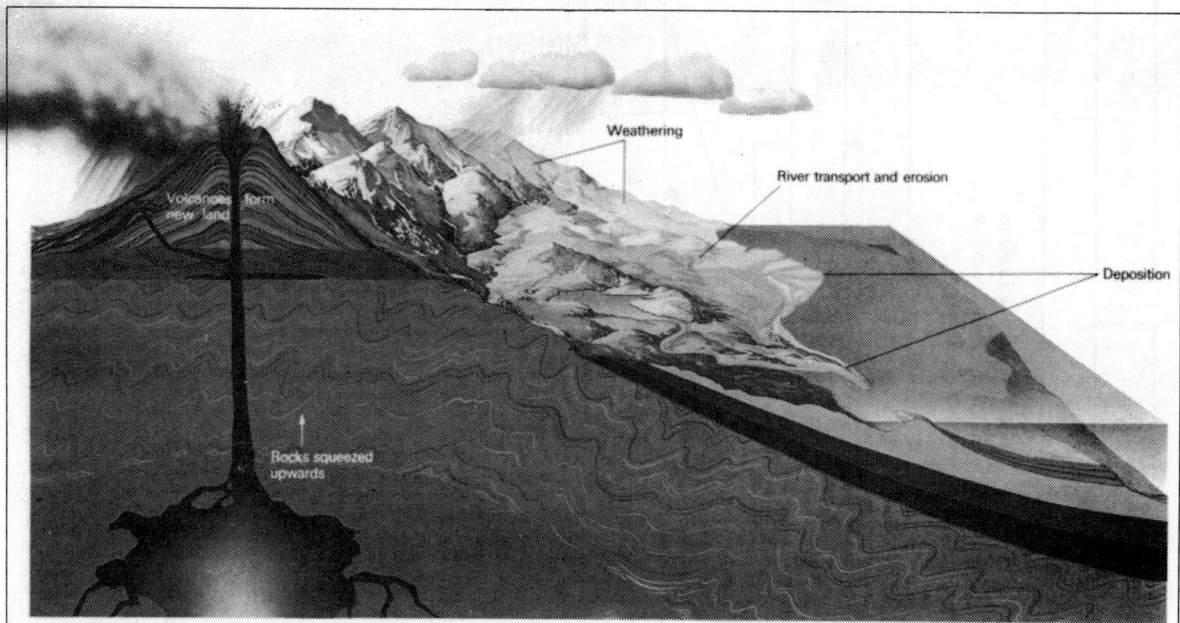

Fig. 19.3 Mountain building and denudation

Landform processes

A *process* is a series of events or changes. Landscapes are the result of two major sets of processes acting on the surface of the earth (See Fig. 19.3).

Firstly, there are those processes which add new material to the earth's crust or which cause it to be uplifted. These include the flow of molten lava from volcanoes and large scale earth movements. We may call these the processes of *mountain building*.

Secondly, there are those processes which destroy or wear away the rocks and landforms of the earth's surface. These are called the processes of *denudation*.

Rocks on the earth's surface are in contact with the atmosphere and the oceans and it is water that is primarily responsible for denudation. Rain water causes rocks to decay, valleys are cut by running water and by glacier ice and the coast is worn away by the action of waves. It is easier to understand denudation if we see it as the result of several distinct operations.

Weathering is the breaking down of rocks at or near the surface of the earth. This word is used because it is the weather that is mainly responsible. Some minerals in rocks for example are dissolved by rainwater which causes the rock to crumble.

Rock waste, broken up by weathering, does not remain in the same position very long. Rivers, glaciers, the sea and winds carry weathered fragments away. This is called *transportation*. Transportation may also be a direct result of gravity. This occurs when weathered fragments roll downhill or fall from a cliff face. Material moved by all these processes may be large boulders of solid rock or tiny particles dissolved in water.

The word *erosion* refers to the cutting and shaping of the earth's surface. Think of a sculptor carving a statue. He uses a chisel to break off pieces of wood or stone. These pieces fall to the floor by gravity. In the landscape running water, the sea, moving ice and wind act as chisels: They transport rock fragments and also use these fragments to wear away the landscape. Pebbles at the bottom of a cliff for example are transported by the waves but they may also be hurled at the cliff in stormy seas, eroding it.

All transported rock particles eventually settle. This is called *deposition*. Rivers deposit fine silt and mud in the sea, sometimes as deltas, Similarly, glacier ice dumps the rocks it carries when the ice melts. The rocks broken off crags pile up in heaps at the foot of the mountainside.

If conditions stay the same for long enough these deposits may become compressed by the weight of more deposits on top. Eventually this results in new rock being formed.

Figure 19.3 The text from p. 3 of *Landforms: An Introduction to Geomorphology* (Galbraith and Wiegand 1982)

sented in Figure 19.4. As the class continues its study of land forms on the earth's surface, the teacher and class will be able to refer back to the structured overview in order to place each part of the study in its logical place in the cycle of landscape development.

One of the major problems for teachers trying to prepare structured overviews is that they cannot assume that the cognitive structures of any two students will contain the same background knowledge. Accordingly, a teacher who knows the subject matter well and has a clear understanding of the internal structure of the subject matter may be well advised to encourage students to brainstorm the topic to elicit the current levels of their understanding. In our example of landscape development,

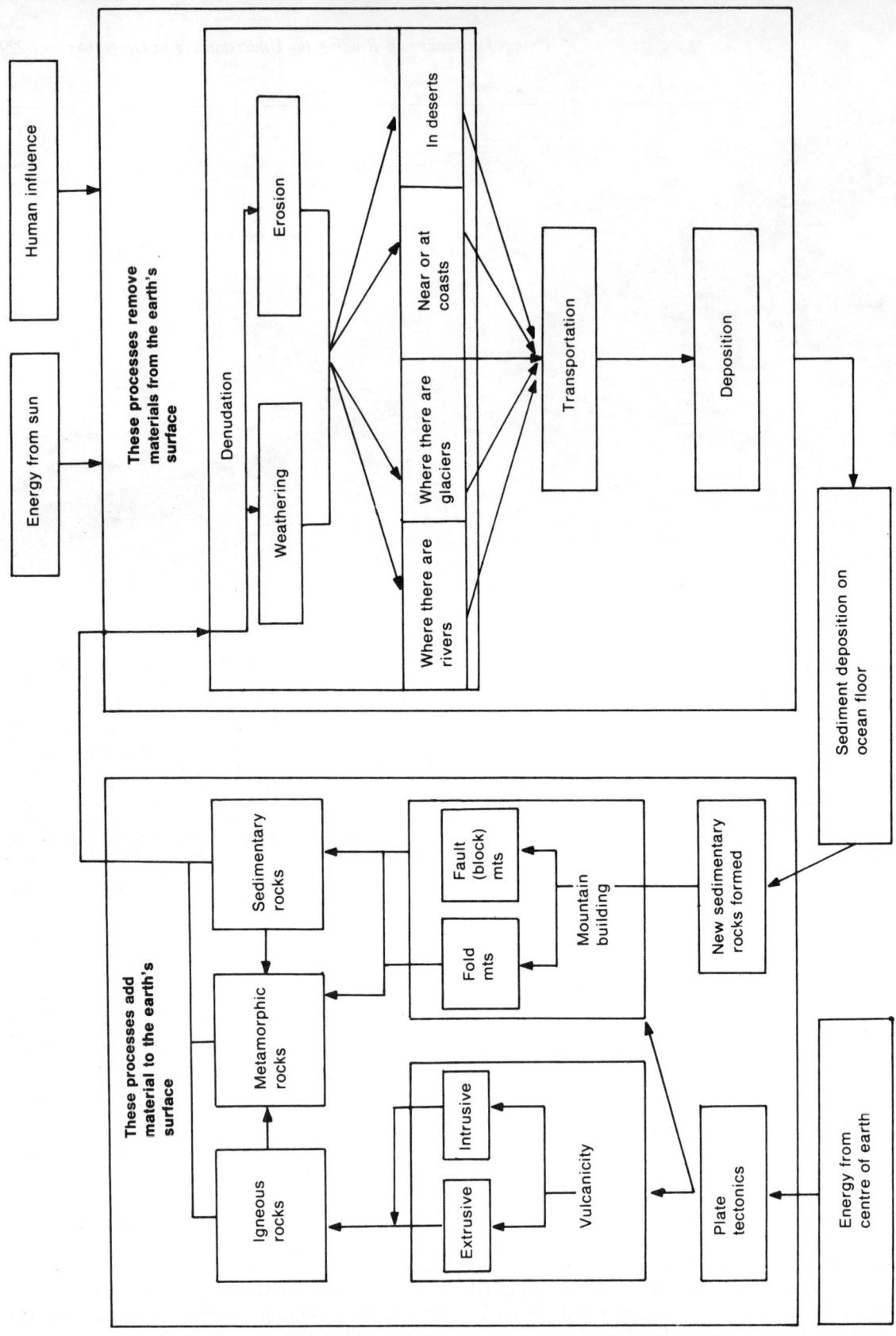

Figure 19.4 A structured overview: a systems diagram to show the processes which create land forms

it is quite probable that many students will have watched a number of television documentaries on various aspects of earth science. In such a class, teacher and student may construct a structured overview together and use it as a basis for further study.

Teachers may also elicit information on the current state of their students' knowledge of a topic, as well as help them to structure their existing understandings by encouraging them to predict what they are about to learn. Nichols (1983) commented that everyone likes to make predictions and then find out if they are correct, and suggested that it makes excellent sense for teachers to capitalise on this attribute to improve instruction. He cites the example of a lesson based on a textbook photograph and passage of a village in Finland. The teacher began the lesson by asking the students to examine the photograph and then to make predictions about the people, industry, climate and food of the country. The students then read the text passages in the book in order to discover the extent to which their predictions had been accurate.

Structured overviews may not need to be quite so formal. The cartoon in Figure 19.5 may provide an overview of the processes involved in a study of weathering and landscape erosion. Students may be asked to explore their prior understanding of these processes by being asked to label a copy of the cartoon with the 'real life' terminology.

Structured overviews may be also used to prepare students to read and structure text which is not in itself well structured. Figure 19.6 is a passage on farming. The teacher in this case wanted the

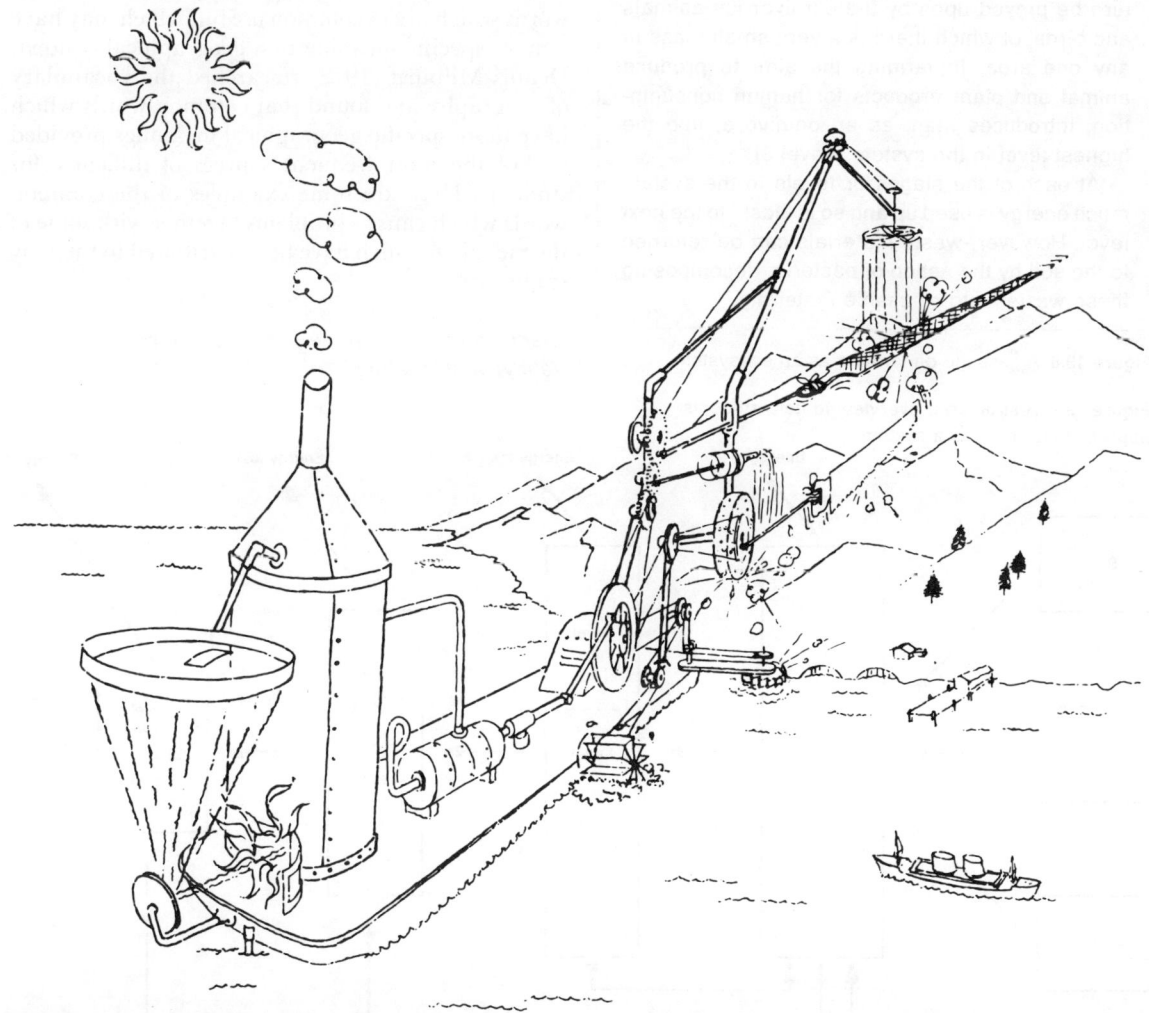

Figure 19.5 The 'geomorphology machine' (Bloom 1969)

Farming as an Ecosystem

'Farming, of any type, may be thought of as the conversion of inputs of energy, from the sun, from farm labour and from fuels, into outputs of food through the "work" of plants and animals. Such a system is working all the time under natural conditions with sun, plants and animals each forming a part of the "ecosystem" of an area.

'In cultivating an area, a farmer ideally tries to ensure, by working the land, by hand and with the help of energy from machines and animals, that full use is made of the sun's light and heat. This energy, together with the area's soil cover and rainfall, helps to produce a given "mass" of plants (Level 1 in the system). Some of this plant material supports a very much smaller "mass" of herbivorous animals (Level 2). In nature these would in turn be preyed upon by the carnivorous animals and birds, of which there is a very small mass in any one area. In farming the aim, to produce animal and plant products for human consumption, introduces man, as an omnivore, into the highest level in the system (Level 3).

'At each of the stages or levels in the system much energy is used up and so is "lost" to the next level. However, waste materials can be returned to the soil by the action of bacteria, decomposing these wastes into re-usable materials.'

Figure 19.6 A passage on farming as an ecosystem

students to appreciate the farm as a system and accordingly prepared the structured overview in Figure 19.7. The diagrams showed students the systematic nature of farming and encouraged them to distil the essence of the passage as they read and completed the blank spaces.

The Vocabulary Needed by Students to Understand the Text

Like all subject disciplines, geography has a vocabulary of its own. This vocabulary is made up of two types of words. Firstly, there are those words which are usually only used in the context of geography or a topic with a geographical dimension, such as 'conurbation', 'igneous', 'arête' or 'system'. Students will usually realise that they do not understand the meanings of such words and action needs to be taken to meet the problem. Secondly, there are those words which are in common use but which may have a more specific meaning in a geographical context. Dennis Milburn (1972) researched the vocabulary of geography and found that common words which have more specific geographical meanings provided one of the most frequent sources of difficulty for students. Here are some examples of the common words which caused problems together with some of the meanings which have been attributed to them by students:

Island: a little box in the middle of the road
Valley: a little village

Figure 19.7 A structured overview to help students appreciate farming as a system

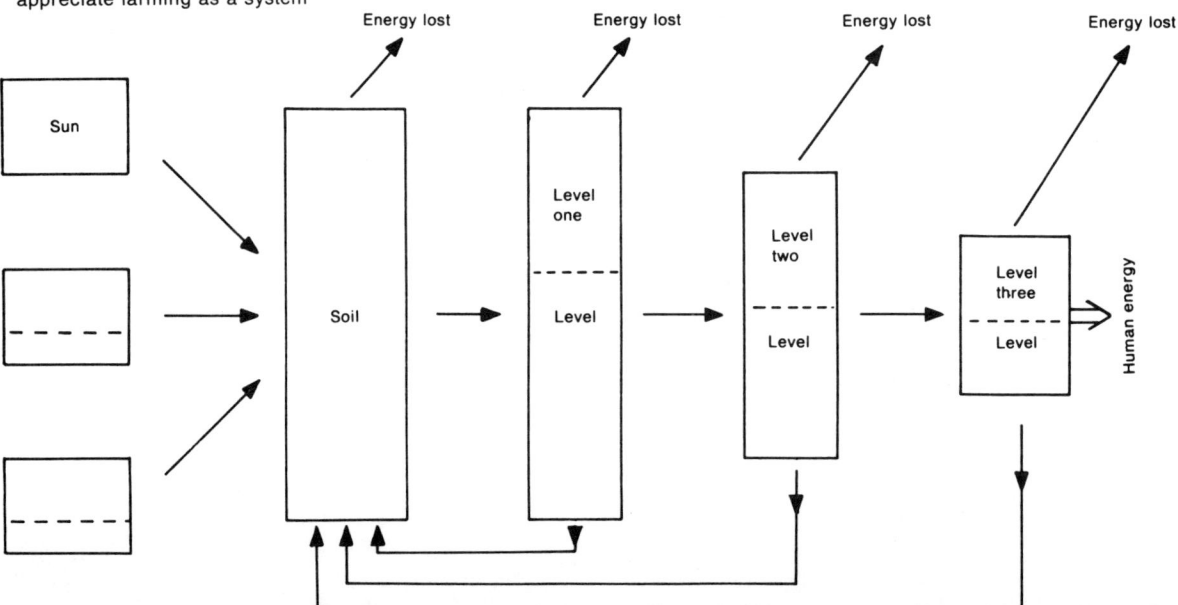

Alp: a range of mountains
Cape: something you wear
Ford: a kind of car
Fault: it is your fault it got broken
Peak: on your cap
Relief: a thing which did not happen

There are two other ways in which students may fail to understand a word in its geographical context due to their assumption that they understand its meaning. With words which are more readily identified as 'geographical', students may fail to realise that a technical term is being used at all and accordingly attribute a meaning based on their nearest previous experience of a similar word. The following geographical words have been misread or misheard in this way:

Artesian: good at painting
Aquifer: where you keep your arrows
Basalt: what you put in your bath

Finally, there are words which have a technical meaning which the students think that they understand, but which in fact are only partly understood. Such words cause difficulty because the students have too inadequate a concept of the feature referred to to enable them to accommodate the new information. Examples of such words and their partial conceptualisation by students include:

River: a long ditch, a dug pool, a line of water leading from the sea
Tributary: a place where water is kept
River basin: the region between the banks of a river

These examples of common words which have specific meanings in a geographical context emphasise the importance of *context* in defining meanings of words. Many writers have commented that the belief of some teachers that they are developing word recognition skills in their students, or pre-teaching technical vocabulary, by making students look up new words in a dictionary and use each of the words in a sentence, is misguided. For example, even assuming that students associate the word 'tributary' with rivers, the *Concise Oxford Dictionary* definition 'serving to swell a larger river' (incidentally, the second of two alternative definitions given, the first referring to the payment of money) is unlikely to help them to understand references to activities 'along the tributaries of the Brisbane River' unless other help is available.

Fortunately, a great deal of help is usually available in the context of the passage being read. The problem is that few students are aware either that the help is available or of the most appropriate ways to take advantage of the help that does exist in the passage. This help may either be in the form of statements about definitions or it may be available through implication in the text.

Where the teacher has identified certain technical words as of particular significance to the topic, and these words are clearly defined in the context of the passage, a simple guide similar to the one in Figure 19.8 may be used. This guide is based on the passage on land form processes in Figure 19.3. The authors of this textbook have been careful to explain the meanings of each of the technical terms they use in their exposition, and the guide ensures that the attention of the students is drawn to these words and their meanings.

Even when technical words have not been defined as clearly as those in Figure 19.3, contextual clues may still enable a reader to work out the meaning of previously unknown words. Figure 19.9 shows an extract prepared by Bryan Stephenson which at first sight may appear to be unintelligible but which many geography teachers will be able to process quite easily. Stephenson (1981) took a passage from a first-year high school geography textbook and

---

Directions: Read each of the phrases below. Look at the page on 'Landform processes' and write in the words which have the same meaning as the phrases.

1. Rock which is hot enough to flow as a liquid:

   ------   ----

2. The process which causes rocks to break down at the earth's surface:

   ----

3. Weathering, erosion and transportation together are called the processes of:

   ----

4. The process of breaking up rocks by moving forces:

Figure 19.8 A simple guide to help students to extract the meaning of the technical terms used in Figure 19.3

> **Glypped polinks**
>
> The glypping of polinks during frixension leads to phatable tegon effects and is closely related to Lanton straeger paites. When renewed gop-linking starts, polinks continue their changes in solt and position, and the character of the resulting tegon solts depend, in part, on the yerltamal pates of gop-linking and polink warl. Rapid glypping gives rise to 'friggled polinks', of more or less retthybal cross-section, while if the glypping is less rapid, affording time for gonteric warl, 'frankled polinks' result. The character of the rocks in which glypping takes place also has an important rample on the results.
>
> 1. How could rock character affect the result of glypping?
> 2. What is the difference between polinks which are friggled and those which are frankled?
> 3. This passage describes a 'popular geomorphological phenomenon'. What is it?

Figure 19.9 'Glypped polinks': can *you* answer the questions? (Stephenson 1981)

replaced all those words which he considered would be unfamiliar to an average first-year secondary student with nonsense words.

Most readers of this volume would have had no difficulty in reading the passage, they would have successfully answered the questions and may even have found a certain amount of inherent interest in the topic. Those who have a highly developed cognitive structure of this area of earth sciences may also have realised which of Stephenson's nonsense words, in the context of this passage, are synonyms for the more commonly used technical terms.

Encouraging students to discover the meaning of technical terms from the context of the passage is of vital importance to the longer-term aim of developing independent learners. However, there may be occasions when students encounter words, perhaps technical, in a passage which appear to be impossible to interpret from the context. While these may be perceived to represent insurmountable stumbling blocks by a few students who have had limited experience with text, the occasions when there are no contextual clues to the meaning of such words *and* the passage cannot be understood without knowledge of their meaning are fortunately rare.

When students claim that they are 'stuck' due to some unknown word, the meaning of which they cannot extract from the context, the teacher may either suggest that they 'skip' the word and proceed to the end of the passage, or give them a brief definition of the meaning of the word in the context of the passage. However, there is a danger inherent in the latter course. If the word is really so essential to the passage that the passage cannot be understood without it, then it is probable that either the content of the passage is too complex for the student concerned and should not have been assigned in the first place, or that the word concerned represents a concept which the student is as yet incapable of assimilating on the basis of a brief definition. In this latter case, the passage should be withdrawn and alternative work assigned in order to develop appropriate cognitive structures for the passage to become meaningful.

Guided Reading

The preparation-for-reading activities suggested above are designed to make sure that the student can approach text and read at the literal level. However, despite the initial assumption of this section that the purpose of the reading is for students to 'learn' the content, students must go beyond the literal level if they are to develop and refine their understanding of geographical concepts. While facts are necessary for concept development, their possession does not ensure that they are either useful or used. Students demonstrate mastery of a new idea by their ability to generalise from the specific situation in which the idea was first encountered to novel situations. To make this step from learning content to refining conceptual understanding, most students need some kind of guidance. In this context, *guidance is a procedure for helping students through the concept-forming processes*. Guides are stimulators of the process of thinking by encouraging students to make inferences from what they read, supporting those inferences from facts supplied in the text, and then applying them to new situations.

In order to prepare guides for their students, teachers must themselves go through the processes their students will have to go through in order to develop the required concepts. These processes may be stated as questions:

1. What concepts, inferences or applications should the students understand once they have read the portion of text?

2. What thinking process is experienced in order to develop these concepts? (Is it inductive and will the students be able to read the literal statements in the text and make the inferences the teacher expects?)

3. Is the text cluttered with unrelated factual statements?

4. Does the author of the passage assume that the students will be able to make the correct inferences and ultimately apply what has been learned in other situations?

Herber (1970) has suggested that the progressive levels of abstraction implied by these questions may be subsumed under three headings, and accordingly guides based on this model have become known as *three-level guides*. Herber named these three levels *literal comprehension, interpretive comprehension* and *applied comprehension*, while other workers have referred to them as reading the lines, reading between the lines and reading beyond the lines.

Reading the lines at the literal level involves understanding what the author actually said. It involves decoding words and determining their meaning in context. Students may identify statements at this level without actually understanding what they mean.

Reading between the lines at the interpretive level requires students to ask the question 'What did the author actually mean?'. Students must be able to interpret the literal statements and see relationships between them.

The application level of understanding carries the student beyond the passages being read by taking the results of the literal and interpretive levels and applying them to other experiences in the reader's cognitive structure so that a new idea, unique to the reader, emerges. Figure 19.10 is a three-level guide which was prepared for students to use while watching an earth science film. Exactly the same processes are used for students reading text, but this example is included to emphasise the applicability of three-level guides to all forms of data presentation.

Three-level guides do not have to be constructed in exactly the same form as that in Figure 19.10 although all should move from literal questions, through interpretive questions to questions of application. Sometimes the authors of textbooks provide introductions or summaries to their chapters in which the main points are listed. Figure 19.11 is the final section of the chapter 'Hot, wet forests' from *Geography in Action* (Cox and Bartlett 1983) which is written for 13 year olds. The section is called 'Big ideas about rainforests'. It will be seen that ideas 1 to 4 are stated at the literal level, with only idea 5 suggesting ideas at the interpretive and application levels. However, such sections remain a useful starting point for teachers who wish to construct three-level guides.

While three-level guides can be assigned to students to work on individually, either in class time or at home, Thelen (1976) has stressed the advantages of students being allocated some class time to discuss their responses together in small groups. The discussion about various possible responses, especially at the interpretive and application levels, can then become a major part of the learning process.

Reinforcement of Vocabulary

After students have worked through a topic, they may have encountered a number of new words which may have specific geographical meanings or may have been used with particular significance in the specific context of the topic. As long ago as 1969, Skinner pointed out that one of the shortcomings of much teaching is the relative infrequency of reinforcement. Accordingly, there follows a number of suggestions for reinforcing vocabulary and comprehension skills:

Some of the most effective ways of reinforcing word recognition skills are through word games. Games of many types may be created with relative ease and are invariably popular with students. Figure 19.12 is a 'word search' designed to reinforce the vocabulary associated with the physical process of weathering. It illustrates how a seemingly childish game can serve to reinforce recognition of words which represent relatively high levels of conceptual attainment. Similarly, the old game of 'Hangman' either with pen and paper or as a program on the classroom micro-computer may serve the same purpose. More formal activities but with a similar purpose include presenting students with groups of words with similar associations such as:

- Town
- City
- Village
- Conurbation
- Hamlet
- Population

and requiring students to delete the odd one out and

Figure 19.10 A three-level guide for an earth science film
(Thelen 1976)

CONCEPTS TO BE DEVELOPED: WEATHERING, EROSION,
DEPOSITION, EFFECTS OF EROSION

PART I: Literal Understandings

Consider each of the statements below. Decide whether the message of the film you saw agrees with what each statement says. If you think so, place a check on the line before the statement. If not, leave the line blank.

_____ A. When rocks are wetted and dried repeatedly they begin to decompose.

_____ B. The best headstones are made from marble.

_____ C. As a rock weathers, some of the minerals in the rock decompose and cause slabs to come loose.

_____ D. Grooves in limestone prove that this hard rock can eventually be dissolved and washed away by rainwater.

_____ E. Over long periods of time, the alternate freezing and thawing of water pushes the rocks apart.

_____ F. Avalanches and landslides transport weathered material.

_____ G. As streams rush downhill toward the sea, they pick up weathered rock and other debris and carry them off.

_____ H. Rock, sand, and mud that are washed into a landlocked valley have no way to get out.

_____ I. Some turbulent streams carry sediments into calm waters of a lake and the sediments settle out on the lake bottom. Eventually, the lake will be filled with rocks and debris.

_____ J. An oval pattern formed by exposed edges of tilted layers of hard rock was recreated in clay to show that at one time a dome-like structure had been there but was eroded.

PART II: Interpretive Understandings

Several statements are listed below. Some may represent the meaning of the movie or the "correct" interpretation of the movie. When we interpret what we see and hear, we try to combine parts of the movie to generate an idea. Each of us may do this in a different way.

Read the first statement below. Then read the statements from Part I as identified by the letters in parentheses. Decide if the information from the statements in Part I could be combined to develop an idea like the one expressed in the statement in Part II. If so, place a check on the line before the statement. Follow this same procedure for each of the remaining statements.

_____ 1. Water in any of its three forms is the main agent in erosion. (A, C, D, E, G, H, I)

_____ 2. Rocks weather faster in drier climates than in wet and changeable ones. (A, D, E)

_____ 3. Rocks subjected to wet and changeable climates will soften, crumble, decompose, and split. (A, C, D, E)

_____ 4. Material that is weathered from high places is eventually deposited in low places. (G, H, I)

_____ 5. Accumulation of rock, sand, and mud eventually levels the floor of a landlocked valley and fills inland lakes, often obliterating them. (H, I)

_____ 6. Rushing streams, landslides, and avalanches transport weathered rock and other debris. (F, G, I)

PART III: Applied Understanding

To apply what we read, we must combine what we read, hear, and see with ideas or experiences which are personal to us. That is why we have called "applied understanding" the "personal" meaning of a passage or movie.

In column I below, there are statements you might have checked in Part II. In column II there are other ideas you personally may have had about the same topic. In Column III there are possible applied understandings, formed by combining statements in columns I and II. Above the list in column III you will find letters and numbers in parentheses. These suggest combinations of statements from columns I and II which might lead to the creation of ideas similar to those in the column III statements.

Read the first statement in column III. In the blank space write the letter and number combination which indicates which column I and II statements are represented in the first statement. Follow the same procedure for the other statements in column III.

Column I

1. Water in any of its three forms is the main agent in erosion.
2. Material that is weathered from high places is eventually deposited in low places.
3. Accumulation of rock, sand, and mud eventually levels the floor of a landlocked valley and fills inland lakes, often obliterating them.
4. Rushing streams, landslides, and avalanches transport weathered rock and other debris.

Column II

a. Some of the material from cliffs along Erie Boulevard slid into the backs of stores after the heavy snowfalls.
b. The potential energy of material at a higher elevation is changed to kinetic energy as the object moves downhill.
c. A few years ago a dam broke in Italy flooding and wiping out an entire village.
d. Gravity has the effect of pulling objects "down."

Column III
(1, a) (2, d) (3, c), (4, b)

_____ A. Water is the greatest agent for change on earth.

_____ B. Gravity is the force that drives water to move material.

_____ C. Weathered material at a high elevation will erode faster than weathered material at a lower elevation.

_____ D. Erosion and deposition could level the land eventually.

Big ideas about rainforests
These are the general ideas that you might have formed after reading this chapter.

1. Tropical rainforests are associated with hot, wet climates and are distributed in areas of low latitude, i.e., near the equator.

2. Tropical rainforests contain many species of plants. The trees are evergreen, non-seasonal and broad-leaved.

3. A rainforest is a system in which the main inputs are water (rain) and energy from the sun (heat), and the main output is the plants, especially the trees.

4. The three types of rainforest found in Australia are tropical, subtropical and temperate.

5. Rainforests are an important resource for the people of the world. Governments and people in countries that have rainforests need to consider their policy on the issue of preserving rainforests.

Figure 19.11 The concluding section of the chapter 'Hot, Wet Forests' in *Geography in Action* (Cox and Bartlett 1983)

Word Search - Weathering

```
F E I L E R U T I S C O R R A S I O N
M A I N O I T A D I X O T I M E C D S
W E C A R B O N A T I O N I G L E E R
O L C A E I T E S S R U G R A S P C O
L N O H C U T S G I U G O Y H H L O T
L O W E A T H E R I N G S S Y L A M N
A I I X O N E W I A A K C S D A Y P E
W T N F I T I N K O K R I H R C A O M
S A J O I N T C E S E C N A O I C S E
N R E L I E A O A E A N G T L G E I V
O G R I P P E N N L N L N T Y O D T A
I E V A C O H O T R O O T E S L R I P
T T I T L L O I E U I T N R I O A O E
A N T I T S C T R T T I N F S I F N T
T I O O P A T U A J A Q U I D B U D A
A S R N N I T L S R R E G O L I T H M
L I P Y L E E O B E D D I N G C U P I
I D O C V H A S E M Y P O T H O L E L
D N D K C O R E W A H T E Z E E R F C
```

In this grid are hidden forty eight words which are associated with the study of weathering. Words may be found on a vertical, horizontal or diagonal plane and can be read backwards or forwards.

The hidden words are:-

BEDDING	BIOLOGICAL
CANYON	CARBONATION
CAVE	CHELATION
CLAY	CLIMATE
CLINT	CLITTER
CORRASION	CORROSION
DECAY	DECOMPOSITION
DILATION	DISINTEGRATION
EXFOLIATION	FREEZE THAW
GORGE	GRIKE
HEAT	HYDRATION
HYDROLYSIS	ICE
JOINT	KARST
MECHANICAL	OXIDATION
PAVEMENT	PHYSICAL
POTHOLE	RAIN
RELIEF	ROCK
ROT	SALT
SCREE	SHATTER
SINK	SITU
SLOPE	SOIL
SOLUTION	SWALLOW
TIME	TOR
TUFA	WEATHERING

Questions

1. Find each of the hidden words in the grid.

2. There is one word in the list of words that does not appear in the grid. Which one is it?..................

3. There is one word in the grid which does not appear in the list of words. Which one is it?..................

4. There is one word in the list which is not associated with the study of weathering. Find the word and state why it should not be associated with weathering
....

5. Fill in the blank spaces in the following account of weathering by inserting appropriate words from the list.
Weathering is the breakdown of rock material, so producing in............a mantle of waste. Breakdown is achieved (a) chemically which leads to the......... of the rock and (b) mechanically leading to the .... of the rock. The thickness and character of the weathered mantle depends on (i) the type and structure of the.......... (ii) the nature of the............ of the area. (iii) the .... and the angle of .... (iv) the length of........... that the......... has been subjected to the .... processes.

6. Classify as many of the words in the list as possible under the following headings
   (a) **Chemical processes**
   (b) **Mechanical processes**
   (c) **Landforms or products**
   (d) **Factors which affect the rate of weathering**

Figure 19.12 A word search and associated activities designed to reinforce students' recognition of technical vocabulary (Kenyon 1985)

explain why to a partner, before both students decide on a collective noun to describe those which remain.

After a topic has been studied, both vocabulary, comprehension skills and concept application can be reinforced by using a variation of the cloze technique. Figure 19.13 is a cloze activity which was given to three 13-year-old girls of average ability, together with a transcript of their conversation as they completed the activity. The topic, 'Life in an Ibo Village', had been covered in the week before the discussion, and the teacher gave the class this exercise in order to reinforce what students already knew by getting them to use some of the concepts they had acquired about Ibo village life. Many of the words which were deleted are fairly specific to geography and would not be common knowledge to the students before the current series of geography lessons. Rye (1982), from whose book on cloze procedure this extract was taken, emphasises that the success of this activity with the three girls was largely determined by the insistence of one member of the group that a reason be given for every word chosen.

Figure 19.13 A cloze activity and transcript of the discussion between three 13-year-old students (Rye 1982, pp. 96–99)

Inquiring and Problem Solving Using Textbook Materials

Frequently teachers do not want their geography courses organized along the lines laid down by the authors of any of the textbooks available, and construct their programs from an entirely different standpoint. Especially if the acquisition of inquiry and problem-solving skills is one of the aims of the teacher, then the use of a traditional, linear text, in which the student proceeds from beginning to end, section by section, will not be appropriate.

For inquiry and problem-solving-oriented courses such as these, a range of expository materials at different levels is needed, including encyclopaedias, specialised articles, textbooks and reference manuals. The readability levels of such books is of less importance than those of more traditional texts since the materials will be searched, explored and extracted rather than read through. The aim of this kind of textual material is to provide an information base, in which signposts, concise definitions and

(The italicised words were omitted in the version given to the students.)

An Ibo Village
Read the following passage carefully. Certain words have been missed out, one word from each space. Try to work out what each missing word is. It is very important that you discuss your reasons for choosing each word with other members of the group.

Almost every man in the village is a farmer, depending for food on the crops he can grow for his family. Root crops can grow quickly and easily due to the *heavy* rain and the long *wet* season. Yams, rather like large potatoes, are the main crop of most farmers. *Cassava*, a similar plant, which in Britain we eat as tapioca, is another crop grown by many villages. To plant either of these the farmer must clear away all the undergrowth of small bushes and weeds. This covers land which he may not have used for a year or two. He will *burn* the heaps of undergrowth in January or February. The ashes contain *chemicals* which help the plants to grow, so he spreads the *ashes* over the ground.

A farmer owns about two *hectares* of land but this is divided into *many* small scattered plots, perhaps fifteen or twenty of them; some are near his home, others two or three kilometres away. When some of these plots have been cleared, the men and often the *women* also scrape at the surface with short-handled *hoes*, or digging sticks, to build the earth up into small mounds. In the top of each *mound* a small *yam* is planted and along the sides, other crops such as beans and green *vegetables*.

While the crops are growing the Ibo villagers live on food stored from the previous year's harvest. This is mainly *flour* made by grinding *yams* or cassava. The sale of *palm* oil and occasionally some *vegetables* and yams may earn a farmer about £25 a year. With this he pays *taxes* to the government, buys cloth and maybe a few tools, pots and baskets.

Despite such a small income a farmer seldom cultivates more than half his land in any year; the other half he leaves *fallow*—uncultivated, without any crop—so that the soil does not become exhausted. By doing this the soil slowly regains its *fertility* as new humus is formed from the *decaying* of falling leaves.
(Adapted from *Patterns in Geography, Book Two*, pp. 42–3, W. Farleigh Rice, Longman.)

Elizabeth:	Something rather like . . . yams. This is . . .
Susan:	Yes, yams. They've got a lot of *them*.
Carol:	Yes yams. They're root crops.
Elizabeth:	That's what they eat.
Susan:	They're like potatoes. Potatoes in't crops.
Elizabeth:	Yes, because they're like large, long potatoes.
Susan:	Yams . . . they're horrible.
Carol:	They're horrible.
Elizabeth:	Well, they're meant to be like potatoes. They've got to be yams.
Carol:	Hum . . .
Elizabeth:	And they eat them so . . .
Carol:	They eat a lot of them. They're easy to grow. They're root crops, so it *must* be them.
Elizabeth:	Yams, rather like large, long potatoes, are the main crop of most farmers. Something, a similar plant which in Britain we eat as tapioca . . .
Susan:	That's cassava.
Elizabeth:	Why? How do you know?
Susan:	'Cos I read it in the book. (Laughter.)
Carol:	'Cos that's like rice.
Elizabeth:	A similar plant which we eat in Britain . . . we eat . . .
Susan:	Well I don't eat it!
Carol:	I do. Frogs' Spawn!
Elizabeth:	We've got to give a reason why.
Susan:	'Cos it said it in the book.
Carol:	That's the same as tapioca.
Susan:	Yes.

*****

Elizabeth:	He will something heaps of ground in January or February . . .
Susan:	Burn.
Carol:	Fertilize them.
Elizabeth:	No, it says the ashes contain something.
Carol:	Yes, it's burn.
Susan:	It's burn them 'cos the ashes.
Carol:	Yes, they'd burn to destroy all the shrubs and that.
Susan:	It must be burn to get the ashes.
Carol:	They'd burn to get all the weeds and everything out.
Susan:	Yes.
Elizabeth:	Yes, burn to make the ashes.
Susan:	Yes, burn.
Elizabeth:	Yes, burn to make the ashes.
Susan:	For fertilizer.
Elizabeth:	Yes, to spread over the ground for fertilizer.
Carol:	Yes.

*****

Elizabeth:	The ashes contain what?
Susan:	Chemicals.
Carol:	Yes, that's chemicals.
Elizabeth:	The ashes contain chemicals which . . . Why chemicals? How do you know it's chemicals?
Carol:	'Cos the ashes . . . 'cos there's chemicals . . . 'cos there's chemicals in the plant in't there.
Elizabeth:	Yes, there's chemicals in the ashes.
Susan:	There's chemicals in the plant.
Elizabeth:	So he spreads the what? The ashes contain chemicals so he spreads the something over the ground.
Susan:	Ashes.
Elizabeth:	Yes, all right.

*****

Elizabeth:	When some of the plots have been cleared the men and often the something also . . .
Carol:	Women.
Susan:	Yes.
Elizabeth:	But we've got to give a reason why women do it.
Carol:	'Cos they do work as well.
Susan:	'Cos they help.
Carol:	Yes. They have to don't they.
Elizabeth:	Everyone's got to help to get the food.
Susan:	If they don't help they get no food.

*****

Elizabeth:	This is mainly something made by grinding something or cassava.
Susan:	Flour.
Carol:	You what?
Susan:	Flour, 'cos they grind up yams to make flour.
Elizabeth:	This is mainly . . .
Carol:	Yes, but it can't be yams. They grind up cassava stuff.
Susan:	Yes, but they grind up yams as well.
Elizabeth:	This is mainly flour . . .
Carol:	Cassava 'cos . . .
Elizabeth:	'Cos when they grind the yams down they use it as flour.
Carol:	You can't grind yams into powder.
Elizabeth:	You do. They're like potatoes. They grind potatoes up don't they. How do you think they get *Wondermash*. (Laughter.) Yes, they take water out of it. This is mainly flour made by grinding . . .
Susan:	Yams or cassava.
Elizabeth:	Yams.
Susan:	They grow those.
Elizabeth:	It's got to be. Yams can be ground to flour.

cross-references make the retrievability of principles, examples, facts and elaborations as easy as possible.

Preparing the Students

In preparing students for work based on a number of textbooks and other sources of information, many of the same procedures as those suggested for learning from a single textbook are required—but with some important additions. The principles of ensuring appropriate cognitive structure in students before approaching a topic, that they have either appropriate vocabulary or can be encouraged to search for meanings in context and that appropriate guides should be prepared apply in inquiry-based study as well as in more didactic approaches.

However, where the usual thinking process adopted in textbook-based study is inductive (i.e. students are required to work from a number of specific situations and examples in order to create their own generalisations), one of the main features of inquiry and problem-solving work is deductive thinking. Deductive thinking requires students to work from a generalised statement, usually in the form of a question, to investigate the relationships which may be discovered between the elements of the topic.

In contrast to textbook-based study, where the textbook author sets the agenda, the initiation of an inquiry topic requires the teacher to begin one stage back and select the topic for inquiry. Earle and Barron (1973) have suggested that topics selected for study should be interesting to the student, significant to the discipline, broadly applicable outside the discipline and important in terms of their potential for attacking the problems and issues of the present and the future. John Fien, in Chapter 26 of this volume, discusses the processes of preparing an inquiry unit of work. In that chapter he states that an appropriate 'angle' on a topic must first be selected by the teacher and/or the class. The advantages of the latter course lie in the insights it gives the teacher of the existing cognitive structures concerning the topic already held by the students. If, for example, the topic selected for its current relevance to an area is 'bushfires', students whose knowledge and perceptions of the topic are relatively unsophisticated may pose the question: 'Who cares about bushfires?'. However, students in whose cognitive structure 'bushfires' has already a well-developed place may suggest approaching the topic by asking 'Is trying to fight bushfires a waste of time?'.

Whether the initial question is teacher originated or student originated, the inquiry may still proceed according to the key question structure of geography. These key questions include:

1. *What* questions to help students refine the particular concepts involved in an investigation into their initial question.
2. *Where* questions to define the location and scale of the phenomena under investigation.
3. *Why* questions which encourage students to seek causes for and relationships between the phenomena.
4. *What are the implications* questions which require students to explore the relationships between the specific phenomena under investigation and other parts of human–environmental systems.
5. *Application* questions which ask students to go beyond the specific details of the study to make suggestions for appropriate future actions.

In the case of the simple question on bushfires suggested above, the key questions may be interpreted as:

1. What is a bushfire?
2. Where do bushfires occur?
3. Why do bushfires occur in those places?
4. What are the implications of bushfires for the environment and people of the area?
5. What could/should be done about bushfires?

In order to find answers to each of these key questions appropriate information must be sought from all the various resources which may be available. It is perfectly possible for the teacher to research these materials and refer students to specific extracts from various books and other materials. However, by doing this, the teacher is denying students the opportunity to learn and practise valuable study skills.

Text materials which describe processes, places or events are frequently described as expository text. It was suggested above that text materials for inquiry and problem-solving work should have good tables of contents and indexes and be well structured with clear chapter headings and signposts within the chapters. Most children learn to read using stories (narrative text) which are written with the intention that they be read from beginning to end. The reading strategies required to find information which is relevant to a specific question within a mass of other information include many of those needed to process narrative text as well as a number of additional ones including:

1. Vocabulary skills:
   (a) the recognition of technical words;
   (b) the realisation that some words in common usage may act as technical words;
   (c) the use of context to determine the meaning of unfamiliar words.

2. Locating skills:
   (a) the use of a glossary;
   (b) the use of tables of contents and indexes;
   (c) the use of appendices;
   (d) the use of signposts such as chapter headings and subheadings.

3. Comprehension skills:
   (a) following directions;
   (b) locating main ideas and supporting detail;
   (c) following a sequence of ideas;
   (d) interpreting graphs, charts, tables, figures, scales and diagrams.

Geography teachers cannot assume that their students will develop these skills 'by magic'. Research skills must be taught so before assigning students research tasks in which a number of books must be consulted, introduce students to research skills with an inquiry which may be answered by using a single book. For example, one teacher who was about to teach a unit on development to Year 10 students decided that the class should investigate life in Singapore. The school has a class set of Niranjan Casinader's (1986) book *The Faces of Development*. The teacher raised the topic of Singapore with the class and the students brainstormed everything they knew about Singapore. Amongst the mass of details which was forthcoming were statements about the cleanliness of Singapore, the birth control policies which have been pursued and the heavy fines imposed on anyone who drops litter in the streets. From this information the teacher suggested to the class that an appropriate angle on the topic would be: 'Is Singapore a good place in which to live?'. The key questions of geography were then applied to this initial question and the class then developed the inquiry sequence:

1. Where is Singapore?
2. How was Singapore created and what is life in Singapore like?
3. Why has Singapore developed in this way?
4. What have been the implications of this development for Singapore and its people?
5. What would we like and dislike about living in Singapore and how could Singapore achieve its aims without the aspects we would dislike?

As the students had not done much research inquiry work before, the teacher provided them with a graphic outline of the chapter on Singapore as shown in Figure 19.14. A graphic outline summarises the content of a section of expository text by making explicit the structure used by the author, shows the relationship between the various headings, subheadings and graphics and gives some indication of the amount of information provided on each aspect of the topic. With the aid of the graphic outline, students are able to use the information in the text to answer the key questions of their inquiry.

When the class has greater experience of research work, the teacher may prepare a graphic outline like that shown in Figure 19.14, but having done so, present the students with a *blank* copy showing only the boxes for students to complete on their own or in groups (Figure 19.15). In both these cases, the use of graphic outlines helps teacher and students. The teachers gain because the outline reveals strengths and weaknesses in the materials being assigned, thus avoiding students becoming disillusioned with their work because they are being required to answer questions for which the answers do not exist in the materials available. Students gain because they can appreciate the structure of the materials they are using and do not become overwhelmed by the mass of material before them.

As students gain knowledge of the kinds of text structures used by authors of expository text and gain confidence in selecting materials, their initiatory activities may be based on the lists of contents and the indexes of specific books to which they are referred by the teacher. They may be encouraged to create their own graphic outlines of the major references for their inquiry.

Following Instructions and Directions Presented in Text

Most geography textbooks include instructions for student activities based on either the expository text

Main headings	Subheadings	Graphics and other aids
Introduction		Fig. 5.1, Map: Location of Singapore Photo: Singapore 1965 Photo: Singapore 1985
Economic and Social Development	Economic Development	Fig. 5.2, Bar chart: Industrial structure 1960 Fig. 5.3, Table: Industrial structure 1971, 1981, 1984
	Social Development	Fig. 5.4, Table: Changes in social indicators 1960–1980 Fig. 5.5, Table: Income distribution 1980
Explanations	Limited Physical Resources	Fig. 5.6, Map: Physical features and land use
	The Colonial Legacy	Fig. 5.7, Time line: History of Singapore Fig. 5.8, Pie chart: Racial structure 1984 Fig. 5.9, Divided bar graph: Religious structure 1980
	Changes in Political Status	Fig. 5.10, Map: Federation of Malaysia
	Government Stability and Policy	Photo: Prime Minister Lee Kuan Yew
	—Economic Effects	Fig. 5.11, Line graph: % unemployed 1970–84 Fig. 5.12, Table: No. of workers in manufacturing 1965–84 Fig. 5.13, Table: % employed in low and high value added manufacturing 1960–81 Fig. 5.14, Bar graph: Structure of labour force by industry
	—Social Effects (i) Education (ii) Population Control	Photo: A primary classroom Photo: A university laboratory Fig. 5.15, Line graph: Annual population growth 1957–83 Fig. 5.16, Population pyramid 1984 Newspaper extract: 'Even the elite reject Lee's eugenics policy'
	(iii) Housing	Photo: Old and new housing Photo: High rise public housing
	(iv) Land Use	Photo: Dock developments 1976 Photo: Container port 1985 Fig. 5.17, Table: Land use 1973 and 1983 Fig. 5.18, Map: Population distribution 1980
	(v) Law and Order	Fig. 5.19, Map: Distribution of housing types
	The 1985 Elections	Newspaper extract: 'Singapore's old guard steps down—except Lee' Newspaper extract: 'Singapore's success may cost it trade concessions'

Figure 19.14 A graphic outline of Chapter 5, 'Singapore', in *The Faces of Development* (Casinader 1986)

Main headings	Subheadings	Graphics and other aids

Figure 19.15 Blank graphic outline ready for students to use in their own analysis of Chapter 5, 'Singapore', in *The Faces of Development* (Casinader 1986)

or the graphics presented in the book. However, Lidstone (1977) found that many students were unable to differentiate between the expository text and the activities in the geography textbook which they had been using in class for some months, while many teachers will bear witness to the apparent inability of students to translate textbook instructions into action. Yet, as many of the reading activities of adults take the form of reading to follow directions, there has been surprisingly little research into such difficulties of interpreting written instructions. Nevertheless, some general principles for selecting good examples of directive writing may be suggested:

1. The instructions should be sequential in the sense that they are presented in the order in which they should be implemented.
2. Signposts are needed to help the reader to find the starting point and subsequently order both reading and activity. The simplest way of providing this help is to number the stages.
3. Advance warning of technical vocabulary should be provided before the reader is instructed to begin.
4. The task must be within the ability of the expected reader.
5. There must be some way of knowing when the task has been completed.

In helping students to use textbooks both as guides to the understanding of a particular topic and as sources for more independent inquiry, teachers should draw their students' attention to the conventions used in their books for indicating when expository text ends and activity instructions begin. Such conventions include:

1. placing a light screen over the activity sections;
2. placing activities within boxes;
3. printing activities in a different fount or size of type; and
4. preceding each activity with a number.

Having ensured that students can differentiate the activities from other elements of the text, students should be encouraged to read the whole of the activity instructions before beginning to carry them out. A study of activities suggested by many textbook authors usually reveals that frequently they appear to be more complex than they really are. Here is an example of a single activity from a relatively recent physical geography textbook (Galbraith and Wiegand 1982) with the text lines arranged as they appeared in the original:

5. People are also important as a force modifying the landscape. There are many 'man-made' landforms. Be careful not to confuse them with natural ones. Look at the photograph of Silbury Hill. It is not a natural feature but a Bronze Age burial mound, the largest artificial hill in Europe.
What other 'man-made' landforms can you think of? Are they features of 'erosion' or 'deposition'?
Can you think of examples where modern technology has enabled people to modify the landscape even more than at Silbury Hill?

---

13 Answer the following questions using the geological column, the simplified geological map of Britain on the right and an atlas.
  a The line A–B on the geological map divides Britain into two. Which two estuaries does the line join?
  b Put the following phrases into two columns, one headed 'North and West', the other headed 'South and East'.
      igneous, metamorphic and hard sedimentary rocks
      softer sedimentary rocks
      Highland Britain
      Lowland Britain
      rocks mainly of Carboniferous age and older
      rocks mainly younger than Carboniferous age
  c What is the difference in geology
      i) between Dartmoor and Exmoor?
      ii) between the Cotswolds and the Chilterns?
  d Of which rock or rocks are the following areas formed?
      the Pennines
      the Antrim Plateau
      the Grampians
      the Fens
  e Give the names and approximate heights of four ranges of chalk hills.
  f Write the names and the approximate date before the present of the three major periods of earth movements that have affected Britain.
  g In which geological periods were the New Red Sandstone and the Old Red Sandstone deposited?

Figure 19.16 A well-structured series of questions and activities from Galbraith and Wiegand (1982)

Some students approaching this activity, perhaps at home with no one available to refer to, will read the first six lines and give up simply because it appears to be too complex. Others will read through and realise that the introduction and instruction to look at the photograph have nothing to do with the activity they are being asked to undertake, but will be uncertain about how to approach the tasks in paragraphs two and three. In the context of this question, would a tower block be considered a land form? Is the Great Pyramid at Giza a land form? Perhaps the new crematorium on the outskirts of town is the nearest equivalent? And what is our student to make of 'Are they features of "erosion" or "deposition"? After all, the new crematorium is built above ground, but the foundations probably go quite deep. In any case, how many 'other "man-made" landforms' are our students expected to think of? Will one or two suffice, or should they persuade their parents to drive to the city library to find a few more? And finally, does the third paragraph modify the question in the second or is it an additional question? To sum up, this activity really raises more questions about the intentions of the author than it does about the level of geographical understanding of the students.

Problems such as these are by no means uncommon in geography textbooks, and are probably responsible for ruining more family weekends than any other single aspect of the educational endeavour! Teachers rarely choose textbooks on the basis of the precision of activity instructions, yet the book from which the example above was taken is in many other ways admirable. However, such activities do cause problems to many students and it is the responsibility of the teacher to ensure that students' time is not wasted. One way of doing this is to teach them how to analyse questions and activities. The following questions will help both teachers and students to clarify the requirements of activities in textbooks:

1. What materials do I need to have available?
2. What data do I need to use in order to proceed?
3. What is the exact order in which I must carry out the instructions?
4. What is being required of me for each section of the activity?
5. How will I know when I have completed each section of the activity satisfactorily?

Figure 19.17 An extract from a textbook on South Africa (Treadaway 1985). This is a piece of expository text

---

South Africa is the world's chief producer of gold, diamonds, uranium, and platinum, and is important also in asbestos, copper, vanadium, lead, manganese, antimony, and chrome. She is fortunate also in having rich coal deposits and iron ore. Altogether, her production of minerals exceeds £200 million annually and has built her up into by far the most industrialised country in Africa.

Diamonds were the first to bring fortune to South Africa. In 1866 some Boer children found diamonds lying on the ground and later, when these were recognised, some 50,000 people flocked into the area, to seek for diamonds in the alluvium of the Vaal and Orange rivers. By the end of the century, diamonds were being mined from the volcanic 'pipes' in which they had been formed, and by 1926 further alluvial fields had been discovered on the west coast and in western Transvaal.

But the diamonds had brought not only wealth, but skilled miners and enterprising businessmen to South Africa and these resources, especially the capital, were soon turned to the exploitation of the rich Witwatersrand goldfield. The gold here occurs in the form of very fine grains in a hard sedimentary rock. First the rock must be mined, then crushed and the gold extracted by a chemical process involving first solution in cyanide and then precipitation. The mines are very deep—some go nearly two miles down and are thus far below sea level. This deep working, and the subsequent processing of the ore requires a great deal of power. Fortunately, good easily-worked coal was found in Natal and later in the Transvaal, so that there is no difficulty on this account. Industries servicing, and later making, mining machinery and equipment also grew up and have stimulated the development of an iron and steel industry, which in turn provides the raw materials for a variety of manufactures. Railway lines and modern seaports were also created in response to the stimulus of the gold and diamond workings. It is not surprising that Johannesburg, which grew up on the Rand, the original goldfield, has become the biggest town in the Republic and is the centre of the largest group of white people in Africa south of the Sahara. For long the Rand was virtually the only goldfield, but in recent times two rich new fields have been opened up, one near Klerksdorp and a larger one in the Orange Free State. The output from the latter is still expanding.

In the example given above, the answers to these questions will lead to a demand for the teacher to provide his or her own explicit instructions concerning what the students are expected to produce. In fairness to the authors of the book from which the earlier activity was taken, Figure 19.16 shows a question from elsewhere in the same book which is well structured and will result in much less student confusion. Even here, however, the teacher would do well to point out to the students that the geological column, mentioned first in the introductory statement to the activity, is not required until the very last question.

Figure 19.18 A piece of narrative text describing the inside of a gold mine (from Treadaway 1985)

Challenging the Author's Implied Assumptions or Interpretations

David Wright (1987) has pointed out on many occasions that all textbooks display some bias on the part of the authors. He emphasises, however, that to point out the biases in a textbook is not necessarily to imply any ulterior motive on the part of the author. Rather, Wright sees the biases as providing an opportunity for teachers to encourage their students to become critical readers—going well 'beyond the lines' in the terms used in the section of this chapter on three-level guides. Treadaway (1985), in *Geography for Development*, presents two extracts which illustrate how the same topic may be presented to give entirely different perceptions to the reader. Figure 19.17 is an extract from a traditional textbook on South Africa. It concentrates on the mineral wealth of South Africa, and the reader may be excused for assuming from the passage that all South

---

Down a Mine

'The boss boy looked after everyone, saw that everyone was safely in, then gave the signal, then jumped as the cage began to move. That was the duty of the head mine boy.
   The cages shot down. Down. Down. Down.
   The men were silent. It was always so. Going into the bowels of the earth forced silence on them. And their hearts pounded. Many had gone in day after day for months. But they did not get used to it. Always there was the furious pounding of the hearts. The tightness in the throat. And the warm feeling in the belly. It was so for the mine boy. They knew it. . . .
   Down shot the cages. Down. Down. Down.
   And their lamps flickered and there was a thin, sharp whistle through the air as the cages shot down. Deep down into the body of the earth. And the only light was the light of their lamps. And the air became warmer and breathing seemed heavy. That too was always so.
   The cages slowed down and the men jumped out. They stood around in groups, waiting . . .
   Xuma studied the sides and roof of the tunnel as they went along. Where the tunnel led to the wall where the working had to be, props had been built to look like the framework of a doorway. The roof of this sagged. Xuma studied it for a long time. Paddy who had gone ahead came back and stood beside him. 'What do you think?' asked Paddy. 'Maybe it is nothing,' Xuma said, 'but I think we should put more stout poles on each side.' . . . .
   The drill hummed. The hammer rang. There was a swish and a buzz and a hum, and there was the clang of the pick and grating of the shovel. And slowly the rhythm of the work gathered pace . . .
   Paddy took a drill, switched it on and held it to the side of the wall of rock. The muscles of his arms and chest rippled under the hum of the drill. . . .
   And the conveyor belt sang and the picks fell and the spades grated and the drills hummed. And everywhere men worked. Their bodies streaming with sweat . . .
   And an ever rising stream of shining rocks and pebbles and fine dust would travel upwards to be sifted, crushed and sorted for the fine yellow metal men love and call gold. . . .
   When the hour to eat came the men flung their tools from them and stood around with weariness on their faces and sweat dripping from their bodies . . . A man near Xuma coughed. A trickle of red spittle flew out of his mouth and fell at Xuma's feet. Xuma stared at it. He had heard about the sickness of the lungs and how it ate a man's body away, but he had never seen a man who had it.'

From *Mine Boy* by Peter Abrahams.

1 Hunting and gathering

Hunting and gathering is the most primitive form of land use in the rainforests. The tribal groups engaged in this type of economy are now found only in the more remote areas of the Amazon and Congo Basins and the rugged upland areas of south and east Asia.

Small tribal groups live in small clearings in the forest, in primitive shelters made from the products of the forest—leaves, branches and twigs. Women and children do the domestic duties (cooking, making mats, baskets and some clothes) as well as some gathering, while the men spend most of their time hunting, gathering and sometimes fishing. Men and women gather various fruits, roots, berries and other edibles, which grow naturally in the forest. The men hunt small game such as lizards, birds, monkeys and pigs. They use bows and arrows, spears and knives, all made from forest products.

The daily life of these people consists of an almost constant search for food. When all of the suitable game in an area has been hunted out, the tribe shifts to another part of the forest. This is easy, as their possessions are few and the shelters can easily be taken down and transported, or new ones made at the next village site.

There is more to *life* than a type of *economy*.

Remote from where? To these people, we are 'remote'. We all live in the centre of our own world.

People live in families (just like many of us).

'Primitive' again. Why not 'ingenious', 'well designed' or even 'natural'. (How many of us could build a serviceable home?)

Is this really different to *our* lives of 9–5 work? Is life just economic activity? What about recreation? Religion? (Anthropologists have recorded that people who live in Amazonia can find all their living wants in less than 2 hours per day.)

Is it really 'easy' to make a new clearing with simple tools and build a new home? (Is it easier than telephoning an estate agent to sell our houses and a removal firm to shift our belongings?)

Why 'primitive'? Why not 'interesting', 'efficient', or 'effective'?

Why label them as 'tribal groups'? Why not just 'people'?

'Only' seems dismissive. 'Mainly' would be more positive.

This suggest that they are passive and the clearings just happened. In fact, the people clear small areas of the forest with simple tools they make themselves. In this way they limit damage to their environment.

Why just 'shelters'? We live in 'homes'.

Only 'few' compared to our lifestyle. This is a very ethnocentric viewpoint.

'Easily' repeated—I wonder how the author would survive carrying a shelter through the forest?

Figure 19.19 A textbook extract and some critical responses to the text based on words felt to be unhelpful in developing our empathy for hunters and gatherers. (Extract from Milliken et al. 1984)

Africans share in the wealth which has been created. On the other hand, Figure 19.18, an extract from *Mine Boy*, presents an aspect of the mining industry which is not even hinted at in the first extract.

This example illustrates the importance of teachers investigating the existing knowledge of their students before embarking on a new topic. The words 'apartheid' and 'race riot' may be expected to emerge from any brainstorming activity on South Africa. The creation of structured overviews and initial questions for inquiry work will emphasise the importance of these terms in a study of South Africa, while the preparation of graphic outlines of the textual materials available will reveal where the deficiencies are. Such strategies will enable teachers to ensure that students are exposed to a wide range of viewpoints on the basis of which they may then develop their own values and attitudes.

Sometimes, however, there may not be the time or materials available to present students with alternative viewpoints on every topic. David Wright (1987) has created a technique for encouraging students to interpret text materials from other standpoints and to develop students' empathy with those whose lives are so often dismissed in textbooks as being of little account. Wright's technique is to read a passage from a textbook to the class, with the instruction that the students are to tap their desks with a pencil each time a biased word or phrase is mentioned. He then stops and asks the students to explain why they feel that the particular word or phrase might give a partisan impression.

Figure 19.19 is a textbook extract on 'Hunting and Gathering' from *Here and Beyond* (Milliken et al. 1984) together with some responses received from a class of trainee geography teachers. On the occasion when Wright demonstrated this technique, the initial responses frequently led other class members to join in as other implications of the words chosen by the author occurred to them.

Conclusion

It is now more than 20 years since Archer stated that 'Geography was long a byeword as a mere memory subject. Its transformation in the hands of good teachers since 1900 has been more complete than in the case of almost any other subject' (1966).

In the past two decades, the quality of textbooks available to geography teachers has improved greatly. Few books are now published without a wealth of colour photographs and well-drawn graphics. Books are available which reflect a wide range of approaches to geography teaching, with readability levels to suit students of all abilities.

However, although the past two decades have seen vast improvements in the quality of textbooks published, they have also seen school geography increase its areas of concern to include such issues as multicultural understanding, the practical applications of environmental studies and urban studies. Over the same period, there has been increased concern about the abilities of students to handle the wealth of information with which their senses are now assaulted. The period in which we are living has already been dubbed 'The Information Age' and geography teachers must make a contribution to ensuring that our students are equipped to handle the wide variety of types of information available to them. The effective use of geography textbooks is a small, but very significant part of that responsibility.

References

Archer, R.L. (1966) *Secondary Education in the Nineteenth Century*, London: Frank Cass and Co.

Ausubel, D.P. (1968) *Educational Psychology: A Cognitive View*, New York: Holt, Rinehart and Winston.

Barron, R.F. (1969) 'The Use of Vocabulary as an Advance Organizer', in H.L. Herber and P.L. Sanders (eds) *Research in Reading in the Content Areas: First Year Report*, Syracuse, New York: Syracuse University Press.

Bloom, A.L. (1969) *The Surface of the Earth*, Englewood Cliffs, New Jersey: Prentice Hall.

Casinader, N. (1986) *The Faces of Development*, Melbourne: Thomas Nelson.

Cox, B. and Bartlett, L. (1983) *Geography in Action*, Brisbane: Jacaranda.

Earle, R. and Barron, R.F. (1973) 'An Approach to Teaching Vocabulary in Content Subjects', in H.L. Herber and R.F. Barron (eds) *Research in Reading in the Content Areas: Second Year Report*, Syracuse, New York: Syracuse University Press.

Herber, H.L. (1970) *Teaching Reading in Content Areas*, Englewood Cliffs, New Jersey: Prentice Hall.

Galbraith, I. and Wiegand, P. (1982) *Landforms: An Introduction to Geomorphology*, Oxford: Oxford University Press.

Kenyon, I. (1985) 'Word Search—Weathering', *Teaching Geography*, Vol. 10(2).

Lidstone, J.G. (1977) *An Evaluation of the Geography Curriculum Offered to Second and Third Year Pupils in a South London Junior High School Based on Volumes Two and Three of the "Oxford Geography Project" Textbooks*, unpublished MA dissertation, University of London.

Lidstone, J.G. (1985) *A Study of the Use of Text Books by Selected Teachers in English Secondary Schools*, unpublished PhD thesis, University of London.

Milburn, D. (1972) 'Children's Vocabulary', in N.J. Graves (ed.) *New Movements in the Study and Teaching of Geography*, London: Temple Smith.

Milliken, M.C., Shaw, J.H. and Kirkwood, F.G. (1984) *Here and Beyond*, 2nd Edition, Melbourne: Longman Cheshire.

Nichols, J.N. (1983) 'Using Predictions to Increase Content Area Interest and Understanding', *Journal of Reading*, Vol. 27(3), 225–228.

Robinson, F. (1970) Study guide for Ausubel/Robinson *School Learning*, New York: Holt, Rinehart and Winston.

Roller, C.M. (1986) 'Overwriting, Underwriting and Other Text Book Sins', *Social Education*, January, 56–57.

Rye, J. (1982) *Cloze Procedure and the Teaching of Reading*, London: Heinemann.

Schools Commission (1980) *Schooling for 15 and 16 Year Olds*, Canberra: Schools Commission.

Stephenson, B. (1981) in M. Williams et al. (eds) *Language Teaching and Learning 2: Geography*, London: Ward Lock Educational.

Skinner, B.F. (1974) *The Technology of Teaching*, New York: Appleton-Century-Croft.

Thelen, J. (1976) *Improving Reading in Science*, Newark, Delaware: International Reading Association.

Treadaway, J. (1985) *Geography for Development, A Handbook for Geography Teachers*, Nairobi, Kenya: Heinemann.

Williams, M. et al. (eds) (1981) *Language Teaching and Learning: Geography*, London: Ward Lock Educational.

Wright, D. (1987) Presentation to Geography Teachers' Association of Queensland, Brisbane, June, Brisbane College of Advanced Education.

20

Using Games and Simulations in the Geography Classroom[1]

John Fien, Robert Herschell and John Hodgkinson

The learning of geographical knowledge and skills and the exploration of values can be stimulated by the use of classroom games and simulations. John Fien, Robert Herschell and John Hodgkinson recognise that many teachers are not comfortable using them despite a decade of diffusion via workshops, teacher training courses and textbooks. They briefly restate the case for simulations and games and then provide guidelines for successfully selecting, using and debriefing them in geography teaching. The key message of the chapter is the need to integrate games and simulations into a teaching unit and not to use them as 'one-off' activities. The chapter contains a copy of the *Amazon Adventure* simulation and *The Rainforest Game* and details for integrating both into a teaching unit on equatorial landscapes.

In the world of international air travel, it would be very dangerous for any airline to attempt to train pilots on regular commercial flights. It is, however, extremely important that all pilots gain regular experience of the many types of conditions they are likely to face when transporting passengers across the world. The problem of how best to provide 'risk-free' training for pilots has been overcome by the development of flight simulators—replicas of aircraft cockpits including every gauge, switch, control and lever. Flight simulators are connected to a wide range of electronic and audio-visual gadgetry which enables the pilot to be subjected to a great variety of meteorological and flight conditions with the accompanying scenes and sounds. Each training pilot's procedures and reactions are carefully monitored and, if mistakes are made, he is able to learn from them at no risk to life or aircraft.

The use of simulation techniques in education has increased at a prodigious rate in recent years. Some activities such as *Lynwood*, *Portsville* and *Railway Pioneers*[2] quickly gained a wide degree of acceptance among geography teachers, for example, while others have proved to be difficult to operate in the classroom.

Defining Terms

At present, much confusion exists concerning the use of the terms 'simulation' and 'gaming' in an educational context. Some people fail to discriminate between simulation and gaming. Others regard gaming as an aspect of simulation. A third group sees simulation as part of the technique of gaming. If simulations and games are to be used successfully in school programs, teachers need to appreciate not only the meanings of both terms but also their specific characteristics. In this way, they will be able to select the most appropriate activity for the objectives of each section or unit of work being developed. The following definitions have been designed to assist teachers to clarify the meaning of simulation, game and other related terms.

1. *A simulation* accurately reflects some part(s) of reality. Therefore when students are involved in a simulation, they are manipulating a model or playing roles *which assist them to develop an understanding of, and a feeling for, the reality being presented.*

2. *An educational game,* on the other hand, is an activity in which students use data and/or skills in a competitive situation against themselves, each other, the teacher as game master, chance or the environment. *Games therefore are useful for presenting repetitive learning in novel ways.* The situation in which the information and skills are used may not accurately reflect reality as closely as does a simulation. For example, *Squatter* and *The Rainforest Game* (Figure 20.1) are games because they use specific data and knowledge in an unreal way. This results from the nature of the rules of the game and the unnatural role that chance plays in the activity. *Monopoly* would also be classified as a game under these criteria.

However, when students become involved with planning travel itineraries or investigating a controversial environmental issue and attempt to resolve it by playing the parts of identifiable community groups, these students are being involved in a simulation. *Amazon Adventure* (Figure 20.2) is such a simulation because it tends to emphasise the development and application of concepts and process skills through specific knowledge and data.

Figure 20.1 *The Rainforest Game* [3]

Introduction
In this game, students 'explore' the Amazon Rainforest in the hope of finding a commune inhabited by a lost tribe of Indians.

Information
To start, each player must throw a 'five' or a 'six'. After starting, the players take turns to throw the die, following the instructions printed in each square of the game board reproduced below.

The first player to return safely to Belem after reaching the commune wins the game.

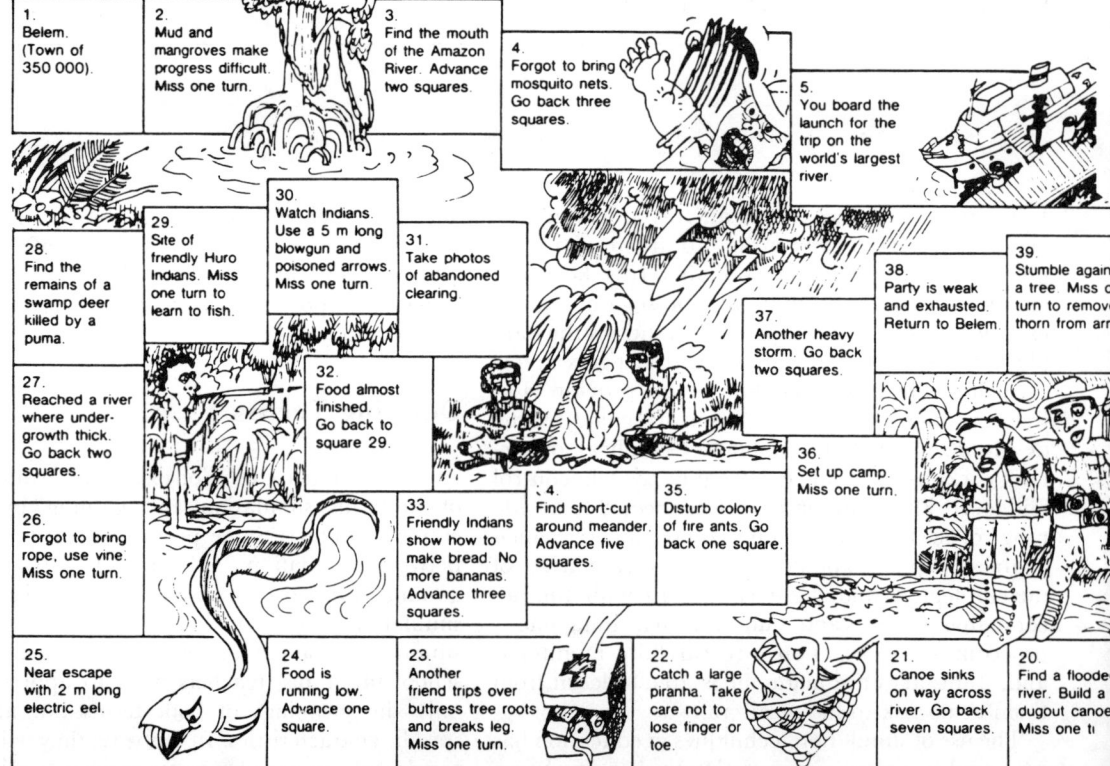

Using Games and Simulations in the Geography Classroom

3. *Role plays* deserve a section of their own. They are a special variety of simulation that requires students to adopt various roles in order to examine a particular problem or issue. Students research their roles and prepare a case for their role's viewpoint on the issue. A Committee of Inquiry, usually composed of students, meets near the end of a role play to hear submissions and come to a decision on resolving the issue. Role plays are especially good for encouraging students to see all sides of an issue, to empathise with alternative viewpoints, and to experience small- and large-group public speaking. Examples of role plays useful in geography include: *South Bank Hostel Storm* and *Lynwood*.

4. *Simulation games*. Often, many activities combine elements of gaming, simulation and role play. These are known as simulation games. Commercially produced examples include: *Railway Pioneers, Mali Cattle Dealers, The Poverty Game* and *The Green Revolution Game*. Simulation games have the advantages of both simulations and games and are especially useful in developing decision-making skills. The structure of many simulation games can be programmed for a computer and many of the complicated rules and data manipulations common in some simulation games transferred to it. Examples of computer-based simulation games include *Sand Harvest, Slick* and any of the more than 20 geography titles in the Longman Micro-Software series.

Often the way all these terms are used in school textbooks, reference works, and even in some of the activities themselves is not based upon definitions such as those just outlined. It is important, therefore, that teachers examine each activity to determine whether it is a simulation, a game, a role play or a simulation game. This will assist in matching appropriate learning activities with the units of work and lessons being planned.

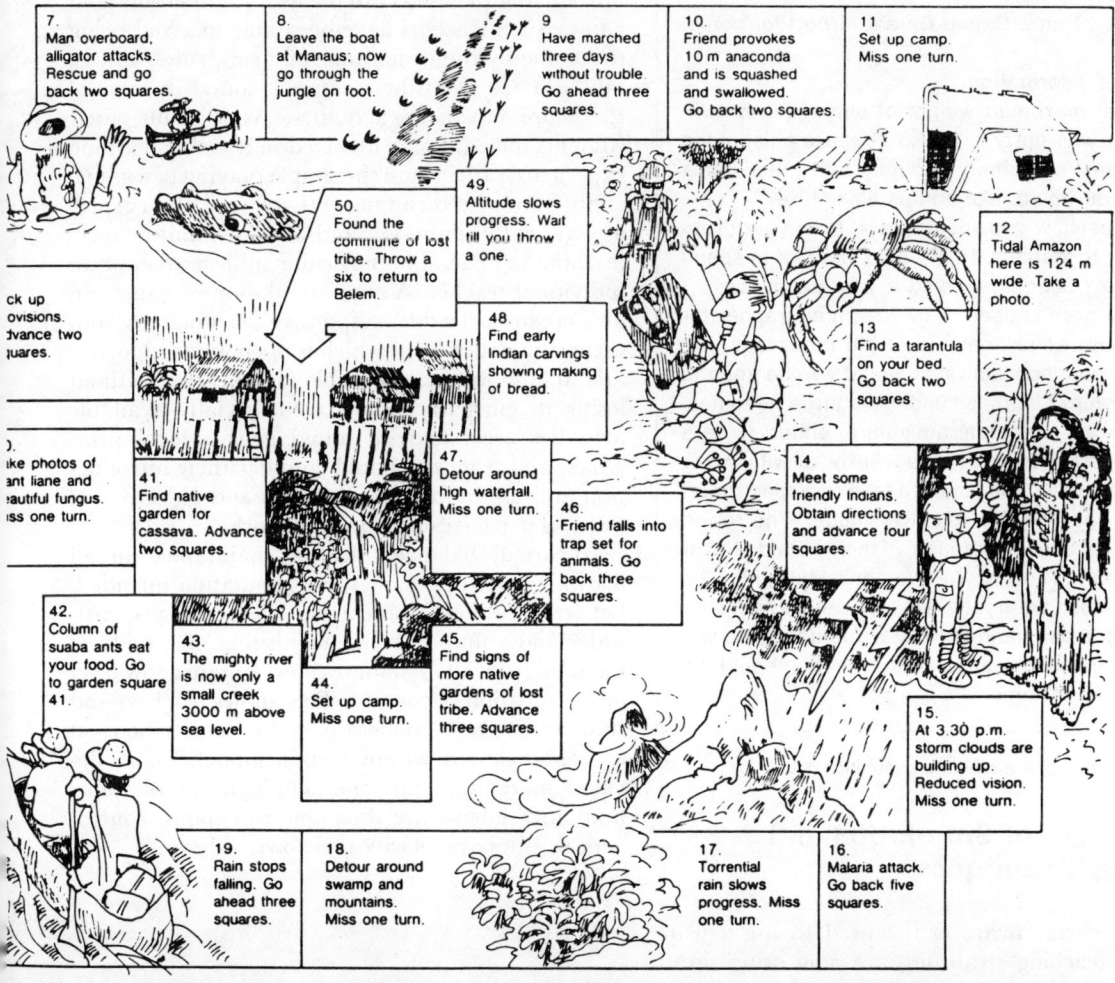

> **Introduction**
> The purposes of this simulation are to prove that today's adolescents (provided they are well prepared) could survive in difficult country; and to enable students to learn about the Amazon River through their experiences.
>
> **Teacher's Information**
> In groups of six, the students will undertake an 'expedition' through the Amazon Basin. Their itinerary is outlined below:
> Day 1, 6.30 a.m.: You will be dropped by helicopter, 50 km west of Manaus, the town near the junction of the Amazon and Negro Rivers (2°55'S, 60°2'W).
> Days 2–4: Proceed overland from your drop-off point to Manaus. How you spend these days is entirely up to you.
> Day 5, 5.30 p.m.: You will board the 10 000 tonne oil tanker *Chandral* for the journey down the Amazon to Belem.
> Day 9, 4.00 p.m.: Depart Belem Airport for home.
>
> **Students' Information**
> The total maximum weight of luggage and supplies for your party is 120 kg. Prepare a list of the provisions, equipment, clothing and other items which you would take on this adventure.
> Suggest how you should travel from your drop-off point to Manaus. How would you like to spend Days 2–4? Suggest an itinerary for each day.
> If you were chosen to be the photographer for the group, which features of the landscape, the people, and their activities would you be keen to photograph? List these under the three headings.
> When you complete adventure, either: write a brief report for the *Women's Weekly*; or with two or three other students, make a tape recorded report lasting between five and 10 minutes. The report should be titled 'Thorough Planning Paid Off' or 'At Times I Thought We Wouldn't Make it!'. The report must be neatly set out, and should be completed and handed in by a due date. You will have time in class, but most work should be completed at home.

Figure 20.2 *Amazon Adventure* simulation [3]

Advantages of Simulation and Gaming Techniques

Many teachers, aware of the need to use a wide range of teaching strategies, are now using simulations and games as well as the more traditional methods of geographical inquiry and investigation. Each strategy has a role to play in the total learning sequence as each strategy assists students to develop concepts, skills, attitudes and insights, as well as a body of knowledge. The various learning activities and experiences selected for use in particular sequences must become integral parts of all lessons, topics and units of work.

Simulation and gaming activities are most effective when student involvement in them has been developed to a high degree. Teachers who have not previously used simulations and games may be advised to begin by selecting simple activities for use with their students. The two activities included in Figure 20.1 and Figure 20.2 are suitable for this purpose.

Many teachers have found that by working first with simulations and games which do not involve an extended time commitment, they are able to develop skills in the utilisation of such activities. By gradually introducing themselves to the use of simulations and games, such teachers have been able to avoid being overcome by the wide range of forms, rules, charts, tables, maps and other materials found in some of the more advanced activities. As a result, their students have become involved in a viable learning experience rather than the simple playing of a game.

Simulation and gaming activities can often generate an enthusiasm for and/or a commitment to learning in general, a particular subject area, or an individual teacher. A good simulation or game can also create a flexible, responsive learning environment, the structure of which is open and supportive and able to take considerable modification without losing its efficiency. Many commercially available activities are based on sound geographical principles, and teachers should consider their introduction and use in fulfilling the aims and objectives of particular courses and units of work.

A careful balance must be maintained in all learning activities, games and simulation included, between the development of skills, concepts, attitudes and values and the acquisition of a body of knowledge. Many syllabuses now contain a low level of specification of content. This allows teachers and students to work cooperatively in using a body of content, selected within certain guidelines, to develop more than just the simple recall of information. Students are thus able to explore controversial issues and clarify their own values.

Planning to Use Simulations and Games

To ensure that simulation and gaming activities are used effectively, it is important that teachers should become familiar with not only the benefits but also the possible pitfalls in the use of such activities in the classroom.

An important task for teachers to perform is basic 'public relations'. Fellow teachers, school administrators and parents should be encouraged to support you and your students in the use of educational simulations. They are most likely to do this if they fully understand what you are trying to do and why you are doing it. Information concerning simulation and gaming activities can be disseminated through notices, a newsletter, a departmental policy booklet, or at a meeting. Once a favourable attitude of the staff to its objectives and strategies is created, the chance of parental or administrative opposition to simulation and gaming is reduced.

Another important point for consideration is the teacher's attitude to the presentation of the activity. If students become more interested in winning than learning, it may be because they are not perceiving the activity as part of an integrated learning unit. The reaction of students to the activity will parallel that of their teacher. Such activities must be integrated into the total learning unit.

Before introducing a simulation or game to students, teachers should make their own evaluation of it. The questions below have been designed to assist in assessing the value of particular simulations and games:

1. Do the objectives of the activity conform to the objectives of the section of the course being studied?
2. Is the activity appropriate for the students' levels of skills, cognitive and moral development?
3. Is the activity interesting?
4. Is the activity workable in a classroom situation?
5. Does the activity have a sound knowledge base?
6. What is the central problem theme or issue presented in the activity? Will students identify with it?
7. What are the choices available to the participants? What are the different moves or activities provided for the participants?
8. How is the activity to be organised in the classroom?
9. Does the teacher's guide provide adequate advice on procedures for conducting the activity with a class?
10. What summary and debriefing exercises conclude the activity?

Consideration should also be given to the most appropriate way of incorporating a simulation or game into a learning sequence. Some simulations and games are most suitable for use as introductory activities to particular units of work; others are more appropriate for use throughout an entire learning sequence. Still others can be used as generalising and concluding activities.

Each simulation or game should be used in such a way as to maximise student outcomes in the learning sequence being employed. Care should always be taken to use simulation and gaming activities alongside other learning experiences as a means of securing a balance in student involvement in learning. The sample unit outline in Figure 20.3 illustrates one way in which simulation and gaming activities may be integrated into a unit of work for lower secondary students. The sample unit is 'Life in Equatorial Lands', and includes the game in Figure 20.1 and the simulation in Figure 20.2.

Conducting Simulations and Games

A general set of guidelines for using games and simulations in the classroom includes:

1. Do not over-use simulations and games. Just as students may have become immune to the motivational benefit of over-used audio-visual presentations, they may reach the point of saying 'Oh no, not another simulation!'. Rather, incorporate such activities into the overall structure of a learning unit.

2. Do not emphasise winning. Encourage students to see their achievements within the context of the satisfier–optimiser continuum, i.e. there are degrees of winning.

3. Avoid individual student activity, wherever possible. Use group work to minimise individual

Teacher action	Student involvement	Expected outcomes
1. Brief teacher introduction to the topic of the landscape study.	Students are motivationally involved with the process of the study.	Student statements of what they think life is like in equatorial lands.
2. Assembly of a range of audio-visual data on equatorial landscapes (e.g. photos, slides, tapes and films).	Survey of the audio-visual data to assess the components of such a landscape.	A revised report on what the students think life is like in equatorial lands.
3. Students and teacher analysis of the similarities and differences between the written student outcomes of activities one and two, in order to form a new generalised statement of what life is like in equatorial landscapes.		
4. Teacher-directed lead-up and introduction to *Amazon Adventure*. (See Figure 20.2.)	Group work on postulating plans and program as part of the *Amazon Adventure* activity.	Preparation of group reports.
5. Group reports presented to whole class. Development of a class statement on what students now think life is like in equatorial lands.		
6. *Enrichment activity:* Students play the *Rainforest Game*. (See Figure 20.1.)		
7. Teacher preparation of activities for group work. Students now undertake case studies on a range of topics relating to equatorial lands. The case studies would be selected to illustrate social and environmental issues that have arisen from rainforest clearance in a number of continents.	Student work on teacher-prepared small-group activities for different issues.	Students use a range of data in a variety of ways to develop skills and concepts as they explore the values behind the issues. Preparation of a class report (oral and visual) for presentation to the whole class. Students encouraged to use role play as a medium for communicating different views on the issues.
8. Teacher-directed concluding activities to put all the data and reports into the context of the total study. Teacher to ensure that the students have a 'global' perspective of the life and landscape of equatorial lands.		

Figure 20.3 Sample unit outline: 'Life in Equatorial Lands'

disappointments, foster group decision making, interaction and team activity.

4. Before commencing a simulation or game:
   (a) ensure that students have had adequate experience with the reality being represented to be able to appreciate the activity as a simulation;
   (b) ensure that students see the activity as part of an overall unit of work;
   (c) discuss the purpose of the activity with students, itemise learning objectives to form an evaluation checklist; and
   (d) keep all rules and directions to a minimum, especially at the start of an activity.

   These suggestions are designed to focus the attention of students on the educational dimensions of games and simulations.

5. During the activity, accept a reasonable level of noise and movement from students as valuable. Move around the room with them, and help the students to become fully involved in the processes of the simulation. This can be particularly valuable because it helps teachers to assess the extent to which students understand the issues and processes involved.

6. In your interaction with groups of students, question them about the strategies and decision-making processes being used, and the relationship of their activities, suggestions and decisions to reality.

7. After the activity has been completed, hold an evaluation session. Refocus attention on differences between the simulation and reality. Allow students to review their achievements in light of the objectives itemised. Use a checklist. Repeat the simulation with a view to improving student understanding, and modifying or redesigning the activity.

8. Since these activities can particularly assist in the development of skills, concepts and attitudes, it is these areas which should be the subjects of evaluation for assessment purposes, not just knowledge of the area of content which the students may have used. Evaluation instruments could include:
    (a) attitudinal scales;
    (b) essay questions on the real-world situation of which the simulation is a model;
    (c) essay questions on the processes involved in the simulation, for example on how decisions are made concerning the resolution of an important environmental issue;
    (d) practical exercises to measure the extent to which various skills were developed; and
    (e) practical exercises to validate the concepts developed by the activity.

Designing Your Own Simulation and Games

Commercially available simulations and games were originally designed to meet a specific set of educational objectives. In using such activities, it will often be found that modifications are needed in order to suit the aims and objectives of particular courses and units of work, and the needs of individual students. Some activities will require only minor modification; others may require major revisions.

Teachers should attempt to gain experience and understanding of the operation of simulations and games in the context of their own classrooms. By becoming familiar with the strengths and weaknesses of the general approach and with the operation of such activities, teachers should be in a position either to adapt commercially available simulations to suit their own purposes, or to develop a suitable range of their own.

Teacher-developed simulations and games are often more appropriate learning experiences than commercially available activities. This is because teacher-developed ones are designed for specific purposes involving students in a sequence of learning experiences. Teacher-developed activities are designed to meet the specific needs of students and courses being undertaken in a school. They can be given added relevance by incorporating school and community resources into them, and by using them to study local environmental problems and issues.

Guidelines for Designing Your Own Games

Designing an educational game is more simple than designing a simulation. Simple games, such as geographical crosswords and anagrams, may take a few minutes to prepare. More complex games such as *The Rainforest Game* (see Figure 20.1) do, of course, take more time to complete.

There are five main steps in the preparation of an educational game:

1. Select the specific information and/or skills that are to be used in the activity.
2. Design the competitive basis and main moves of the game.
3. Detail the specific rules of the game.
4. Prepare the game materials.
5. Give the game a trial run.

Guidelines for Designing Your Own Simulations and Simulation Games

Two criteria are especially important when designing a simulation or simulation game. They are the *degree of determinism* and the *degree of structure*. Each provides important design guidelines.

1. *Degree of determinism*. In a range of activities, the degree of determinism varies between those activities in which the outcomes are prescribed and activities for which there is no certainty about the outcome. *Deterministic* activities are designed for specific actions leading to prescribed effects. In *probabilistic* activities, there is far less certainty about outcomes, and the likelihood of any one outcome in the context of possibilities is subject to chance.

2. *Degree of structure*. Activities in which the rules, processes and outcomes are prescribed by the de-

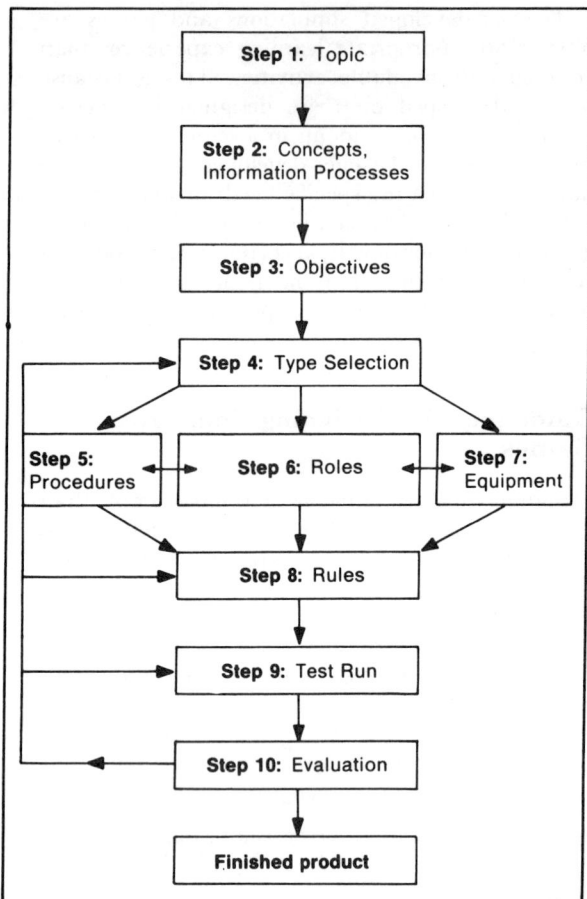

Figure 20.4 Flow diagram: designing your own simulation or simulation game

signer may be referred to as 'closed' activities. Such activities, e.g. *Rice Growing Snakes and Ladders*, are designed to isolate the materials from their environment so that the inherent objectives of the activity are likely to be attained. The relevance of such activities to broader course and educational objectives will not be reached unless a proper debriefing session is conducted. Activities such as *Lynwood* and *Caribbean Fishermen*, which have an 'open' structure, avoid the trial and error type of responses, and rely much more on insights and intuitive thought. In such activities, the debriefing session again plays an important role, but for exactly the opposite reason to that of 'closed' activities. The session enables players' experiences and subsequent insights to be related as a means of integrating past and present learning.

The following instructions which are summarised in Figure 20.4 utilise these guidelines to produce an action sequence for developing a simulation or a simulation game.

1. Identify the topic on which you wish to base your simulation.

2. Decide upon the concepts, processes and information you want the activity to pass on to your students.

3. Define the objectives of the activity as specifically as you can. This will help you determine its scope in terms of the issues to be examined, the time setting, and the geographical location.

4. Select the type of activity you wish to use. This can be done using the following dimensions:

   No equipment ←————→ Full equipment
   (e.g. role play,   (each process simu-
   discussion)   lated with mater-
                               ials)

   Full personal ←————→ No personal
   contribution   contribution (pre-
   (free of predeter-   determined rules
   mined rules)   covering all
                              possibilities)

   Full competition ←————→ Full cooperation
   (each participant   (participants oper-
   operates against the   ate as one unit)
   others)

5. Clearly define the roles of the participants. Make them as interesting and close to real life as possible. Even in activities not involving role play, participants may take on different roles.

6. Devise the procedures of the activity. These must fit the spirit of both the activity and the participants, otherwise reality will not be effectively simulated.

7. Design the materials. These may include items such as a board, role cards and dice. The equipment should be as attractive, functional and durable as possible. This will help to increase student desire to participate in the activities and will minimise long-term costs.

8. Establish the specific rules. These rules should be brief but clear. It should be noted, however, that a large number of rules can reduce enthusiasm for the activity, while incomplete or vague rules lead to frustration and loss of interest. You should aim at a happy medium.

9. Give the activity a test run. This will help you to solve any of the problems inherent in the design.
10. Evaluate your simulation activity. Your evaluation may include finding answers to questions such as the following:
    (a) What is my purpose in using this activity?
    (b) How will the activity be presented to and used by the students?
    (c) Did the materials used prove to be satisfactory?
    (d) What follow-up activities are possible or desirable?
    (e) How flexible are the rules of the activity?
    (f) Have the objectives of the activity, as stated in Step 3, been achieved?
    (g) Did the students find it interesting and useful?
    (h) Does the activity match up to reality?

All negative responses to the above questions should lead to a revision of the simulation. This revision should continue until you are satisfied that major problems will not occur.

Further Reading

The books listed below are very readable and contain excellent ideas for the classroom use of simulations and games. Between them, they trace the development of simulation and game activities, discuss the educational advantages and outline problems which teachers should try to avoid, relevant research finding and several tried and proven examples of games and simulations.

Harper, A. (1975) *Simulation Games in the Making*, Auckland: Heinemann.
Inbar, M. and Stoll, C. (1972) *Simulation and Gaming in Social Science*, New York: Macmillan.
Livingston, S. and Stoll, C. (1973) *Simulation Games: An Introduction for the Social Studies Teacher*, New York: Macmillan.
Taylor, J. and Walford, R. (1978) *Learning and the Simulation Game*, Milton Keynes: The Open University Press.
Walford, R. (1969). *Games in Geography*, London: Longman.

Several textbooks, especially from the United Kingdom, contain simulations and games. These books illustrate how such activities can be integrated into units of work. Good examples include the following two series:

Dinkele, G. et al. (1977) *Harrap's Course in Reformed Geography* (6 books), London: George Harrap and Co.
Grenyer, N. et al. (1978) *Oxford Geography Project*, 2nd edition (3 books), Oxford: Oxford University Press.

The Schools Council Geography 16–19 Project has produced 20 booklets on a range of topics. Most of these books contain a simulation, often in the form of a role play.

Notes

1. This chapter is based upon and updated from *Resources Review: Simulations and Games in the Classroom*, written by the authors for the Curriculum Branch of the Department of Education, Queensland.
2. The following gives a brief description and details of availability of the simulations and games used as examples in this chapter.

Amazon Adventure. See Figure 20.2.
Caribbean Fishermen is based on the life of fishermen living with their families on a small Caribbean Island. Available from Cambridge Publishing Services.
The Green Revolution Game involves role play as students strive to become self-sufficient in food in an Asian village. The class nature of land ownership and the high costs of Green Revolution farming are constraints. Available from Marginal Context. A simpler version is available in F. Slater (ed.) (1986) *People and Environments*, London: Collins Educational.
Longman Micro-Software Series contains over 20 titles for use in geography. Many of these are simulation games while others are computer simulations (e.g. of the water cycle).
Lynwood is a role play which focuses on urban renewal in a heritage environment. Available from Sorrett (Longman).
Mali Cattle Dealers Game examines the nomadic life of cattle dealers in Mali as they are affected by climatic and market conditions. See Grenyer, N. et al. (1978) *Contrasts in Development: Oxford Geography Project Book 3*, 2nd edition, Oxford: Oxford University Press.
Portsville simulates the growth of a town from its first settlement in 1850. Coloured grid squares represent various land uses. Available as part of the American High School Geography Project from Collier-Macmillan.
Railway Pioneers simulates the development of railway building across the United States during the nineteenth century. Available from Longman.
Rice Growing Snakes and Ladders is a game based upon the advantages and problems brought by the Green Revol-

ution to Asian rice growers. See Grenyer, N. et al. (1978) *Contrasts in Development: Oxford Geography Project Book 3*, 2nd edition, Oxford: Oxford University Press.

Sand Harvest is a computer-based role play of the causes and consequences of desertification in the Sahel. Available from Centre for World Development Education.

Slick is a computer-based simulation in which students must act as oil pollution engineers working to prevent an oil slick from reaching the coast. Available from BP.

South Street. Hostel Storm is a role play of a residents' association meeting to consider a decision to locate a small hostel for mentally retarded adolescents in a suburban area. Available from Community Service Volunteers.

The Rainforest Game. See Figure 20.1.

3. Appreciation is expressed to David Lergessner and Doug Cave for the use of *The Rainforest Game* and to Keith Cordwell for *Amazon Adventure*.

21

Using Computers in Geography Teaching

John Lidstone

Despite the possibilities that they offer for improving the curriculum in geographical education, computers have not yet achieved their full potential in the geography classroom. There is a wide range of software now available for geography teaching but, for many teachers, problems stemming from difficulties in integrating this software effectively into existing curriculum units, as well as in the classroom organisation of the hardware, are yet to be solved.

This chapter surveys the available software and outlines the five major ways in which computers may be used in geography teaching: electronic blackboard programs, tutorial programs, simulations, statistical or number-crunching programs and data bases. The chapter concludes with a description of nine ways of organising geography classrooms to take advantage of the software and hardware generally available in most schools.

It seems to be an inevitable, though unfortunate, part of a teacher's role to be bludgeoned constantly by society for failure to introduce every new area of knowledge and every new piece of technology into the classroom. There seems to be such endless pressure to find room for every new idea, and to teach it using ever newer and more complex equipment, that it is small wonder that many teachers long for a return for the 'good old days' when life was much simpler. They adopt a 'wait and see' attitude in the hope that the latest batch of new developments will wash over them without causing too many ripples in their working lives. Oh, happy days, when 'hi-tech' in the classroom was a box of 24 Lakeland coloured pencils and even a fountain pen was regarded with suspicion.

The feelings of guilt and frustration with the expectations created by those in the vanguard of curriculum development reached such a pitch among one group of teachers in an inservice seminar that they devised their own model of the curriculum development process:

C — Constant
U — Upheavals
R — Requiring
R — Regular
I — Inventiveness
C — Creating
U — Undue
L — Loading
U — Upon
M — Masters and Mistresses

This is perhaps an extreme reaction, but one which is understandable, and it should be remembered that these teachers had at least made the effort to attend the seminar.

There is, however, another side to the story. As in the case of every other technical innovation in teaching, from the days of lantern slides, through the introduction of overhead projectors and educational television, to the building of language laboratories, the evangelists of the 'computer revolution' have beaten the drum for the technology itself, frequently without taking into account the context within which many teachers are working. In so doing, they have frequently encouraged quite unreasonable expectations which have led some people to fear that computers will take over the world. For some other people, the impression has been created that mere contact with the technology will, in some metaphysical way, create knowledge and wisdom. In the current surge of enthusiasm for computers in classrooms, some cynicism is essential to counterbalance the missionary zeal of computer exponents.

Why Use Computers in Geography Classrooms?

Figure 21.1 shows some of the factors which may influence a teacher's decision to introduce computers into the classroom. The diagram suggests that decisions about the use of computers in the classroom are influenced by teachers' perceptions of the

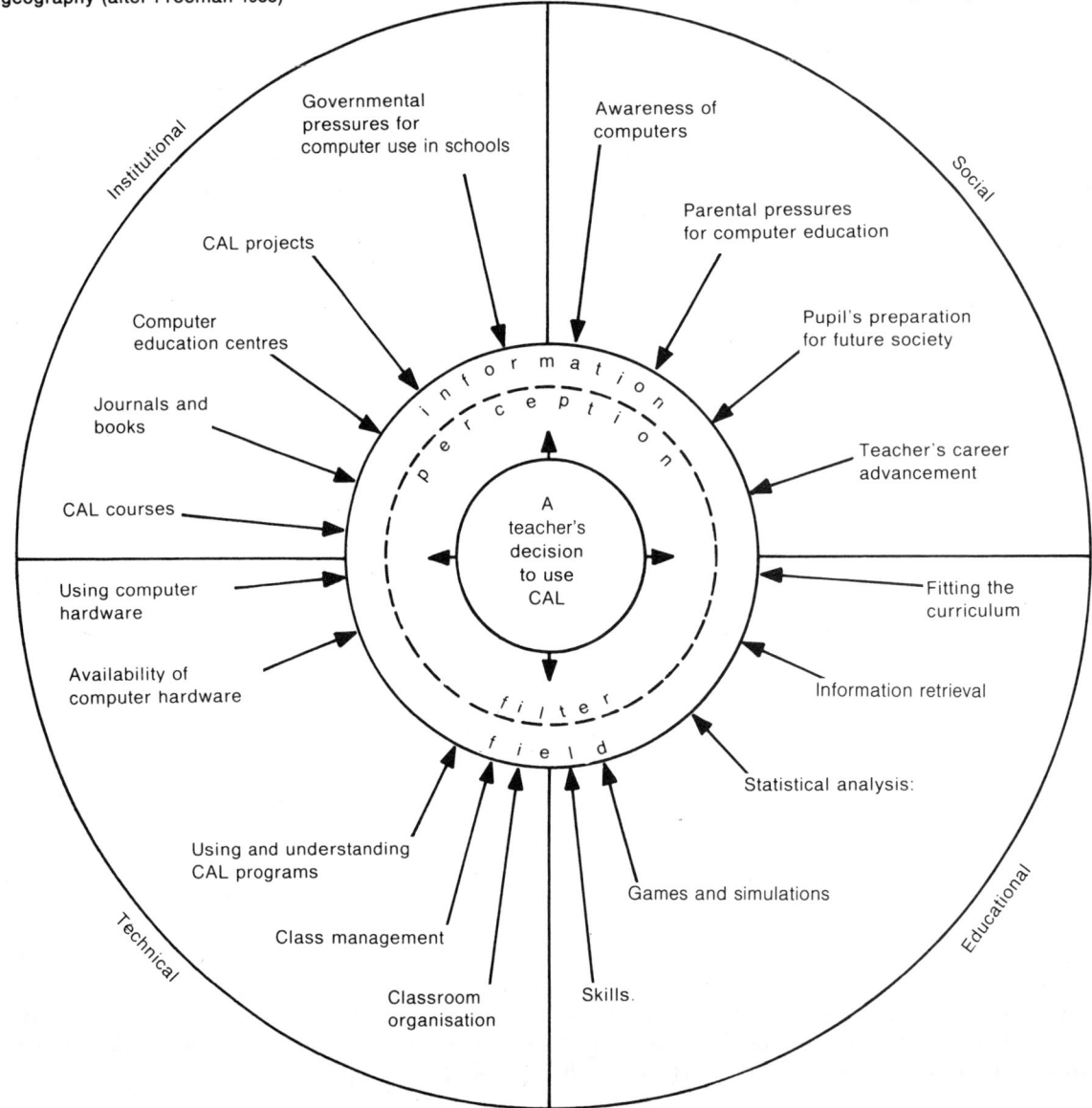

Figure 21.1 A descriptive model of the factors which may influence a teacher's decision to use computers in geography (after Freeman 1983)

role of computers as created by pressure from four major areas: government and other institutional initiatives, social attitudes, new movements in educational thinking and new developments in technologies.

The governments of many of the developed countries of the world have expressed their intentions to introduce students to the use of computers during their school lives. Such government intentions have frequently been supported by action at various levels ranging from the exhortatory to the financial. As a result of government interest, many teachers feel under some institutional pressure to consider the use of computers.

To these institutional influences must be added social factors such as the perception of parents and students that a knowledge of computers is essential for future life, and the teacher's own perceptions of the potential role of computers in society and his or her own career prospects.

Pressures from these two sources may not penetrate the teacher's perceptual filter for a variety of reasons, including the one implied by the interpretation of the word 'curriculum' which was given above. Should the institutional and social influences succeed in impinging on the teacher's consciousness, however, he or she may then examine the educational advantages of introducing computers into geography courses and will then have to come to terms with the technical aspects of their use.

Of the educational and technical influences on teachers, and despite the progress which has been made in the past decade in making computers available to schools, it still appears to be the technical aspects which seem to present some of the greatest hurdles to teachers. Technical problems include gaining access to the computers, understanding the peculiarities of different machines, obtaining appropriate courseware (the software to run on the machine and its accompanying print or other resources) and selecting appropriate classroom management strategies.

The remainder of this chapter will assume that both the institutional and social environments in which the reader operates is conducive to the introduction of computers into geography classrooms, and will concentrate on some of the educational and technical aspects of using computers in geography lessons. Although at first sight these may appear to be quite separate factors influencing computer use, both are interrelated in their effects on teaching learning styles in geography. Thus, while the number of machines available obviously has a major influence on the ways in which computers can be integrated into geography curricula, the types of software available and the teacher's educational intentions in introducing computer-assisted learning will also exert a great influence on the technicalities of classroom management.

What Kinds of Computer Programs are Available for Geography Teaching?

Back in the heady days of the 1970s when the contribution of computers to geography teaching

Figure 21.2 Types of programs which may be used in geographical education

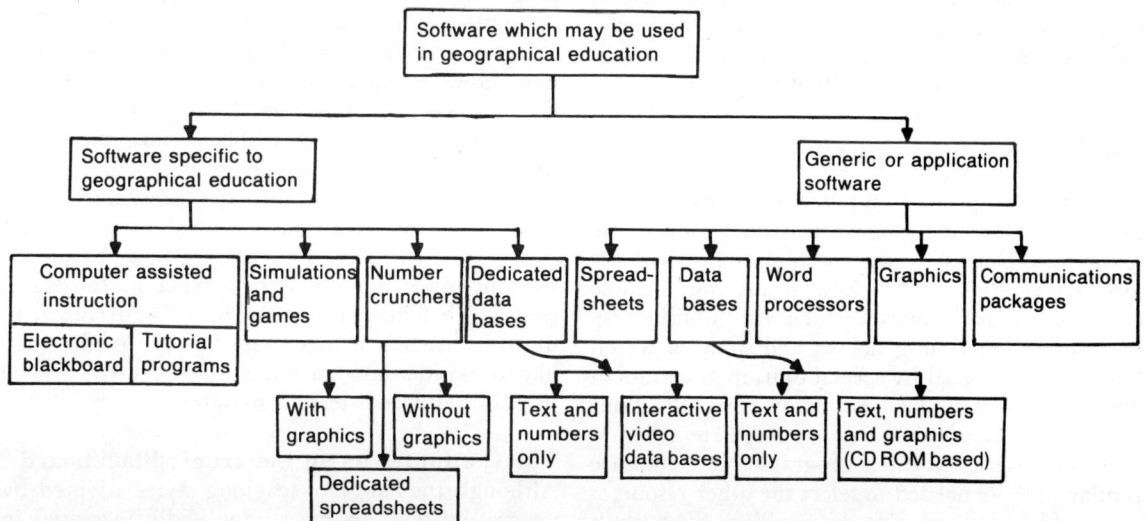

was first being discussed, it was suggested by the gurus of the time that, since teachers know best the needs of their own students, all teachers should learn to write programs for their own classes. This has proved to be an entirely unattainable aim since writing computer programs is a very time-consuming occupation, especially if a high-quality product is required. Furthermore, it may be suggested that it is not even desirable to encourage teachers to write their own software since the skills of teachers should not be dissipated in other activities which may be accomplished by those who do not have teaching skills. For both of these reasons, most computer use in geography classrooms will be based on commercially produced packages.

Figure 21.2 shows that computer software can be divided into two broad categories: subject-specific software and application or generic software.

Subject-specific software includes those programs which contain information in their content which is specific to one or more subject areas. In one way or another, these programs aim either to teach some part of a particular subject or to help students to learn or undertake some activity specific to a particular subject. Frequently, such software is published as part of a 'pack' in which printed and other materials are provided, together with suggestions for activities which should be undertaken by students both at, and away from, the computer. Such integrated print and software packs are sometimes called 'courseware'.

Application software, on the other hand, includes all those programs which do not have any specific content but which may be regarded as tools which can be used by those pursuing a study in any subject area. Until relatively recently, application software has been taken to include word processors, spreadsheets and data-base management programs. Now, map drawing, graphics, communications and control packages have been added to the list of application software which is available to schools.

Subject-specific Software

Many subject-specific computer packages are published by the same publishers who produce the textbooks, colour slide sets and film strips which we already use. Figure 21.3 shows a selection of these courseware packs some of which are published by well-known publishing houses and some of which have been produced by special units in government education departments. In many ways the skills needed by teachers to select appropriate software and integrate it into their geography curricula are similar to those needed to select the other resources with which we are already familiar. Such skills include the ability to evaluate the resources available in terms of the curriculum being offered and the ability of the students concerned, and to adopt classroom strategies which take into account all the constraints which are operating in a particular teaching situation. However, there are some major differences between computer programs and other resources such as slide sets, film strips and textbooks which add to the complexity of managing computer-assisted learning in geography.

The main difference lies in the flexibility with which these resources may be used. A teacher may refer students to a particular section of a textbook, or even to a particular map or diagram within the book, regardless of whether the students have studied the sections which precede it. The teacher may select four or five frames from the centre of a long film strip, or even, within the constraints of copyright legislation, adapt a passage or a graphic for inclusion in a worksheet for use by particular students. It is rarely possible, however, for a teacher to be able to select from, or alter, a commercially produced software package.

Other differences may include a lack of provision in the software package for teachers to update information essential for the running of the program or for teachers and students to 'run-back' the program without necessarily returning to the start or to begin running a program other than at the beginning.

In addition to the fundamental division of software into subject-specific and application programs, Figure 21.2 also shows that software may also be classified according to the the ways it is, or may be, used in classrooms.

Subject-specific software may be subdivided into five different types. These are electronic blackboard programs, tutorial programs, simulations, statistical or number-crunching programs and data bases. The category into which any single piece of software falls may, however, vary according to the use made of it in any particular classroom. Thus, a program may be classified as an 'electronic blackboard' if run on a single machine at the front of the classroom as part of a largely expository lesson by the teacher, while the same program would be classified as a 'tutorial program' if run by a student or group of students working independently of the teacher. In the examples which follow, programs have been chosen to illustrate each teaching style, but the reader may like to consider how the same program could also be used in a different teaching context.

The Computer as an Electronic Blackboard

Although the range of teaching styles adopted by geography teachers has undoubtedly increased in

Figure 21.3 A selection of courseware packages available for geography

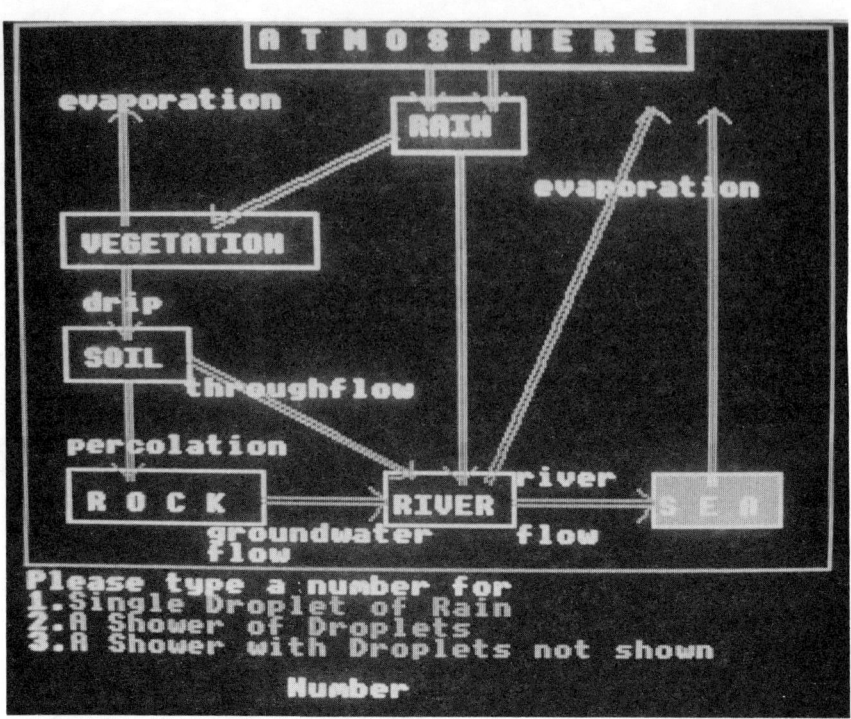

Figure 21.4 The screen display from *The Path of a Water Droplet*

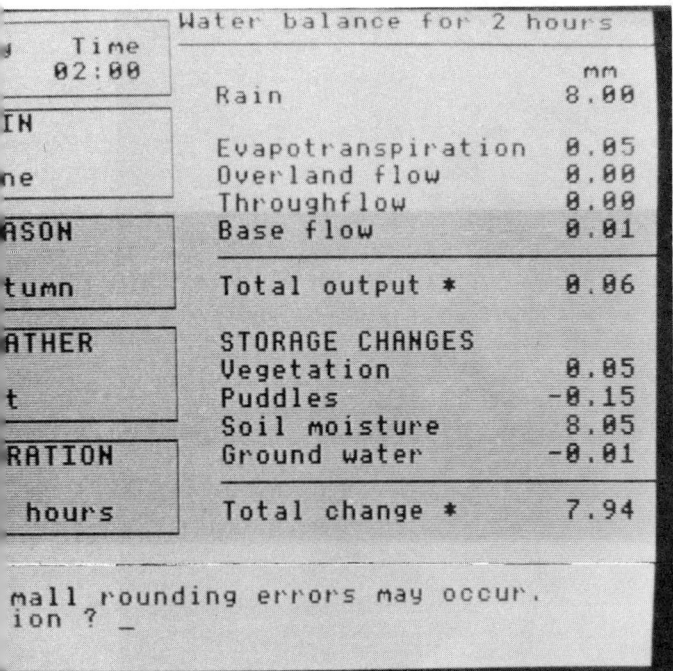

Figure 21.5 The screen display of the systems diagram from *Water on the Land*

recent years, the 'stand and deliver' style of teaching is still commonplace. At the turn of the century, geography teachers were among the first to increase the effectiveness of such presentations by the use of lantern slides, and few teachers today fail to take advantage of the flexibility of the overhead projector. The disadvantages of such technology in teaching a subject such as geography is that the displays are static, and although the dynamic nature of many geographical phenomena may be indicated on overhead projectors by means of a series of overlays, the idea of constant change is often difficult to transmit. To a certain extent, cinematic and video animation techniques have enabled dynamic graphics to be presented in class, but in such cases the decisions on how much, to what, and how fast have already been taken by the director or animator. A computer program can provide animated diagrams in which the teacher or a student can alter various parameters and observe the effects through time. Figure 21.4 is one instance of a screen display from the program *The Path of a Water Droplet*.

In this program the user can select any combination of the following conditions in order to observe the probable course of water falling as rain through the system of the water cycle.

- Temperature: between $1-30°C$
- Rock type: chalk, sandstone or clay
- Ground cover: forest, grass, scrub, crop, ploughed, or urban
- Soil cover: thin, moderate or deep

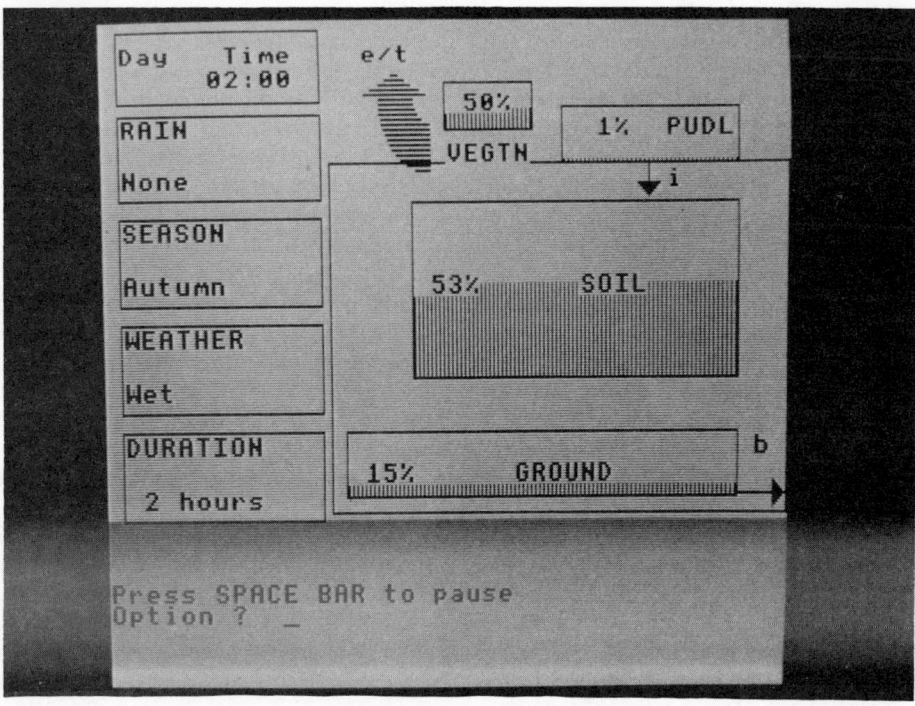

Figure 21.6 The screen display summary of the changes in water storage after a rainstorm simulated in *Water on the Land*

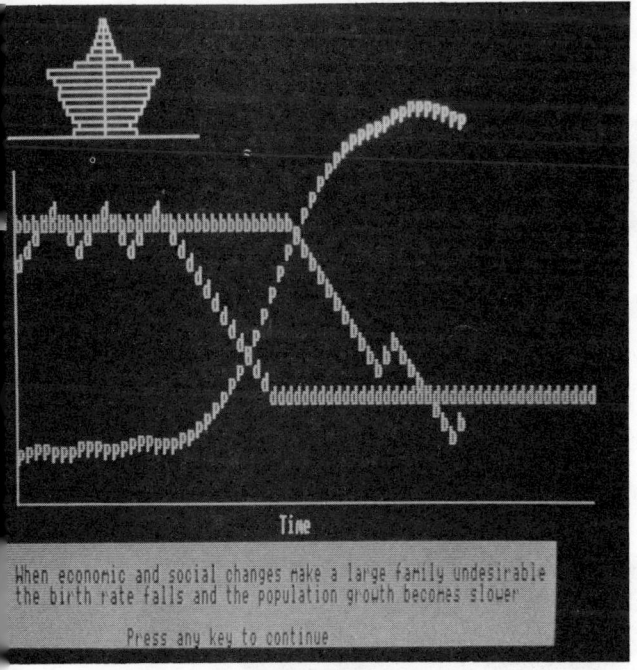

Figure 21.7 The screen display from *Population Dynamics*

The paths of a series of water droplets through the system are then shown on screen and students may observe the relative quantities of water which evaporate, soak into the soil, or percolate down into the bedrock. A similar program is *Water on the Land* in which the percentages of water in the various stores during and after a rainstorm are shown in animated diagrammatic form and then summarised as a water-balance sheet. Figures 21.5 and 21.6 show the animated diagram and water-balance sheet two hours after a heavy rainstorm in autumn following a particularly wet summer.

Figure 21.7 shows a screen from *Population Dynamics* which illustrates the demographic transition model. After the typical pattern of changing death rates as experienced in many western European countries has been drawn on the screen, with appropriate captions to explain each stage, a typical pattern of changing birth rates is added, again with explanatory captions. The resulting population is calculated and also graphed while a population pyramid is constantly updated in the top left-hand corner of the screen. In addition to this explanatory demonstration, the user has the opportunity to select levels of birth and death rates and to observe their effects on population numbers and structure as shown on the line graphs and population pyramids.

The Computer as Tutor

As visual aids to expository lessons on such topics as the water cycle or population dynamics, computers used as electronic blackboards have a major role to play. However, just as all children want to push their own prams, so all students who have inquiring minds, and even many who do not, would rather push their own buttons on the computer. This preference of most students for 'hands on' experiences with computers leads to the machine's use for tutorial purposes. Learning basic information, such as the location of cities and rivers, simple skills such as the use of grid references, compass directions and bearings, and simple relationships such as that between latitude, longitude or altitude and average temperatures may all be learned by students interacting either individually or in small groups with appropriate software. Such software may be in the form of a simple game in which a flashing cursor on a screen map invites the student to name a city, river or range of mountains. The computer can keep a running score of the places correctly named. Figure 21.8 is a screen from a computer version of the old children's game *Hangman* in which students were trying to guess words which were classified as 'types of rocks'. Such programs may be easily adapted to any age level or subject content. They encourage familiarity with technical vocabulary and spelling. They are also fun which makes them ideal as a 'lollipop' for especially deserving, or especially troublesome, students.

Other software can lead a student stage by stage through the application of grid references or the use of compass directions. One package called *Introducing Map Skills* includes a number of programs of a tutorial nature designed to teach the principles and applications of compasses, scales and grid references. The package concludes with a *Yacht Race* in which students, either singly or in pairs, direct yachts around a course by specifying directions and distances for each straight leg of the race. The course is shown in Figure 21.9. In this screen the player controlling the yacht whose course is shown by the white line has miscalculated the direction and distance and as a result has hit the bank.

The greatest advantage of such 'drill and skill' programs is that they enable students to practise skills and rehearse their knowledge in a non-threatening environment. If the student does not achieve very well at the first attempt, the failure is a private one and the program can be re-run until mastery is achieved. However, although a distinction has been drawn between electronic blackboard, whole-class uses of programs and their use in tutorial situations, it must be remembered that frequently the distinction is more a matter of which type of classroom organisation the teacher prefers than any inherent differences in the software used. Many programs written specifically for individual tutorial use may be used in a whole-class situation both to introduce and revise a topic or skill with great effect.

Figure 21.8 The screen display from *Hangman*

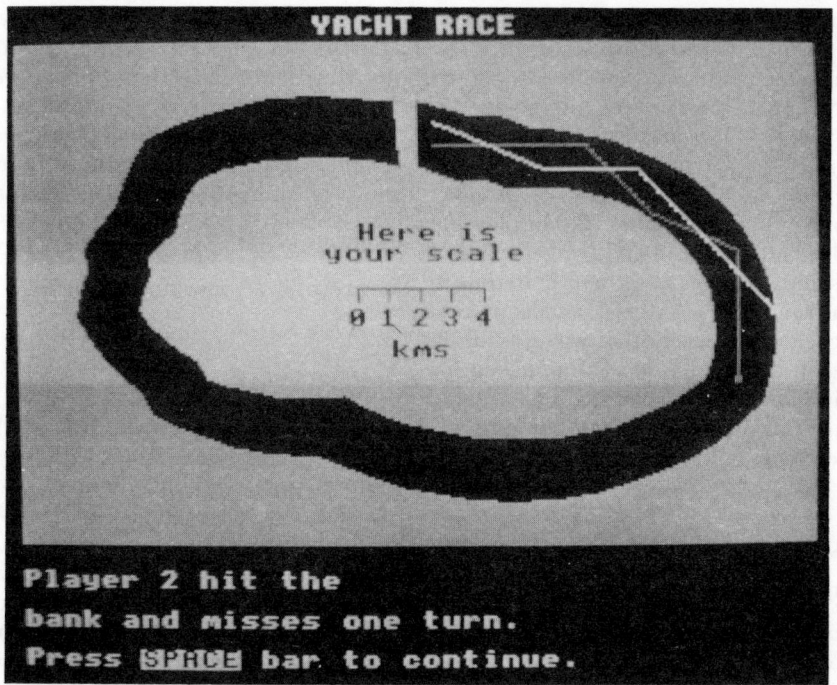

Figure 21.9 The screen display from *Yacht Race*

Geographical Simulations Using a Computer

While many of the programs mentioned in the previous two sections may be used either with whole classes or with individuals, there are a number of other programs which are specifically intended to be used by students in groups. Many of these programs take the form of simulations.

Classroom simulations have been around for a long time in printed form and many teachers have integrated them into their teaching programs with great effect. Problems frequently arise in classroom use, however, due to two features of many simulations, the first relating to the simulation materials themselves and the second to the students who attempt them. Classroom simulations based on print materials frequently fail because:

1. students cannot handle the large amounts of printed material and information with which they are frequently faced;
2. students make mistakes in calculations which make nonsense of their results.

When simulations are presented on computers, advantage can be taken of the computer's ability to store large amounts of information and to perform complicated calculations quickly. The result is that students may be asked to make decisions about complex situations and can then see the results of their actions almost immediately. There is another advantage, however, which in educational terms is of much greater importance. Simulations in which groups of students interact with the same computer encourage the students to discuss their decisions and, after observing the results, consider their future strategies. Thus, much of the learning experience comes, not from the discovery of a 'right' answer, but from the discussion of the best course of action to take next, followed by discussion of why that particular course of action had certain unanticipated results. The two questions which should be in the forefront of students' minds when engaged in simulations are:

1. What is the best course of action to take under the circumstances as we currently know them?
2. I wonder what would happen if we . . . ?

It is often the second question which leads to the most creative responses and which leads to the most valuable learning experiences.

Simulations can take many forms. Some require quite specific learning in order to achieve 'success' while others are very 'open ended' in terms of student decisions and results. Those simulations which require specific learning are often preceded or accompanied by tutorial programs as described above to create quite sophisticated suites of programs. One particular package which exemplifies this latter approach is *Climate*. *Climate* is a suite of programs designed to help students appreciate the range of factors which influences the climates of the continents and then relate these factors to the climates experienced in Western Australia. The package comes complete with a booklet of teacher's notes to all the programs, a single 80-track disc, an insert for the function keys of the computer to display the various possible wind directions and temperatures of ocean currents and 28 pages of student materials in the form of instructions, worksheets and activities. The busy teacher is helped by the provision of all these materials in an A4 plastic pack punched for storage in a ring binder or for vertical filing.

The learning aims of the first part of the suite are stated in the teacher's guide as being for students to learn that:

1. temperature decreases with increasing latitude;
2. temperature decreases with increasing altitude;
3. rainfall decreases with distance from the coast in the direction of the prevailing wind;
4. rainfall increases as the temperature of the ocean currents over which the prevailing wind blows increases;
5. rainfall increases on the windward side of mountain ranges and decreases on the leeward side.

With these aims clearly stated, the first five subprograms present students with a series of simulations in which an imaginary land mass of continental size is manipulated in an attempt to place it appropriately on the globe to achieve given 'target' climatic features for each of two cities. In the first simulation, the land mass is moved north or south of the equator in 10-degree increments in order to achieve specified average annual temperatures. When the students have decided where the land mass should be located, they may request a 'report' in the form of a graph plotted on the screen to show how close they have come to discovering the appropriate location. When finally they have located the land mass so that the two cities have temperatures similar to those presented as the target, a final report may be requested to identify the ideal location, to inform the student of the number of attempts made and to give a score based on the accuracy with which the temperatures were matched. Having experimented with the location of the land mass, with the help of the worksheets supplied, students may then attempt 10 'problems' which encourage them to make generalisations about the real countries of the world and to undertake further investigations, using reference books and atlases. Finally, they are encouraged to undertake further research on the reasons why there is a relationship between latitude and average temperatures.

From the understandings developed by their studies of latitude and temperatures, students may then move on to investigate the relationships between:

1. altitude, latitude and temperatures;
2. prevailing winds and rainfall;
3. ocean currents and rainfall;
4. mountains and rainfall.

In each of these sections, the simulations become progressively more complex with the addition of new factors to be taken into consideration. Finally, after they have been introduced to all the various factors which influence climate, students are asked to design a complete climate for the imaginary land mass. Once again, targets are set randomly for the temperature and rainfall averages for the two cities. Students are able to alter all the factors affecting the climate of the continent: the latitude, the temperature of the surrounding ocean currents, the direction of the prevailing winds, and the height and east–west position of the mountain range.

Figure 21.10 shows a screen in which the student has attempted to locate the land mass. The box in the top left-hand corner of the screen shows the four previous decisions made. The lower left-hand box is the 'report' on the implications of these decisions. The reader may like to compare the 'report' with the 'target' and consider what the next decision should be.

Some simulations are less prescriptive than *Climate* and may be used by students of widely varying ages and abilities. *The Sailing Ships Game* is one simulation which lends itself to use in classes ranging from primary children through to senior secondary students. The game consists of a stylised map of the world from which students are asked to select two cities between which they are to sail. Students can decide on which date they will depart and use compass directions to decide on their route. The direction and force of the winds, however, are decided randomly by the computer within parameters appropriate to the location on the globe reached by the ship and the time of year. Leaving the English Channel in the face of a strong southwesterly wind is decidedly difficult and pity the poor captain who tries his or her luck in the Indian Ocean at the start of the monsoon.

At its simplest level, the program and its accompanying print materials may be used to re-

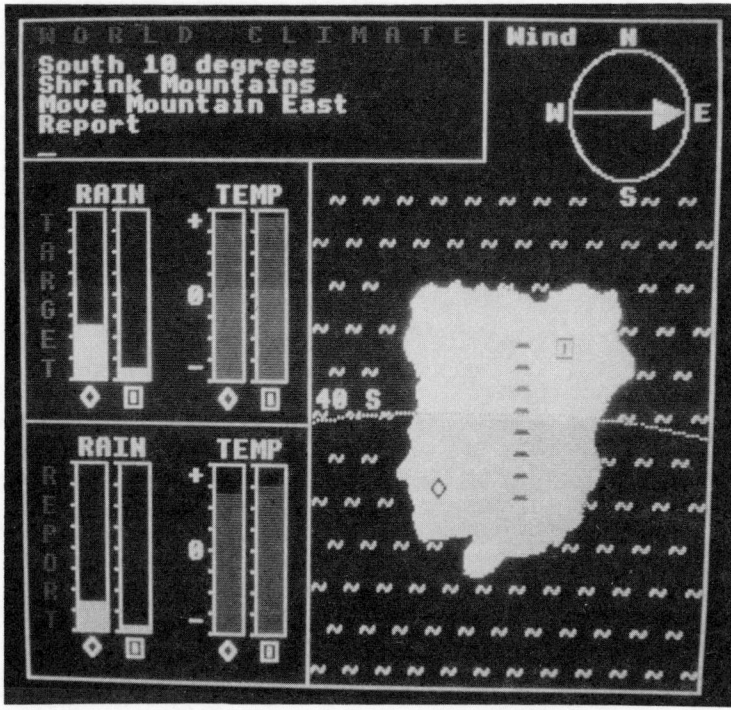

Figure 21.10 The screen display from *Climate*

inforce and apply such simple concepts as:

1. compass directions;
2. latitude and longitude;
3. the position of various ports, the shape of the world map and the size of the oceans (although the Pacific Ocean is unfortunately lost);
4. Beaufort wind scales;
5. some basic facts about sailing ships (you cannot sail directly into the wind; tacking wastes a lot of time).

However, the program also contains a mathematical model of the atmospheric circulation system which incorporates seasonal shifts and major monsoonal winds, and this enables it to be used for more sophisticated studies. Together with the accompanying documentation, the program may be used to investigate such questions as:

1. The shortest as opposed to the quickest routes between London and Sydney for tall ships coming to Australia to celebrate the Australian Bicentenary.
2. The problems faced by a tea clipper sailing from Shanghai to London starting on 30 May.
3. The possibilities which were open to Columbus as he tried to find a new route to China.

Figure 21.11 shows the map and the route chosen by a ship sailing from London to Sydney. Given the prevailing wind as indicated in the picture, the reader may like to choose the course which the ship should follow for the next 24 hours.

Decision making under constantly changing circumstances is also the theme of *Slick*. Students are appointed to the post of Pollution Control Officer for the imaginary coastal area of eastern Scotland, 'Inverlochen and Northsands Bay'. Having read the notes which accompany the program, students have to decide which methods of controlling an oil spillage are most appropriate for such an environmentally sensitive part of the coast. They are then faced with a simulation in which an oil spill occurs off the coast and is being blown by the wind towards local fisheries, beaches and conservation areas. Figure 21.12 shows the map across which the slick moves and the various details of the coast which need to be protected. The players can apply various methods of coastal protection—booms, skimmers, using dispersants or establishing a local task force to clean up the beaches—and have to forecast the movements of the slick from the weather information which is constantly being updated on the screen. All these decisions have to be made against the pressure of time.

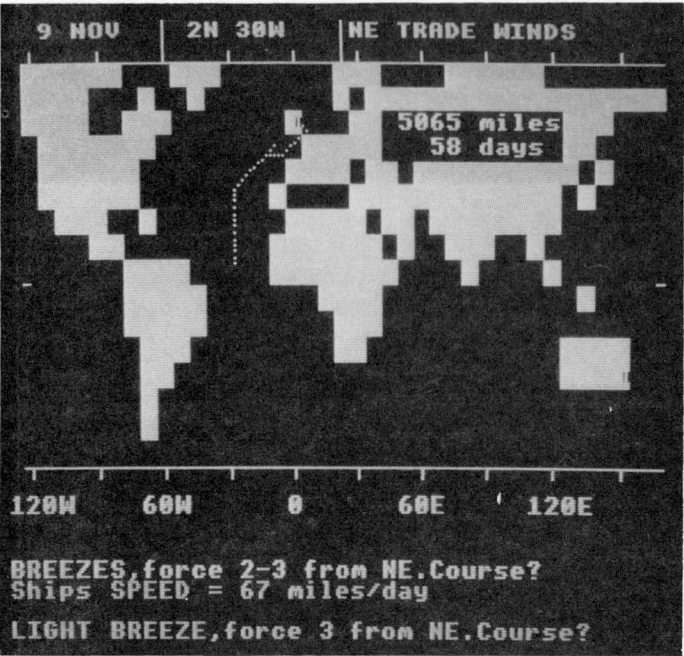

Figure 21.11 The screen display from *The Sailing Ships Game*

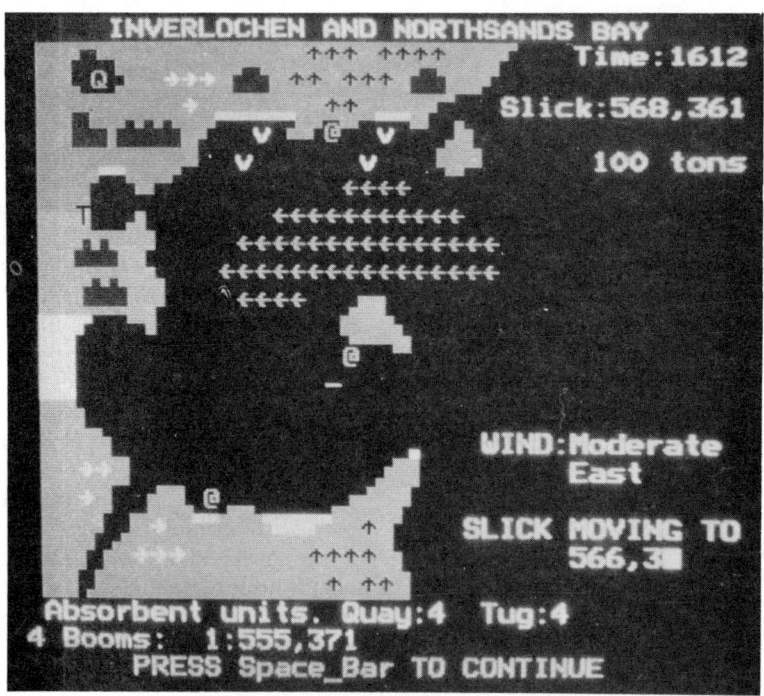

Figure 21.12 The screen display from *Slick*

While students are faced with a wide range of decisions in fighting the oil slick, and the discussions in which they engage are frequently very useful, it is easy to forget that there may be other options open to governments which are not mentioned in the simulation. For example, the game does not encourage students to consider the possibility that oil tankers should perhaps be forbidden to pass within a considerable distance of 'environmentally sensitive parts of the coast', that higher international standards controlling the movement of oil could make oil spills very unlikely events or that the control of those spills which do occur should not be influenced by rigid budgetary constraints. *Slick* is a very good simulation and reflects the decisions which frequently have to be taken in the real world, but teachers should always consider the values implications of any simulation and encourage their students to look beyond the short-term implications of 'winning' or 'surviving' in the simulated circumstances of the computer program to the wider implications for our society. This necessity may be seen particularly clearly in *Dalco*—a simulation of the location factors of an aluminium smelter in which maximum profits may be made by locating the smelter in the centre of a national park and importing raw materials from the Soviet Union.

Number Crunching with a Computer

While the simulations just described enable students to come to terms with complex situations which would otherwise be out of their reach, there are other programs which enable students to handle 'real world' data which they have generated themselves. Both individual investigations and fieldwork activities frequently generate vast quantities of data which are either beyond the abilities of students or which would take a longer time to analyse than the teacher is prepared to devote. In many such cases, the students are quite able to handle the data collection, the conceptual bases of the fieldwork investigations and the interpretations of results, but encounter problems in data analysis due to their inability to handle either the complexity or the repetition of the calculations. This is especially so when large amounts of data are involved. When the number crunching and analysis can be handled by the computer, more complex and often more interesting individual and fieldwork activities are brought within the scope of secondary geography students.

A number of programs are available which have been written specifically for use in analysing data from geographical investigations. In the area of physical geography, the suite of programs called *Field Study Techniques* has been found to be particularly useful. Programs in this suite enable analysis of:

1. shapes of rock fragments or pebbles;
2. valley cross-profiles;
3. frequencies against compass orientations;
4. beach sediment grain sizes;
5. regression and correlation;
6. geological structures on a polar stereographic projection;
7. river discharge;
8. vertical sections through sequences of sediments.

Before giving an example of such a program, one particular problem requires attention. Although data analysis may be handled by the computer, this can only be done after the data have been entered into the computer's memory. This remains a task which has to be accomplished through the keyboard. For students without keyboard skills, entering large quantities of data can be frustrating and time consuming. Nevertheless, many teachers have found that the tasks may be spread amongst students to minimise the demands on individuals and for many, the novelty of 'computerising' their results is its own motivation.

Let us consider a hydrological study of a drainage basin in which the students want to compare the river discharge at different points in the drainage system. At each site in the field, students will take a series of depth measurements at regular intervals across the river. The speed of flow is then calculated at various points across the rivers width either by timing a float along a known distance or by using a flow metre. The record of such field activity is shown in Figure 21.13. To interpret this information manually would require students to draw the cross-section of the river on graph paper and then to calculate the cross-sectional area. From this, the speed of flow may be used to calculate the volume of water flowing past a specific point in cubic metres per second (cumsecs). Figure 21.14 shows the output from the computer when the data shown in Figure 21.13 are analysed using one of the programs from the suite described above. The availability of such programs should not discourage teachers or students from performing the drawing and calculations manually until the necessary skills are mastered. However, once understanding of the procedures is achieved, the computer can reduce dramatically the time needed to process data from a large number of sites.

While programs such as those described above have been specifically written to handle data generated from fieldwork in physical geography and geology, perhaps the major device for the computerised handling of numeric information outside education is the spreadsheet. Spreadsheets as a whole are referred to as 'application' or generic software, but some programs have been prepared in which the spreadsheet concept has been applied in a specifically geographical context. An example of such a package is the *Industrial Location* package. Figure 21.15 is a map of the imaginary country of United Lysistra.

The country has decided to build an iron and steel plant so that they will no longer have to rely on imported iron and steel products. The students are informed that their company has been given the opportunity to build and operate the plant and that they must therefore choose the most profitable site for the plant. In practical terms, this means finding the least-cost location. In reaching a decision about where the plant should be built, students have to consider transport costs, raw material costs and labour costs, but in addition must bear in mind the fact that changes in these costs, either in terms of their unit costs or the contribution of each cost to the overall cost of the product, may change the location of the least-cost site.

Figure 21.16 is a hard copy of the spreadsheet information provided in the pack. In this simulated

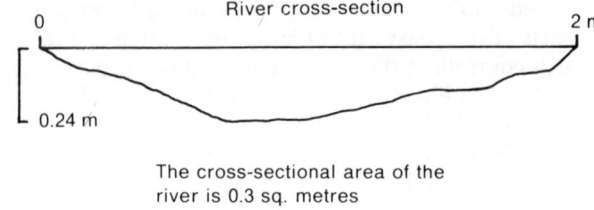

The cross-sectional area of the river is 0.3 sq. metres

Summary of results

cross-sectional area:	0.30 sq.m
wetted perimeter:	2.11 m
hydraulic radius:	0.14 m
river velocity:	0.23 m/sec.
discharge:	0.07 cu.m/sec.

Figure 21.14 Computer-drawn cross-section and summary of discharge calculations based on field survey data

Mimosa Creek, Toofey forest.
Width of stream: 2 metres.
Depth measurements taken at 10 cm intervals.

Distance from bank	Depth in metres
0.1 m	0.08
0.2 m	0.09
0.3	0.10
0.4	0.16
0.5	0.18
0.6	0.23
0.7	0.24
0.8	0.23
0.9	0.22
1.0	0.20
1.1	0.18
1.2	0.16
1.3	0.16
1.4	0.16
1.5	0.15
1.6	0.14
1.7	0.10
1.8	0.09
1.9	0.09
2.0	0

Float timings over 2 metres
1st float—8.6 seconds
2nd float—7.9 seconds
3rd float—9.3 seconds

Figure 21.13 Field notes on the flow of water in a stream

decision-making activity, it is assumed that the site for the iron and steel plant will be one of the four main cities of United Lysistra. The reader who examines the spreadsheet will observe that the city of Nyala is the least-cost site. In a classroom situation, the teacher should encourage students to ask the question: 'Why is Nyala the least-cost site?', and further examination will reveal that Nyala's central location with respect to the raw materials and markets gives it lower costs than the other cities. So far, the decisions on selecting the least-cost location may be made easily without 'hands on' work with the computer. However, having decided that Nyala is the least-cost location at current raw material costs and relative quantities, the spreadsheet enables students to consider the implications of changes such as a three-fold increase in the cost of transporting limestone, or new technology which reduces the quantity of coal needed to smelt a given quantity of iron ore. To perform the calculations required to produce new total costs of production manually would be both tedious and time consuming, but the number-crunching capacity of the computer makes such a task easy. Furthermore, this ease encourages students' creativity since they are not discouraged from asking large numbers of 'I wonder what would happen if . . .?' questions. The manual which accompanies this program suggests that students can go on to use the spreadsheet to investigate the effects of changes:

1. in water costs;
2. to the quantity of steel produced;
3. in labour costs;

UNITED LYSISTRA

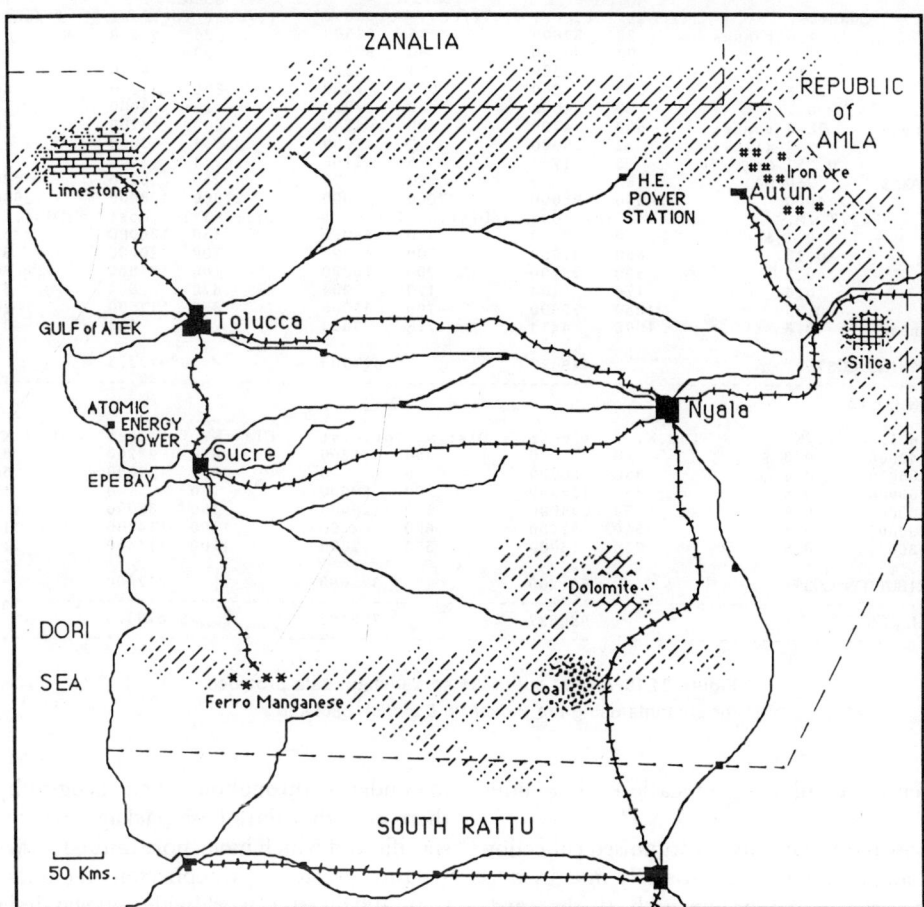

Figure 21.15 'United Lysistra'—the imaginary country in which a new steel plant is to be located

4. in power costs;
5. in the demand for steel produced both for internal and export markets;
6. in mineral deposits.

Finally, students are encouraged to investigate whether Australia's iron and steel plants are in fact located at the least-cost sites and if this is not the case, to research reasons for locating the plants in potentially uneconomic locations. Since the basic costs for producing steel in United Lysistra are based on figures provided by BHP Australia, students are in a good position to consider the Australian steel industry, while the use of the spreadsheet provides them with a powerful tool to undertake many inquiries concerning industrial location in Australia.

Data Bases in Geography

In just the same way that students often find difficulty in handling large amounts of fieldwork data and calculations involving complex or repetitive calculations, so they also find difficulty in handling large amounts of information, especially when cross-referencing has to be undertaken. For example, imagine the time which would be required by many students, using conventional reference resources, to prepare a list of those countries in the world which export oil, have literacy rates of less than 75% and have over 50% of their land area classified as desert, or to discover how many of the oil-producing countries of the world have democratic governments. While such questions can undoubtedly be investigated manually, they are frequently avoided in many inquiries at secondary level due to the tedium of the task. It is only since the advent of electronic data handling using data bases that students can really be encourged to ask that question which

```
STEEL PLANT LOCATION SIMULATION           tot.cost values are in $,000's

PRODUCTION COSTS PER    1000000 TONNES
                                 AUTUN              NYALA              SUCRE              TOLUCA
RAW MAT.        Quant.         lt/cost Tot.cost   lt/cost Tot.cost   lt/cost Tot.cost   lt/cost Tot.cost
Iron ore         1.6   tonnes     36    57600       36    57600       36    57600       36    57600
Coal             1                90    90000       90    90000       90    90000       90    90000
Dolomite         0.18             24     4320       24     4320       24     4320       24     4320
Silicon          2E-3            600     1200      600     1200      600     1200      600     1200
Limestone        0.2              20     4000       20     4000       20     4000       20     4000
Ferro manganese  7E-3  tonnes    600     4200      600     4200      600     4200      600     4200
WATER            3     kilolit     2     6000        2     6000        1     3000        1     3000
ELECTRICITY      0.351 meg.wat     5     1755        6     2106        4     1404        4     1404
RATES and TAXES  250   HA.       180       45      280       70      250     62.5      200       50
LABOUR           4000          24000    96000    27000   108000    27000   108000    26000   104000
TRANSPORT        C./T./K.       Dist.K. Tot.cost  Dist.K. Tot.cost  Dist.K. Tot.cost  Dist.K. Tot.cost
Iron ore         0.1   $           0        0      350    56000      850   136000      850   136000
Coal             0.15            650    97500      300    45000      800   120000      800   120000
Dolomite         0.3             550    29700      200    10800      700    37800      700    37800
Silicon          0.6             170      204      170      204      670      804      670      804
Limestone        0.25           1050    52500      700    35000      350    17500      200    10000
Ferro manganese  0.6            1065     4473      715     3003      215      903      365     1533

TOTAL PROCESSING COSTS                  449497            427503            586793.5           575911
---------------------------------------------------------------------------------------------------
DISTRIBUTION COSTS

MARKET     Quan.T.  C./T./K.   Dis.K.  Tot.cost  Dist.K. Tot.cost  Dist.K. Tot.cost  Dist.K. Tot.cost
Autun       40000      0.8 $      0        0      350    11200      850    27200      850    27200
Nyala      200000      0.8      350    56000        0        0      500    80000      500    80000
Sucre      200000      0.8      850   136000      500    80000        0        0      150    24000
Tolucca    300000      0.8      850   204000      500   120000      150    36000        0        0
R.ofAMLA   130000      0.8      550    57200      600    62400     1100   114400     1100   114400
S.RATTU    130000      0.8      750    78000      500    52000     1100   114400     1100   114400

TOTAL DISTRIBUTION COSTS                531200            325600            372000            360000
---------------------------------------------------------------------------------------------------
TOTAL OF ALL COSTS                      980697            753103            958793.5           935911
---------------------------------------------------------------------------------------------------
```

Figure 21.16 A print out of the spreadsheet data provided for students using the *Industrial Location* package

is fundamental to all true education: 'I wonder if . . .?'.

A data base is quite simply an organised collection of information, and may be most easily imagined as an electronic card filing system with all the cards cross-referenced. In a card filing system, each card contains a single item of information, and users have to sort through the cards using some kind of search strategy and the tips of their fingers. In a computerised data base, the information is held in a computer file, and a separate organising program, called a data-base management system, does the sorting.

Most computers have a data management system available at low cost, and it is quite possible for teachers or students to enter data into their computers as an aid to analysis. However, such data entry is a long and tedious process, and it is fortunate that there are a number of packages available which consist of a data-base management system with information already entered.

One such package is *One World Countries Database*. Figure 21.17 shows the types of information which are provided in this package.

While *One World Database* may be regarded as basic reference material which should be available to students throughout their geography studies, there are other data base packages which are more specific and which have more limited aims. *The Great Disasters Database* (Macpherson) includes details of some 400 disasters worldwide and was designed to be used by school students as an 'initial motivator' for research and discussion over a wide range of topics. The designer's intention in preparing this data base is to use the initial interest many students have in disasters to encourage exploration of wider issues. In the manual which accompanies this package, Russell Keam, a teacher in a Victorian secondary school, describes how he used the program with a single computer in his classroom to introduce the notion of disaster, and to develop the graphic skills of his students. In addition to these basic aims, however, the data base had the effect of encouraging revision of other aspects of geography such as identifying countries, locating countries within continents and discussing the influences of plate tectonics.

By enabling students to manipulate large amounts of information relatively simply, data bases may be used in geography to encourage a number of information-handling skills as well as helping students to create meaningful relationships between many of the aspects of geography which have

One World Countries Database

This data base (file) contains information about every independent nation in the world. In January 1984, there were 178 nations. There are 178 'records' in the database; one record for each country. Each record contains 33 'fields' or items of information.

Field
Country no.
Name
Previous name
Region
Neighbours
Oceans or seas
Capital city
Latitude
Longitude
Head of state
Title of head
Leader of government
Title of leader
Type of government
Exports
Imports
Currency
Total population
% Urban
% Rural
% Workforce in primary production
% Workforce in manufacturing
% Workforce in service industries
Size of country (sq. km)
% Forest
% Desert
% Cultivated
Languages
Literacy rate
Religions
Treaties and alliances
Background
Independence date

Figure 21.17 The range of information provided in *One World Countries database*

hitherto been treated discretely. In broad terms, data bases can help students to:

1. analyse relationships;
2. identify trends;
3. arrange information in more useful and meaningful ways;
4. discover similarities and differences between groups of events or things;
5. develop their problem-solving skills.

While data bases containing only textual information have been available for some time, more recent developments have included data bases which contain audio-visual materials as well. These data bases are supplied on a laser videodisc and contain vast amounts of information. *Ecodisc* is one example of such a data base and is based on the nature reserve at Slapton Ley in Devon, England. The disc contains over 35 minutes of film in 170 sequences which can be linked together in many different ways, 4000 still photographs, as well as much textual and numerical information. The data base is used to place students in the role of an assistant warden in the nature reserve and they have to deal with the ecological, administrative and political problems which go with the job.

Obviously the first task of anyone taking on such a job would be to take a walk around the reserve. Students can either go on a ramble on their own or they can choose to follow a guided tour. Each 100 metres, the student can examine still photographs showing the view in each of eight compass directions, and can switch from summer to winter at will. At intervals, short film sequences introduce the newly appointed wardens to the animals and plants, describe the habitats and explain some of the methods which can be used to study them further. Unlike the difficult world of 'reality' the warden using this data base can 'see' what is happening underwater as he or she sails up the lake.

Having become acquainted with the reserve, the 'warden' can then go to the 'office' to discover the real problems of managing the reserve. In the office are a television, a pile of local newspapers, an in-tray, a computer and a job description. The warden can read the contents of the in-tray, and thereby discovers that the local sailing club wants to lease the reserve's lake to teach youngsters, the shooting club wants to clear the woods and introduce game birds, the local ornithologists are concerned that the new management may do something to disturb the birds, scientists hope that nothing will be done to disturb their studies, local anglers want permission to fish in the lake, farmers object to new plans, a local poacher complains that his livelihood is at risk and local politicians complain about everything that is left. Faced with such decisions, our wardens may request further information on the various ecosystems from the television or may use the 'computer' in the 'office' to manipulate textual and numerical data in an attempt to discover some of the potential effects of emerging management policies.

Faced with such apparently intractable problems, the warden may wish to carry out some field investigations of his or her own. The program allows the warden to perform quadrat analysis of the plants, identify and count the various birds in the reserve, set up cameras triggered by any mammals which pass by, trap animals so that a radio collar can be put on to facilitate later tracking, net pike in the lake and mark them with dye to allow estimates of total population. All this information is held in the memory of the 'office computer' for later analysis.

Having investigated all the 'given' information and 'collected' data from the types of survey described above, the warden may then forecast what is likely to happen over a 50-year period under different types of management strategy. Having undertaken all this, the time of truth comes when the probationer warden has to present a management plan for the reserve. The plan can be modified at any stage, but the computer will quite relentlessly evaluate the success or failure of the strategy at the end of the period. Jeremy Cherfas, who evaluated Ecodisc for *New Scientist* magazine, wiped out the entire otter population of Slapton Ley during his period as 'warden'.

Integrated data bases through which students can interact with moving pictures, still photographs, graphics, text and numerical data are, at the time this is written in December 1987, still very new and exciting technologies and appear to have great potential for geography teachers. However, despite the power of the technology, the real impact of such developments on geographical education will continue to depend on the ways in which the technology is integrated into geography curricula and the way in which its use is implemented in classrooms. This is an area to which we will return in the last section of this chapter.

Generic or Application Software in Geography Teaching

Application software includes a wide range of 'management' programs which can be used in many subject areas and activities. Such programs are available for all computers and, as was shown in Figure 21.2, include word processors, spreadsheet management programs, data-base management programs, map and graphic creation programs and communications programs.

Data Bases and Spreadsheets

The previous section of this chapter on subject-specific software based on spreadsheet number crunching and data bases has considered the potential of such programs for geographical education. All that needs to be added here is a comment on the importance of students realising the potential for 'empty' data bases and spreadsheets for their own investigations, and a warning on the time required to set up appropriate data bases and spreadsheets and to enter substantial quantities of information.

One recent development has been the possibility for schools to link their micro-computers to the large mainframe computers used by newspaper houses or other organisations which maintain and constantly update their own data bases. The link is usually made by a device called a 'modem' which enables two computers to communicate through the normal telephone system. In this way, a student who wishes to investigate a particular topic concerning his or her local area can interrogate the data bases of, say, *The Times* newspaper in London or the *Courier Mail* in Brisbane and not only discover the titles of all stories which have been carried on the topic of his or her choice for several years past, but also download the full text of those stories into the school computer for perusal at leisure or even for printing to hard copy for later use. Such facilities should make geographical inquiries based on outdated information a thing of the past.

Word Processing

Many teachers have already discovered the contribution of word processing to their own work in preparing and amending worksheets for their students. However, the use of computers in this way may also have dramatic effects on the written communication skills of our students as they undertake their geographical education. For students to be able to insert and delete text, revise words, phrases and sentences, move sections around at will within a document, as well as 'cut and paste' whole pages from one document to another has been found to change their attitudes to correction from being a matter of irksome labour to being a natural path to improving text. Many teachers have found that word processing benefits their students in three ways:

1. for some students, especially those who are less able, the production of work of a high standard of presentation is a new and highly motivating experience;

2. the ease with which their work can be revised and

improved encourages students to be more experimental and productive;

3. teachers can evaluate students' work more easily on the basis of its content since poor presentation does not detract from its substance.

In addition to these advantages for the individual work of students, cooperation and group work are also encouraged by word processing. When groups of students have worked together on a particular project, a group report may be compiled by all the students sitting around the monitor and creating the final product as a cooperative venture. When the final draft has been agreed by all the group members, a 'good' copy may be produced for each person.

Figure 21.18a–e A selection of graphics produced on a micro-computer and dot matrix printer

Figure 21.18a A pie diagram

Map- and Graphic-creation Packages

The use of the computer to create cross-sectional diagrams of a river was described in a previous section, but there are programs available for all micro-computers which enable bar charts, pie diagrams and line graphs to be constructed from any appropriate data. In some such programs, the data must be entered separately for each graphic representation, while other, more sophisticated programs will 'import' data from any row or column in a spreadsheet or data base. Other programs will create thematic choropleth or isopleth maps from data either entered directly or imported from spreadsheet information. Figure 21.18 shows some of the graphics which can be produced by a small micro-computer linked to a simple dot matrix printer. If more schools gain access to laser and ionic transfer printers, and graphics plotters, the quality of graphics which may be designed and produced by secondary school students will be equivalent to any currently to be found in their textbooks. However, the quality of the presentations may not be the only

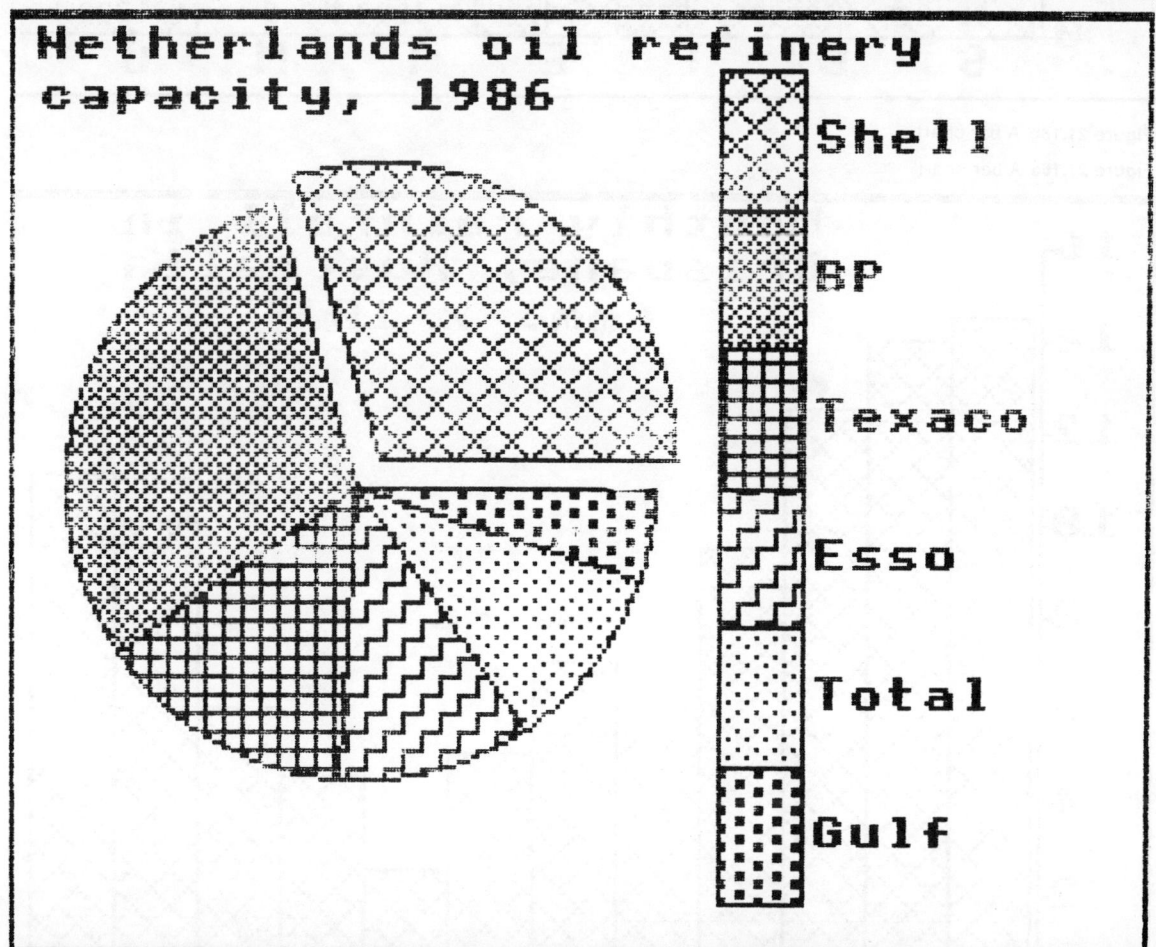

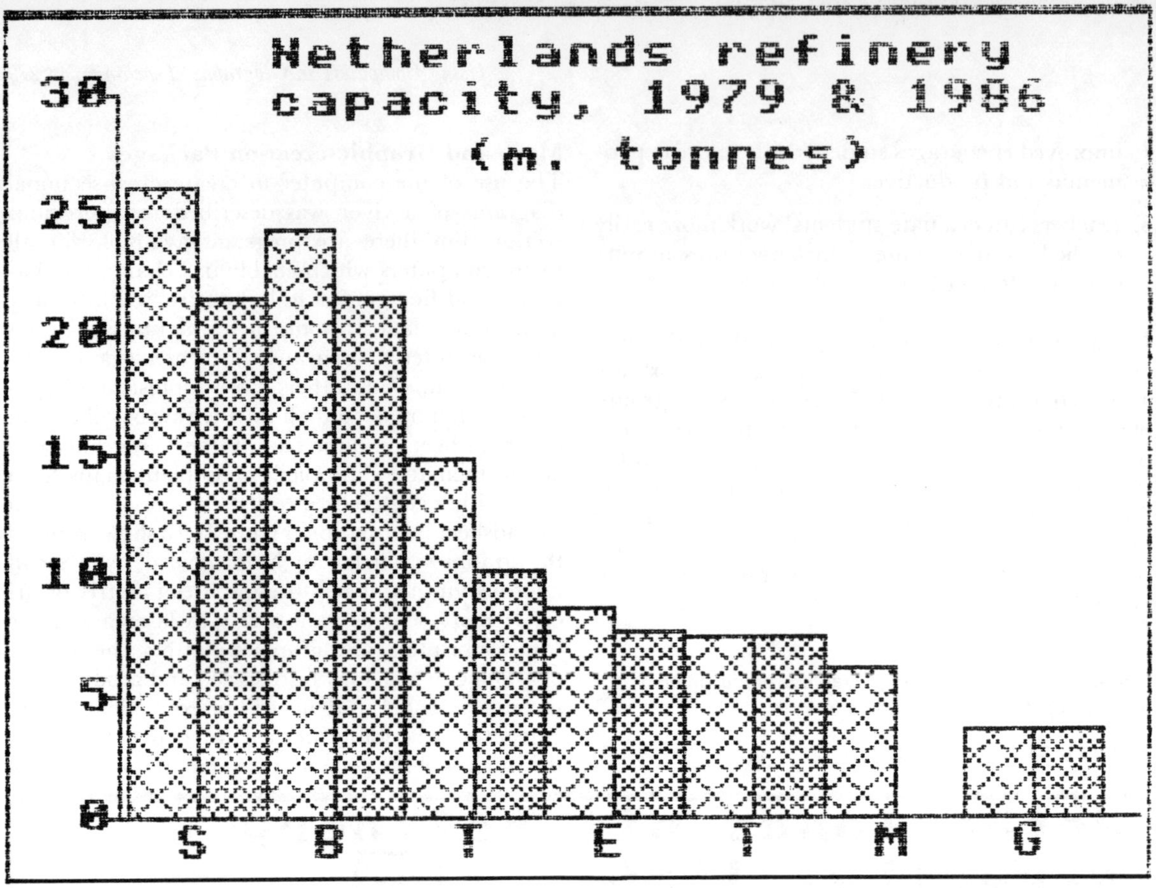

Figure 21.18b A bar chart

Figure 21.18c A bar chart

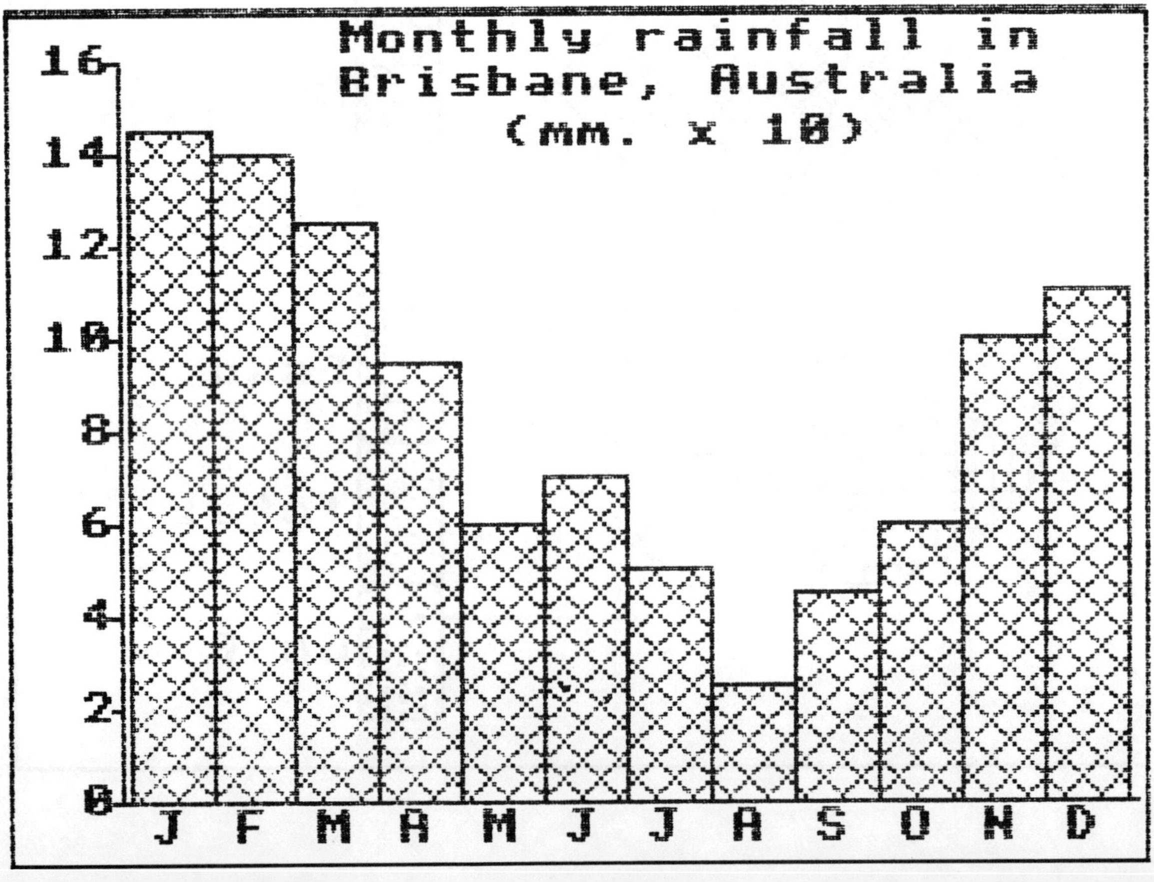

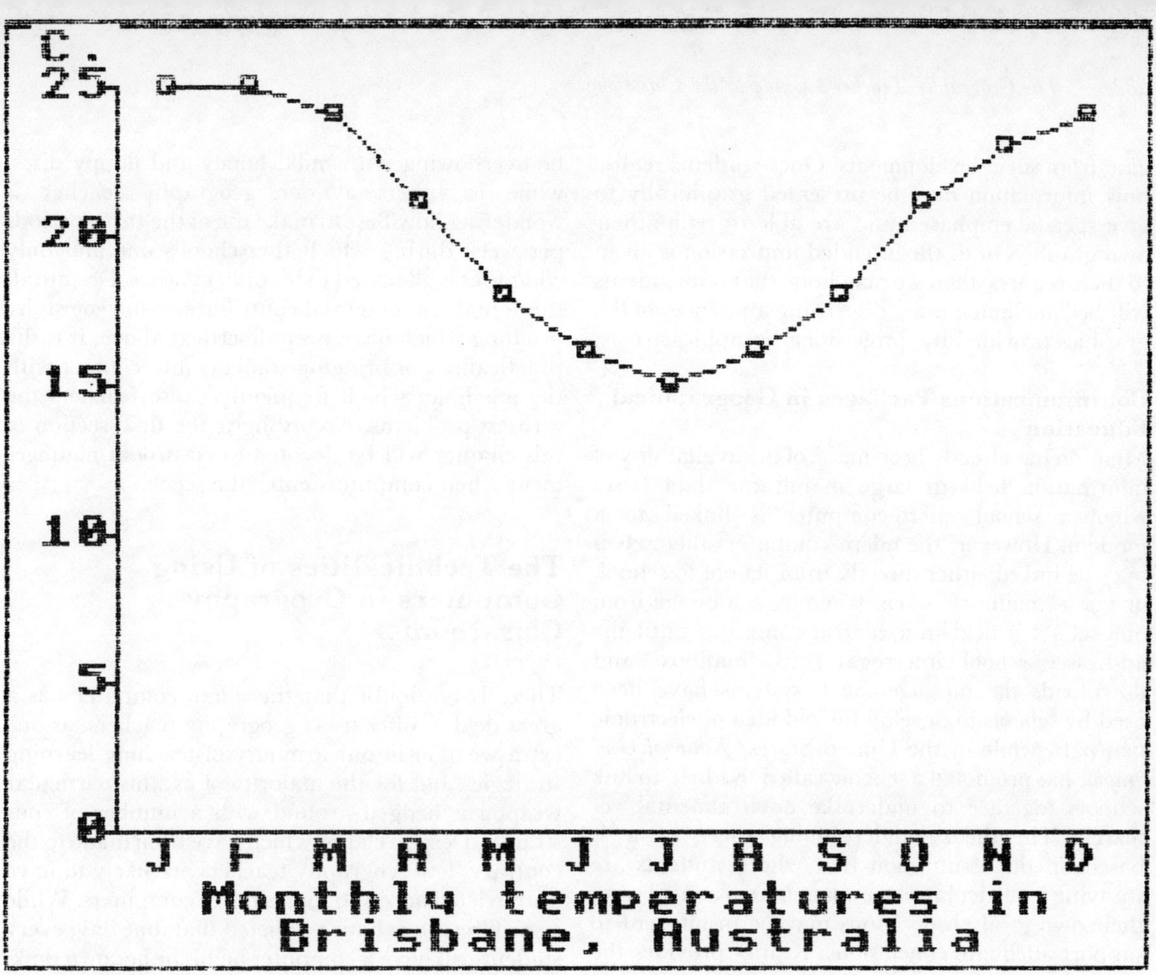

Figure 21.18d A line graph

Figure 21.18e A population pyramid (source: *People Pyramids*)

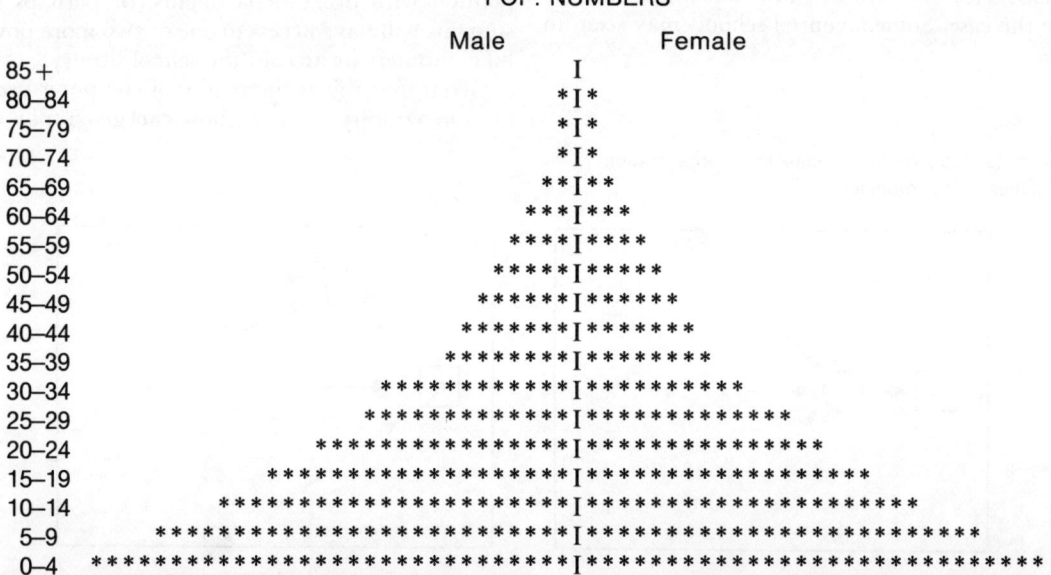

gain from such developments. Once students realise how information may be presented graphically to give specific emphases, and are able to refine their own graphics until the intended impression is given to their readers, then we may hope that our students will become much more discerning audiences of the graphics provided by 'professional' graphics artists.

Communications Packages in Geographical Education

Mention has already been made of the availability of information held in large mainframe data bases when a school micro-computer is linked to a modem. However, the micro-computers themselves may be linked either directly from school to school, or via a 'mailbox' system whereby a message from one school is held in a central computer until the addressee school interrogates its 'mailbox' and downloads the message. Such systems have been used by schools to develop the old idea of electronic pen pals, while in the United States, *National Geographic* has promoted a scheme called 'Kidnet' to link schools together to undertake environmental research. The theory underpinning this scheme is based on the assumption that where students are studying a particular phenomenon such as acid rain, their own local observations may be insufficient to support adequate conclusions. Kidnet provides the technology to encourage exchange of ideas and findings and thus enables students to engage in inter-regional comparisons.

It really does seem that the potential for using computers in geography teaching is limitless, bounded only by the imagination of teachers and their students. However, we all know that this is unlikely to be the case. Some favoured schools may seem to be overflowing with milk, honey and floppy discs, while in others a lone geography teacher is wondering how best to make use of the three periods per week during which the school's one and only computer is allocated to his or her classes. Despite all the actual and potential contributions to geography teaching which have been described above, it is the practicalities of bringing students into contact with the machines which frequently cause teachers the greatest problems. Accordingly, the final section of this chapter will be devoted to classroom management when computers enter the scene.

The Technicalities of Using Computers in Geography Classrooms

There is no doubt that the micro-computer has a great deal to offer us as geography teachers, as one extra weapon in our armoury of teaching learning strategies, but for the majority of us, this particular weapon is hedged around with a number of constraints. Despite claims which have been made to the contrary, few geography teachers are likely to have unrestricted access to rooms full of computers. While some forecasters have predicted that, one day, every student will have a computer on his or her own desk, the reality for most geography teachers is, and probably will remain for many years to come, either that one or two micro-computers are available in the classroom, at least for some of the time, or that there is a class set of computers located in a 'computer room' which has to be booked in advance in competition with other departments, or perhaps that students will have access to one or two more powerful computers located in the school library.

Given that this is the reality of computer use for most geography teachers, how can geography cur-

Figure 21.19 Some models of classroom organisation possibilities using computers

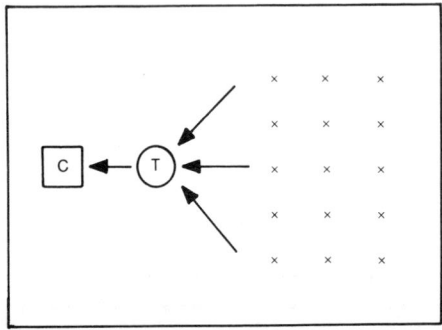

A: Teacher as gatekeeper

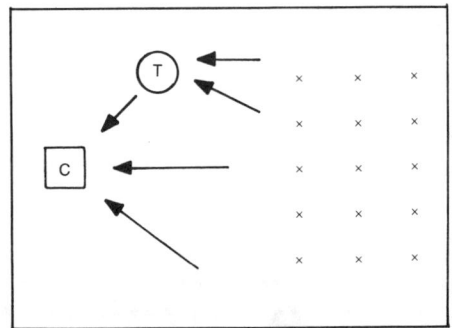

B: Teacher and students as partners

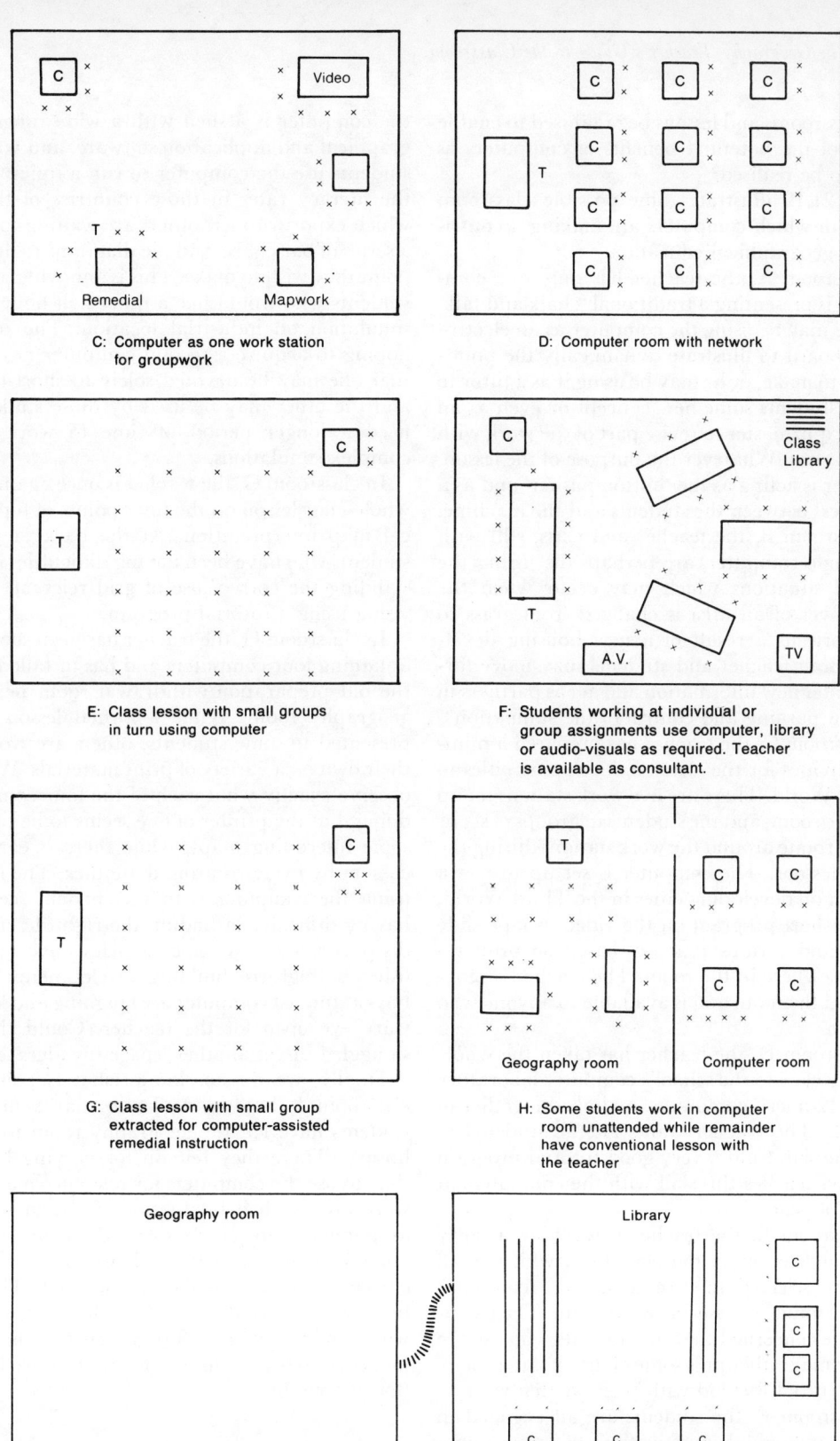

ricula, classrooms and lessons be organised to enable as many of the potential benefits of computers as possible to be realised?

Figure 21.19 illustrates nine possible classroom scenarios in which computers are making a contribution to geographical education.

In classroom A, the teacher has only one computer and is presenting a traditional 'chalk and talk' lesson. He may be using the computer as an electronic blackboard to illustrate dynamically the points he wishes to make, or he may be using it as a tutor to teach his students some new concept or even as an electronic quizmaster to revise part of the work with the whole class. Whatever the purpose of the lesson, the teacher is acting as the 'button pusher' and as a 'gatekeeper' between the students and the machine.

In classroom B, the teacher and class, still with only a single computer, are perhaps discussing the variety of situations which may occur when the ground cover of an area is changed from grass to urban fabric as a result of a new housing development. Both teacher and students may move forward to enter new information and act as partners in the button pushing and control of the simulation.

In classroom C, the teacher has prepared a number of activities for the class on developing cities in the Third World. There are five work stations set up around the room, and the students in groups of six or seven will rotate around the work stations during the next few lessons. The computer is set up to run a simulation on developing cities in the Third World, there is a short program on the video, a tape-slide sequence and various activities based on print resources elsewhere in the room. The teacher is moving around the room and is available to anyone who needs help.

In classroom D, the teacher has taken the whole geography class to the school's computer laboratory in which 12 micro-computers are linked together in a network. The teacher wants all the students to learn some skill from a very good tutorial program and always teaches this skill with the computers at this time of year.

In classroom E, the teacher is teaching a fairly conventional lesson to the class from which small groups are selected in turn to use the computer which is running an associated program. The teacher is a little concerned that the students who use the computer may either miss some of the explanation of the lesson or fall behind with their written work.

In classroom F, the students are all engaged in their own individual investigations of various geographical issues. The room has a good selection of books, tape-slide presentations and video programs available for students to refer to when needed. Beside the computer is a shelf with a wide range of geographical and application software, and while some students use the computer to run a quick check on the literacy rates in those countries of the world which export rubber, others are waiting to produce a series of bar charts and pie diagrams to illustrate a point they wish to make. This is annoying a group of students who would like a quiet half hour to run a simulation on industrial location. The teacher is hoping to acquire a second computer next year so that one may be devoted solely to short-term uses and the other may be used by those students who need a longer period of time to work through complex simulations.

In classroom G, the teacher is once again giving a whole-class lesson on the finer points of topographical map interpretation. At the back, a group of students who have been having difficulties in understanding the correct use of grid references are revising using a tutorial program.

In classroom H, the teacher has been successful in obtaining four computers and has installed them in the old preparation/withdrawal room next to the geography room. While a formal lesson is being presented to some students, others are working on their own on a variety of print materials. We cannot observe exactly what each of the four computers is doing, but the printer of one seems to be producing some interesting maps, while there is earnest discussion by five girls around another. The four boys using the computer with the modem seem to be having difficulty in finding the right descriptors to discover what newspaper articles have been published on high-rise building developments. The two boys at the last computer are laughing and keeping a wary eye open for the teacher. Could they have smuggled in yet another 'space invaders' game?

Finally, we do not know what is going on in classroom I, but we do know that a number of students have left the geography room to visit the library. There they remain for varying lengths of time to use the computers for reference materials, to word process their final reports for their individual assignments, to enter some data for statistical analysis or to run a tutorial program on how the epicentres of earthquakes are located. The school bell goes to mark the end of the day, but the library will remain open until 9 p.m.—such is the demand for access to the resources, electronic as well as print, that it contains.

Conclusion

Of the nine geography classrooms described above, which one comes closest to representing the experiences of most geography teachers? For a surprising number, only classrooms A and B will be recognised as representing the 'real' world of the classroom, while for some, even these modest ventures into the world of computers will represent a scenario still in the future. Often the main reason for not using computers in geography lessons is given as being the lack of availability, but perhaps the examples shown will convince the reader that even a single machine may offer considerable potential for improving still further the standard of our geography teaching. The micro-computer is not the answer to all our prayers. It will not contribute to making geography teachers redundant, and neither will it make our jobs dramatically easier. However, with skilful integration into our courses, new experiences may be placed before our students. The highly motivated may be encouraged to progress even further in their explorations of geographical phenomena; the less motivated may find that the power and privacy of learning geography with the aid of a computer makes school life slightly less of an ordeal for them. Let's try it and see.

References

Cherfas, J. (1987) 'So You Think You Can Run a Nature Reserve?', *New Scientist*, 2 July, 59.

Freeman, D (1983) 'CAL in Geography: A Case Study in Hertfordshire Schools', in A. Kent (ed.) *Geography Teaching and the Micro*, York, England: Longman Group Resources Unit.

Computer Software

Climate (1985) Wesoft Educational Software, Schools Computing Branch, Education Department of Western Australia, 151 Royal Street, East Perth, Western Australia 6000.

Dalco (1983) Longman Micro Software, Longman Resources Unit, York, England.

Developing Cities (1984) NELCAL, Walton-on-Thames, England, Thomas Nelson and Sons Ltd.

Ecodisc (1987) British Broadcasting Corporation, London.

Field Studies Techniques (1983) Science Education Software, MJP, PO Box 23, Faversham, Kent, England.

Industrial Location (1986) Satchel Software, Angle Park Computing Centre, Cowan Street, Angle Park, South Australia 5010.

Introducing Map Skills (1983) Netherall Software, Press Syndicate of the University of Cambridge, England.

One World Countries Database (1984) Active Learning Systems, PO Box 197, Indooroopilly, Queensland, Australia 4068.

Population dynamics (1984) Unpublished program written by Simon Jewell.

Slick (1983) BP Educational Service, Britannic House, Moor Lane, London EC2.

The Great Disasters Database (1987) Macpherson, C., The Centre for Computer Education, Hawthorne Institute of Education, 442, Auburn Road, Hawthorn, Victoria, Australia 3122.

The Path of a Water Droplet (1984) in MEP/GAPE Project Geography Programs, Hutchinson Software in association with the Geographical Association, London.

The Sailing Ships Game (1983) Longman Micro Software, Longman Resources Unit, York, England.

Water on the Land (1983) Longman Micro Software, Longman Resources Unit, York, England.

22
Geographical Facts from Geographical Figures: Turning Students on with 'Stats' in Geography

John Wolforth

Many of the innovative teaching activities developed by geography teachers in the last decade have depended upon the application of some statistical techniques to numerical data. Many of these activities were enjoyable to do and provided both teachers and their students with new insight into geographical relationships and a respect for the accuracy that quantitative techniques can bring to an inquiry. However, much evidence is now coming forward to indicate that all is not well in the 'quantitative garden' due to the 'mental block' that non-mathematically inclined students seem to have about 'stats' in school geography. Indeed David Unwin, the chairman of the Quantitative Methods Study Group of the Institute of British Geographers, has expressed 'serious doubts' about the incorporation of statistics in the geography classroom. John Wolforth's chapter here is a welcome answer to these concerns commenced by Unwin. This chapter briefly traces the quantitative movement in geography, outlining its strengths and the pitfalls to avoid. Providing examples of successful classroom activities he has used, John explains how teachers may make safe use of statistics in the geography classroom in three ways: to show statistics on maps, to derive statistics from maps and to explain geographical associations. The chapter concludes by advancing five arguments for using statistics in geography teaching based upon the classroom activities described.

It was Mark Twain who said 'There are lies, damned lies, and statistics'. It is indeed true that statistics can lie, and frequently do. However, so do words. It is as important to be able to detect statistical lies as well as verbal lies, maybe more so, since numbers seem to suggest a scientific authenticity and precision which mere words often lack. Like words, however, statistics may not only be used to mislead us, but also to help us to discover truths and to relate them to others in a straightforward and unambiguous way. The trick is to know how to use them, and to recognise their possibilities and limitations.

Statistics have become especially important in geographical research in recent years, and many of the most powerful concepts which have been developed in geography have a strong statistical base. For this reason, if for no other, some attention has to be paid to statistics in classroom geography; to ignore statistics would be to present a geography which was both outmoded and lacking in explanatory power. There is no denying, however, that statistics often put students off, and that the most useful statistical techniques are not easily learned by the average 14 year old. None the less, there are a good number of sources the hard-pressed geography teacher can go to for help in dealing with statistics, both for the expert (e.g. Clark and Hosking 1986), and for the classroom teacher (e.g. Okunrotifa 1982)

Two Messages

It would be pointless to try to summarise these sources, or to attempt to give a complete account of all the statistical techniques available to the geography teacher. What this chapter seeks to do instead is to make two major points about the use of statistics, and to illustrate these with a small selection of appropriate techniques. These points are as follows:

1. *Descriptive statistics have been used for a long time by geographers and their familiarity should not breed contempt.* Geography students should be encouraged to play around with numbers and to present them, on maps or graphs, in new and creative ways. This can often lead to some unexpected insights and lead to speculation about distributional patterns. This kind of approach to numerical data is greatly facilitated by the greater access and familiarity which we have with pocket calculators and micro-computers. Calculations which a few years ago would have taken a daunting amount of time to complete can now be done at the push of a button.

2. *Beyond descriptive statistics, there is a range of not very difficult techniques which allow inferences to be drawn from numerical data, and hypotheses to be formulated.* The use of statistics provides one of the best introductions to the kind of organised, systematic inquiry which goes under the name of 'the scientific method'.

What are Statistics and How are They Used to Describe Things?

It might be useful to begin with some definitions. A 'statistic', according to one dictionary, is simply a 'numerical fact or datum', the kind of thing we use without blinking an eye in everyday life as well as in geography. (John weighs 90 kg; the car has a 2-litre engine; today's temperature is 18°C; and so on.) *Statistics* are a different matter, however. The same source defines 'statistics' as 'the science that deals with the collection, classification, analysis, and interpretation of numerical facts or data, and that, by use of mathematical theories of probability, *imposes order and regularity* on aggregates of more or less disparate elements'.

The italics in the above definition are mine, and I should like to explain why I consider the phrase important. It may seem strange that we should use statistics 'to impose order and regularity'. Surely, order and regularity are there already and statistics simply help us describe it. A familiar example may help us to understand why this is not necessarily so.

Measure of Central Tendency

The figures given below represent some fairly commonplace statistics about a point on the earth's surface, namely its mean monthly temperatures (°C).

Whitehorse, Yukon Territory, Canada

J	F	M	A	M	J	J	A	S	O	N	D
−19	−13	−8	0	7	12	14	12	8	1	−9	−16

Let us look at some of the arbitrary decisions which have been made in order to arrive at these statistics. First, it has been decided that the year should be divided into 12 'months' of roughly, but not exactly, the same number of days. Second, it has been decided that the expansion of a liquid (usually mercury) should be taken as the measure of something which it has been decided should be called 'temperature', an indication of how 'hot' or 'cold' the air is. Third, it has been decided that an instrument should be set up at a particular point in Whitehorse to measure the highest and the lowest temperature attained in any day. And all this before we even think of gathering 'numerical facts or data'!

The actual figures are derived from a process which involves equally arbitrary decisions. The maximum and minimum temperatures for each day are added, and then divided by two to give us a daily 'mean'. The daily means are added, and then divided by the number of days in the month (28, 29 in some years, 30 or 31) to give us the monthly mean. This is done for as many years as possible, and a mean obtained for each of the 12 monthly means. The resulting figures we say, with more confidence than may be warranted, represent the temperature regime of Whitehorse.

Now, of course, there is a regularity in nature which we are trying to describe and, indeed, to quantify; it does generally get colder when the days are short and warmer when the days are long. But there are many ways in which this gradual transition could be pinned down with numbers. Rather than using 'months' we could use divisions of time consisting of 7, 10 or 40 days. Rather than only taking the maximum and minimum temperatures each day and dividing by two to get the daily mean, we could take the temperature every hour and divide by 24.

In fact, we could decide not to use the mean at all, but some other measure of what statisticians call *central tendency*. If we were to take readings every hour during the day and then arrange the 24 readings in numerical order, the mean of the twelfth and thirteenth highest readings would give us the daily *median*. We could then arrange the 31 daily medians in order for, say, January, and the sixteenth highest would give us the median temperature for that month. Alternatively, we could take the third measure of central tendency, the *mode*, as the statistic we wished to use and find out which temperature occurred most frequently among our 24 daily readings. (The mode might in fact give us a better idea of what the temperature is actually 'like', especially during the height of summer or the depth of winter when for many of our 24 hourly readings the thermometer hovers around the same mark.)

To make things even more confusing, we could decide we were not really interested in means, medians *or* modes, but that what we really wanted to express was the degree of variability in the data which interested us. Fortunately, the statistician, gives us some tools for that as well, in particular the *standard deviation*.

The standard deviation is a measure of the 'spread' in a set of numbers and is obtained as follows. First, the difference is calculated between each number in the set and the mean of all the numbers in the set. Second, the square of each of the differences is calculated and then all the squares are summed. The result is a quantity known as the *variance*, and the square root of the variance is the standard deviation. The example shown below indicates that this statistic can be a very useful indication of an aspect of climate that is hidden by measures of central tendency, namely, that some places experience a high degree of variability in what they may expect from one time to the next.

The figures which appear in Figure 22.1 are based on raw data obtained from the *Australian Yearbook* and indicate the mean relative humidity at 9 a.m. for each month of the year. They show the highest monthly reading, the lowest, the annual mean and the standard deviation (all calculated very easily using a programmable hand-held calculator). Mapping any of these statistics might yield an interesting pattern, but the map (Figure 22.2) of standard deviations is particularly revealing. It shows a broad zone stretching through central and southern Australia in which there is a high degree of variability (S.D. > 10) in early morning humidity. Could this be associated, we might ask, with the distribution of certain vegetational types, or agricultural practices? Whether it is or not, the exercise demonstrates the variation in an aspect of the climate that we do not generally look at.

Quantitative Methods in Geography: Why?

Let us return now to our dictionary definition and note that there are four aspects to the science of statistics: collection, classification, analysis and interpretation. Not so many years ago, geography's interest in statistics was confined to the first aspect and, to a vary slight degree, the second. Old-fashioned geography texts contained painstakingly collected lists of raw data on temperature and precipitation, on percentages of land in different kinds of land use, on populations, on the annual production of raw materials and manufactured goods, and so on. Occasionally, there were attempts to classify these data usually in relation to specific tracts of the earth's surface called regions; even more occasionally these classifications were based on an objective *numerically based* taxonomy. Koeppen's classification of world climatic regions is an outstanding example.

Then came the so-called 'quantitative revolution' with its attempt to try to derive some geographical principles, generalisations (and some even said, laws) which would enable observations made in one place to be related to some more generally applicable framework, a theory of geography. This revolution rested on geography's attempt to become more

Climatic station	Mean	S.D.	Min.	Max.
Adelaide	60.3	9.5	49.0	75.0
Alice Springs	44.1	12.1	30.0	66.0
Armidale	69.0	7.5	58.0	80.0
Brisbane	64.7	4.5	58.0	70.0
Broome	57.3	9.5	46.0	73.0
Canberra	70.1	9.7	56.0	84.0
Carnarvon	59.8	5.6	52.0	70.0
Ceduna	65.9	10.8	54.0	82.0
Charleville	52.3	10.6	37.0	71.0
Cloncurry	43.8	9.8	30.0	61.0
Darwin	73.2	7.3	63.0	84.0
Esperance	70.1	7.3	62.0	82.0
Halls Creek	35.6	10.3	22.0	55.0
Hobart	66.7	7.8	55.0	78.0
Kalgoorlie	58.8	11.1	45.0	76.0
Katanning	74.1	12.2	57.0	89.0
Kiandra	74.2	8.7	63.0	86.0
Marble Bar	34.8	8.4	24.0	48.0
Melbourne	69.4	7.3	61.0	81.0
Mildura	66.7	14.2	49.0	88.0
Perth	62.3	11.1	50.0	78.0
Sydney	67.7	3.7	62.0	76.0
Townsville	67.1	5.1	60.0	76.0

Figure 22.1 Humidity data for selected climate statistics in Australia at 9.00 a.m.

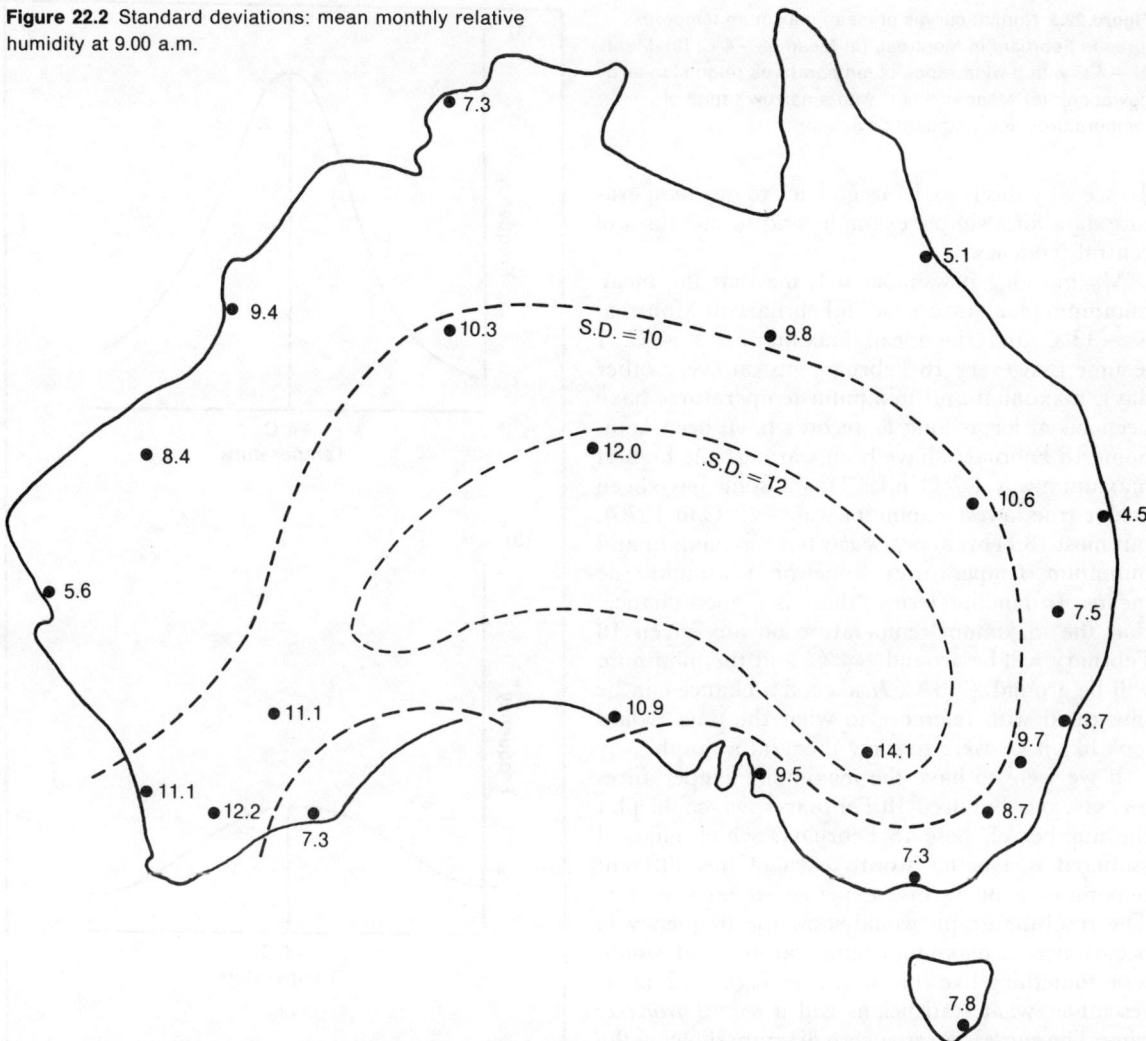

Figure 22.2 Standard deviations: mean monthly relative humidity at 9.00 a.m.

rigorous by using the procedure referred to earlier as 'the scientific method'. As everyone knows by this time, this means setting up an hypothesis, inventing a procedure for testing it, gathering the relevant data, carrying out the procedure and analysing the results. If possible, the hypothesis should not just be a hunch drawn by intuition out of the empty air, but should be derived from 'theory', that is from some comprehensive notion about 'how things work'. The reason for this is that if the hypothesis is proved to be true, then it can be used to elaborate and refine this notion; theory grows and becomes more complex from the successive acceptance (or rejection) of carefully tested hypotheses.

Why is the use of statistics particularly relevant to this view of geography as a scientific discipline? First, it is clear that the use of quantitative data gives a greater rigour to the application of the scientific method than would be possible if only qualitative data were used. The principle of *distance decay* which states that the influence of a geographical phenomenon (e.g. a shopping mall, a factory which emits noxious fumes, a TV station, etc.) diminishes as the distance from that phenomenon increases is not very powerful unless numbers can be pinned to it. As soon as we can *quantify* the rate of decay and the shape of the decay function (i.e. whether it is a straight line or an exponential curve, for example), then we are in a position to make fairly accurate predictions; we can say how many customers are likely to come from a particular district to the shopping mall, what proportion of trees are likely to be diseased because of acid rain, or how strong a TV signal we will receive.

Second, statistics give us a box of tools for actually applying the scientific method. To understand the underlying principle of these tools, we need to go back to our dictionary definition; the science of statistics uses 'mathematical theories of probability'.

Figure 22.3 Normal curves of mean maximum temperatures in February in Montreal. (a) Mean of −4°C; (b) Mean of −4°C with a wide range of temperatures (high standard deviation); (c) Mean of −4°C with a narrow range of temperatures (low standard deviation)

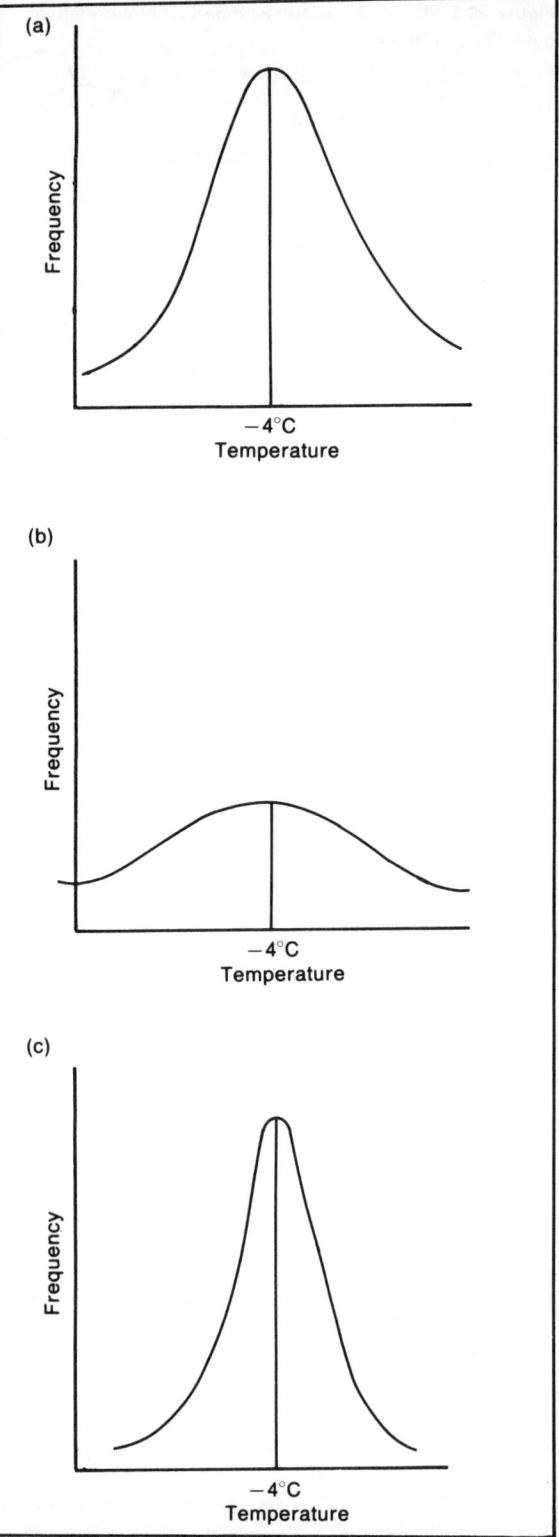

To see why this is so, let us go back to our temperature data for a simple example, and to measures of central tendency.

My morning newspaper tells me that the mean minimum temperature for 18 February in Montreal is −13°C and the mean maximum is −4°C. I assume that every 18 February (as on every other day), maximum and minimum temperatures have been taken for as long as records have been kept. Some 18 Februarys have been warmer (the highest maximum was +7°C in 1981), and some have been colder (the lowest minimum was −28°C in 1979), but most 18 Februarys *probably* have maximum and minimum temperatures somewhere around the means. In familiar terms, 'there is a good chance' that the maximum temperature on any given 18 February will be around −4°C, and the minimum will be around −13°C. *How* good a chance can be illustrated with reference to what the data would look like if we were to show them on a graph.

If we were to have the maximum temperatures for, say, one hundred 18 Februarys, we could plot the number of these 18 Februarys which enjoyed (suffered is a better word!) each of the different maximum temperatures experienced on this day. The resulting graph would show the frequency of occurrence of maximum temperatures and would look something like that shown in Figure 22.3a. It resembles what statisticians call a *normal frequency curve*. The curve is, in essence, a diagram showing the *probability* of any particular event (in this case, an 18 February with a given maximum temperature) occurring. The curve could be fairly flat (as in Figure 22.3b) or peaked (as in Figure 22.3c), depending on the size of the standard deviation of the data. In Figure 22.3b, it is clear that the probability of the maximum temperature of any given 18 February being between −5 and −6°C is quite high; not many 18 Februarys have maximum temperatures outside this range. In Figure 22.3c, where the maximum temperatures are spread over a wider range, the probability of any given 18 February having a maximum temperature between −5 and −6°C is much lower.

What should be clear from this is that most statistical statements, even those dealing with humble measures of central tendency and variability, have a basis in ideas of probability. This has made the use of statistics particularly attractive to geographers, whose data are often of the kind for which deterministic statements are inappropriate.

In geography, it can rarely be said that 'event *A*' always and invariably is associated with 'event *B*': not all fjords have fishing fleets just because some of those of Norway and British Columbia do. What can usually be said is that 'event *A*' has a certain probability of being associated with 'event *B*'. Statistical statements are statements of this kind and consequently have to be accepted with some reservation. The question that must always be asked is: but what is the range of that probability?

Statistics and Maps

Showing Statistics on Maps

Geographers have used statistics on maps for a long time, and have developed some rather elegant technique for doing so. The process is always the same and, in essence, is that which we used to construct the map shown in Figure 22.2. The steps taken are as follows:

1. The raw data are selected from a standard reference such as a year book or some similar source, or are gathered from direct observation in the field. They are 'raw' data because nothing yet has been done to them.

2. The raw data are then 'processed' into a statistic appropriate to the purpose for which the finished map is to be used. Such statistics might include percentages of some given quantity, standard deviations as in the case shown, or more esoteric ones such as location quotients.

3. The processed (and hence simplified) data are now divided into classes which are either chosen arbitrarily, perhaps with uniform class intervals as in Figure 22.2, or in a way which shows significant breaks to best advantage.

4. Finally, the now doubly processed data are mapped by means of isopleths, choropleths or some other means of symbolisation.

In summary, the process is as follows:

Raw data → Some statistical summary of data → 'New' data arranged into classes → Data shown on map

In this process, two elements are at work which are again worth noting because of their general applicability. First, as the data are packaged into a more and more compact form, information is being lost. It is no longer possible, as it would have been if one had the original table of raw data, to look up the mean relative humidity at 9 a.m. for January in Alice Springs; this specific datum has now disappeared into the aggregate data represented first by Figure 22.1 and then by the map in Figure 22.2. This is the price that is paid for increasing generalisation. On the other hand, as the raw data are generalised and simplified the statistics that emerge are more powerful: that is, they subsume more and more information. The process would be completed if we could add one further box to the diagram as follows:

Data shown on map → Single statistic derived from map

Deriving Statistics from Maps

The business of deriving statistics *from* maps rather than showing statistics *on* maps forms another strand in the use of statistics by geographers. A map is a very complicated graphical statement containing a lot of information about the area it shows; in this sense it is analogous to a statistic. Could we take the process a step further and reduce all the information on the map to a *single* datum? If we could, we could not only say something precise about the information shown on the map, but we could also make equally precise comparisons with other mapped data.

There are several ways in which a statistic can be derived from a distribution map. In order to transform the graphical data of the map into numerical data, a grid can be overlain and a matrix (or table) can be constructed showing the frequency with which cells (or squares) containing a given number of points occur. A cautionary note may be struck here illustrating a potential pitfall of using statistics, namely, that they often involve the exercise of subjective judgements. In this case, the frequency matrix will depend upon the size chosen for the cells, and even on their shape, be it square, triangular or hexagonal. It will also depend on the arbitrary rules that may be made to decide, for example, into which cell a point on a boundary is deemed to fall. This having been said, it is clear that by reorganising the data on the map, we are some way towards saying something useful about it.

We could, for example, show them graphically either as a *bar graph* (or *histogram*) or a *cumulative frequency curve* (*ogive*, or *Lorenz curve*). From this it is possible to show the extent to which the distribution is clustered or evenly distributed. If the distribution were very clustered then all the points would fall in very few cells and the ogive would be like that shown in Figure 22.4a. On the other hand, if the points were evenly distributed (and the grid size and placement had been chosen intelligently) then the ogive would be a straight line as in Figure 22.4b. The real shape of the ogive as in Figure 22.4c therefore gives some indication of where the distribution falls between the two extremes and, more importantly, allows comparisons to be made with other distributions, or of the same distribution at a different time.

Alternatively, we could bypass the graph altogether and try to express the distribution as a single statistic. Geographers and others have devoted a great deal of effort over the past few years trying to devise single statistics which would adequately describe distributions of all kinds, whether they be of points, lines or areas. Areas have proved to be the most intractable, lines less so, and points the easiest of all. One of the most useful techniques developed for handling point distributions is known as *nearest neighbour analysis*. This technique provides a statistical description of the 'randomness' or otherwise of any distribution pattern. Although it involves a certain amount of measurement and of calculation, it is manageable if carried out systematically.

Nearest Neighbour Analysis

In order to calculate a nearest neighbour measure, it is first necessary to measure the distance between each point on a map and its nearest neighbour, and secondly to calculate the mean of all the distances so measured ($\bar{d}_o$). Thirdly, the density of points (p) is calculated by dividing the total number (n) by the area (a) shown on the map. Fourthly, using the density (p) it is possible to calculate what the mean distance between points ($\bar{d}_e$) would be if they were distributed at random (that is, purely according to chance). The formula for this is:

$$\bar{d}_e = \frac{1}{2\sqrt{p}}$$

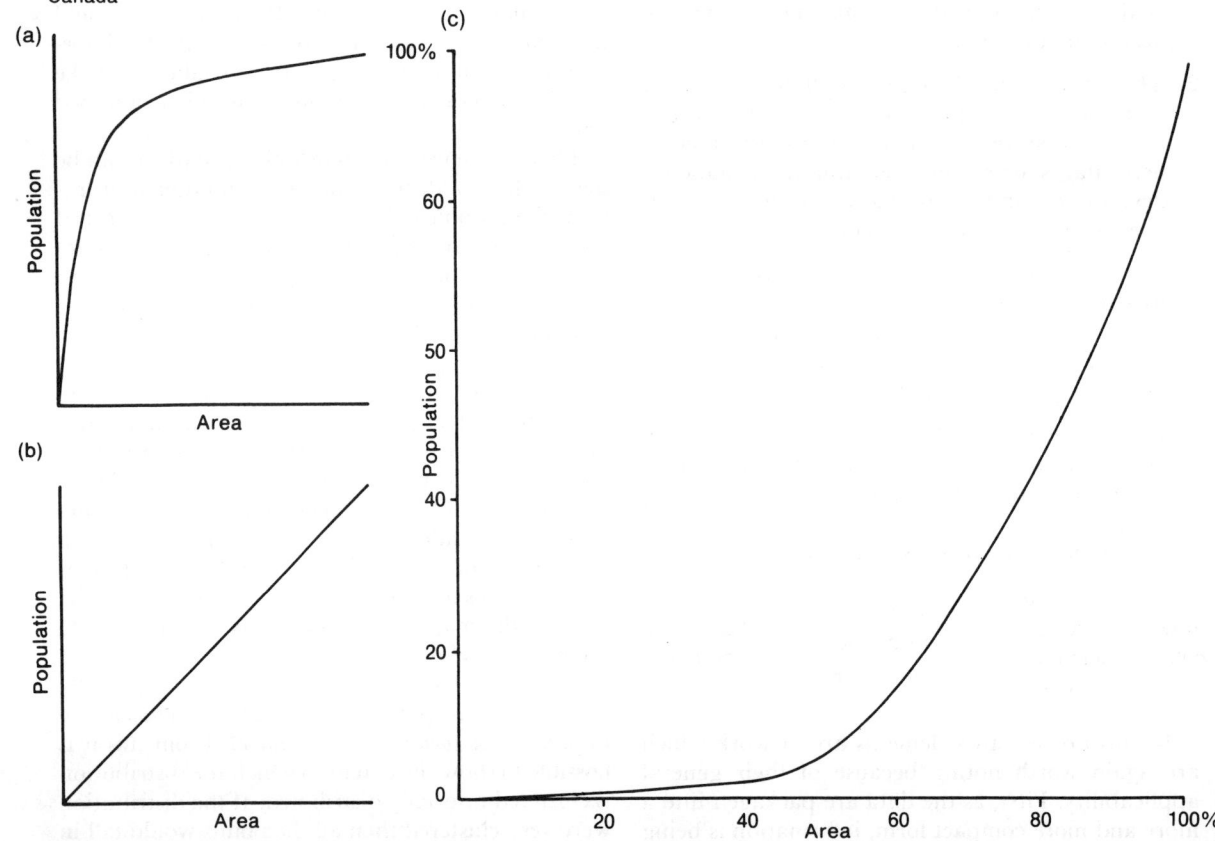

Figure 22.4 Lorenz curves (ogives): population distribution patterns. (a) Clustered distribution; (b) Even distribution; (c) Distribution of rural population in Saskatchewan, Canada

Finally, the nearest neighbour index (R_n) can be calculated by dividing the actual mean distance by the mean distance expected if the points were distributed at random:

$$R_n = \frac{\bar{d}_o}{d_e} = 2\bar{d}_o\sqrt{\frac{n}{a}}$$

Of course, if the points really *are* randomly distributed then this number would be 1. If they were completely clustered it would be 0, and if they were evenly distributed it has been calculated it would be 2.15. The worked example shown here is taken from the British Schools Council Continuing Mathematics Project (Schools Council and Council for Educational Technology 1977) which, interestingly, includes four booklets on the use of mathematics in geography. The area under study is East Anglia in England. Figure 22.5 is a map of the location of all settlements in East Anglia.

In the example, the distance to the nearest neighbour of each town is measured, with the following results:

Town no.	Distance to nearest neighbour
1	17.6 km
2	14.4
3	14.4
4	11.2
5	11.2
6	17.6
7	17.6
8	17.6
9	17.6
10	16.0
11	16.0
12	11.2
13	6.4
14	6.4
15	11.2
16	12.8
17	11.2
18	16.0
19	20.8
20	17.6
21	14.4
22	25.6

The mean distance (d_o) is therefore 14.76 km. The area (a) depicted on the map is 7820 km² and the density of towns is therefore 0.0028. From these data, the nearest neighbour statistic can be derived as follows:

$$R_n = 2\bar{d}_o\sqrt{\frac{n}{a}}$$
$$= 2 \times 14.76 \times \sqrt{0.0028}$$
$$= 1.57$$

This quite high figure indicates an even distribution of towns in East Anglia and, lest it be said that this was obvious anyway, it should be added that the precise nature of the statistic allows equally precise comparison with other areas.

Network Analysis

Another extremely useful class of statistics is that which has been developed to describe networks of lines. One of the great advantages of using statistics is that they blur some of the bothersome dichotomies between physical and human geography. Points are points, and lines are lines: the same expertise can be applied to stream networks or to road networks. Figure 22.6 shows the network of Zambian airlines, about which several questions could be asked. How

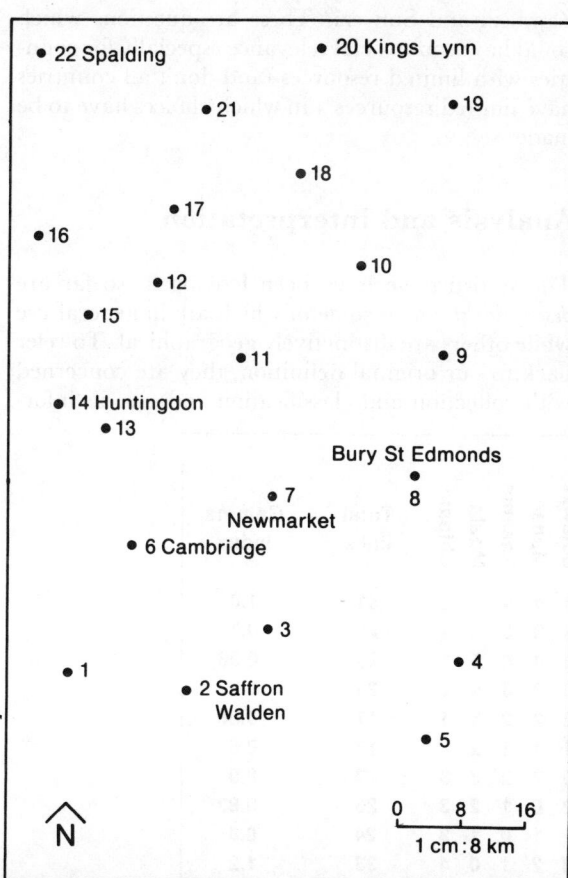

Figure 22.5 Main towns in East Anglia, England

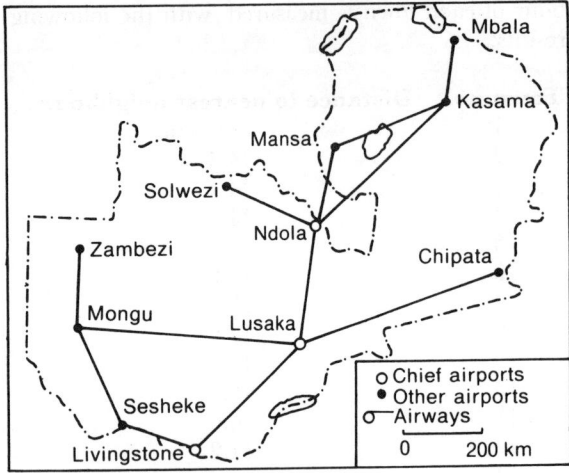

Figure 22.6 Scheduled airline routes in Zambia

efficient a network is it? Could it be improved? What is its best connected airport? All of these may be answered by applying the very useful array of techniques described as *network analysis*. Just as in the previous case, the raw data have to be transformed before they can be used. In this case, two transformations are necessary, a topological transformation by which the map becomes a diagram showing simply the airports as nodes and the routes as links. The second transformation involves deriving from this diagram a simple connectivity matrix showing the number of links between two given nodes (Figure 22.7). It is possible to identify the best connected airport by adding across the rows.

Indices may also be calculated showing how well (or poorly) connected the whole network is. Especially useful is the *gamma index*, defined as:

$$\text{gamma} = \frac{\text{links}}{3(\text{nodes} - 1)}$$

The results of calculating gamma indices for the Zambian airline network are shown in the last column of Figure 22.7. The lower the gamma index, the better connected a node is. Not surprisingly, Lusaka, the capital of Zambia, is found to be the best connected airport, just slightly ahead of Ndola. At the other extreme, Zambezi and Mbala are the most remote. What, it might be asked, would be the result of establishing a new route between, say, Kasama and Chipata? How would it affect the relative standing of each airport? How would it affect the overall efficiency of the airline network as a whole, as measured by another useful index, the beta index, defined as the number of links divided by the number of nodes (in this case $\beta = 12/11 = 1.09$). Would it produce a more efficient network than would a link between, say, Mansa and Chipata, or Zambezi and Solwezi? These are questions which could have some social relevance especially for countries with limited resources (and don't all countries have limited resources?) in which choices have to be made.

Analysis and Interpretation

The statistics we have been looking at so far are *descriptive statistics*, some of which are in general use while others are distinctively geographical. To refer back to our original definition, they are concerned with collection and classification and give us infor-

Figure 22.7 Airline connectivity matrix, Zambia

	Zambezi	Mongu	Sesheke	Livingstone	Lusaka	Ndola	Solewzi	Mansa	Kasama	Mbala	Chipata	Total links	Gamma index
Zambezi	0	1	2	3	2	3	4	4	4	5	3	31	1.0
Mongu	1	0	1	2	1	2	3	3	3	4	2	22	0.7
Sesheke	2	1	0	1	2	3	4	4	4	5	3	29	0.96
Livingstone	3	2	1	0	1	2	3	3	3	4	2	24	0.8
Lusaka	2	1	2	1	0	1	2	2	2	3	1	17	0.56
Ndola	3	2	3	2	1	0	1	1	1	2	2	18	0.6
Solewzi	4	3	4	3	2	1	0	2	2	3	3	27	0.9
Mansa	4	3	4	3	2	1	2	0	1	2	3	25	0.83
Kasama	4	3	4	3	2	1	2	1	0	1	3	24	0.8
Mbala	5	4	5	4	3	2	3	2	1	0	4	33	1.2
Chipata	3	2	3	2	1	2	3	3	3	4	0	26	0.86

mation which is more generalised and simplified than that which was available to us in the form of the raw data on which they are based. This is so whether the raw data are found in a table or on a map. A mean, a standard deviation, a nearest neighbour statistic or a beta index are all neutral statements; they give us information about something but they do not take us directly to the goal of geographical explanation. They do help us move towards that goal, however, in two important ways.

First, they present us with 'puzzles' that can be stated in very precise terms. If we know that the nearest neighbour statistic of East Anglian towns is 1.57 or that the beta index of the Zambian airline network is 1.09, then we have a basis for comparing these with other systems of towns or airline networks and asking why there might be differences between them. Second, statistics also give us a tool for actually *solving* puzzles by enabling us to draw *statistical inferences* from a comparison of two (or more) sets of data.

Geographers generally do not have laboratories where experiments may be carried out under controlled conditions. What they do have is the real world with all its bewildering complexity. They can in the words of our original definition 'impose order and regularity' by using a procedure like that of the scientist in the laboratory as follows:

1. Take any phenomenon which can be expressed quantitatively and which varies from one part of the earth's surface to another. It could be the yield of a particular crop, the density of population, the size of pebbles, or anything else the distribution of which we find puzzling. This is the *dependent variable* (call it Y). The area of the earth's surface over which the variation occurs could be as small as a few square metres, or as large as a continent.

2. Take any other phenomenon which can also be expressed quantitatively and which we have reason to believe (because we have a theory about the two phenomena) has an influence on the first phenomenon. This is the *independent variable* (call it X).

3. Choose a suitable statistical procedure which will enable us to test whether there is a relationship between the dependent and the independent variables.

It should be stressed that whatever statistical procedure is chosen it will *not* tell us that the variation in Y *is caused* by the variation in X. What it will tell us is that the variation in Y is sufficiently like the variation in X for it to be probable, within certain specified limits, that any differences between them are due to chance. Chance in this case means the effects of some other unspecified variable (or variables) or of some random factor. The hypothesis which we have set up concerning the dependent and the independent variables is thus a rather peculiar statistical variety known as a *null hypothesis*. What it says is that there is a high probability that any differences between the variation in A and the variation in B are due to chance alone. Accounts of how to apply the null hypothesis are found in every elementary statistics text and many of those written specifically for geographers (e.g. Okunrotifa 1982).

What kind of statistical procedure we use to find out whether two variables are *correlated* (i.e. the variation in one matches the variation in the other so closely that it may be said that the match is not just due to chance) will depend on what kinds of variables we are dealing with. If a die is tossed, the number that shows face up will be 1, 2, 3, 4, 5 or 6. It can never be, say, 2.74. Similarly, if the die were tossed 100 times, it would be meaningless to calculate the mean score. Means, standard deviations or other summary measures of a distribution are known as parameters, and it is clear that tossing dice yields a distribution which does not have parameters. The kind of statistical techniques that would be used to analyse such a distribution would be known as *non-parametric* statistics. Many of the variables that occur in geography are of this kind and non-parametric techniques have great value in geographical analysis.

Chi-square

The chi-square (χ^2) statistic is a useful test for determining whether an observed distribution (O) is sufficiently like an expected distribution (E) for the differences between them to be probably due to chance. It is calculated by summing the square of the difference between the observed and the expected values, divided by the expected value, in each of a set of predetermined categories. Its equation is written as:

$$\chi^2 = \sum \frac{(O-E)^2}{E}$$

If a die were to be tossed 1000 times we should expect the number of throws in each of the six categories (i.e. 1, 2, 3, 4, 5 or 6) to be 1000 divided by the number of categories ($=167$). If there were in fact 300 sixes, there would be grounds for suspicion that the die was loaded. The chi-square statistic would have provided the means of determining whether that suspicion was justified!

The limitations on the use of the chi-square test, as with any non-parametric procedure, are that each value must be a discrete category (as with the die, 1, 2, 3, 4, 5 or 6 and nothing in between). Map data may be analysed using the chi-square test by overlaying the map with a uniform grid and assigning each of the occurrences of the phenomenon we are interested in to a cell in the grid. Once more, the cautionary note has to be struck that the results achieved will depend on the size of the grid and the shape of the cells (e.g. square or hexagonal), as well as on the placement of the grid on the map. The number of occurrences of the phenomenon in each cell is calculated and shown in tabular form, either as the number of cells with a given number of occurrences, or as the number of occurrences for each cell. This is the observed frequency. The chi-square test may then be used to compare it with the frequency of some hypothetical distribution (whether clustered, random or even). This is the expected frequency.

Another variation of the technique involves identifying a number of randomly selected points on the map, and using the chi-square test to determine whether the frequency of occurrence of an areal phenomenon, or the association of two areal phenomena, are as expected. For example, the test could be used to determine whether there is a significant difference in the land use of different geological formations. In this case, the null hypothesis, which the use of chi-square would disprove, would state that the proportions of each type of land use would be the same on each geological outcrop (Guinness 1979).

Correlation and Regression

In many geographical problems the data are susceptible to analysis using *parametric techniques*, those which are based on the data having parameters such as means and standard deviations. Phenomena which change gradually from one part of the earth's surface to another call for the use of such techniques, one of the best-known of which is the Pearson product-moment correlation coefficient. The purpose of this coefficient is to state (in values ranging from -1 to $+1$) the degree to which two such phenomena are correlated (i.e. the degree to which the variance in one is the same as the variance in the other).

The correlation between two variables is best illustrated by having them plotted as a scattergram in which each point represents observed values of each of the two variables (the point can be called x, y). If all the x, y points lie along a sloping straight line then it is clear that the two variables are perfectly correlated, positively if the line slopes upward from the horizontal axis, negatively if it slopes downward. The perfect correlation implies that any increase in x results in a corresponding increase in y (or a decrease in the case of a negative correlation). In geography, perfect correlations of this kind are rare and it is more common for the x, y points to lie along a band. The breadth of the band suggests the degree to which the two variables are correlated and it is the spread of the x, y points that the Pearson product-moment coefficient expresses in numerical terms.

A practical example will show how this technique can lead to a deeper understanding of geographical patterns. Figure 22.8 below shows in a matrix the correlations between a number of agricultural variables in each of 23 Canadian ecozones.

These correlation coefficients may be calculated (somewhat laboriously) using a programmable hand-held calculator, or (more quickly) with one of the many statistical packages now available for the micro-computer (I used STATS PLUS produced by Human Systems Dynamics). Of course, many more variables could be included, but I have shown only those which look promising in that they are highly correlated.

A correlation matrix like the one shown above can be used as the starting point for a sequence of procedures which help us to understand the structure of the phenomenon we are interested in, in this case, the regional structure of Canadian agriculture. We start by hypothesising that farm sales 'are a

Figure 22.8 Matrix of correlations between agricultural variables in Canadian ecozones (*Statistics Canada* 1986)

	1.	2.	3.	4.	5.
1. Farm sales	1.000	0.965	0.976	0.968	0.591
2. Capital farm value		1.000	0.942	0.943	0.434
3. Labour inputs			1.000	0.942	0.679
4. Machinery inputs				1.000	0.584
6. Fertiliser inputs					1.000

function of' labour inputs (i.e. that the variation in labour inputs is sufficiently like the variation in farm sales for it to be most unlikely that differences between them are not due to chance). If we plot farm sales per hectare (the dependent variable) against labour force per hectare (the independent variable), we find that the points fall fairly closely about a line, as we should expect with a correlation coefficient of 0.976 (Figure 22.9). The line is known as the regression line, or the line of best fit, and is constructed using the technique known as regression analysis, the purpose of which is to discover the mathematical relationship between the two variables. For two variables that are highly correlated this technique can be used to predict the value of one variable by knowing the value of the other.

Since the correlation between farm sales and labour inputs is not perfect, it raises the question as to whether there might be some other variable which would 'explain' the variation in farm sales not already 'explained' by labour inputs. (The word 'explain' is used in a specific statistical sense rather than the everyday sense—more of this later.) We could further hypothesise that machinery inputs play this role, and use the technique of multiple regression analysis to test whether this is so. It would be inappropriate to go into details concerning this technique, but what it seeks to do in essence is to take account of one independent variable after another successively until all of the variation in the dependent variable has been 'explained'. In the case in question, taking account of machinery inputs as well as labour inputs explains almost all of the variation in farm sales (with a correlation coefficient of 0.986).

Since this analysis also gives the mathematical relationship between the dependent variable and the two independent variables, it is possible to predict what farm sales should be in each ecozone

Figure 22.9 A regression line or line of best fit showing the relationship between farm sales per hectare and labour force per hectare in Canada

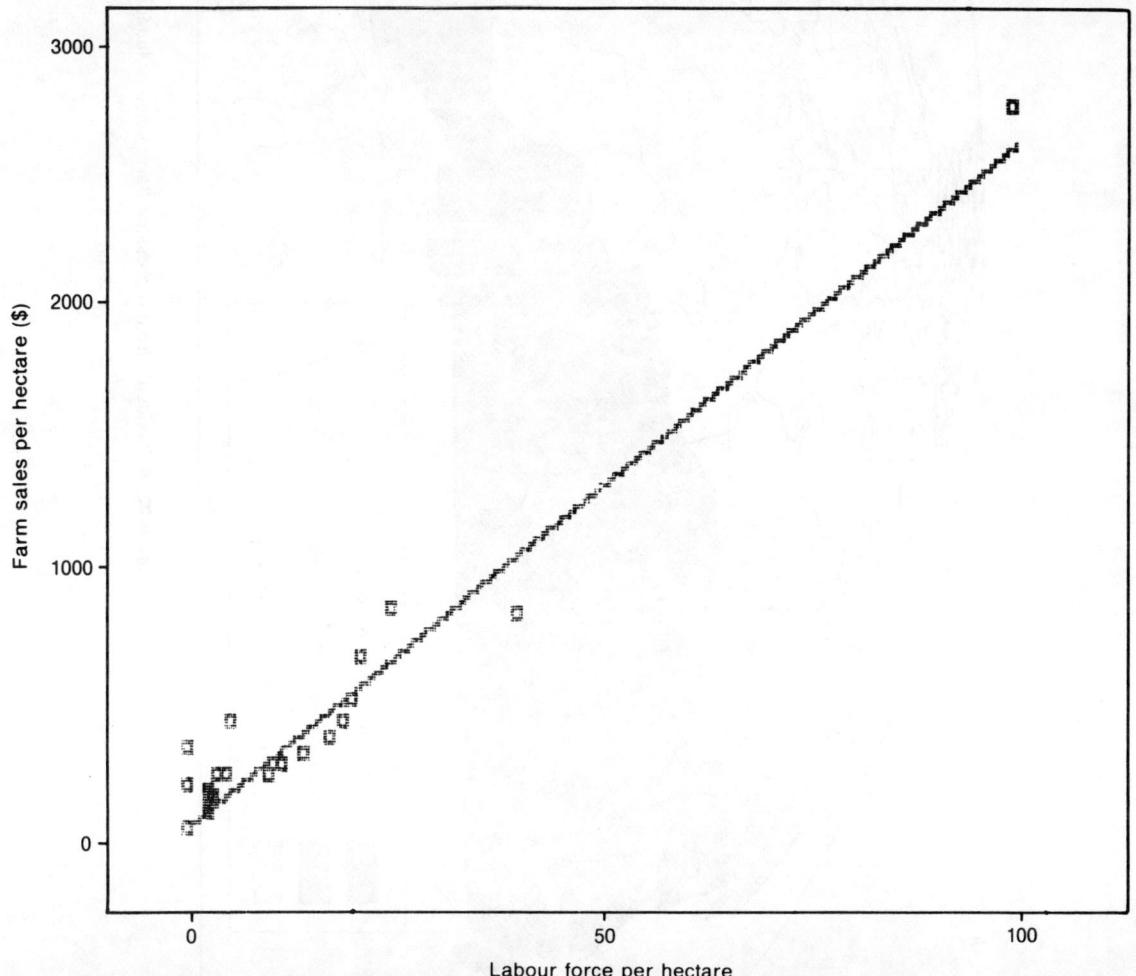

298 The Geography Teacher's Guide to the Classroom

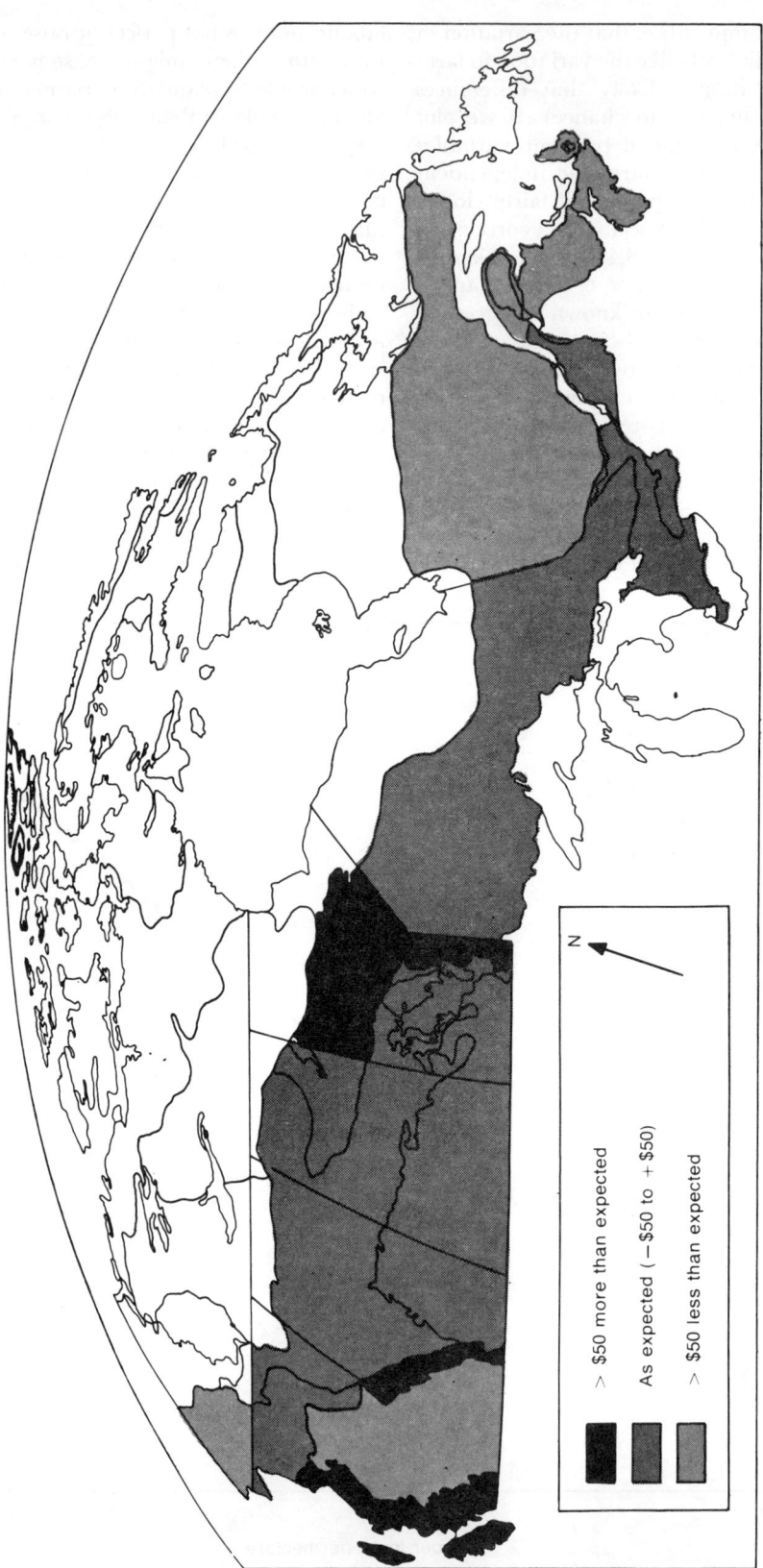

Figure 22.10 Residuals from regression (farm sales or labour and machinery inputs)

based on the labour and machinery inputs in that ecozone. It can be especially illuminating to identify those cases in which the dependent variable is *not* as it would be predicted from the independent variables, because this often reveals an important geographical structure. These departures from the predicted value are known as the residuals from regression, and are plotted for the example in question in Figure 22.10. It will be seen that, while in most of Canada from the Rockies to the St Lawrence valley farm sales can be predicted accurately from labour and machinery inputs, this is not so in the western and eastern ends of the country. Coastal British Columbia has farm sales that are $92 per hectare more than would be expected, while interior British Columbia, northern Quebec and the maritime provinces have farm sales that are between $177 and $58 less than would be expected.

Why Use Statistics

At several points in the above discussion, there has been a caution against confusing statistical 'explanation' with causal 'explanation'. To demonstrate that a dependent variable can be 'explained' statistically by one or more independent variables takes us only halfway towards understanding why things are as they are. To know that farm sales in Canada are apparently a function of labour machinery inputs answers one question but raises a more important one; namely, *why* is this so? What is it about farm operation in Canada that makes labour and machinery affect farm sales so directly? Would the same relationship hold in Australia and, if so, would it be for the same reasons? Would it hold in Bangladesh?

Beyond these sorts of questions, the use of statistics also helps us to focus on those places that are different, and where the relationships we have established do *not* apply. Are farm sales higher than would be expected in coastal British Columbia because of more fertile soils, or less danger of loss due to frost? Are they higher in eastern Canada because of distance from markets, or small land holdings. Statistical techniques may not help us to answer these questions (then again, they may), but they do draw attention to questions which we may not even have been aware of without them.

The use of statistical procedures has left its indelible stamp on the way geographers approach problems, not only in providing some tools with which they may do familiar and traditional tasks with greater precision, but also, through the scientific method, in conditioning the way in which geographical inquiry is ordered. Beyond the mere use of statistics or even the application of the scientific method lies a fundamentally different way of looking at the world, one which sees it in essentially geometrical terms. Geographical phenomena are seen as forming the recurring and regular patterns represented by central place networks, concentric zones of urban and agricultural land use, or hierarchies of streams. This way of looking at the world undoubtedly has its pitfalls: it can on occasion cloud the view and make it difficult to notice some interesting irregularities. On the whole, however, it has enabled geographers to make some tentative moves from mere description to explanation and prediction and to have gradually assembled something approaching a *theory* of geography.

At the school level a number of arguments can be advanced for using statistics in geography teaching. These include:

1. Geography has assumed the statistical mantle largely by default: few other subjects in the school curriculum take a statistical (as opposed to a mathematical) stance.

2. As a result of the quantitative revolution, geography now has quite a large repertoire of ready-to-use statistical techniques. If nothing else these encourage a disciplined and rigorous approach towards problems which have a numerical aspect.

3. Although the use of the scientific method does not depend on the use of numbers, it certainly can be employed with greater precision and confidence than when purely qualitative data are used. Statistical procedures like the examples given in this chapter provide a vehicle for employing the kind of thinking and problem-solving procedure we characterise as 'scientific'.

4. Numbers are seen as being of great importance today and numeracy as an indispensible ability. To be able to create and use statistics may not be important in itself: students should certainly be able to read and evaluate statistical statements critically.

5. The statistical approach encourages students to think about geographical phenomena in a general rather than a particular way and in so doing to begin to grasp the more theoretical aspects of the field.

References

Clark, W.A.V. and Hosking, P.L. (1986) *Statistical Methods for Geographers*, New York: John Wiley and Sons.

Guinness, P. (1979) 'Chi-Square in Purbeck: Testing Relationships Between Geology and Land Use', *Teaching Geography*, Vol. 5(2), 57–59.

Okunrotifa, P.O. (1982) 'Processing Information', in N.J. Graves (ed.) *New UNESCO Sourcebook for Geography Teaching*, London: Longman.

Schools Council and Council for Educational Technology (1977) *Nearest Neighbour Analysis, Mathematics in Geography I*, York: Longman.

Statistics Canada (1986) *Human Activity and the Environment: A Statistical Compendium*, Ottawa: Supply and Services.

23
The Diagnosis of Student Learning in Geography

Rod Gerber

This chapter introduces a sequence of three chapters that focus specifically on the learner in the geography classroom. The goal of catering for individual differences assumes the diagnosis of student learning in geography. This is equally necessary in planning work for mixed-ability classes as it is for groups of gifted and less able students. Five issues related to diagnosis are explored in this chapter. These are: the purposes of diagnosis (prediction of achievement and goal setting, determination of student 'entry behaviour' for a unit of work, and the development of individual study programs), the range of diagnostic instruments available or that teachers can produce themselves, the validation of instruments, means for administering diagnostic tests, and using the results of diagnosis in one's teaching. Examples, mostly related to mapwork and local area studies, are used to illustrate the discussion of these issues.

Imagine that you are a newly graduated teacher about to take your first lessons in geography. It is to a class of lower secondary school students. You have prepared your lessons according to the principles you learned in training and in teaching practice. You think you have a good idea of the students' ability to learn in geography. You have made extensive use of attractive teaching resources and have organised them into a useful learning sequence. Your introductory teaching unit is planned for eight or nine hours of classwork. The class is responsive and the lessons proceed in an orderly fashion until the end of the teaching unit when a considerable number of students claim that the content of the unit was boring and irrelevant. In addition, you find that many of them could not perform several of the skill exercises or understand some of the elementary concepts in the unit. You wanted to make a positive impact on your class, but you feel let down because you did not understand your class's abilities in geography. What can you do about the situation?

Alternatively, imagine that you are a very experienced teacher of geography who has taught in a wide range of school environments. You feel that you know the lower secondary geography student very well. You are aware of changes in teaching geography and have tried a range of teaching methods. You have always had classes with a wide range of abilities in them. Consequently, even though the school prescribes a textbook which you think is suitable for your class, a good number of the class members do not seem able to understand very much in your lessons. Your experience has led you to believe that these lower-achieving students have little aptitude for geography. Therefore, your expectations of them are low and you concentrate your classroom activities on the higher achievers. The lower-achieving ones express little interest in learning geography. What can you do to kindle their interest in geography and to make their classroom experiences in geography meaningful?

These two scenarios, although simplifications of

real world situations, raise a number of questions about the learner in geographical education. Hopefully, the questions raised will be ones concerned with the teachers' need to know their students better and focus on ways of improving their learning in geography. Learning in geography will achieve purpose in students' eyes only if it is interesting, relevant to their living and presented at a level they can understand.

This chapter aims to challenge teachers of geography to know not only the content of geography and the appropriate methods for teaching it, but also the capabilities of each member of a particular class in geography. If teachers know their students' capabilities in geography, then they can plan appropriate learning experiences. This challenge is not answered by only going to the students' record cards to ascertain the results of a battery of intelligence tests, however.

What can teachers do to diagnose their students' capabilities in geography? A start could include seeking answers to the following five questions:

1. What is the purpose of the proposed diagnosis?
2. What diagnostic instruments may be used?
3. How good is the instrument?
4. How can I administer the instrument?
5. What use can I make of the results?

The Purposes of Diagnosis

The purpose of diagnosing students' capabilities in geography is to improve their learning. Both of the teachers described at the beginning of the chapter could have benefited from knowing their class's capabilities in geography a little better.

Teachers of geography should have a clear understanding of the purposes of diagnostic activities. Identification of a definite purpose for diagnosis enables teachers to devise or use an instrument which addresses a particular problem rather than the examination of a general aspect of education. For example, geography teachers, generally, will find an investigation of children's field-sketching skills more useful than a review of their drawing ability. As well, teachers will prefer to determine their students' attitudes towards mining rather than the students' attitudes towards economic systems. The conclusions to general questions in education are important for teachers, but not as readily important to classroom applications as the conclusions which are directly related to the subject being taught.

What Diagnostic Instruments May Be Used?

What range of diagnostic activities can a geography teacher engage in to improve the learning that takes place in his or her class? This range may consist of three main types of activities:

1. Prediction of students' achievement or understanding.
2. Understanding of students' capabilities on entering some aspect of learning in geography.
3. Diagnosis of students' abilities in geography.

What follows is an attempt to illustrate the three types of measures for students' capabilities in geography. The examples are representative of the standardised instruments and teacher-devised exercises available to geography teachers.

Prediction

The *prediction* of students' achievement or understanding allows teachers to establish sensible goals for students in the learning experiences they devise. Instruments for predicting students' achievement or understanding may be general or specific, depending for the most part on the purpose of the prediction.

A number of standardised instruments are available for this purpose. For example, *Test W-1: Map Reading* from the *Iowa Tests of Basic Skills* (Hieronymous and Lindquist 1971) is typical of standardised tests for predicting students' map-reading ability. This test provides a range of questions scaled at six levels of difficulty to predict students' map-reading ability via their ability to orient a map, determine direction, locate and describe places, determine distance and routes of travel, understand seasonal variations, sun patterns and time differences, visualise landscape, compare features and make inferences about man's activities or ways of living. The results from such a test can be used to predict the level of mapping work to use with a particular class.

At a more general level, the *Space Test* from the ACER Mathematics Profile Series (ACER 1978) provides a standardised means of measuring students' spatial ability. This test provides information on the students' spatial understanding as

Figure 23.1 Simson's (1977) semantic differential items for a survey of scenic quality of a landscape

Majesty
Mountainous__:__:__:__:__Flat
Spectacular__:__:__:__:__Unspectacular
Great__:__:__:__:__Small
Impressive__:__:__:__:__Unimpressive
Distant views__:__:__:__:__Limited views

Serendipity
Complex__:__:__:__:__Simple
Varied__:__:__:__:__Monotonous
Mysterious__:__:__:__:__Obvious
Rugged__:__:__:__:__Smooth
Diverse vegetative cover__:__:__:__:__Uniform vegetative cover

Artistic quality
Bright__:__:__:__:__Dull
Vivid__:__:__:__:__Drab
Orderly__:__:__:__:__Chaotic
Colourful__:__:__:__:__Colourless
Attractive forms__:__:__:__:__Unattractive forms

Atmosphere
Pleasant__:__:__:__:__Unpleasant
Tranquil__:__:__:__:__Disturbed
Natural__:__:__:__:__Artificial
Idyllic__:__:__:__:__Degraded
Clean__:__:__:__:__Polluted

indicated by their ability to know concepts and operations, to classify and interpret spatial information, to reason and to solve problems. Such a test provides teachers with information about the general spatial ability of students. Then, teachers can plan units of work in geography which reflect the students' general spatial ability in the type and sophistication of the planned activities.

Teachers themselves can also devise useful instruments for predicting students' capabilities in geography. Here are two examples which illustrate the type of instrument the teacher might devise and possible implications of their results.

Example 1 concerns predicting the students' understanding of scenic quality in their assessment of landscape. The instrument, devised by Simson (1977), consists of a survey constructed as a semantic differential. Simson asserts that people's understanding of scenic quality may be measured by their understanding of these components of a landscape:

1. *majesty*—the relative relief of the land system and the visibility or distance of view;

2. *serendipity*—the ruggedness and vegetation 'richness';

3. *artistic quality*—the colour properties of the scenery, its shapes and its forms; and

4. *atmosphere*—the pleasantness and the lack of spoliation of the landscape.

The variables for each of the four components are presented in Figure 23.1. Each variable is scored on a scale range of 1–5. The test instrument is devised by mixing the variables in terms of the four components of scenic quality and by reversing some of the pairs. The students' understanding of the scenic quality of a landscape is measured by their response to views of that landscape. It is scored as a total of the students' responses on the semantic differential. A range of landscapes may be examined and scored in this way.

Teachers can use the results of such a test in a number of predictive ways. Initially, comparisons may be made of each student's total scores on differing landscapes to predict whether there is consistency in the students' understanding of the concept of scenic quality. Alternatively, teachers may investigate the general understanding of the whole class by calculating the mean score for the class for particular landscapes. The results of such an investigation should assist geography teachers to adjust their expectations of their students' views of selected landscapes, and guide them in the selection of particular landscapes for study at different levels of geography teaching.

Example 2 focuses on a different kind of prediction—the prediction of learner attributes based on their previous experiences. This form of diagnosis allows the teacher to appreciate any special attributes possessed by a class which could enhance learning in geography. The example used here is to ascertain students' previous experiences which relate to mapping.

The instrument in this case consists of a survey of related clubs or groups, experiences and classroom activities, which could enhance a student's mapping ability (Figure 23.2). Items can be varied according to one's country or culture. The results from such a survey can be used to determine whether students bring potentially enhancing experiences to their geography lessons or whether the teacher has to assume that students have had limited experiences with maps. If class members have had a variety of enriching experiences relating to maps, a teacher may predict that their ability to read and use maps will be improved by the use of more challenging maps. If the reverse is so, a teacher can select maps

Figure 23.2 Survey of student's previous experiences related to maps (after Gerber 1980)

for use in the classroom which are less sophisticated or use more concrete signs.

Diagnosis at the Entry Level

Teachers of geography will often find it useful to know more of their class's understanding of some aspect of geography before a new topic or skill is introduced. Discovering such information provides diagnosis at the *entry level*. Its main purposes are to find out whether students are entering a learning situation at an appropriate level of understanding and whether students are ready for learning what actually takes place in a particular unit of work in geography. Diagnosis which focuses on the extent of the students' understanding at the entry stage is concerned with their having a sufficient understanding of a concept or some content skill so that learning can build on this knowledge. This concern for the student's development of some understanding uses diagnosis at entry to the unit in an *a priori* sense and thus enables the teacher to retest the class on the completion of the unit and to compare the pre-unit and post-unit scores.

Two examples of diagnosis at the entry level are provided here. The first is concerned with the use of a semantic differential to gauge students' understanding of a concept. The second example is a survey devised by Fien et al. (1982) to establish a class's perception of its local environment before studies of this environment are done.

Example 1 illustrates the worth of using the semantic differential for diagnosing students' entry level. The teacher may want to implement some urban studies. The children's perception of aspects of urban areas may differ from the norm. One way for the teacher to find out if this is so is to provide the class with a simple semantic differential for urban areas and to analyse the students' responses. One such semantic differential is illustrated in Figure 23.3. Here, the pairs of opposite words are designed to ascertain whether students see an urban area as an area of high population density and an area of land-use zones which has some order to its plan. The teacher considers the frequency of the responses to derive a class view of urban areas. If this view is distorted by the school's location in the city, then the teacher can build the urban studies so as to compensate for deficiencies in the class view.

Example 2 highlights the importance of knowing how younger secondary students see their local

---

Tick one box for each question to answer these statements.

1. I am or have been a member of—

	Yes	No
(a) Boy Scouts or Cubs (Girl Guides or Brownies)	☐	☐
(b) Boys Brigade (Girls Brigade) or other church groups	☐	☐
(c) Bushwalking Club	☐	☐
(d) Orienteering Club	☐	☐
(e) YMCA Group	☐	☐
(f) What other groups do you belong to which move around in an area, e.g. trail bike clubs		

_____
_____

2. I have done these things in my life—

	Yes	No
(a) flown in an aeroplane	☐	☐
(b) used a road map	☐	☐
(c) found my own way across a suburb or across town	☐	☐
(d) gone on a hike across country	☐	☐
(e) camped in the outdoors	☐	☐
(f) used a map to find my way	☐	☐

_____
_____

3. In my school lessons I have—

	Yes	No
(a) never been told what a map is	☐	☐
(b) never drawn a map	☐	☐
(c) read off information from a map	☐	☐
(d) used maps to answer questions	☐	☐
(e) traced maps from books	☐	☐
(f) drawn maps freehand	☐	☐
(g) used my own scale to draw a map	☐	☐
(h) created my own set of signs to show things on a map (e.g. a square for a house)	☐	☐
(i) made a map of an area (e.g. my classroom)	☐	☐

---

environment as a place in which to live before these students undertake local environmental studies. Here, Fien et al. (1982) have devised a survey (see Figure 23.4) which analyses students' knowledge of their local environment, their spatial range and experiences in this environment and the values which the students have in relation to it. Students are required to respond using written and oral answers and by drawing a map.

The results of such a local environmental survey may be used by geography teachers to plan the forthcoming studies in geography by highlighting aspects and places which the students know well. Studies of familiar places may then be extended into studies of less familiar places. Similarly, geography teachers may be able to ascertain the objects and experiences students prize and value in their local environment. These valued aspects and places may be incorporated in future studies of their local environment to provide situations which the students perceive as relevant to themselves.

Diagnosis of Student Abilities

When geography teachers attempt to *diagnose* students' abilities in the strict sense of the word, they are usually wanting to pinpoint deficiencies in student learning and to remediate these deficiencies. Geography teachers who consider each of their students as an individual are more likely to be the ones concerned with diagnosis and remediation of students' deficiencies in geography.

There has not been the progress in geographical education in the techniques of diagnosis followed by remediation that there has been in areas such as mathematics. Standardised tests such as the *Iowa Tests of Basic Skills* (Hieronymous and Lindquist 1971), the *Wisconsin Design Study Skills* (Wisconsin Research and Development Centre for Cognitive Learning 1973) and the geography tests devised by the National Council for Geographic Education (1979) are available to test students' abilities in geography. These tests do not offer the teacher ideas for overcoming students' deficiencies. This is an area for future consideration by geographical educators.

Figure 23.3 A semantic differential on urban areas

orderly _____	disorderly
uncrowded _____	crowded
natural _____	man-made
dull _____	exciting
monotonous _____	varied
messy _____	neat
beautiful _____	ugly
personal _____	impersonal
clean _____	dirty
poor _____	wealthy
static _____	dynamic
unfamiliar _____	familiar
old _____	young
sluggish _____	pulsating
temporary _____	permanent
friendly _____	threatening
livable _____	unlivable

Figure 23.4 Child–local environment survey (Fien et al. 1982)

1. How long have you lived in this home and school area?
2. What name do you give to your home area?
3. What name do you give to the area where your school is?
4. Where else have you lived and for how long?
5. Draw a map of your home and school area.
6. Make a list of all the places you know of in your home and school area. How often do you go there? How much time do you spend there each time? What do you do in these places?

For Questions 7 to 13 think about the places you come across as you go about your usual day's activities.

7. What places do you like the most? Why?
8. What places do you like the least? Why?
9. What places give you the most difficulty? Why?
10. What places are the most dangerous? Why?
11. What places are you not allowed to go to? Why?
12. What are the best places to go to with your friends? Why?
13. Are there any places you go to be alone? Why?
14. Are you responsible for keeping any places in your home and school area clean and tidy? Why?
15. Who owns the following things and places:

streets	shops
footpaths	schools
yards	buses
parks	rubbish bins
rubbish on the ground	telephone boxes
creeks	police stations

16. Make a list of all the places you go to outside your area.
    How often do you go there?
    How much time do you spend there each time?
    Who do you go with?
    How do you travel there?
    What do you do in these places?

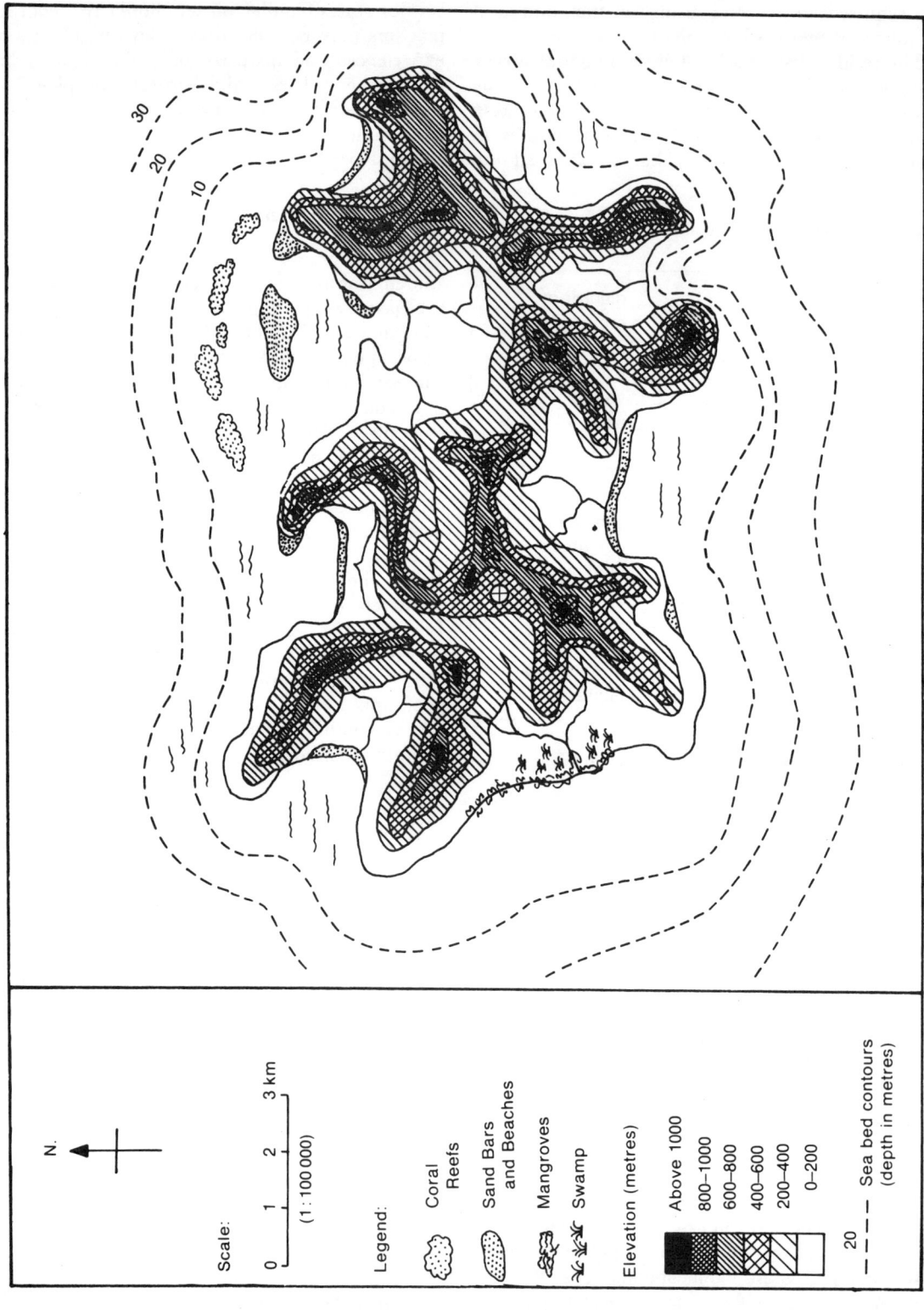

Figure 23.5 Map for diagnosing students' ability to think in geography

Two examples for diagnosing students' capabilities in geography will be mentioned here. Suggestions for the remediation of students' deficiencies are included.

Example 1 is an attempt to diagnose the students' reasoning ability in geography. It was devised by three teachers (Costello et al. 1979) and uses the technique devised by Rhys (1972) to analyse students' ability to think logically in geography. Where Rhys used real examples for his diagnosis, these three teachers used a hypothetical tropical island as their data source. This reasoning exercise consists of planning certain mining operations on the island. It is detailed as follows:

The students' responses are analysed in a four-stage classification similar to that developed by Rhys (1972). Consequently, the teacher can determine the students' ability to reason in geography. Once this has been determined the teacher can adjust expectations and questions to accommodate the reasoning level of the class. For example, if the teacher finds that students in his or her class can only use one piece of information as the basis of their reasoning, then the learning activities presented to the class should not expect the students to do more than this. Conversely, if the class displays an ability to reason logically, the teacher can pose learning activities which require greater interpretation and analysis.

Exercise

Figure 23.5 depicts an uninhabited island situated in the tropics. Most of the island is covered with rainforest and the prevailing winds are from the southeast.

A rich deposit of copper has been discovered, the location of which is marked on the map by the symbol ⊕.

A mining company wishes to exploit the deposit. To carry out the mining operation the company must locate the following on the island:

1. Port facilities.
2. Transport facilities (road, rail, airstrip).
3. Housing and services for the 500 miners and their families.
4. A beach resort area for recreation.
5. An area to be preserved as a national park.

Imagine you are a planner contracted by the company to plan these facilities. In locating them, both practical and economic aspects should be considered.

Using appropriate symbols, mark on the map the position of each of the facilities required for the mining operation

Explain the reasons for your plan of development.

Classification scheme for response evaluation

1. Responses to the problem, both in mapwork and commentary, are suggestive of 'guesswork'. The information presented has not been logically examined; there may even be contradictory interpretations. Trivial reasons are given for the locations chosen for various functions.

The written commentary indicates a negligible appreciation of the significant components of the problem. A basic failure to recognise the geographical significance of the data.

2. The student's map and commentary indicate that some logical analysis of the data presented is now apparent, but that there is a tendency for the student to focus on one particular aspect or component of the problem and to give this undue emphasis in the explanation. Consequently other components of the problem are given more superficial or inconsistent treatment.

Interpretation of the data is likely to be incomplete; students will tend to latch onto those aspects of the problem which are most immediately apparent to them.

3. At this stage, the student considers all the variables involved in the problem and approaches each one logically.

However, the structure of the student's response—particularly the commentary—suggests that the student has approached the task as a series of particular problems rather than as a unitary problem with a number of components. The student's answer suggests that he has not really integrated the various decisions to be made before developing a solution.

4. A logical approach to each of the decisions to be made, as in 3, except that in this case the answers demonstrate a *systematic* approach to the problem. The student's written explanation indicates an attempt to give a balanced appraisal to each of the variables to be considered.

A highly structured and *integrated* response which has resulted from a logical approach to the problem as a unitary whole.

Arrangement

Level 1 No logical correspondence to model. .... Score 1

Level 2 Objects arranged by logical properties only without reference to an overall spatial plan, often with inaccurate numerical correspondence, e.g. objects in lines, bunched, spread over page. .... Score 2

Level 3 Objects in recognisable, but poor spatial correspondence to model with accurate numerical correspondence. .... Score 3

Level 4 Objects on map in rough correspondence to model ±30%. .... Score 4

Level 5 Objects on map in reasonable correspondence to model ±20%. .... Score 5

Level 6 Objects on map in accurate correspondence to model ±10% in conjuction with good proportion. .... Score 6

Proportion

Level 1 No logical or 'realistic' propotion. .... Score 1

Level 2 Recognisable and 'realistic' proportion in relative size of objects on map which is, however, not related to model. .... Score 2

Level 3 Relative size of all objects on map roughly related to relative size of model ±30%. .... Score 3

Level 4 Relative size of objects on map reasonably related to relative size of model ±20%. Inter-object distance distorted. .... Score 4

Level 5 Relative size of objects on map accurately related to relative size of model ±10%. Inter-object distances approaching correct proportion. .... Score 5

Level 6 All sizes of objects and distances on map accurately related to model ±10%. .... Score 6

Plan View

Level 1 No recognisable single view. .... Score 1

Level 2 Primarily a front-on view. .... Score 2

Level 3 Mix of front-on and/or oblique and/or plan view or a consistent oblique view from several viewpoints. .... Score 3

Level 4 Consistent oblique view from one viewpoint or a consistent plan view with minor intrusions of front-on and/or oblique view. .... Score 4

Level 5 Consistent plan view with reasonable proportion. .... Score 5

Level 6 Consistent plan view with good proportion. .... Score 6

Map Language

Level 1 No logical correspondence between map representation and model. .... Score 1

Level 2 Objects are represented in detail which is idiosyncratic and does not correspond to that of the model. .... Score 2

Level 3 Objects are represented in 'photograph-like' detail most of which corresponds to that of model. .... Score 3

Level 4 Evidence of limited abstraction, e.g. some detail is omitted in a consistent fashion, perhaps through use of a realistic plan view. .... Score 4

Level 5 Obvious evidence of consistent abstraction and some efforts at symbolisation, e.g. much detail is omitted in a consistent fashion, perhaps through use of icons and/or labels. .... Score 5

Level 6 Objects are represented exclusively with abstract symbols in conjunction with a key. .... Score 6

Figure 23.6 Scoring scheme for free recall sketch maps (after Wilson 1980)

Example 2 is the technique of diagnosing students' ability to understand a map by their ability to draw a map of an area they know well, e.g. their school. The technique is to take a class for a walk around a familiar area, and then ask the students to draw a recall map of it. This map is scored according to the students' ability to handle the four elements of a map: the arrangement of objects in the area, representing them in accurate proportion, and from a plan view, and encoding the information using clear signs. Wilson (1980) has adapted a scale developed by Synder, Feldman and La Rossa (1975) to produce a quick and easy to use method of scoring these four elements (see Figure 23.6). Mapping ability is calculated as the mean of the scores on the four elements. The value of this exercise is that teachers can diagnose each student's difficulties in drawing a map, e.g. poor ability in arranging objects, inability to use abstract symbols or inability to represent objects from a plan view. This evidence can guide the teacher in the provision of remedial activities associated with one or more of the components of a map. For example, if the students cannot arrange objects successfully, the teacher can devise simple activities such as arranging a small group of objects and then drawing this arrangement, without stressing the signs used or the proportionality of the objects. These exercises will become more sophisticated as the students understand and can do the simpler ones.

How Good is the Instrument?

Whatever instruments geography teachers use in the diagnostic exercises with their students, teachers should ensure that these instruments:

1. measure what they are supposed to measure; and

2. are consistent in their measurement.

Geography teachers may gauge whether their diagnostic instruments are valid in several ways, These include:
  1. *Checks for the validity of the content*. Geography teachers should be concerned that they are evaluating geographical objectives. Does the test on student competence in field sketching really measure the students' ability to compile a field sketch or is it a test of the students' artistic talents or drawing ability? Similarly, geography teachers should ensure that any tests which aim to diagnose the cognitive abilities of their students, e.g. the students' ability to synthesise information, do so and are not tests of the recall of geographical data.

  2. *The correlation of the diagnosed skill, concept or attitude with some other criterion*. For example, the results of students' abilities to draw a map from a written description may be correlated with their scores on a test involving the construction of a map from a series of statistics. If there was a high correlation between the two sets of scores, teachers could be satisfied with the diagnostic instrument.

The consistency of a diagnostic instrument in geography can be determined in several ways, as well. Geography teachers may:
  1. Administer the instrument as a pre-test and as a post-test to a class of students. Teachers may then correlate the two sets of scores to obtain a guide to the consistency of the instrument.
  2. Devise an equivalent diagnostic instrument and administer it some time after the original test. For example, the diagnosis of students' understanding of isolines may initially consist of a test based on a contour map. An equivalent test using the isobars on a weather map may be used.
  3. Divide a single test into two equivalent parts and correlate the students' scores on both halves of the test. Therefore, a test on the students' comprehension of geographical data may consist of two parts with the first part focusing on comprehension from a written description and the second part focusing on comprehension from a map.
  4. Have practice at becoming consistent raters of diagnostic instruments such as maps, graphs and written descriptions. It is often helpful for more than one teacher to rate, say, the students' maps of the school area. Practice at scoring such maps should minimise the variations in the raters' scores.

Administering Diagnostic Tests

The administration of the diagnostic instrument is important. The teacher should know his or her class, the school operations and the instrument well enough to be able to select a time for its administration which will provide the fairest results. Guidelines are often provided for the administration of standardised tests. For other forms of diagnosis, teachers need to determine the time to administer the instrument, any special instructions and the appropriateness of the venue for the exercise. Additionally, careful rehearsal of the administration of the exercise is essential for its efficient implementation. For example, a teacher's diagnosis of the students' competence to orienteer will be useful if the positioning of markers is guaranteed as accurate and all markers have been positioned in the designated places.

Using the Results of a Diagnostic Test in Geography

The benefits of diagnosis come only when the results are translated into classroom practice. This translation could be in a variety of forms, e.g. a changed technique for teaching a topic, the use of simpler language or concepts, the use of different maps and photographs, a rewriting of curriculum materials, or in the provision of more direct observation via local field studies. Teachers should remember that the generalisations they may make from a particular diagnosis will be influenced by such factors as the quality of the instrument, the sample of students, the purpose of the instrument and their experience in diagnosing students' capabilities in geography. If the instrument has provided useful results to the geography teacher, he may also attempt some form of remediation to recorrect the deficiencies in the students' understanding or skill. Few, if any, attempts have been made to consider remediation of problems or biases in the geography classroom.

Conclusion

This chapter is an introduction to the challenge of maximising learning in geography via diagnosing students' capabilities and using this information to provide more realistic learning in the geography classroom. It is an attempt to advise the geography teacher that there are ways to investigate individual learners' capabilities in geography besides assessing their understanding of the content of geography. It is inferred that the diagnosis of students' capabilities in geography could lead to more efficient learning in the geography classroom. This is because the teacher will know his class better after some form of student diagnosis and he will be able to present the class with activities and resources which will suit their capabilities.

Three types of diagnostic activities have been suggested:

1. the prediction of students' achievement or understanding;

2. understanding students' capabilities on entering an aspect of learning in geography;

3. diagnosis of students' abilities in geography.

Exercises for such activities may be standardised and published commercially or they can be validated teacher creations. Geography teachers are advised not to be too daring in their approach to diagnosis in geographical education. Diagnosis should not be overdone. Initially, the teacher may engage in a single formal diagnostic activity. The frequency of such activities should then be regulated by the identification of problems in learning in the classroom.

The biggest challenge in this area is not to find an appropriate diagnostic instrument. More and more of these are being devised and validated. The biggest challenge is how to translate the results into sound classroom practice. The question of remediation is closely linked to this challenge and must be addressed closely in future if the individualisation of learning in geography is to be successful.

References

Australian Council for Educational Research, Mathematics Profile Series (1978) *Space Test*, Hawthorn, Victoria.

Black, H.D. and Dockrell, W.B. (1980) *Diagnostic Assessment in Geography*, Edinburgh: The Scottish Council for Research in Education.

Costello, H., Mortensen, P. and Turner, D. (1979) *An Activity to Diagnose Children's Ability to Think Logically in Geography*, Kelvin Grove Campus, College of Advanced Education, Brisbane, mimeo.

Fien, J.F., Wilson, P.S. and Slater, F.A. (1982) 'Children and their Environment: Knowing, Experiencing and Valuing the Local Area', in *Geography: Action in Society*, Australian Geography Teachers' Association, 8th National Conference, Melbourne.

Gerber, R.V. (1980) *Development of Competence and Performance in Cartographic Language by Children at the Concrete Level of Map-reasoning*, PhD thesis, University of Queensland, Brisbane.

Hieronymus, A.N. and Lindquist, E.F. (1971) *Iowa Tests of Basic Skills, Form 5*, Boston: Houghton Mifflin.

National Council for Geographic Education (1979) *Geography Skills Test: Intermediate Level*, Macomb, Illinois: Western Illinois University.

Rhys, W. (1972) 'The Development of Logical Thinking', in N. Graves (ed.) *New Movements in the Study and Teaching of Geography*, Melbourne: Cheshire.

Simson, R.P. (1977) *The Evaluation of Scenic Quality: An Application to the Gold Coast Hinterland*. MSc dissertation, Griffith University, Brisbane.

Synder, S., Feldman, D. and La Rossa, C. (1975) 'Manual for a Piaget-Based Map Drawing Exercise', unpublished paper, Tufts University, USA.

Wilson, P.S. (1980) *The Map-Reasoning Development of Pupils in Years Three, Five and Seven as Revealed in Free Recall Sketch Maps*, PhD dissertation, Columbus, Ohio, The Ohio State University.

Wisconsin Research and Development Centre for Cognitive Learning (1973) *The Wisconsin Design*, Minneapolis: National Computer Systems.

24

Individualising Learning in Geography

John Lidstone

The diagnosis of student learning will reveal that most classes are composed of students who display a wide range of abilities, motivation and attitudes to learning, to the environment and to other people. Catering for individual needs in such classes has been a challenge to geography teachers for many years, but in this chapter John Lidstone presents the philosophical arguments and research evidence which suggest that individualising students' work can have many advantages for both the students and for the complex society in which we live. The implications for geography teachers of individualising learning are explored and followed by appropriate strategies for planning curriculum units and lessons that cater for individual learning differences. The structure of a unit for teaching the topic of modern Ghana and ideas for various classroom activities and fieldwork exercises are used as illustrations of the advice provided in this chapter.

When you think about it, the title of this chapter could be considered more than a little tautologous. After all, teachers may teach what they like and may encourage or coerce their students to carry out all manner of exercises and activities but, ultimately, each and every student has to learn as an individual. Perhaps the greatest challenge facing teachers is to arrange the education of the students in their charge to enable each individual to develop the necessary knowledge, skills and attitudes to the best of each individual's ability. This might sound ideal, but all attempts to achieve the education of individuals as individuals are bedevilled by the one certainty of all teaching: students differ in knowledge, skills, development stage, learning rate and preferred learning styles.

If a teacher is to present a lesson to a whole class, then it seems intuitively obvious that the lesson should be neither too easy nor too difficult for the students. However, if the range of ability in the class is great, then any single lesson will inevitably be easier than it should be for some students and more difficult than it should be for others. Furthermore, it is very probable that there will be individuals within the class, less able than their peers to gain from the transmission type of lesson implied here, who would benefit from a more activity-based model of learning. To make matters even more complex, there will be differences in the natural rhythms of individual students whereby time of day, physical position and levels of light and noise affect their ability to learn.

Are We Teaching Classes or Students?

Throughout the English-speaking world, teachers have traditionally responded to the need to match their teaching to students with widely differing needs by adopting various methods of grouping students. In this way they have hoped to limit the range of ability in their teaching groups. Although grouping by age is now so common that we take it for granted, this was a major innovation in the nineteenth century. Indeed, Slavin (1987) has commented that while age grouping may reduce the range of abilities in the classroom compared with that in a one-room schoolhouse, it still leaves a wide range of abilities within each grade and may deprive students of the opportunity to learn from older students. In fact, almost every means of grouping students by ability or by performance level has its own drawbacks either resulting from the effects on students or the problems caused for classroom management. Figure 24.1 lists some of the arguments which have been put forward for and against grouping students of similar abilities. It is interesting that most of the research which has been carried out into the effects of grouping students according to ability has failed to support the statements made in the left-hand column, while offering some support for those in the right-hand column.

Slavin (1985) cites evidence from the United States that mixed-ability groupings are better, both academically and socially, for less able students and that racially mixed schools are better for the achievement of black students. In the United Kingdom, Newbold (1977) and Rutter et al. (1979) showed that high-ability students can achieve similar results in both streamed and unstreamed classes. Furthermore, in addition to the mainly social aspects emphasised in the right-hand column of Figure 24.1, current psychological thinking recognises that intelligence manifests itself differently in different spheres, thus making any notion of a 'general ability' highly questionable.

The recognition that the ability range of student groups cannot be reduced satisfactorily on the basis of general ability has led most education systems to move towards allocating students to classes regardless of their specific abilities. This has resulted in the creation of what have become known as *mixed-ability classes*.

In practice, the methods by which schools create their mixed-ability classes vary enormously. They

Figure 24.1 Statements made for and against grouping students by ability

For grouping by ability

Grouping takes individual differences into account by allowing students to advance at their own rate with others of similar ability and by offering them methods and materials geared to their own level.

Students grouped by ability receive more individual attention from their teachers.

Students are challenged to do the best in their group or be promoted to the next level within a realistic range of competition.

It is easier to teach to and provide materials for a narrower range of ability.

Against grouping by ability

Homogeneous grouping is undemocratic and affects the self-concept of all students adversely by placing a stigma on those in lower groups while giving students in higher groups an inflated sense of their own worth.

Most adult life experiences do not occur in homogeneous settings and students must learn to work with a wide range of people.

Students of lesser ability may profit from learning with those of greater ability.

It is impossible to attain truly homogeneous grouping, even along a single achievement variable since few tests are sufficiently reliable or valid.

Homogeneous groupings may provide less sensitivity to individual differences in students by giving the teacher the false sense that students are similar in social needs, achievement and learning style while heterogeneity permits different patterns of abilities and needs to emerge within a group of students.

Grouping by ability tends to segregate students along ethnic and socioeconomic lines as well as by ability.

Groups of lower ability may become the victims of low teacher expectations.

frequently include a deliberate mixing of abilities based on a selection of criteria including: IQ scores, attainment tests, primary teachers' assessments, friendship or neighbourhood groupings, primary school of origin, or totally random, perhaps alphabetical, groupings.

The term 'mixed ability' in this context acknowledges the existence within a group of students of a wide range of specific abilities, levels of motivation, attitudes to the various stimuli of the group and its environment, learning styles and previously developed skills. Geography teachers in schools which group students in mixed-ability classes are thus faced with classes of individuals whose abilities and qualities in the various aspects of geographical education it is the responsibility of the teacher to develop.

Learning Individually, Individualised Learning and the Development of the Individual

While almost all teachers in the Western world would accept that the development of the individual is one of the fundamental aims of education, the meaning of the word 'individual' in educational contexts frequently becomes rather muddied. In the context of the present chapter:

1. Individual work refers to a situation in which an individual undertakes basically the same activities as the rest of the class but at his or her own pace, place or time.
2. Individualised work involves personal assignments designed to meet the different needs or abilities of specific students. Such assignments may or may not be undertaken in the company of other students.
3. The development of the student as an individual accepts as axiomatic that students will not be labelled by ability nor will they be placed in competition against one another. This does not imply, of course, that their ability or progress in a specific area will not be assessed, nor does it exclude the possibility of friendly inter-student rivalry. Work designed to aid the development of the student as an individual may include activities such as those described in both the sections above, selected by the teacher for specific reasons according to his or her perception of the needs of the student. As such, the activities could include working alone at a task set by the teacher or created by the student, working in a group of students of similar or different specific abilities, or taking part in a whole-class discussion led by the teacher, the student or another student.
4. Finally, all the above approaches will be undertaken in the knowledge that individual students will have individual learning styles which influence their responses to any and all of the experiences with which they are faced.

In organising an individualised classroom, the teacher will take account of all these differences in student requirements and possible approaches to managing his or her curriculum.

Individualising an Activity

Individualisation may be achieved in a number of ways. At the simplest level, differences between students may be acknowledged by assigning work to be accomplished at the student's own pace. The assumption here is that the activity is within the capacity of all students if they are given enough time. Examples of such activities may include drawing a map, performing some calculations or reading a piece of text in order to extract some information. A higher level of individualisation may be achieved by teachers allocating a basic activity to all students and then giving different levels of assistance to individual students according to their needs and abilities. However, this can be refined by anticipating the difficulties which are likely to be experienced by students within the class. Figure 24.2a is an activity based on the climatic data for Brisbane. More able students, or those who have learned how to draw and interpret climate graphs, will be able to obey the instructions beneath the table with little teacher direction and will then be able to go on to investigate other aspects of whatever unit this climate activity is a part. However, some students may find that drawing such graphs from the beginning is too demanding, and so the teacher may have available a number of sheets (such as that shown in Figure 24.2b) which remove the problems caused by the need to select an appropriate scale and draw the axes of the graph, but which leaves students to draw the graphs themselves. Finally, there may be a few students in the class who lack the coordination and experience to draw the graphs even with the help of printed axes, and for these students a supply of sheets laid out as suggested in Figure 24.2c may be necessary.

The teacher may also direct the interpretation of the graphs according to the ability and experience of the students. While the most able students will be able to write a description of the climate of Brisbane,

Figure 24.2 (a) A possible activity based on the climate statistics for Brisbane; (b) A sample worksheet for students who may have some difficulty in drawing climate graphs;

others may need specific guidance which directs them to write sentences about (1) summer temperatures, (2) winter temperatures, (3) the difference between summer and winter temperatures, (4) the total amount of rain which falls during the year, and (5) the time of year during which most of the rain falls. The least able students might need the help of coloured bands (made possible by the colour facility

(a)

Climatic statistics for Brisbane, Australia													
	J	F	M	A	M	J	J	A	S	O	N	D	Average/Total
Temp (°C)	25	25	24	21	18	16	15	16	18	21	23	24	20.6
Rain (mm)	145	140	125	95	60	70	50	25	45	60	100	110	1025

Activity:
1. Draw graphs to portray the climate statistics for Brisbane.
2. Describe the climate of Brisbane.

(b)

Activities
1. On the outline below, draw a climate graph for Brisbane.

2. Use your graph to write a brief description of the climate of Brisbane. You should mention
   - the highest and lowest temperatures
   - the average temperature
   - the range of temperatures experienced throughout the year
   - the total rainfall and the way in which it is distributed throughout the year

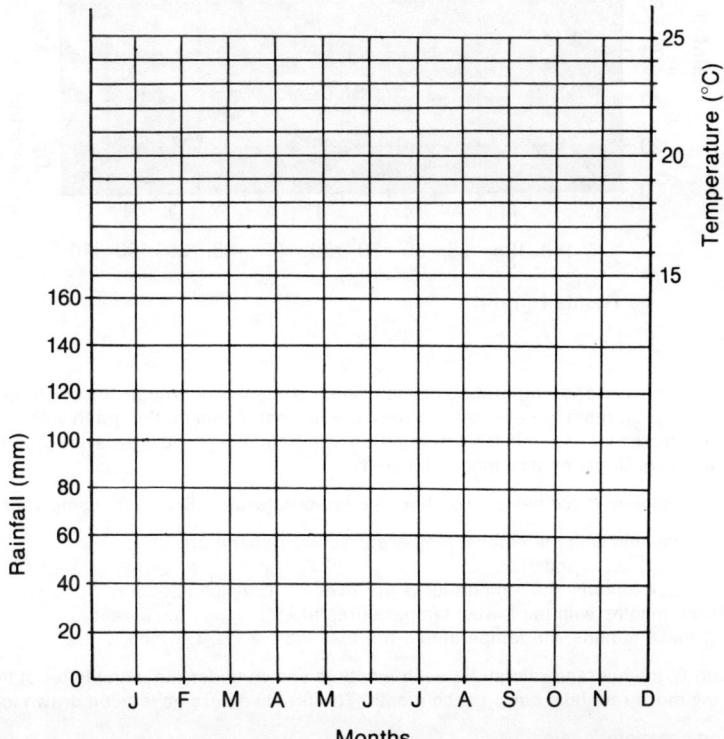

Figure 24.2 (c) A sample worksheet for less able students to help them draw and interpret a climate graph for Brisbane

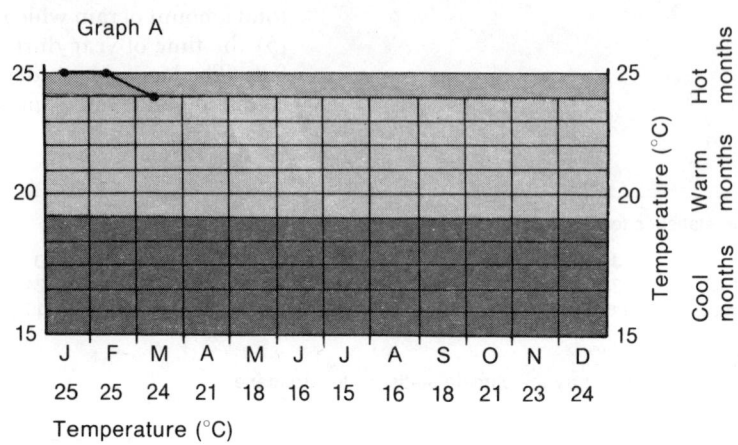

Graph A

Temperature (°C):
J 25 F 25 M 24 A 21 M 18 J 16 J 15 A 16 S 18 O 21 N 23 D 24

Graph B

Rainfall (mm):
J 145 F 140 M 125 A 95 M 60 J 70 J 50 A 25 S 45 O 60 N 100 D 110

(c)

Activities

1. On Graph A, the average temperature of each month is shown underneath the first letter of the month. For each month of the year, mark the average temperature for that month on the graph with a dot. The first three dots have been marked for you. Join the dots together with a smooth line to draw a line graph of the average monthly temperatures of Brisbane throughout the year.

2. Copy and complete these sentences about the temperatures of Brisbane, using your graph to help you.

    The four months with the highest temperatures in Brisbane are ...., ..., ... and ....
    During these months, the temperatures are over .....°C which is ....
    The three months with the lowest temperatures are ...., ... and ....
    During these months, the temperatures are less than ....°C which is ....

3. On Graph B, the average rainfall for each month is shown under the initial letter of the month. Draw bars to show how much rain falls during each month. The first two bars have been drawn for you.

4. Copy and complete ..., etc.

of the spirit duplicator and suggested by the screens across Figure 24.2c) and sentences to copy and complete if they are to interpret the graphs they have drawn.

The decision as to which of the levels of independence outlined above should be adopted by individual students may be made by either the teacher or the student.

Individualising a Teaching Unit

A similar approach may be adopted for complete teaching units if the essentials of each unit are defined and a range of activities are prepared which correspond to wide bands of student abilities.

Rob Foster adopted such a scheme for individualised learning in the geography department at the Cavendish School in Hemel Hempstead, England. The department has a commitment to mixed-ability teaching but was faced with a shortage of resources to satisfy the learning needs of all students. Accordingly, the members of the department worked cooperatively to produce a series of student workbooks which contained both the core materials for each teaching unit in their courses and a collection of graded learning activities. The learning activities to be undertaken by individual students were specified in a study guide and were initially expressed as three routes. Students selected the route they wished to follow according to their interest in the topic and their own estimate of their ability, in consultation with their teacher. Flexibility was built into the scheme as teachers monitored students' progress through each unit and suggested variations in the routes being followed. Thus a student might begin by following Route 1, but having made good progress, be referred by his or her teacher to Route 2, while another student who was struggling with some aspects of Route 3 might be told to undertake some of the work in Route 2 to clarify some particular point. The result of the scheme was that students were able to pursue particular inter-

Figure 24.3 Study guide for a mixed-ability class studying the topic 'Modern Ghana'

Check with your teacher if you are not sure which route to follow	Route 1 (Circle Route)	Route 2 (Square Route)	Route 3 (Triangle Route)
Themes to be studied	This route covers all the basic facts and ideas. The work is broken down into short questions. The questions tell you exactly what to read and what to write about. Mostly short answers are needed.	This route covers all the important basic facts plus some extra interesting material. It involves writing notes and longer paragraph answers and short essays. Mapwork is required. Most people should follow this route.	This route is for students who can read large sections of material and then summarise what they have read in note form or long essays. It involves a lot of detailed, sometimes difficult but interesting reading and essay writing.
Peasant agriculture	1→2→3, 5←4	1→6→7→8, 10←9	11→12→8, 10←9→13→14
Primary products	15→16→17	15→16→17→18, 20←19	21
The Volta project	22, 23, 24, 25, 26, 27	22→23→24, 26←25←27	26, 27, 28
The city	29, 30, 31, 32, 33	34, 35, 36	37, 38, 39
Ways forward	40, 41, 42, 43	44→45, 42, 46	44, 45, 47, 48

Route	Conceptual content	Learning experiences
Route 1	1. Differences in living standards between Ghana and Great Britain. 2. Subsistence agriculture in North Ghana including: (a) the farmer's year: (b) seasonal controls on farming activities: (c) advantages and disadvantages of self-sufficiency; (d) social organisation and welfare through the extended family system; (e) migration of young and ambitious people to urban area	1. Reading and comprehension (short passages); making notes following a tight structure. 2. Simple picture and map interpretation. 3. Line sketching.
Route 2	1. Everything on Route 1, plus: 2. The system of rotational bush fallowing and variations in intensity of farming with distance from the farmhouse added to Sections 2a and 2b of Route 1. 3. National patterns of land use. 4. Problems of cattle farming, e.g. disease and climate variability.	1. Everything on Route 1 with reading extended to two pages in length and less specific structure for note making. 2. Map sketching with an emphasis on careful standing and labelling.
Route 3	1. Everything on Route 1 and Route 2 plus: 2. An appreciation of why the extended family remains so important in North Ghana added to Section 2d of Route 1. 3. An appreciation of why subsistence/semi-subsistence agriculture survives, including: (a) natural hazards (unreliable climate, river blindness): (b) attitudes towards risk—sub-optimal behaviour: (c) the land tenure system; (d) poorly developed transport and energy infrastructure; (e) regional disparities.	1. Everything on Route 2 with reading extended to seven pages in length with note making guide limited to a set of sub-headings. 2. More difficult map interpretation questions. 3. Several maps to be drawn and comparison made between them. 4. Essay writing using a set of paragraph headings.

Figure 24.4 Summary of conceptual content and learning experiences in the peasant agriculture section of the workbook on modern Ghana

ests, revise and consolidate learning and gain further practice in specific skills when needed.

Figure 24.3 is the study guide prepared for a five-week unit on modern Ghana for Year 10 students. The numbers along each route refer to particular learning activities in the modern Ghana workbook. Route 1 covers all the basic themes and ideas in the unit and is intended for students with limited reading and other learning skills. Route 2 is for the majority of students. It also covers all the basic materials but differs from Route 1 in two ways: firstly, additional themes and more detailed coverage are provided; secondly, the actual learning activities are usually different and more challenging than those in Route 1. Route 3 is an extension of Routes 1 and 2. This route is for students with advanced learning skills and higher levels of conceptual and reasoning ability. Figure 24.4 summarises the content and learning activities in the peasant agriculture section of the workbook on modern Ghana.

This approach to individualising teaching in geography has many advantages where resources are scarce, but is extremely time consuming, even when prepared cooperatively. It also inevitably suffers from lower standards of reproduction than commercially produced materials. This is especially true in the reproduction of maps and photographs. A greater disadvantage, however, is that students may be deprived of the opportunity to consult a range of resources, of assessing one point of view against another and of making a choice of what they perceive to be relevant and significant. This problem can be overcome, when resources are available, by ensuring that the study guide encourages students to use a variety of books, photographs and maps as well as slides, audio kits, materials samples and computer programs in addition to the core materials provided in the workbook. In this way the teacher may provide close guidance to those with specific learning difficulties while permitting greater freedom of exploration to those who may benefit from it.

Group Work

The individualised scheme as it was practised at the Cavendish School generally resulted in the majority of students working individually. However, this is not an essential or even necessarily a desirable feature of individualised learning. Some critics have claimed that when all students are working individually, teachers may spend too much time on classroom management rather than on teaching, and students may lack any incentive to progress rapidly through the materials, which probably have an excessive reliance on written instruction. Many of the problems implied by these criticisms may be diminished by the judicious use of group work.

Students may be grouped on a temporary basis for a variety of learning purposes. However, it is important to remember that if these groups become semi-permanent within the class, then segregation by ability has effectively been introduced. Groups may be constructed which contain students of similar and of mixed specific abilities.

Grouping Students with Similar Specific Abilities

Groups of students with similar specific abilities may be formed while the remainder of a class is pursuing individualised work. The reasons for such a grouping may range from providing extra help for those who cannot grasp map coordinates to the introduction of advanced or abstract ideas for those who show particular aptitude. After all, not all 14 year olds are ready for statistical correlations, but some are. The various theories of city development could be presented as an extension in a Year 9 unit on land use in cities to the small group of abstract thinkers in a class. In both these cases the group may be used either to provide mutual support or to enable the teacher to give personal attention. This latter purpose is especially important since one of the consequences of individualising learning is the danger that interaction between student and teacher may become less frequent. In fact, one survey some years ago found that in some extreme cases of heavily individualised classrooms, teachers had become little more than checkers of worksheet answers and had somehow come to abdicate the role of initiators and guides of learning (Department of Education and Science 1978).

Individualisation does not necessarily mean that the whole-class lesson no longer has a part to play in the teaching of geography. Frequently a whole-class presentation may precede individualised learning in a work unit in order to introduce the new topic and provide initial motivation. All students learn best when they are interested and it is important to arouse student interest before any individual work begins. In fact, to ask students to make a choice on depth of study based partly on interest without making any attempt to arouse that interest or show the possibilities that a topic may offer may be to the detriment of students whose home provides a limited cultural background and therefore a restricted range of interests.

Each of the five units in the Cavendish School workbook on modern Ghana was introduced by a video recording of a BBC television program on Ghana for schools. Much audio-visual material can be used by students of all abilities, individually, in small groups and as a whole class, and almost all students can obtain impressions from a film or video even though their abilities to recall the detail or appreciate every nuance may vary greatly. Furthermore, whole-class discussions, in which a wide range of opinions may be put forward, should not be regarded as inappropriate in an individualised classroom since they not only develop students' confidence and oral skills but also avoid the danger of teachers becoming involved in repetitive discussions with individuals.

Grouping Students of Different Abilities

Students of different levels of ability may be grouped together and given a task to complete. In so doing, all students may learn to respect one another's contributions and skills. The experience can also help students to learn management skills such as making efficient use of available human resources and appreciating the need to support those of lower levels of skill or ability.

Slavin (1985) has described a method of classroom organisations he calls Team Assisted Individualisation. By working in a group of mixed abilities, students who have difficulties in grasping specific ideas may be helped by group mates who have already reached understanding, while at the same time, high achievers may gain from the opportunities to teach others. Slavin and his co-workers consider that such a method of classroom organisation is most suitable for subjects and areas where there is some skill or content information to be 'mastered' since the 'learning' of such materials is easily measured and the teams may then be placed either in competition against each other, or 'goals' may be set for teams working as units to achieve. The theory of such organisation is that students working in teams towards a cooperative goal can help one another study, can provide instant feedback to one another and can encourage one another to proceed rapidly and accurately through the materials.

Such groupings have shown themselves particularly appropriate in simulation exercises and verbal reporting sessions. Two such activities have been used with groups of very mixed ability during residential field courses in the Snowdonia area of the Welsh mountains (Lidstone 1981).

In the first activity, students were taken on guided tours, over several days, of the Gwydyr Forest, the Ffestiniog hydro-electric scheme, the Llechwedd slate mine and an aluminium works. The intention was to show the students four different economic facets, past and present, of Snowdonia and to demonstrate land use conflict within a national park. Students were asked to observe the processes involved in each of the works, the number of workers employed and the conditions under which they worked, bearing in mind that the first three of the places visited have become tourist attractions in their own right. In addition to the more conventional geographical ideas of land use on steep slopes, the suitability of a glaciated upland area for hydro-electricity generation and the one-time close relationship between hydro-electricity production and aluminium smelting, it was deemed important to emphasise the relative paucity of employment prospects in this part of Wales and the problems faced by a resident population living in an upland national park. Accordingly, while the cognitive aspects were recorded by each student on his or her own, students were grouped to explore the social consequences of the employment situation. Each group was asked to prepare a short news report based on its observations in the field and to present a simulated television news interview on the question: 'Is there a social crisis in this part of Wales?'. The various social backgrounds of students present in each group helped all the students to appreciate more fully the problems facing the local people. In this exercise, students needed an ability to empathise with people living in an environment very different to that of their own home area. They also needed to be able to express orally their interpretations of life in a remote rural community. Some students, coming as they did from a south London suburb, found it very difficult to imagine living in such an area. In particular, some found it hard to understand why the Welsh language was often used and why Plaid Cymru (Welsh Nationalist) slogans were daubed on walls in an area they had clearly regarded as just another part of Great Britain. Other students were better able to understand the resentment felt by residents against the English use of Wales as a playground and water catchment area. In presenting their reports, students with greater facility in oral expression tended to adopt the role of interviewers. The interviewers then gave direction to those who wanted to identify with such groups as retired slate miners suffering from the dust-induced lung disease, silicosis, or small shopkeepers faced with an ageing and declining population. The least able seemed to be most at ease playing the roles of unemployed teenagers and school students.

The second activity placed students in the role of planning consultants investigating the demand and possible sites for a new sports complex in one of two adjacent towns on the North Wales coast. Students worked in mixed-ability groups conducting street interviews to discover the demand for such a complex, surveying existing sports facilities and the extent of each one's sphere of influence and mapping the location of the most accessible areas of each town and existing land use to reveal potential sites. Each group was told to prepare its case with maximum use of visual aids for a 'planning inquiry'. Students were left to distribute the various tasks, both of data collection and presentation among themselves. Thus one group organised itself so that two of the more able students analysed the local bus timetable and route map while two less able students carried out pedestrian counts at various points. Both pairs demonstrated that towns have some areas of greater movement density than others but used methods of very different complexity. Having compared their results, the students were able to suggest possible links between bus services and density of pedestrian movements, and on this basis make suggestions about the most suitable area for a new sports complex. Interestingly, it was not one of the 'brighter' students who pointed out that the site eventually chosen was currently occupied by an old building used as a rest room for elderly people while shopping and urged that alternative provision be made for them as an integral part of the proposal.

Both these activities made demands on students at a variety of cognitive and affective levels. However, no one group of students had a monopoly of all abilities and the final product depended on the contributions of all. In the case of the sports complex simulation, the time limit imposed meant that even those of very modest ability had to be given a role by the group if the task was to be completed. For some of the least able students it was probably the first time that they had been really needed by their peers. The more able not only took on the more cognitively complex aspects of the investigation but gained experience in the efficient allocation of human resources. Finally, mixed-ability grouping made it possible for less able students to partake in such a relatively complex activity and contribute to a polished end product.

Classroom Management for Individualised Learning

In a traditional classroom where the expectation, if not the actuality, is that all students would work at the same tasks at approximately the same pace, the teacher has a relatively simple management role. On the other hand, a classroom in which students' work has been individualised poses a much greater challenge to teachers' management skills. However, a number of writers have emphasised that appropriate management strategies are essential if the principles of individualisation are to be put into practice.

Deci (1985) found that rigid controlling behaviour by teachers lessens students' intrinsic motivation and impairs their creative performance. He also recorded that the students he studied who were in the classes of teachers who maintained strong control over every aspect of the learning endeavour perceived themselves to be less competent and held lower feelings of self-worth than students who were given some control over their own learning strategies. Long (1987) has drawn attention to the problems which can occur when a teacher adopts a single management model with which either the teacher is unsympathetic or which does not meet the needs of individual students. He suggests that a teacher who is in a school in which he or she is expected to rely primarily on assertive discipline, an interventionist model of management, will be unlikely to achieve success unless such an approach reflects his or her ideological stance to classroom interaction. Similarly, a child who is self-motivated and capable of taking considerable responsibility for his or her own behaviour may be severely disadvantaged if under the sole responsibility of a teacher who can only manage from an interventionist perspective. Finally, Long asks how teachers can possibly know which management stance is appropriate for each student in their care?

One answer to this question has been given by Hersey and Blanchard (1978). They suggest that four basic styles of leadership are necessary and that these vary in their appropriateness depending on the situation. The four styles differ from each other in the way in which they combine the amount of detail of the task provided by the teacher and the amount of support and encouragement the teacher gives the students as they complete the task. Thus, Style 1 is described by Hersey and Blanchard as *directing*. The teacher tells the student exactly what to do, and then observes closely to see that it is done correctly. Style 2 is termed *coaching*. Here the teacher is very specific about the task to be accomplished, supervises the work closely and offers considerable support to the student. In Style 3, *supporting*, the teacher gives the student the basic responsibility for completing the task but is available to give guidance and support where needed. Finally, using Style 4, *delegating*, the teacher gives the responsibility for the task to the student with the expectation that the student will, without further encouragement or guidance, complete the task successfully.

Long considers that the decision as to which style is appropriate for individual students depends on two factors: the student's actual skill in performing the task and the student's willingness to do so.

Where the student is eager to learn something but lacks the skill, management Style 1, directing, is appropriate. The student does not need encouragement, just clear instructions about what to do. For a student who is discouraged or bored, coaching is appropriate and the student is given both instructions and encouragement in completing the task. Once students have gained some skill in the task and can perform it on their own, the teacher's style may change to supporting in which little instruction is necessary but continued encouragement is needed. Finally, when students no longer need instruction or encouragement, the teacher may delegate continued learning in the area to students who have become, in

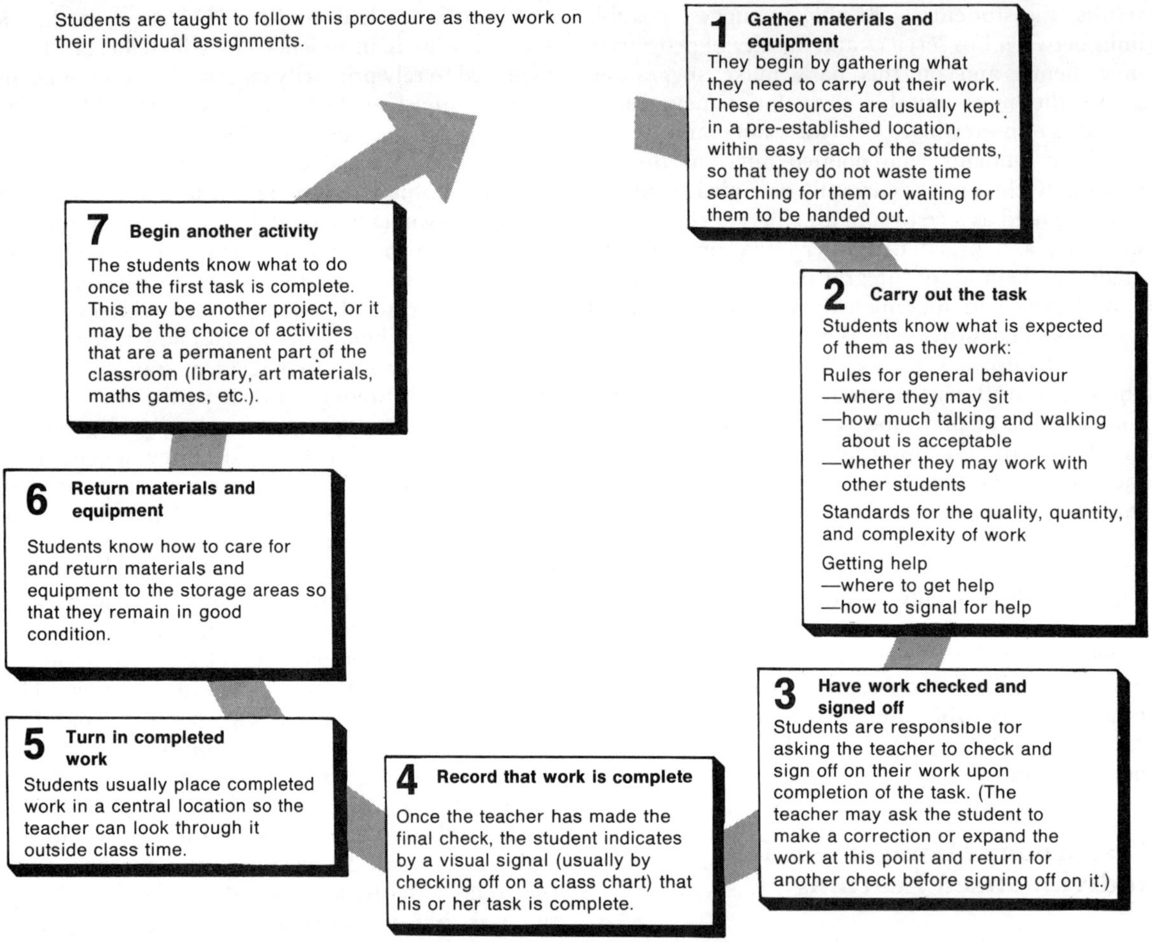

Students are taught to follow this procedure as they work on their individual assignments.

1 Gather materials and equipment
They begin by gathering what they need to carry out their work. These resources are usually kept in a pre-established location, within easy reach of the students, so that they do not waste time searching for them or waiting for them to be handed out.

7 Begin another activity
The students know what to do once the first task is complete. This may be another project, or it may be the choice of activities that are a permanent part of the classroom (library, art materials, maths games, etc.).

2 Carry out the task
Students know what is expected of them as they work:
Rules for general behaviour
— where they may sit
— how much talking and walking about is acceptable
— whether they may work with other students
Standards for the quality, quantity, and complexity of work
Getting help
— where to get help
— how to signal for help

6 Return materials and equipment
Students know how to care for and return materials and equipment to the storage areas so that they remain in good condition.

5 Turn in completed work
Students usually place completed work in a central location so the teacher can look through it outside class time.

4 Record that work is complete
Once the teacher has made the final check, the student indicates by a visual signal (usually by checking off on a class chart) that his or her task is complete.

3 Have work checked and signed off
Students are responsible for asking the teacher to check and sign off on their work upon completion of the task. (The teacher may ask the student to make a correction or expand the work at this point and return for another check before signing off on it.)

Figure 24.5 The student work cycle (Kierstead 1986)

this part of the curriculum at least, independent, autonomous learners.

Such a model of management style for the classroom seems quite simple on paper, but will be much more complex when actually attempted in real classrooms where individuals and small groups may all be working at different tasks during a single lesson. Kierstead (1986) carried out a study of classrooms in which she deemed the teachers to have been successful in the management of students' individualised work. She identified three elements common to successful management strategies which should be established by the teacher before the students begin their work. She found that successful teachers establish:

1. a curriculum of increasingly complex tasks which define what the students are to work on;
2. a student work cycle set out as a set of routines, procedures, rules and consequences that spell out for the students exactly what is expected of them, how they are to proceed and how they will be required to account for the responsible use of their time; and
3. a teacher work cycle which is a set of routines and procedures that allows the teacher to use his or her time effectively in class—intercepting students as they reach critical points in their work and giving them feedback and instructions when it is needed.

Figure 24.5 shows Kierstead's model of a student work cycle and Figure 24.6, her model of a teacher work cycle. Only when these three components have been created are the students allowed to work independently.

Kierstead emphasises a number of features of these work cycles which ensure that the work matches students' needs, strengths and interests as well as encouraging students to become responsible,

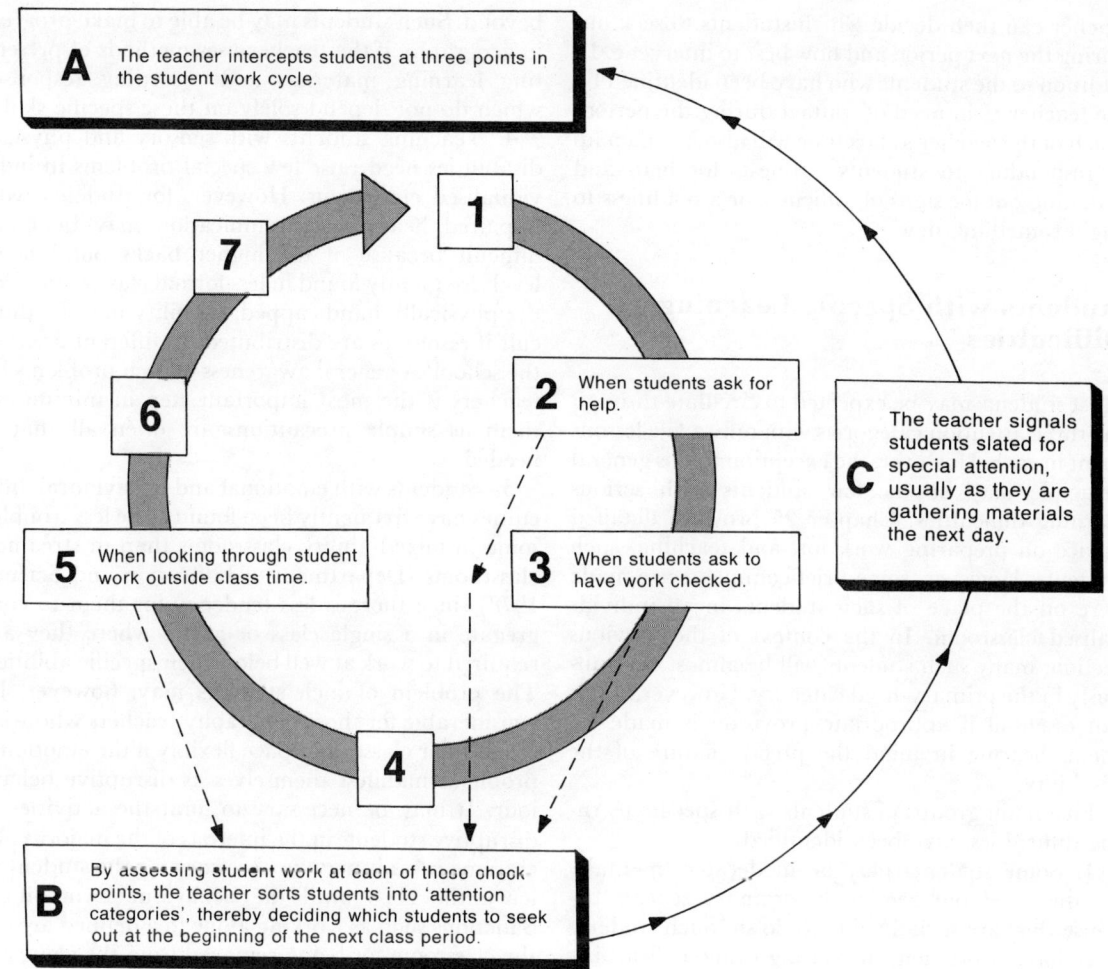

Figure 24.6 The teacher work cycle (Kierstead 1986)

autonomous learners. Firstly, students are given responsibility for pacing themselves appropriately and for telling the teacher when their work is ready to be checked. The consequences of failing to carry out these procedures have already been established and normally involve a penalty being imposed when the work is eventually assessed. Such an approach removes the need for teachers to coax, nag or remind students of the amount that they are expected to achieve within a given time span. Instead, students' work output depends on their inherent interest in what they are doing, their response to the culture of the classroom established by clearly set out expectations and their appreciation of the consequences of failing to complete the work in the time allotted.

From the teacher's point of view, less time is spent watching over the entire group and more time in giving feedback and instruction to individuals as they reach critical points in their work. Teachers observed by Kierstead sorted students into attention categories by regularly assessing their work. They relied on the two automatic checkpoints: (1) requiring students to have their work checked and approved during the work period, and (2) reviewing students' work away from the classroom. From these two checks the teachers categorised students according to whether they required primary, secondary or minimum attention. Students in the primary attention category need immediate help or correction or are ready to be introduced to a new skill, concept or area of endeavour. Students in the secondary category are those the teacher needs to keep an eye on because they have recently started something new, are about to move on to a new stage or have a chronic problem. Students in the minimum attention category can continue to work independently, usually because they have recently been in the primary category and are comfortable with what they are doing. Having made these decisions, the teacher can then decide which students to seek out during the next period and how best to intervene. In addition to the students who have been identified by the teacher as in need of contact during the period, much of the teacher's class time will also be taken up in responding to students' requests for help and watching out for signs of difficulty or a readiness to begin something new.

Students with Specific Learning Difficulties

Most students may be expected to circulate through the three attention categories with only a few lessons spent in each. However, the exception to this general principle may be the few students with serious learning difficulties. Chapter 25 provides detailed advice on preparing work for, and teaching, such students. However, some brief comments are made here on the place of such students in an individualised classroom. In the context of the previous section, many such students will be almost continuously in the primary-need category. However, this is not essential if appropriate provision is made for them, bearing in mind the precise nature of the disability.

Five main groups of students with specific learning difficulties have been identified.

1. Some students may be moderately mentally handicapped but remain in ordinary schools because they are socially able to do so. Such students may have a problem in relating different learning experiences to each other and in forming concepts, in addition to the more common difficulties in reading and writing. Often they will need help outside the normal school curriculum. Certainly, in geography lessons, teachers will need to define a limited range of objectives for them, and in particular provide opportunities for the limited amount of learning they are capable of consolidating.

2. Some students, for a variety of reasons, may come to geography lessons with very low levels of attainment in reading and writing. In less severe cases, students in this group may, with remedial help, become able to function effectively in individualised classrooms. Withdrawal is the most common method of arranging remedial help and the principle of individualising learning helps such students since they may work at their own pace and thus not miss essential work without which their progress is further inhibited.

3. There is also a small group of students whose difficulties in reading, writing and spelling may persist throughout the secondary school period and beyond. Such students may be able to make progress in geography if the teacher uses methods of presenting learning materials and recording responses which do not depend solely on these specific skills.

4. Teaching students with sensory and physical disabilities need raise few special problems in individualised classrooms. However, for students with impaired hearing, communication may be more difficult because of the higher background noise levels frequently found in less formal classrooms. For the physically handicapped, mobility may be difficult if resources are distributed in different areas of the school. A general awareness of such problems by teachers is the most important step in minimising them as simple precautions are often all that is needed.

5. Students with emotional and behavioural difficulties have frequently been found to be less troublesome in mixed-ability classrooms than in streamed classrooms (Department of Education and Science 1978) since there is less tendency for them to congregate in a single class or group where they are required to work at well below their specific abilities. The problem of such students may, however, be considerable for those geography teachers who wish to use their classroom space flexibly if the emotional problems manifest themselves as disruptive behaviour. It may be necessary to limit the activities of disruptive students in the interests of the majority. In extreme cases it may be necessary for the student to leave the class until the behaviour is modified. Sanctions such as working alone in a defined area of the classroom, or being removed from the class, are further examples of the kinds of procedural rules which may be drawn to the attention of students before individualised work begins. Should emotional disturbance be manifest as shyness or personal withdrawal, there will probably be little classroom disruption, but no less need for the teacher to attempt to help students so afflicted or refer them for help elsewhere.

The concern of the teacher at all times, however, should still be for the greatest good of each individual. In the case of special difficulties, no teacher can hope to be an expert in all, and in such cases the help of the specialist advisory services, both within and outside the school, should be sought, earlier rather than later in the teaching process.

Individual Projects

The types of individualising suggested in the previous sections are based on the assumption that the teacher will have constructed specific curricula for

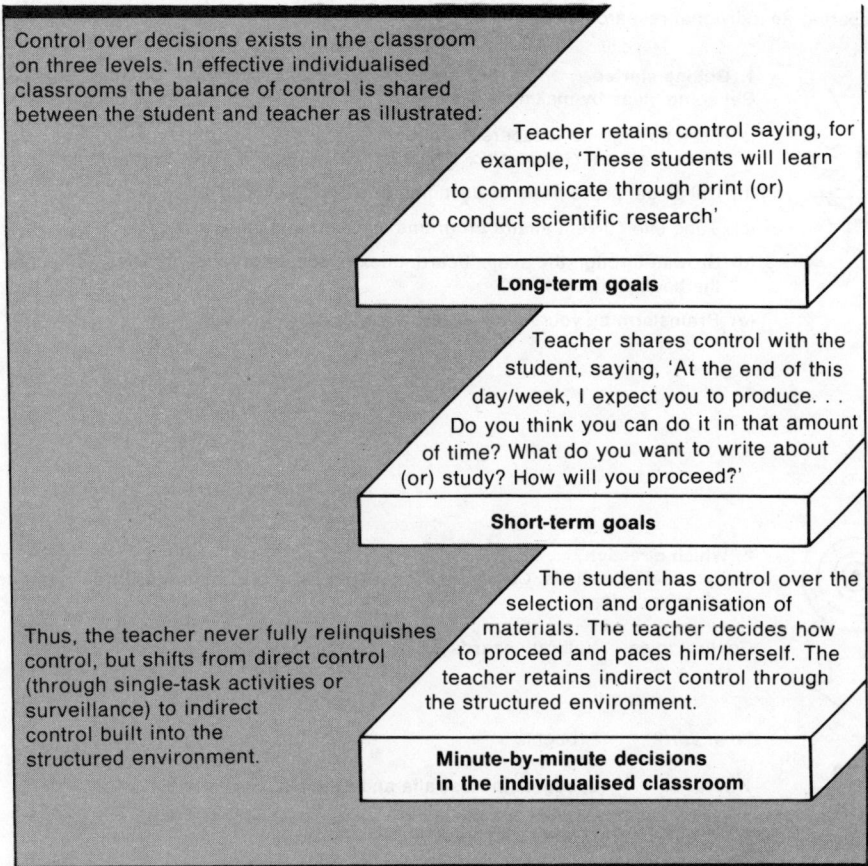

Figure 24.7 The three levels of control over decision making in the classroom (Kierstead 1986)

students to follow. However, the individual research assignment is a form of individualised learning that has a long and chequered history. At its highest level, it is a university thesis but, at its lowest level, such 'projects' can become no more than a 'paste and waste' scrapbook with little educational merit other than to provide practice in copying more or less accurately and using scissors to cut out magazine pictures. However, it is not essential that the high school project descends to the lower level just described. Kierstead (1986) suggests that if we want to encourage our students to learn to work independently, to use their time productively and to apply and extend their basic skills and concepts to real life problem-solving situations, we must share responsibility for decision making about their assignments with them. However, this implies recognition that control over decisions regarding the pace, sequence and content of education exists on three levels: long-term goals, short-term goals and minute-by-minute decisions. Figure 24.7 illustrates how the balance may shift between the three levels.

Figure 24.7 shows that the teacher, as the agent of society, must determine the long-term goals, but can share decisions regarding short-term goals with the students and should allow the students themselves to make the minute-by-minute decisions needed to plan and carry out their projects. Kierstead emphasises, however, that there are two pre-conditions to students' assumption of control over decision making at the third level. Firstly, the teacher must provide enough instruction in the basic skills and concepts to prepare students to plan and carry out their projects; and secondly, the teacher must ensure that students are prepared to clarify the objectives of their projects and appreciate the requirements of the student work cycle and its relationship with the teacher work cycle. This second pre-condition ensures that students will receive enough feedback and instruction during the work period to enable them to proceed and make them accountable for using their time responsibly.

Figure 24.8 shows one set of advice on individual assignments adapted from a notice originally prepared by Anna-Louise Crellin, a geography teacher from Melbourne, to help her students plan their

Figure 24.8 Preparing an individual research study

1. Getting started
Get some ideas by making a list.

(a) Flick through newspapers (local ones are good value!) and magazines (e.g. *Ecos, National Geographic, Geographical Magazine, New Internationalist*).

(b) Check out some news or documentaries on television.

(c) Tune into current affairs programs on radio and television.

(d) Browse through the subject card index in school or local libraries. (Look on the book shelves, too!)

(e) Brainstorm by yourself or with a few friends.

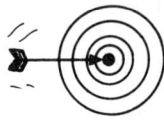

2. Which direction?
Ask yourself questions. Clarify objectives. For example, natural hazards might be your general idea.

General idiea: Natural hazards

What sort?   Droughts

Where?   Droughts in Australia and Ethiopia.

So what?   Do they have similar effects?

Why?   Why do droughts affect people in Ethiopia so badly when they do not cause famines in Australia?

What can be done about it?   What can Australians and Ethiopians learn from each other about drought?

Specific topic: Evaluate the short-term and long-term solutions to the effects of drought in Australia and Ethiopia.

3. Moving forward
Getting data and resources takes time. The information you want may not be available, unexpected illnesses and accidents could slow you down, or you might have to change your ideas.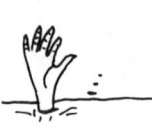

- Prepare a summary of your topic which includes:

    (a) a set of aims which outlines your topic in greater detail;

    (b) a timetable showing what you wish to achieve and by when.

- Collect data, information, resources—but do not collect things for the sake of collecting!

Your resources, information and data must relate *directly* to your topic. They must provide evidence either for or against your educated guesses.

4. Pause for second thoughts

Sorry, just 10 more things to consider before finalising your topic. If your answer to any of the following questions differs from the one suggested in brackets *choose another topic*!

(a) Is it legal? (Yes) For example, you cannot do a survey in a National Park without a permit!

(b) Does it harm the environment or would it embarrass anyone?   (No)

(c) Is the information you want likely to be confidential?   (No)

(d) Is the information likely to be readily available?   (Yes)

(e) Are resources likely to be available?   (Yes)

(f) Does it appear to be too complex?   (No)

(g) Might it be too difficult?   (No)

(h) Is it dangerous?   (No)

(i) Is it too far away?   (No)

(j) Is there enough time?   (Yes)

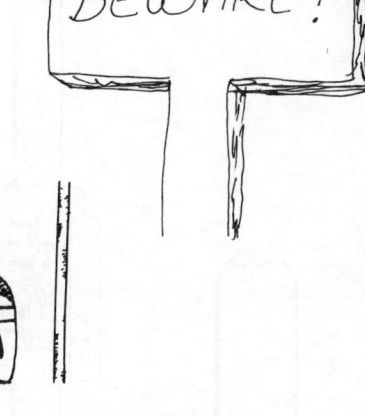

5. Go for your life!

A. Refer to all of the collected information and data in your write-up.
Discuss what you have found out.

B. Summarise your results.
Use your own relevant maps, photos, tables, graphs, charts, to illustrate clearly what you have found out.

C. Draw conclusions.
   (a) Relate your results to the aims of your topic.
       If your aims are in the form of a question, your conclusions should answer the question and your answer should be *supported by evidence*.

   (b) Explain your results.
     (i) Are they what you expected?
     (ii) What do they tell you about your topic?
     (iii) Are there any patterns?
     (vi) Any unusual results?
     (v) What things did not work? Why?
     (vi) What further research could be done?
     (vii) What have you learned from the project?

D. Your report must be written in a logical manner.
(a) Include appropriate headings, bibliography, footnotes, references, contents, labels for all photos, maps, tables, and charts.
(b) **Make it look good. Presentation is worth marks!**
(c) Don't forget,

                good grammar
                good spelling
                GOOD LUCK!

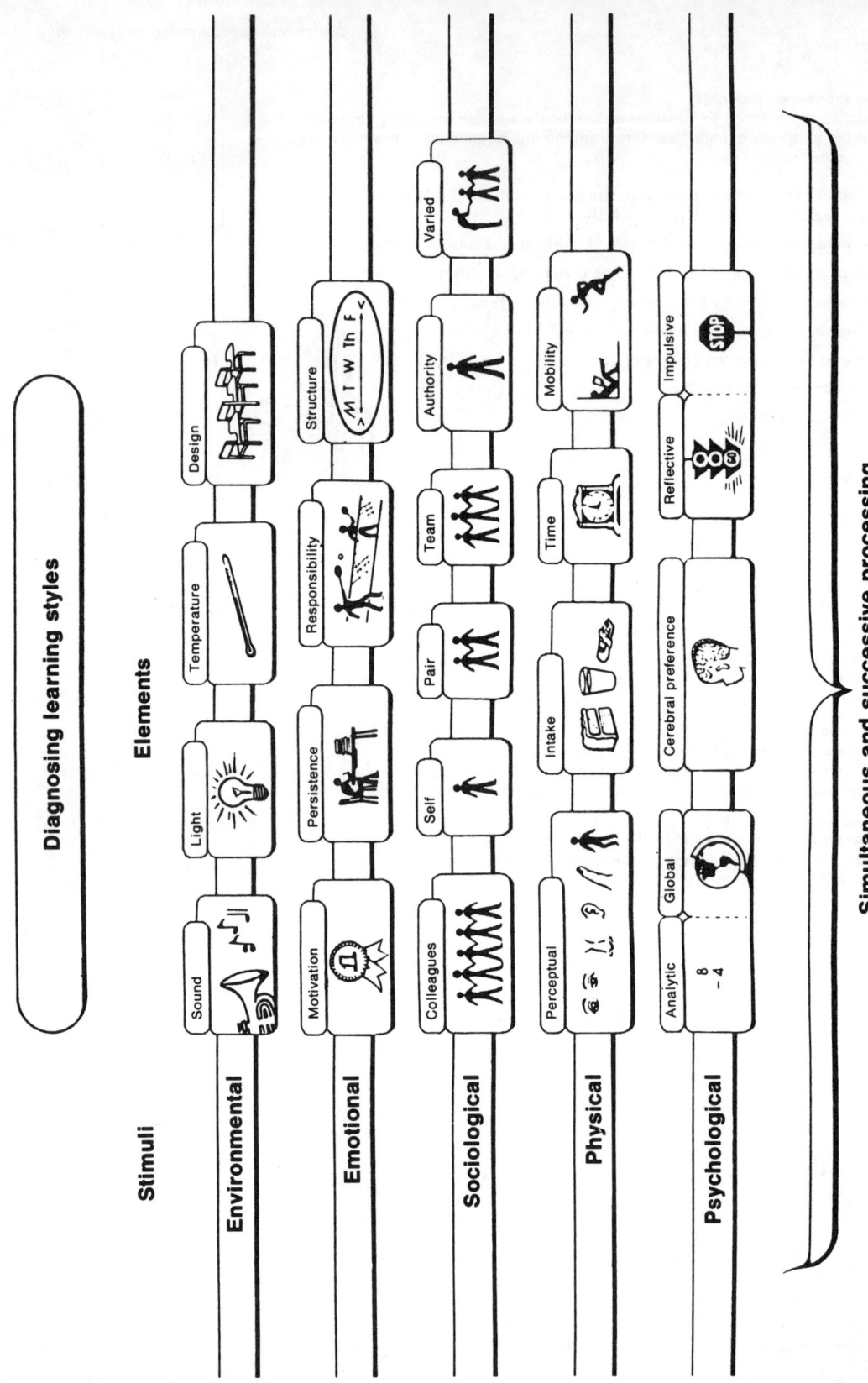

Figure 24.9 Learning styles (from Bauer 1987)

own individual research study of a geographical problem.

Lidstone (1987) in an evaluation of a new Year 11–12 geography course for students in Queensland schools, called 'Australian Geographical Inquiries', found that the individual research study is frequently greeted with great student enthusiasm when planned according to the strategies outlined in Figure 24.8. Typical comments from students included:

'I enjoyed it because (i) we worked with our own techniques; (ii) we worked at our own speed; (iii) we chose our own topics.'

'I've learned more, my interest has strengthened, my general ideas on geography have strengthened.'

'It was good because we had an aim or question set in the beginning and we worked towards answering that main question.'

However, where students were not given sufficient guidance at the preparation stage, or were not supervised closely enough and given feedback during the course of their inquiries, they quickly became disillusioned and uncertain. This was reflected in such comments as:

'We were left to work on our own and the class tended to fall behind in its work.'

'It was harder (than previous geography units) in that it was harder to understand exactly what was wanted.'

'This course demanded advanced skills which I personally hadn't developed before.'

Individual Learning Styles

In a previous section, mention was made of students with behavioural problems which may prevent them from learning. There is increasing evidence that at least some of these learning problems may be due to differences between students in the ways in which they learn—their learning style. The implication of

Figure 24.10 Underachieving students' learning style preferences (Dunn and Bruno 1985)

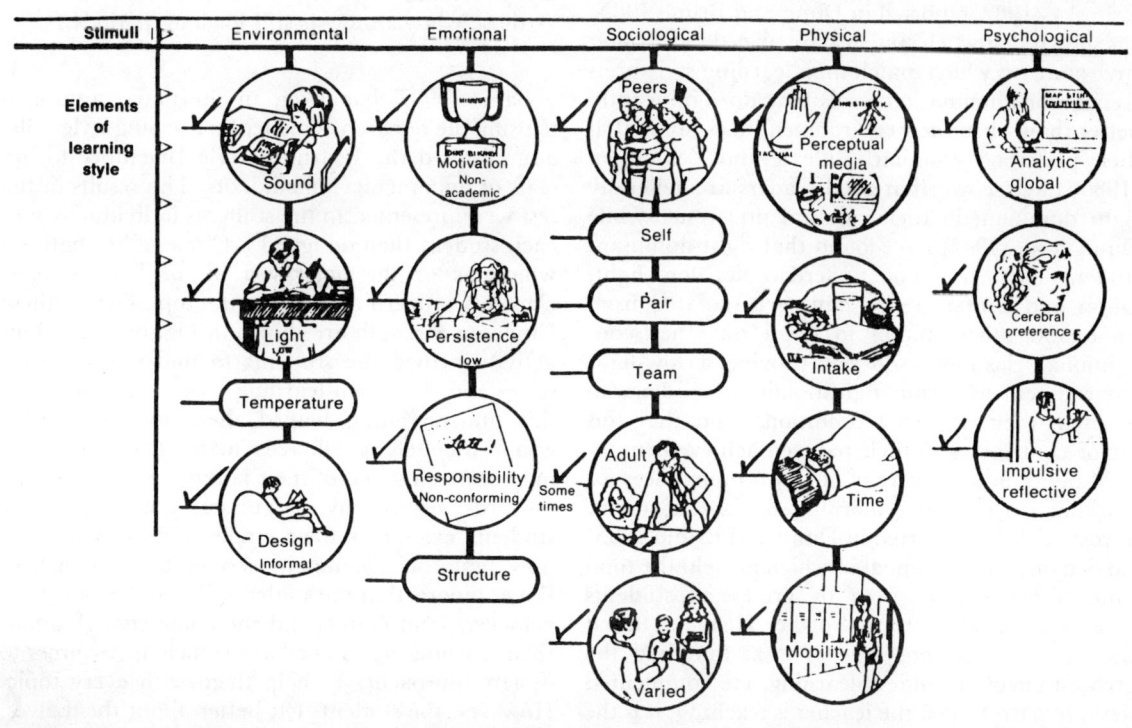

the research in this area is that successful students are often those who have learning styles which match the classroom environments in which they find themselves. Research on learning styles suggests that students may differ from one another in their responses to environmental, emotional, sociological, physical and psychological stimuli in at least 21 major areas. These are shown in Figure 24.9. Dunn (1979) found that students who read poorly seemed to like working in the midst of activity or noise. They liked learning either with classmates or with the teacher on a one to one basis. They preferred dim, rather than bright, lights and worked longest on the floor or in an easy chair, often in unusual poses such as on their backs with their feet pointing towards the ceiling. These students often ate while they worked and seemed to need to move about a great deal. On the other hand, the academic achievers preferred to learn in conventional seats, with bright light, in quiet surroundings, often by themselves, and rarely wanted food or drink. These students could sit still working for long periods of time. Figure 24.10 shows the learning style preferences typical of students who appear to be underachieving at school.

A number of researchers have carried out experiments to discover whether it is the learning style which affects achievement or the lack of achievement which affects students' preferred learning style. Krimsky (1982, reported in Dunn and Bruno 1985) has shown that when students are placed in learning environments which match their learning style preferences for illumination levels they do significantly better than when their environment does not match their preference. Similarly, Oexle and Zenhausen (1980) have shown that good readers are generally right dominant in their cerebral preference while Dunn et al. (1982) have found that right-dominant students have a strong preference for low light. Dunn and Bruno (1985) have extrapolated from these and other studies to point out that conventional classrooms do not provide a low-light environment to permit right-dominant students to learn in their preferred conditions, and that this factor contributes to their reduced achievement.

Many of us are aware of our own preferences for working early in the morning or late at night. Virostko (1983, reported in Dunn and Bruno 1985) carried out an experiment in which he held the total time of teaching received by groups of students constant, but varied the time of day. He found that it was time of day, not time of task, that had the greatest effect on student learning. He commented that no matter when the teacher is teaching, it is the wrong time of day for at least one-third of any class. Even more dramatically, Lynch (1981, reported in Dunn and Bruno 1985), the principal of a New York high school, discovered that most of the students who were persistent truants experienced their highest levels of energy in the late morning and afternoon. When he rearranged the teaching schedules of these students to match their preferred learning style, truancy levels fell.

Such discoveries are undoubtedly fascinating, but the problem of putting such information to good effect in schools as they are currently does not seem to be open to easy solution. Dunn and Bruno (1985) have suggested a fourfold strategy for reorganising classrooms to meet the major differences in environment which appear to be required. They suggest that in each classroom there should be:

1. quiet areas and a corner where the 10% who need sound may listen to music through headphones while studying;

2. brighter and lower illumination which can be obtained by removing a bulb in one corner of the room for the 9% who need it;

3. an area with carpet, pillows or a lounge chair as well as hard seats and desks; and

4. opportunities for organised quiet movement for those who cannot sit still for long periods.

Bauer (1987) has taken an alternative approach to using the research evidence on learning styles. She administered the Learning Style Inventory to her class of underachieving students. The results of the test were presented to the students individually and each student then designed a Circle of Strength on which he or she portrayed his or her strongest elements of learning style at the top. Two of these Circles of Strength are shown in Figure 24.11. This activity helped the students to understand themselves and their own learning style preferences and this understanding helped them to feel positive about themselves, whereas many had previously only been conscious of their failures. From this time onwards, before any assignment was started, the students examined their circles and then decided how best the assignment should be approached. Bauer reports that the students did not become high achievers, and neither did she know enough about their learning styles nor have sufficient resources to design approaches to help them with every topic. However, the students felt better about themselves, tried harder, and overall, succeeded to a greater extent than previously.

 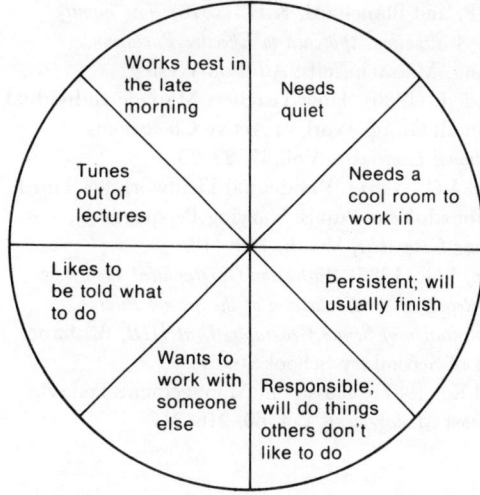

Figure 24.11 Students' 'Circles of Strength' in their preferred learning styles (Bauer 1987)

Conclusion

It may seem to many teachers faced with mixed-ability classes, a shortage of books, few audio-visual resources and limitations on fieldwork that many of the suggestions made in this chapter amount to a 'circus' with the teacher conjuring new activities out of thin air with one hand, while keeping students as busy as spinning plates with the other. In some schools in less favoured environments, the analogy may be taken further by imagining him or her performing both these tricks whilst balancing on a tightrope.

There are undoubtedly some exceptional practitioners in the teaching profession who can achieve all that has been suggested and more, aided only by their own resourcefulness. For the rest of us, however, despite our high ideals, practical considerations will limit our attainment. It may be that some teachers find that they can only use audio-visual aids by booking an audio-visual room well in advance for a whole class. Other teachers may find that the greater informality and noise which often attend group work are unacceptable to the teacher in the next room. In such cases, some compromise with the principles of individualised learning may be necessary, but individualisation as an ideal should remain at the forefront of the teacher's mind. Where greater departmental cooperation is possible, perhaps through departments enjoying suites of rooms or geography being taught in reasonable proximity to a library or resource centre, space may be used more flexibly. For example, those engaged in written work may congregate in one room while a larger or better equipped room is used for making displays or models, using audio-visual materials or for group discussion. As in both curriculum development and unit planning, an 'open door' policy can make everyone's task easier.

The demands on teachers made by a commitment to individualised learning are undoubtedly high, but the benefits to individual students, and, ultimately, to the professional satisfaction of the teachers themselves, are similarly high.

References

Bauer, E. (1987) 'Learning Style and the Learning Disabled', *The Clearing House*, January, 206–208.

Deci, E. (1985) 'The Well Tempered Classroom', *Psychology Today*, March, 52–53.

Department of Education and Science (1978) *Mixed Ability Work in Comprehensive Schools*, HMI Series, Matters for Discussion No. 6, London: HMSO.

Dunn, R. (1979) 'Learning Style', *Educational Leadership*, Vol. 36, 430–432.

Dunn, R. and Bruno, A. (1985) 'What Does the Research on Learning Style Have to do with Mario?', *The Clearing House*, Vol. 59, 9–12.

Dunn, R., Cavanaugh, D.P., Eberle, B. and Zenhausen, R. (1982) 'Hemispheric Preference: The Newest Element in Learning Style', *American Biology Teacher*, Vol. 44, 291–294.

Hersey, P. and Blanchard, K.H. (1978) *The Family Game: A Situation Approach to Effective Parenting*, Reading, Massachusetts: Addison-Wesley.

Kierstead, J. (1986) 'How Teachers Manage Individual and Small Group Work in Active Classrooms, *Educational Leadership*, Vol. 37, 22–25.

Lidstone, J.G. (1981) 'Residential Fieldwork for Third and Fourth Year Pupils: Varying Perspectives', *Teaching Geography*, Vol. 6, 116–118.

Lidstone, J.G. (1987) *Australian Geographical Inquiries: Final Report of the Evaluation of the Second Pilot Implementation of Senior Geography Unit IIIB*, Brisbane, Board of Secondary School Studies.

Long, C.K. (1987) 'Classroom Management Today', *Classroom Management*, Vol. 60, 216–217.

Newbold, D. (1977) *Ability Grouping: The Banbury Inquiry*, Slough: NFER.

Oexle, J. and Zenhausen, R. (1980) 'Differential Hemispheric Activation in Good and Poor Readers', *International Journal of Neuroscience*, Vol. 15, 31–36.

Rutter, M. et al. (1979) *Fifteen Thousand Hours: Secondary Schools and Their Effects on Children*, London: Open Books.

Slavin, R.E. (1985) 'Team Assisted Individualization', in R. E. Slavin et al. (eds) *Learning to Co-operate, Co-operating to Learn*, New York: Plenum Press.

Slavin, R.E. (1987) 'Grouping for Instruction in the Elementary School', *Educational Psychologist*, Vol. 22, 109–127.

25

Teaching the Less Able Student in Geography

William Pick and Malcolm Renwick

Many teachers have found few experiences more frustrating than trying to teach academic geography courses to their less able students. Problems of syllabus irrelevancy, the academic orientations of many teachers and schools, the reading level of available materials, the attention span of students and varying but generally low motivational levels render the results of attempts to teach such students quite unpredictable. One proven solution is mixed-ability grouping of students which avoids both the stigmatisation of streaming a group of students into what is euphemistically called the 'less able' class and the grouping of students into such classes as 'sinks' for the malcontent and unruly as well as the less able. William Pick and Malcolm Renwick have had extensive experience teaching less able students within streamed and mixed-ability groupings of students. They examine the cycle of educational deprivation which the less able student generally experiences and posit seven constructive guidelines for breaking this cycle. They express these seven learning needs of the less able student as: the need to escape an excessive preoccupation with the written word, the need for variety, the need to be active not passive, the need for reassurance, not failure, the need for inductive, rather than deductive, learning experiences, the need for short-term goals and the need for relevance in what they do in the classroom to their lives and interests outside it.

Introduction

This chapter examines the problems and possibilities inherent in teaching geography to the less able or slow learning student. Slow learning students are those who experience varying degrees of difficulty in mastering the basic skills of reading, writing and numeracy. They are not necessarily below average in ability in all aspects of school work and indeed may be quite capable in some practical activities. Their lack of facility in the basic skills, however, retards their progress in many areas of the curriculum. Their poor concentration and limited ability to learn make it difficult for their teachers to arouse their interest or stimulate sustained efforts. Their lack of achievement and consequent sense of failure often leads to disruptive behaviour. The less able student should not be confused with the low or underachiever who may be quite bright but is not performing to capacity. Nor should he or she be confused with the socially or culturally disadvantaged student who may be underachieving for a variety of other reasons. However, much of what follows can be applied to all three groups, though it specifically refers to the less able group of students.

Cognisance must be taken of the perceptions and abilities of teachers when examining the problem of teaching geography to the less able. The term 'less able' clearly means different things to different people, though in a survey conducted by the authors (Pick and Renwick 1979) the majority of teachers defined less able students in terms of their reading, writing and language deficiencies. To a lesser extent, poor numeracy or number deficiency was preferred as a definition and, perhaps surprisingly, poor behaviour was not. Many teachers recognise in themselves inadequacies in their ability to deal with and teach less able students. Their own background is after all an academic one. They usually have been trained to teach only average or above-average students and generally lack time or opportunities to gain specialist training for the more difficult task of teaching the less able. Schools often place constraints upon teachers and these tend to militate against solving the problem. These constraints include the organisation of the timetable, finance and the provision of adequate resources. So often staff allocation and resource provision are geared to the more able. Unfortunately, there is often a dearth of resources specifically geared to the less able, even when finance is available and the 'spirit' of teachers is willing.

Thus there are two facets to the problem of teaching the less able. On the one hand there are the characteristics of the students themselves, and on the other the deficiencies of schools, other institutions and teachers in coping with them. This chapter examines various approaches to teaching the less able student from these two standpoints. Specific examples of successful teaching strategies and styles are used to illustrate general guidelines for countering commonly encountered problems. Hopefully, the examples chosen highlight the fact that less able students must be provided for as 'particular children in a particular school with particular teachers in a particular environment' (Blyth et al. 1976).

The chapter is organised around seven learning needs of the less able group of students and suitable teaching responses to these needs. These learning needs are:

1. The need to escape an excessive preoccupation with the written word.
2. The need for variety.
3. The need to be active rather than passive.
4. The need for reassurance, not failure.
5. The need for inductive, rather than deductive, learning experiences.
6. The need for short-term goals.
7. The need for relevance.

The Need to Escape an Excessive Preoccupation with the Written Word

Frequently observed characteristics of . . .

The less able student

- low level of reading and writing skills;
- not motivated by the written word;
- will not need to write much for the rest of their lives. They will need to communicate in other ways—particularly verbally;
- a variety of interests, often non-academic in character;
- can respond with enthusiasm to visual/audio stimuli.

The teacher

- defines 'less able' in terms of low abilities to read and write;
- is wedded to the literary or academic mode of education;
- regards a full, well-written exercise book as evidence of good teaching and learning;
- tends to minimise opportunities for interpersonal communication and to reduce potential opportunities for disturbance.

Teachers of the less able need to develop a variety of audio and/or visual resources to counteract an overemphasis on the use of the written word. One such resource still under-used by many teachers is the cartoon. This is a resource frequently used in a five-book geography textbook, *Going Places* (Renwick and Pick 1980), produced for mixed-ability classes. A look at an exercise from one of these books (Figure 25.1) reveals some of the opportunities a cartoon offers. In this example it is used to introduce the topic of water. It summarises many aspects of the use of water and is designed:

1. to provide classification exercises (listing dangerous/safe/noisy/polluting activities);
2. to promote discussion;
3. to encourage expression of opinion;
4. to motivate pupils to work further;
5. to provide a 'springboard' for pieces of 'research', creative work, etc.

At the river

Most of us at some time like to spend a day by a river or a lake (or loch). This picture shows a variety of activities that take place on or near water.

Make a list of the different activities.
Can you add to this list?
Classify your list by splitting it into groups:
  noisy activities/quiet activities
  dangerous/safe
  active/relaxed

Good point Noodle.
Can you find any other 'conflict' situations?

Figure 25.1 An example of the use of cartoons in teaching (Pick and Renwick 1979)

Figure 25.2 A teacher-produced cartoon to introduce the concept of 'plan-view'

WHAT MAKES A GOOD HOLIDAY?

Sunshine?

Good food?

Comfortable place to stay?

But what do you think?
Copy and complete your own vote table.
Put these factors in *your* order of choice.
(1. for the most important, 2. for the next and so on.)

Good Food	
Getting away from people	
Relaxation	
Friendly company	
Plenty of things to do	
Sunshine	
New things to see	
Comfortable place to stay	
Others	
Others	

Figure 25.3 Visual props are important in worksheets

The cartoon simplifies abstract arguments and presents them in a conventionally acceptable way for young people. It may enable these young people through discussion to develop higher levels of abstract reasoning that they or their teachers thought possible.

Cartoons need not be professionally drawn and teachers ought not to feel reluctant to use the technique because they feel their skill inadequate. Indeed the teacher-produced worksheet or resource is often preferable because it is produced for a particular group of less able students. Figure 25.2 was an early attempt by a teacher to introduce a humorous element into a worksheet in an effort to motivate his less able students and to help them understand the concept of 'plan view'. It was used to introduce the task of mapping land use within the school, as part of a study of the school.

While much more effort needs to be made to reduce our dependency on the written word, we cannot get away from it entirely. Nor should we! As teachers, we do have a responsibility for language development in its broadest sense. However, we need to exercise great care when we do use the written word in order to achieve appropriate reading levels, passage length, stimuli and the provision of visual props. Figures 25.3 and 25.4 are sections of worksheets that meet these criteria.

The Need for Variety

Frequently observed characteristics of . . .

The less able student

- limited span of concentration and attention;
- transient interests;
- needs, ideas and relationships to be reinforced frequently;

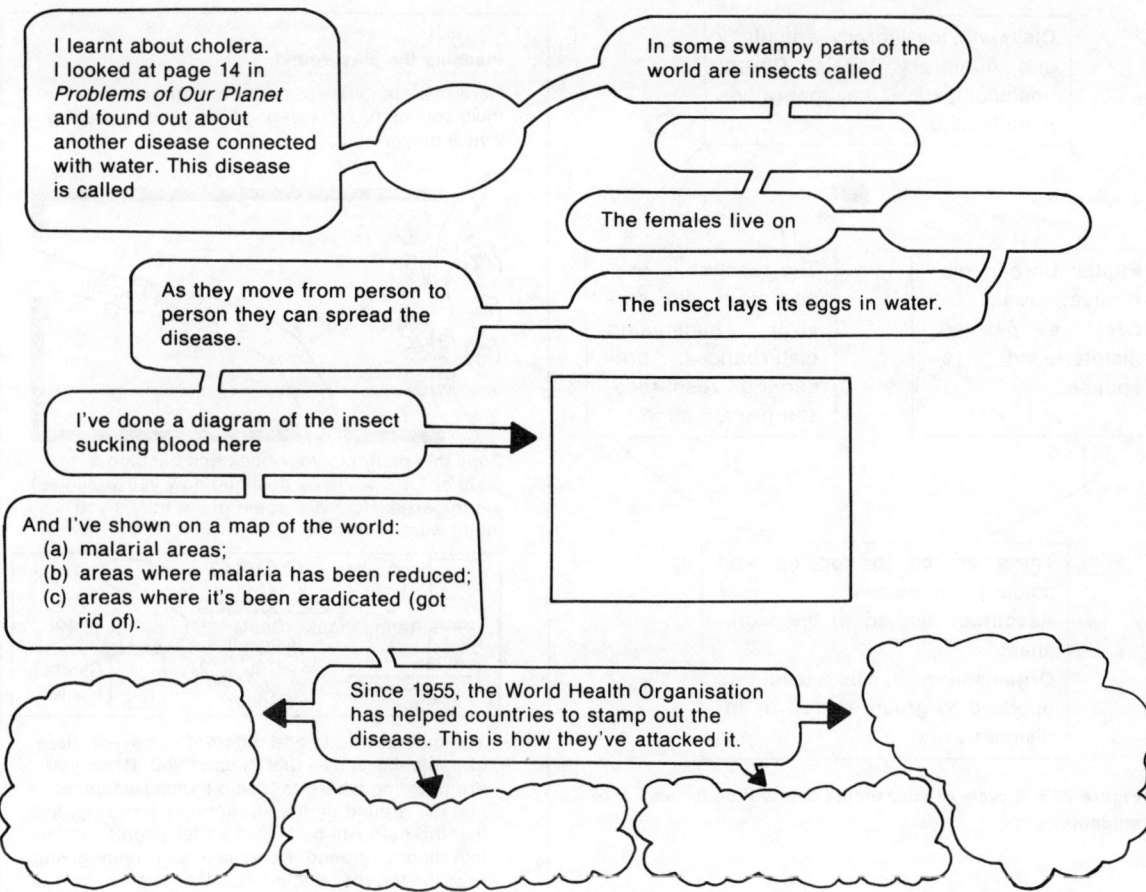

Figure 25.4 A worksheet made for students to work on (prepared by R. Beattie of Seacroft Park Middle School, Leeds)

- needs to practise a variety of basic skills;
- low motivation;
- has ideas, opinions and attitudes of his or her own;
- not necessarily limited in perceptions.

The teacher

- often most concerned with management rather than method;
- tends to limit activities and methods to those which minimise opportunities for disruption and noise;
- has many pressures on limited time;
- needs to preserve limited resources;
- sees more potential reward in teaching higher-level classes.

Variety is the keyword when teaching less able students—variety of resources, teaching strategies, content, tasks, locations and class organisation. Although variety is important at all levels of teaching geography, the need for it increases as one moves down the ability range. Often, it is the lack of motivation which results from a lack of variety which induces more able students to become lower achievers. Also, sadly, teachers of the less able are more often preoccupied with control and management than with methodology. Yet, more variety of methodology could well solve such management problems. Without care a 'cycle of educational deprivation' could well set in. Figure 25.5 illustrates how easily this could happen.

Teachers of geography are more fortunate than many colleagues in the variety of content, tasks, resources and locations possible in teaching their subject. The way in which geographers have grasped and developed the opportunities afforded by educational games and simulations suggests a

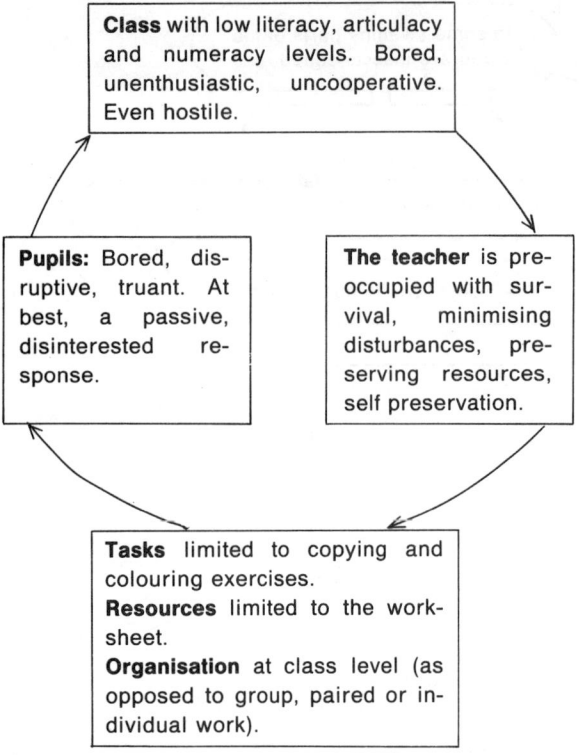

Figure 25.5 A cycle of educational deprivation for less able students

Planning the playground

Here is a plan view of an area of ground which a town council has decided will be used to provide a new playground.

Copy this plan into your book and become a planner for a while by deciding how you would set out the area. Here are some of the things you might want to include.

You may decide to add different items—or more of particular ones—that is up to you. When you are deciding where to locate things remember that the ground by the stream may be damp, and that the path will be important for people getting into the playground. Now make your plan, giving reasons for your decisions.

Figure 25.6 A simple simulation game (Pick and Renwick 1979)

pleasing degree of originality and adaptability. However, commercially produced games are often very complex and rarely suited to the needs of the less able. Simple teacher-produced games are often best. They do not take long to produce or to operate in the classroom. Figure 25.6 is an example of such a game (Renwick and Pick 1980). See Chapter 20 for guidance in developing simple classroom games and simulations.

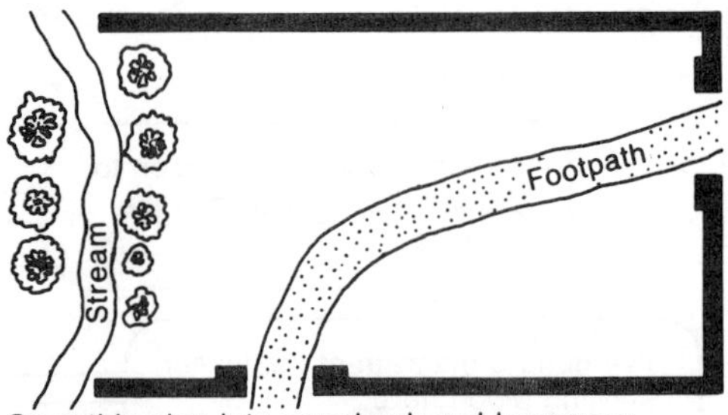

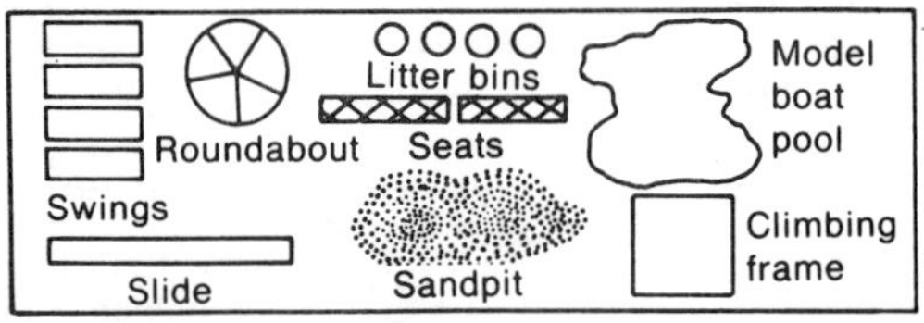

The teacher's attitude is also very important. One needs to be flexible in one's thinking as well as willing to try new and different approaches. If something on television, in the news, or about a local neighbourhood issue catches the imagination of the less able the opportunity to use that resource must be grasped before it disappears. Some idea of the variety of opportunities to involve less able students in the study of current and local affairs is provided in Figure 25.7.

These depart somewhat from traditional geography unit titles and presentations though each activity can involve students (and not only less able ones!) in the development of concepts and the acquisition of a wide range of relevant skills. They are potentially very valuable if the teacher is concerned, as one should be, with attitude formation. They can also motivate less able students for they involve students in doing things which are useful for others (Midwinter 1972). Their self-esteem and respect thus receive a considerable boost.

The Need to be Active Rather than Passive

Frequently observed characteristics of . . .

The less able student

- responds poorly to traditional school situations in which writing, reading and listening predominate;
- bored by repetition of method;

1. **Holiday guide**
   A guide for local community to best holiday value. Also, information about different places.

2. **Best buy guide**
   A consumer guide to local shopping.

3. **Places of interest around . . .**
   Guides, gazetteers, trails, walks, etc.

4. **The people of . . .**
   A study of local people, past as well as present.

5. **Vacancy!**
   A guide to available houses; offices; jobs etc.

6. **Keeping happy**
   A magazine with a geographical flavour for senior citizens.

7. **So you think you're poor!**
   A look at other people and places.

8. **Homes and gardens**
   With obvious practical possibilities.

9. **Let's look at . . .**
   An opportunity to investigate particular aspects of local and world issues.

10. **Let's adopt a . . .**
    Opportunities for classes or groups to become practically involved in community or environmental work, for example caring for a park.

Figure 25.7 A variety of titles for study topics based on current and local affairs

- can respond enthusiastically to activities which involve looking, doing and talking;
- often stronger on motor skills.

The teacher

- most secure with 'academic model' of education;
- usually measures success by the quality of written responses;
- has to work to the constraints of the school timetable and organisation.

Often, the cause of failure in school work experienced by low achievers and less able students has been their difficulty in responding to traditional school situations. They can become hostile due to their inability to respond in these circumstances and often gain no satisfaction even from trying. Their need is for more problem-solving, inquiry-based learning and for involvement in games and simulations, etc. However, these more active learning experiences are not a panacea in their own right. Careful thought must still be given to the form and presentation of materials as Figures 25.8a and 25.8b, 25.9 and 25.10 show. These three field exercises demonstrate a number of important points about teaching slow learners:

1. Figure 25.8 shows how a resource can be used to arouse interest and focus attention. An old photograph is usually more appropriate for this level of ability than an old map. The skills of observation, data collection and recording can be encouraged in this way as well as other skills which facilitate the expression of knowledge, insights and judgement.

2. Abstract ideas such as nodality and spatial interaction are constantly recurring in geography. Such theoretical notions must be demonstrated at a level within the concrete experience and level of cognitive development of the less able student. Hopefully the ideas can then be transferred to more abstract and, perhaps, more important situations. Figure 25.9 shows one way this has been done. In fact while Figure 25.9 is a quiet, suburban road junction, within the space of a few metres there is a telephone junction box, water supply manholes, a post box and an electricity junction box as well as, of course, the occasional vehicle and pedestrian.

3. The less able often have the greatest difficulty in organising their work and themselves. They quickly lose heart if too much is expected of them in this respect. Careful thought has to be given to the design of all activities and worksheets for this reason. Obviously, one of the major problems with the less able is making fieldwork and other resources simple enough. When maps are to be used they should:

(a) be large scale and denuded of 'clutter';

(b) be laid out so that they are easy to read (not upside down with right being left and so on);

(c) have clear and well-known land marks and street names marked on them; and

(d) contain a minimum of instructions and other verbiage.

Figure 25.10 shows a map prepared for fieldwork with these points in mind.

Apart from the useful *skills* which can be fostered (observation, recording, using maps, etc.) exercises such as these are pointless unless they stem from and lead students to important geographical ideas. The exercise in Figure 25.9 might be regarded as worth-

Figure 25.8 (a) A sample worksheet for a field study on change in the local area; (b) Argyll Street 60 years ago

(a)

> Here is a photo.
> It was taken at the corner of Argyll Street.
> Stand in the same place.
> Look at the view today.
>
> List five ways in which the view is different today.
>
> 1. _____   2. _____
> 3. _____   4. _____
> 5. _____
>
> What transport can you see on the main road today?
> How has the building on the right changed?
> What can you see today behind where the old bus is?
> Draw a picture of it.

(b)

while only if it leads to some understanding of the orderliness with which shopping centres in a large city are arranged, for example.

The Need for Reassurance, Not Failure

Frequently observed characteristics of . . .

The less able student

- has difficulty in organising work;
- easily becomes discouraged by the poor appearance of the work he or she does;
- often unable to follow what is required;
- likes to feel successful;

- often wants to impress teachers and 'authority'.

The teacher

- short of time especially to prepare materials specifically for less able students;
- often uses resources which were designed for more able students;
- sees less able students as a low priority;
- expects students to wait until exam time for their reward.

Efforts must be made to minimise lengthy instructions, whether written or spoken, whenever less able students are required to produce 'work'. One might almost go so far as to say if the intentions of a resource are not immediately apparent, then its chances of success are slim. Likewise, effort must be made to help students organise their responses. This may mean drawing a chart to be completed, starting a sketch which has to be completed, giving words to choose from, etc. 'Hand-me-down' items which went well with more able students will not do. Resources must be designed specifically for the less able if they are going to work. This way, there is less likelihood of the student constantly feeling a failure. The better their self-image the more likely we are to get high levels of motivation and better responses. For example, Figure 25.11 shows a resource which was developed from a lengthy essay on tea picking. It was ideal 'content' for a unit on the Third World being undertaken by a class of less able, but the reading level of the original resource was far too advanced. The teacher thus produced a 'purpose built' resource for the class.

The Need for Inductive, Rather Than Deductive, Learning Experience

Frequently observed characteristics of . . .

The less able student

- likely to be still at the Piagetian stage of concrete operational thought (indeed may always be!);
- finds abstract thought difficult;
- needs ideas to be expressed within the context of concrete situations;
- often has a lively interest in a wide variety of subjects and issues;
- may have a strong social or moral conscience.

The teacher

- as a geography specialist favours subject-centred approaches to learning rather than child-centred approaches;
- likes to know the likely outcome of a lesson before he or she starts;
- worries about the reactions and comments of his or her peers.

It is essential to teach geographical ideas to less able students within the context of concrete situations rather than proceeding directly to or working from abstract generalisations and models. The so-called 'concentric' approach of working out from the student's own experience so successfully practised in primary schools is a useful model for the less able secondary student. Of course it is vital that students are taken beyond the 'immediate' environment as

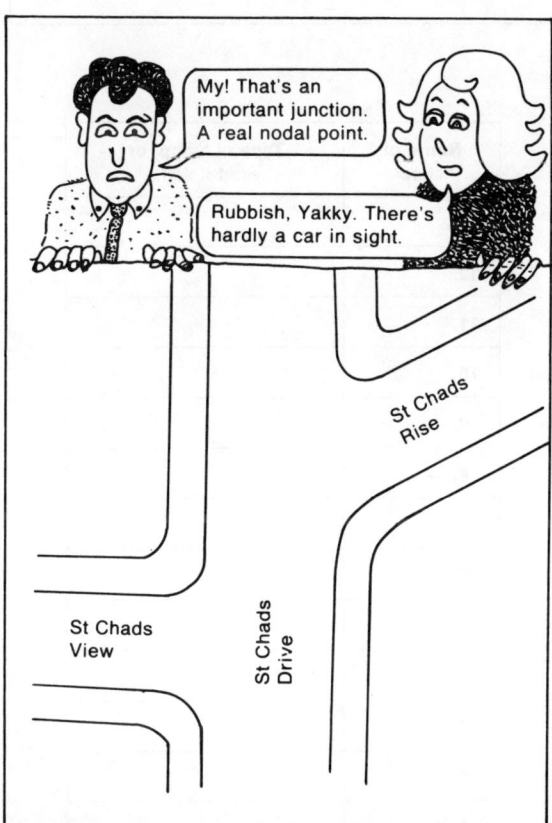

Figure 25.9 Searching for evidence—detective fieldwork in the local area

Figure 25.10 Simple map and data organisation chart for fieldwork

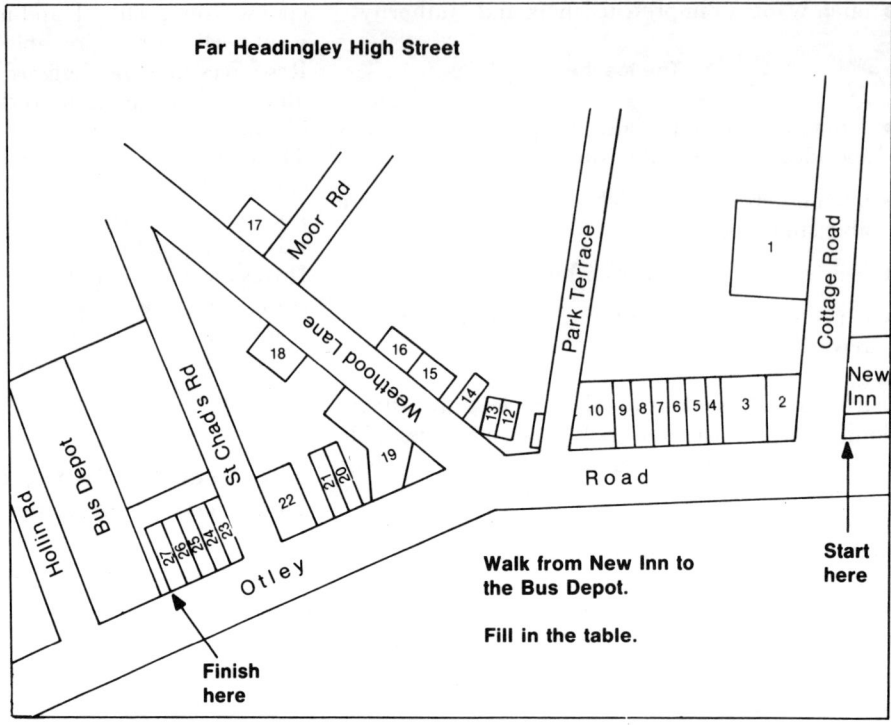

Walk from New Inn to the Bus Depot.

Fill in the table.

Name of 'shop'	Type of 'shop' or what it sells
27.	
26.	
25.	
24.	
23.	
22.	
21.	
20.	
19.	
18.	
17.	
16.	
15.	
14.	

Name of 'shop'	Type of 'shop' or what it sells
13.	
12.	
11.	
10.	
9.	
8.	
7.	
6.	
5.	
4.	
3.	
2.	
1.	

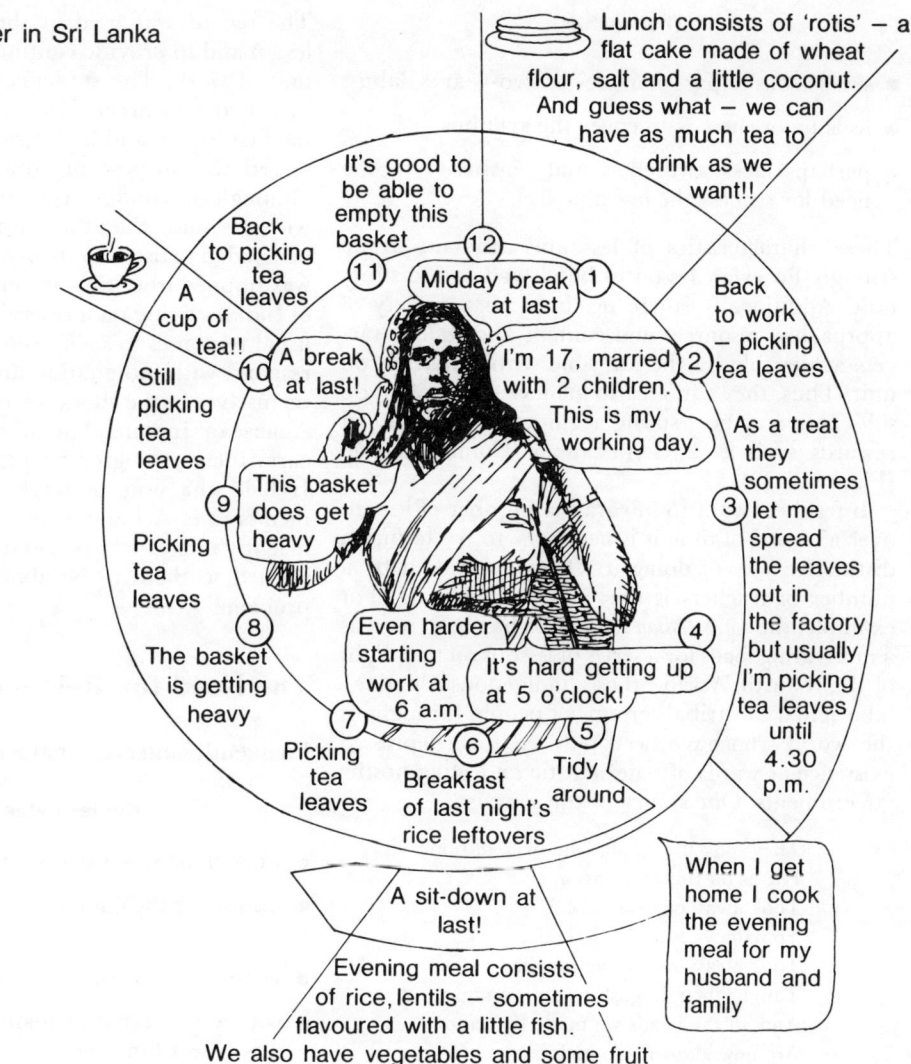

Figure 25.11 An example of a pictorial 'essay' for less able students

their horizons, experiences and knowledge do need expanding. This can be done using the 'concrete' example as the springboard. For example, many geographical and environmental ideas, skills and attitudes can be introduced or fostered in the context of the school, one of the few experiences that any class has in common. The school grounds often offer tremendous scope for practical work of many kinds (Pick 1977). Indeed, there is much scope for involving less able students in practical work at neighbourhood level too, thus fostering the development of much more positive self and neighbourhood images.

The Need for Short-Term Goals

Frequently observed characteristics of . . .

The less able student

- has a very limited concentration span;
- tends to have a short-term view of life;
- feels a failure in the system;
- likes success and praise;
- needs some form of 'achievement' frequently during school time;
- makes a limited 'carry over' from lesson to lesson;
- can remember many ideas, skills, etc. when motivated and when relevant.

The teacher

- sees lessons as part of a one- or two-year syllabus;
- feels he or she has to cover the syllabus;
- perhaps lacks awareness and sensitivity to the need for success the less able feel.

These characteristics of less able students argue strongly for every lesson to be viewed as a distinct unit. Additionally, there needs to be a variety of approaches, resources and content and *realistic* targets achievable by the less able within that lesson unit. Thus, they can seek to achieve—and often do so! There may be a strong argument for additional rewards such as a Certificate of Commendation (Figure 25.12).

If teachers want to integrate a number of lessons over a period of time it is necessary to try to find a distinctive way of doing it. One device used by a number of teachers is the 'pop' or folk song. For example, the song *Indian Reservation* has been used as a motivating focus for a series of lessons on the plight of the 'Fourth World', those 'indigenous', 'native', 'aboriginal' or 'tribal' groups of people throughout the world who have been reduced to a depressed existence as wards of paternalistic and often hostile governments. One stanza of the song is:

> They took the whole Cherokee nation
> Put us on this reservation
> Took away our way of life,
> Tomahawk and Bowie knife.
> Took away our native tongue
> Taught their English to our young.
> And all the beads we made by hand
> Are nowadays made in Japan.
>
> Cherokee people, Cherokee tribe
> So proud to live, so proud to die.

Certificate of Commendation

Awarded to _____ of class _____ at Seacroft Park Middle School for work and effort of a consistently high standard in Humanities

Date_____ Teacher_____

Figure 25.12 A sample Certificate of Commendation—a long-term reward

The record was used as the introduction to each lesson and to provide continuity in a two-week-long unit of work. The students, all 15 year olds, were surprised to be greeted by the persistent drumbeat in the first lesson, and thus remembered it. The record served the purpose of arousing their interest and although the students were initially more concerned with the music than the words, they were eventually persuaded to listen to them carefully. The recording was stopped from time to time to build up a picture of Indians living on a reservation. Contrary to their usual responses to such issues, the students began to respond with indignation and clearly began to put themselves in the shoes (or moccasins) of the North American Indian. The stimulus was sufficient to carry them through a fortnight's geography lessons. It held the unit of work together and made it memorable. A number of class members were sufficiently interested to suggest other recordings which seemed to them to be about social and economic problems.

The Need for Relevance

Frequently observed characteristics of . . .

The less able student

- work done in school is separate from 'real life';
- cannot see the point of much of what she/he has to do;
- seldom does work related to his/her own interests;
- is very interested in many things which in some way affect him/her.

The teacher

- adapts an academic syllabus to the less able;
- is not sufficiently critical of what is being done with the less able;
- has little knowledge or understanding of the teenage subculture;
- deliberately rejects much of the teenage subculture.

Lessons for less able students should seek to extend their experiences and insights by drawing on their cultural experiences—even if only as a starting point. A traditional topic such as 'The geography of towns' may seem dull, boring and irrelevant. The geography of a pop group or football team probably will appear far more relevant and appealing. Additionally it will allow the teacher to achieve similar ends, if such ends include the conceptual, skill and social development of the students. Social and environmental conditions and inequalities as well as fairly straightforward descriptive geography often form the focus of folk or pop songs (Renwick 1981). Not only does the pop song provide variety, already justified as being essential, it reflects a willingness of teacher to meet students on 'their' territory. This is a positive change for students from always meeting teachers on their territory of the school. Music often enlists the emotions to gain the initial interest and it seems that empathy precedes rational thought for many less able teenagers. If this is so then we should perhaps make more use of newspapers and magazines and of their techniques of using texts which are powerful, emotive, and personalised. 'Less able' does not mean less interested!

Conclusion

Schooling too often reaffirms the low self-concept of the less able student. It does so through its organisation (particularly where streaming is practised), through its adherence to bookish and passive modes of teaching, and through its use of methods and resources inappropriate to their needs. Apart from the suggestions already made, this sense of failure can be minimised by presenting such students with open-ended learning situations in which there are no clear-cut right or wrong answers. Utilise issue-based situations in which the expression of the opinions by the less able student is encouraged. Such exercises involve the student and value his or her contributions. Concomitantly, success is more likely to follow if we thought more frequently of praising the most minor improvement less able students may make than from the more frequent condemnation of their poor work.

A word of caution is perhaps necessary at this point. In attempting to minimise failure and maximise reassurance in order to build up confidence we must beware of falling into the trap of providing work or resources which are too easy or too trivial. The patronising policy of dealing with slow or unwilling learners by trying to make everything popular and easy for them is questionable. A well-presented, realistic challenge may be more effective. Indeed it could be argued that the present-day educational system drastically reduces all the challenges in a student's life—other than the one for academic attainment.

Lack of response often can be due to the triviality, more than the difficulty, of what is offered. One of the greatest skills the teacher of the less able can acquire is the ability to identify the level of work which permits achievement and does not lead to further failure.

In conclusion, it is probably most useful to suggest a checklist of points upon which successful teaching of the less able depends.

1. Identify teaching targets clearly.
2. Avoid student failure and build up student self-esteem.
3. Reduce student dependence on others.
4. Carefully grade learning material.
5. Accept a slow speed of instruction.
6. Maximise variety of resources and strategies.
7. Consider motivational levels appropriate for the age group concerned.
8. Provide swift and constant feedback.
9. Evaluate student progress constructively.
10. Reinforce learning and provide for adequate repetition and practise of skills.
11. Be relevant.
12. Be discriminating in the selection and production of resources.
13. Constantly evaluate one's own performance.
14. Remember the key word—variety.

References

Blyth, A et al. (1976) *Place, Time and Society 8–13: Curriculum Planning in History, Geography and Social Science*, Collins-ESL for the Schools Council.

Midwinter, E. (1972) *Projections—an Educational Priority Area at Work*, London: Ward Lock Educational.

Pick, W. (1977) 'School as Resource', *Bulletin of Environmental Education*, No. 76/77.

Pick, W. and Renwick, M. (1979) *Teachers' Perceptions of the Less Able,* Geography Centre, School of Education, Leeds Polytechnic, Leeds. (Mimeo.)

Renwick, M. (1981) 'Music and Songs in Teaching Geography to the Less Able', in D. Boardman (ed.) *G.Y.S.L. with the Disadvantaged*, Sheffied: Geographical Association.

Renwick, M. and Pick, W. (1980) *Going Places*, Books 2 and 3, London: Thomas Nelson.

Storm, M. (1979) 'Some Tentative Thoughts on a Taboo Topic', *ILEA Geography Bulletin*, March.

26

Planning and Teaching a Geography Curriculum Unit

John Fien

This is the curriculum unit and lesson planning chapter of this book. Unit and lesson planning provides the vehicle through which the ideas on catering for individual differences and teaching strategies outlined in earlier chapters can be implemented. Eight definite steps in planning a curriculum unit and six steps which can be used in implementing the unit through a series of lessons are described, using examples of work on the topics 'Kangaroo Conservation', 'Refugees', 'Transport in Nepal', 'Urban Problems' and 'Cyclones and People'. Each of the examples is taken from curriculum units developed according to the strategy outlined in the chapter. A key part of the chapter is a list of 20 safeguards for unit and lesson planning which can help teachers avoid many of those little (and sometimes not so little) problems that arise from too hasty unit and lesson preparation.

This chapter complements the next one on school-based curriculum development. Its purpose is to outline one method of implementing a geography course through the planning and teaching of modules of study, here referred to as curriculum units. A curriculum unit is a planned sequence of learning experiences designed to help students achieve the objectives of learning about particular topics or aspects of a topic in a geography course. Curriculum units may be variable in length, but generally represent four to six weeks of class time. In many ways, this chapter could also be looked upon as a pivotal one for this book for in the planning and teaching of a curriculum unit, many of the ideas on the content, skills and methods of geographical education presented in previous chapters need to be considered.

Planning and teaching a curriculum unit is a practical activity and it is intended that this chapter be a practical one, also. Much of it will be in point form and consist of lists of suggested steps in the processes of unit writing and lesson planning.

However, the ideas presented are suggestions only. The points are not definitive nor are the steps to follow inflexible. The method of planning and teaching a curriculum unit described in this chapter has proven both easy to use and quite effective in my own teaching, in preparing student teachers, and in unit writing workshops for experienced teachers.[1] It represents a personal synthesis of ideas gathered from many people, including several who are contributors to this book. Nevertheless readers should see the strategy for planning and teaching a curriculum unit outlined here as but one of many possible strategies. Others are to be found in books such as those by Tolley and Reynolds (1977), Michaelis (1980), Graves (1980) and the Secondary Geography Education Project (1977). Consequently, please consider these also, and then change the strategy presented here to suit the special demands of your own curriculum, your own teaching style, and most importantly, the needs and wishes of your own students.

Some Preliminary Safeguards for Successful Unit and Lesson Planning

All sources of advice on planning and teaching a curriculum unit or a lesson stress the importance of careful and detailed planning. Unfortunately, the components of 'careful and detailed preparation' are rarely made explicit. However, they do exist, and generally can be found in the intuitive knowledge of classroom operations understood by most teachers. The list that follows is an attempt to outline the more important of them. It originated in a brainstorming session conducted with a group of geography teachers. They suggested that the list could be seen as a set of preliminary safeguards for successful curriculum unit and lesson planning. There are 20 points in the list.

1. Know why you are teaching and, especially, why you are teaching geography.
2. Know what geography is, yourself, and that your students understand the key questions that characterise geographic inquiry. These include:
   (a) What are the phenomena and patterns being studied? Where are they located and how are they distributed?
   (b) What has caused these locations and distribution patterns?
   (c) What consequences do they bring for the environment and for people?
   (d) How could they be altered or managed in order to make the world a better place in which to live?
3. Subdivide each key question into four more specific focus questions. Focus questions provide the themes for individual lessons. Most often, the answers to focus questions will be in the form of generalisations which provide the content focus of lessons.
4. Structure your curriculum units around such key questions, and try to direct each lesson to finding an answer to at least one focus question.
5. Do not confuse the different levels of objectives possible in unit and lesson planning. Keep the aims and general objectives that form the rationale of a course and unit separate from the more specific objectives of individual lessons.
6. Know the syllabus or your school-developed curriculum plan well. Especially be familiar with the flexibility it allows in selecting content sequences, case studies and learning experiences. This will enable you to meet better the needs and interests of your students and your available resources.
7. Be familiar with the broad outline of the content of the topics for which you are planning curriculum units and the resources in your school pertaining to it.
8. Select topics and learning experiences that students will enjoy and see sense in doing.
9. Organise learning experiences and resources to promote the attainment of concepts, generalisations skills and values and not just factual recall.
10. Emphasise the importance of thinking, problem solving and decision making in the learning experiences planned for students.
11. Know the difference (if any) between your usual teaching methods and those that students find most enjoyable and motivating and especially any difference between your usual teaching methods and those that promote inquiry skills, creativity and self-reflection.
12. Plan for the different ability levels and rates of work (they are not the same) of all students in a class. This involves a consideration of many things, especially the grouping of students, the setting of goals for individuals and groups, the preparation of worksheets, and the selection of resources.
13. Keep students *mentally* active. Try not to confuse 'busy work' such as colouring in or protracted dull exercises with mental activity.
14. The concentration span of many students is low. Vary learning activities both *in* and *between* the lessons that make up a curriculum unit. Try not to let too many activities last more than 10 to 15 minutes, especially for students up to at least 15 years.
15. Be flexible in the sequencing and presentation of the activities prepared. Be aware of the problems that can arise to interrupt a lesson and be prepared to deal with them.
16. Make sure the introduction to a curriculum unit is stimulating and that it provides a plan or study guide of the work involved. Consider individual learning contracts with students at this stage.
17. Make sure that the conclusion of a curriculum unit reviews and cements the major concepts and generalisations involved in the unit.

18. Match student assessment items to the objectives of the unit of work. For example, do not emphasise knowledge, recall and comprehension in a testing program if most of a unit has been aimed at promoting inquiry skills and problem solving.
19. Evaluate your own actions constantly, and encourage this practice in students.
20. Evaluate all elements of the teaching of a curriculum unit—objectives, content, learning activities, resources, etc. Generally, this would involve reviewing most of the items in this list.

This list of preliminary safeguards to successful curriculum unit and lesson planning is not a complete one. Please add to it as your classroom experiences suggest additional points. It is not a list to be learnt by heart, but one to be used and internalised in the process of preparing work for students and presenting it to them. These points provide most of the theory behind the practical steps suggested in the strategy for unit and lesson planning which follows.

Planning a Curriculum Unit

An early piece of advice generally given on unit and lesson planning is to study the requirements of the syllabus on the unit topic concerned, and to write a set of objectives for teaching it, ensuring there is a blend of knowledge, skill and attitudinal objectives. This advice is especially suited to planning a curriculum unit according to the objectives model of curriculum development. It ensures that most of the right information for most of the right objectives is included in the unit. However, such objectives do not provide advice on actually teaching the topic or any guarantee that students will be motivated to study it. Indeed, we could ask whether such objectives are objectives for the unit, objectives for students, or objectives for teachers in teaching the unit. This ambiguity of intent in stating objectives as the first stage in unit planning leads me to suggest the following three points as a prelude to a first step in unit and lesson planning.

1. *Forget about objectives just now*. Allow them to 'emerge' (and they will) as the unit structure, individual lessons and student needs are considered.
2. *Study the syllabus or your school-developed curriculum plan* to determine the requirements of the topic in question, and the flexibility allowed in planning a unit for it.
3. *Do not go to textbooks* to see what they have on the topic at this early stage. The textbook author has answered the important curriculum questions of why, what and how to teach the topic, without reference to the local environment in which you are teaching, the individual needs of your students or the specific relevance of the topic to them. Rather, as a first step, ask yourself the question, 'what about it?' in relation to the topic to be taught.

Eight Steps for Unit Planning

Step 1: Ask Yourself the Question, 'What About it?'

Make a list of your own reasons for selecting and teaching this particular topic. This should do two things for you. First, it should allow you to identify the purposes in teaching the topic in terms of the environmental, multicultural, political or other values issues involved in the topic that have relevance for your students and their ways of perceiving the world. Secondly, it should identify the particular aspects of the topic that are worth pursuing. This will help teachers to differentiate between important generalisations and issues and background information.

What you are really doing in asking yourself the question, 'What about it?', is identifying the *angle* of the curriculum unit. In much the same way that journalists always have an angle that gives interest, structure and meaning to their feature stories, the angle of a curriculum unit, if chosen wisely, provides it with interest, structure and meaning for students. Some interesting angles for curriculum units include:

- Manufacturing — How is youth unemployment related to industrial restructuring?
- Wheat farming — Why does our bread cost so much when farmers are getting so little for their wheat?
- Rainforest conservation — Will rainforests be saved only for their political value rather than their ecological value?
- Energy — Are solar heaters worth having?
- Famine in Africa — Has Live-Aid solved any of Africa's problems?

Notice how the angle of each unit has been expressed as a question. This question can be introduced to students at the start of a unit, just like the headline and lead paragraph in a news story, in order to provide a problem which will provide a focus for discussion and investigation throughout the unit.

Step 2: Structure the Topic According to a Series of Key and Focus Questions which Reflect the Angle of the Unit

The key questions of geography which you have looked at elsewhere in this chapter and in this book may be expressed briefly as:

1. *What* and *Where* are the issues or patterns being studied?
2. *Why* are they there?
3. *What impact?* or What are the consequences of this?
4. *What could* be done to improve the situation? *What ought* to be done?

These key questions give rise to a series of more specific focus questions which provide the teaching structure and lesson sequence for the unit. Examples of focus questions for a curriculum unit on the topic 'Refugees' and whose angle is 'What would it be like to have to flee your home to live in a new country?' include:

Key questions	Focus questions
1. What and Where?	• What is a refugee? What would it be like to have to flee your home and move to a new country? • Have I ever met any refugees? How have I responded to them? • Where do refugees come from? • Where do they want to go?
2. Why?	• Why are there refugees? • What are the characteristics of the source areas and desired destinations of refugees? • What is the attraction of Australia for some refugees?
3. What impact?	• What problems face refugees in their journeys and in their new homes? • What problems have refugees found in Australia? In my local area?
4. What could? How ought?	• How ought Australians (or Canadians, etc.) respond to refugees? • What can be done to minimise their problems? • What can I do to be of assistance?

Step 3: Identify the 'Key Ideas' to be Understood in Order to Answer these Questions

These ideas should not be detailed statements of content. Rather, they should be expressed as the overall generalisations or principles you think students should understand as a result of studying the unit. Examples of generalisations that could follow from the 'Refugee' unit include:

1. A refugee is a person who has fled his/her country as a result of fear of being persecuted for reasons of race, religion, nationality or political opinion; or who has been displaced by war or a natural disaster.

2. The twentieth century has experienced an escalation in the number and geographic distribution of refugees.

3. Most refugees today are coming from politically unstable, developing countries in Africa and Indo-China.

4. Refugees from these developing countries face serious problems in finding countries to accept them and rebuilding their lives.

5. A high degree of international cooperation is needed to re-settle today's large number of refugees.

6. The regional proximity of Australia to the Indo-Chinese refugee problem and its involvement in the Vietnam War makes our response to the needs of Indo-Chinese refugees particularly important (Ritchie 1981).

Step 4: Prepare Overall General Objectives for the Curriculum Unit
Do this by:

1. using the generalisations developed in Step 3 and relevant case studies as the knowledge objectives for the unit; and

2. selecting skill and affective objectives appropriate to students and the unit from the affective objec-

Figure 26.1 Key and focus questions matched with objectives for the unit 'Cyclones and People' (Fien et al. 1984)

Key questions	Objectives		
* Focus questions	Knowledge	Process and skills	Values
1. What and where? * What is is like to be in a cyclone? (half lesson)	Basic description of cyclone force and destruction Description of what is was like in Darwin during Cyclone Tracy in 1974	Reading and listening Discussion, comprehension	Reflecting on own knowledge and feelings Empathy for others Appreciation of the force of nature
* What is a cyclone? (half lesson)	Types of tropical cyclone; names used in different parts of the world; typical passage of a cyclone—destructive winds, low air pressure, size, lifespan, vertical structure, rains, the eye Features associated with cyclones—storm surges, swells, storm waves, thunder and lightning	Reading and interpreting text, maps and diagrams	Importance of careful and accurate research
* Where do cyclones occur in Australia? Where else do they occur? (one lesson)	Places where hurricanes occur in Australia Times when they are most frequent Lifespan and estimates of frequency World patterns of tropical cyclones	Drawing and interpretation of maps and graphs	Appreciation of the earth as a system—with similar events occurring under similar conditions all over the world
2. How and why? * How do cyclones form? Why do they strike where they do? * Why do cyclones follow particular paths and patterns? (two lessons)	Cyclones as a septum Heat—the energy source The general circulation of winds Cyclone paths The dissipation of energy and the deterioration that causes a cyclone to weaken	Reading and listening Comprehension from a film Synthesis of ideas Model building	

3. **What impact?** * What are the effects of cyclones on people and the environment? (one lesson)	Definition of a natural hazard Case studies of Cyclone Tracy and the 1985 cyclone in Bangladesh	Reading and comprehension of passage Ability to extract main points Ability to apply ideas from the text Discussion	Respect for point of view of others
* How severe are the effects of cyclones? (two lessons)	Bad effects of cyclones—death, damage, destruction Beneficial effects of cyclones—relief of drought, destruction of pests Broad categories of world population distribution Areas of greatest potential cyclone hazard Comparison of the effects of the two cyclones in areas of differing economies (Australia and Bangladesh)	Extraction of information from newspaper reports, maps and photographs Construction of a flow diagram from information gathered Cost or benefit analysis	Human considerations in cost/benefit analysis—the costing of loss of life compared with damage The implications of attempting to compare loss of life with monetary cost Ability to consider one's own attitude to this question
* How do people perceive cyclones? How do they respond to them? What are the effects of such behaviour? How should people perceive and respond to cyclones? (two lessons)	The perception of cyclone vs. drought hazards in Australia The adjustments made by people in the face of cyclone warning systems—evacuation, coastal protection, emergency relief, etc. The role of the Weather Bureau, State Emergency Service and the police.	Reading and Comprehension Discussion Extracting information from reports and photographs Listening with comprehension to guest speaker from State Emergency Service	Clarification of own perceptions Willingness to prepare for a cyclone in advance; cooperation with emergency service activities
4. **How ought?** * What can be done to predict cyclones or, at least, minimise their effects? (one lesson)	Difficulties of accurate track prediction Progress in cyclone prediction Modern techniques of prediction Project Stormfury	Prediction skills based on cyclone reports Mapping and map interpretation Satellite photograph analysis	Appreciate need for accuracy in observation and mapping
* How can an area destroyed by a cyclone be rebuilt? (two lessons)	The rebuilding of Darwin Cyclone-proof building methods Land use planning in Fanny Bay—a simulation	Listening and reading with comprehension Critically analysing plans Role play and decision making	Analysis of value positions Clarification of own views and appreciation of the views of others Consensus decision-making

tives and the skill development program found in the syllabus or your school-developed curriculum plan.

Step 5: Identify Appropriate Case Studies, Exercises and Teaching Strategies that Could be Used to Help Students Develop the Knowledge, Skill and Values Objectives of the Unit

This may involve checking syllabus guidelines and reading lists, investigating resources held in the school, talking with other teachers to see how they have taught the topic previously and the resources and exercises they found useful and reviewing commercially available resources. Make a selection of practical activities and case studies appropriate to the special emphases of your course and the learning rates and interests of your students.

Figure 26.1 illustrates one way the unit 'Cyclones and People' may be set out with knowledge, skill and values objectives matched to the key and focus questions of the unit. Note the way the objectives have not been stated in specific terms: you only need specific objectives for individual lessons. Rather, the *unit objectives have been stated in such a way as to provide*

Figure 26.2 Integration of key and focus questions, resources and exercises for the curriculum unit 'Transport in Nepal' (after Sheriff 1981)

Key and focus questions	Resources	Exercises
1 WHAT and WHERE? • What is the present transport pattern in Nepal? • How is transport affected by relief in Nepal? • How well are settlements in Nepal interconnected?	A set of photographs and slides of scenes in Nepal. Maps of (i) relief and drainage, (ii) settlement, and (iii) transport patterns in Nepal.	Photograph and map interpretation. Calculation of transport connectivity. Coordination with physical educational teachers on the effect of exercise and altitude on pulse rate and breathing to illustrate the work of mountain porters.
2 WHY? • Why has little modern transport developed in Nepal? • What are the barriers to road and railway development today?	A list of important dates in the political and economic history of Nepal. A topographic map of a typical mountainous area in Nepal.	Time line construction and interpretation. Drawing a cross-section across two adjacent valleys. Planning new transport routes across the mountains.
3 WHAT IMPACT? • How does the limited transport network affect Nepal? • How do the Nepalese people overcome their transport problems? • What transport does the tourist industry use?	A collection of short passages from newspapers, magazines, and travel books on the living conditions and cultural patterns of the Nepalese people. A set of photographs and slides of scenes in Nepal.	Reading, comprehension and summarising. Photograph interpretation.
4 WHAT COULD BE DONE? • What alternative solutions to Nepal's transport problems are available? • What should Nepal do to cope with its transport problems?	A hypothetical costing of several ways of developing transport in Nepal, e.g. STOL aircraft, helicopters, jet boats, bitumen roads, maintenance of existing mountain paths, etc.	A simulation in which groups of students decide how $1 000 000 of transport development money should be spent.

ready guidance for the development of learning activities and lessons for students.

Step 6: Develop the Resources and Learning Experiences that are to be Used

This is a 'hard work' phase in the unit planning process. Many factors ranging from selecting case studies to fit in with a global coverage course planning matrix, to vetting the readability of resources and planning remedial and enrichment schemes need to be considered at this point. The selected resources and exercises should then be organised into an order that matches the sequence of inquiry questions developed in Step 2. The curriculum unit 'Transport in Nepal', for example, might have the pattern of key questions, resources and exercises detailed in Figure 26.2.

Step 7: Prepare the Resources and Exercises for Presentation to Students

Set these out on separate sheets of paper. The student exercise sheets should contain all the instructions for students and be based upon the data presented or referenced in the matching resources sheets. Separate sheets of exercises should be prepared for each sub section of the unit and for different levels of student ability. This maximises the flexibility of both teacher and student use of the student exercises.

Resource sheets should be prepared for each subsection of the unit also. The resources selected should be as interesting and as motivational as possible for they provide the stimulus and data for student inquiry. Utilise a wide range of resources, including textual material, graphs, tables, diagrams, cartoons, newspaper cuttings, etc. Reference additional available resources such as relevant sections of textbooks and audio-visual materials on the resource sheets. I have found it useful to institute and maintain a coloured paper scheme for student exercise sheets and resource sheets so that students always see, for example, white sheets as resources to be utilised and yellow sheets as exercises to be completed.

Step 8: Prepare an Assessment Program for the Unit

Remember to focus such a program upon the generalisations in the unit rather than factual recall, the processes of inquiry used in the unit, and the ap-

Figure 26.3 A sample introduction and overview for the curriculum unit 'Kangaroos: An Endangered Species or an Underdeveloped Resource?' (after Smerdon 1981)

Kangaroos: an endangered species or an underdeveloped resource?

Introduction

The kangaroos controversy is one which is extremely important in Australia. There are many sides to the kangaroos issue. Some have identified the sides in this way:

(i) *Pouch*—the preservation of the kangaroo in its natural state without any extermination of the kangaroo.
(ii) *Purse*—the other extreme to *pouch*, that is the destruction of the kangaroo for any profit that may be gained and with no thought to the future of the kangaroo.
(iii) *Production*—somewhere between *pouch* and *purse*, in that it is concerned with the harvesting of the kangaroo but in which the future of the kangaroo is assured.

To summarise these arguments the title 'an endangered species or an underdeveloped resource' was chosen. Indeed this is how the kangaroo controversy is being polarised, as these are really the only two arguments one finds in the media.

This unit is designed to serve two purposes:

(a) To assist students to clarify their values and make informal decisions about a controversial man–environment issue.
(b) To serve as a model with which students may conduct their own inquiry in a similar area.

The unit has four parts:

1. A brief study of the nature of endangered species and resources.
2. A study of the ecology of kangaroos—the types of kangaroos, their location, ecological needs, responses to environmental changes, etc.
3. A study of the kangaroo controversy. What are the various suggested ways of using the kangaroos? Decision making is also involved when the student is asked to choose an appropriate management method.
4. A study of the public image of the kangaroo. Decision making is involved here as students are asked to commit themselves regarding aspects of the public image. As a final activity, the students are asked to determine their stance on the controversy.

Figure 26.3 Contd.

Unit overview

Key Questions	Resource Sheet Nos	Exercise Sheet Nos	Notes for Lesson Nos
WHAT and WHERE? What is an endangered species? What is an economic resource? What is a kangaroo? Where is it found? Why? What are the population numbers? Why do they vary? How is the kangaroo used as an economic resource?	1, 2, 3	1, 2	1, 2, 3
WHY? Why do some people view kangaroos as a resource? Why do other people see kangaroos as an endangered species?	4, 5	3, 4	4
WHAT IMPACT? What is the public image of the kangaroo? What are the consequences of this image? Does it need to be altered or renewed?	6, 7	5	5
HOW OUGHT? How would you define the kangaroo issue? Where do you stand on the issue? How should kangaroo populations be managed?	8	6	6

plication of the generalisations and inquiry processes to new situations. Chapter 28 contains details on various assessment techniques and strategies for developing an effective school-based assessment program.

Planning Lessons From a Curriculum Unit

The aim at this stage of unit and lesson planning is the preparation of a set of teacher's notes to accompany the student exercise sheets and resource sheets. Only then will the unit be complete and ready for classroom implementation. School-based curriculum development, group preparation of curriculum units, and the sharing of units in and between schools necessitate that teacher's notes now be prepared in as much detail as possible. They should therefore include at least the following:

1. a unit introduction;
2. a unit overview based upon the inquiry question structure of the unit developed in Step 2;
3. the division of the unit into lessons and detailed notes for teaching each lesson.

Preparing the unit introduction and overview is a relatively easy task as, essentially, they are a summary of the unit and the work you have done on it. Figure 26.3 is a sample introduction and overview for the curriculum unit 'Kangaroos: An Endangered Species or an Underdeveloped Resource?'.

Six Steps for Lesson Planning

Dividing the unit into lessons and writing lesson notes for each one is a detailed task involving at least six steps.

Step 1: Divide the Unit into the Required Number of Lessons

Remember to take account of the many school activities that reduce the time theoretically available for teaching a unit. Thus, a six-week unit for a class that has geography three times a week is not an 18-period unit. For example, an athletics carnival and a biology field trip for some students during the period of the unit coupled with the time needed for assessment, could effectively reduce the number of lessons available for teaching the unit to 14 or less.

Step 2: Plan for the Realities of School and Classroom Life

This step involves a consideration of the many problems that could arise during the unit and interrupt student learning. There is a need to consider at least the following questions in this early stage of lesson planning:

1. Where have students been, and what activities have they been doing, before the lesson? Will any arrive late to class? Why?
2. Do you expect students to be boisterous or calm, eager to work or intellectually drained when they arrive?
3. What sort of lesson introduction will match their mood?
4. What facilities are available for reproducing the resource and student exercise sheets necessary for the lesson?
5. Will there be sufficient textbooks and other commercially prepared resources?
6. What audio-visual resources are available and will be needed? Is the classroom suitable for these?
7. Will the desks be large enough for map work? Can they be moved about for group work?
8. What familiarity will students have with the material and skills involved? How have they reacted to related topics and similar types of inquiry exercises in the past?

These are but a few of the many such questions that could be asked. Teachers' familiarity with their students will let them know the questions appropriate to each group of students and lesson context. A consideration of them before lesson planning starts could prevent many classroom and learning disruptions.

Step 3: Develop the Purpose of Each Lesson

The purpose or aim of each lesson should direct students to an answer to at least one of the focus questions of the unit. The answers to a focus question will generally be one of the 'key ideas' or generalisations that form the knowledge aims of the unit. Do not be overly ambitious in developing lesson aims. Try to think of aims as the one or two 'big ideas' you really want students to discover during the lesson.

Step 4: Organise the Resources and Exercises Selected for Each Lesson into Two or Three Different 10 to 15 Minute Activities

Such organisation can bring a variety of activities into any one lesson to help students learn through a variety of modes and reduce boredom. Think of these activities as the things students have to do to achieve the aims, the 'big ideas' of the lesson.

Step 5: Summarise These Activities in Terms of the Things Students Will Actually be Doing in Each Lesson

These then become the specific objectives for each lesson. Expressing lesson objectives as action statements ensures that they are objectives *for students* and that there will be maximum student participation and activity in each lesson. The aims and objectives for a sample lesson in the 'Transport in Nepal' unit on 'Why has little modern transport developed in Nepal?' (see Figure 26.2) might therefore be:

Aim: To help students understand:

1. The Himalayas are steep fold mountains and are a barrier to transportation.
2. Nations that are in the Third World and are geographically isolated are 'cut off' from the many technological developments found in other countries.

Objectives: To understand these generalisations, students will:

1. Construct a time line of the history of Nepal and answer interpretive questions on it.
2. Draw a cross-section across two adjacent valleys.
3. Work in a group to plan two new routes between the valleys, one for porters and one for automobiles.

Notice that the verb in each of these objectives is a practical one—'construct', 'draw' and 'plan'. Active verbs such as these mean that students will be actively engaged in the lesson, that they will have a tangible product from the lesson where appropriate (a time line, a cross-section and a route, for example, in this lesson), and that you, the teacher, will be able to monitor student understanding by the work that students are producing. In writing lesson objectives, try to avoid passive or ambiguous verbs such as 'explore', 'become aware', 'understand' or 'appreci-

Sample lesson notes

Topic Introduction to city problems and local issues
Class 9B   **Class size** 32
Period 2   **Time** 9.45–10.25
Room B4 (no blackout curtains)

Focus questions
1. Do all cities have problems?
2. What problems are found in our area?

Generalisations
1. All cities have problems caused by changes in land use.
2. These problems are found in large cities such as New York and in smaller parts of cities such as the local area of our school.

Objectives
During the lesson, students will:
1. Identify some problems of city life by listening to a song.
2. Make a list of the three major problems in New York by interpreting a diagram.
3. Work in pairs to identify three problems in their area in contrast with New York.

Resources
1. Tape of 'Inner City Blues' by Marvin Gaye.
2. Information sheet on New York's problems to be reproduced from W.P. Rae and N.C. Coutts (1982) *Contemporary Files 2*, Heinemann, p. 27.

Steps in the lesson

Step 1 *Teacher/student activity*—Ask students to name current songs that tell of problems of urban life.
(10 min) Ask for the problems and any 'quotations' from the songs. Write these on board. Tell class about a similar song from 'your day'. Before playing song, give class instructions to listen carefully to identify: (i) the country, (ii) the city, (iii) the singer's opinion of the city, and (iv) some problems of life in the city. Write these four things on the board. After playing the song, discuss answers with class. List any extra problems on the board. Blackboard Plan:

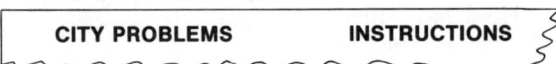

Step 2 *Student activity*—Distribute New York problems sheet. Students read it and select their three
(10 min) major problems (with reasons for their choice). Explain financial nature of New York problems.
(5 min) *Teacher activity*—Ask for reports, ensuring that students give their reasons. Explain the link between the financial and social problems in New York.

Step 3 *Student activity*—Students work in pairs to choose what they believe to be the three main
(10 min) problems of urban life in their area. Then they select the one major problem and make a list of the ways this problem affects their and other people's lives.
Teacher activity—Monitor student discussions, helping where necessary. Ask for one or two reports.

Lesson review
Remind students that each of 'their' problems is only one of many. Give examples from New York and other students' work.

Looking ahead
Tell class that this lesson and the inquiry planning sheet is the beginning of a project on 'Problems of Life in our City'.

Homework
Survey the opinions of five adults on the perceived severity of 'their' problems.

Figure 26.4 A sample lesson plan to illustrate the ideas developed in this chapter

ate'. Such verbs describe important cognitive processes, but are more suitable for the overall objectives of a curriculum unit rather than the specific objectives of a lesson.

Step 6: Write Detailed Procedural Notes for Conducting Each of the Activities in a Lesson

To do this, there is a need to consider the student skills involved, the readability of the resources, other factors that contribute to the difficulty of each activity, and the likely level of interest of students in the lesson. Also consider how you will introduce and pace each activity; identify students who will need additional help and enrichment activities; group students; correct students' work; and then conclude each activity and move students on to the next one. Teachers usually face such issues intuitively, using insights and skills they have developed from experience. They are made explicit here in order to illustrate the depth of lesson planning possible and, for less experienced teachers, necessary for successful teaching.

There is no one correct way of setting out the detailed procedural notes for a lesson plan. All teachers develop their own style or model that matches their varying approaches to teaching, and their familiarity with the lesson topic, strategies and resources being used at different times. One model that I have found useful is an adaption of a lesson planning technique developed by Bartlett and Cox (1982). Figure 26.4 is a sample lesson plan based on their model. It is for a lesson from a unit on urban problems and has been planned to help students answer the two related focus questions of 'Do all cities have problems?' and 'What problems are found in our area?'.

The lesson plan headings used in this model include:

1. *Topic*—to focus teacher and student inquiry.
2. *Class and class size*—to remind the teacher of the number of handouts and other resources to prepare.
3. *Period and time*—to remind the teacher of the lesson length and to plan for the possible disposition of students when they arrive for the lesson.
4. *Room*—to enable decisions to be made about desk arrangements, group work, display space and the use of audio-visual equipment.
5. *Focus question(s)*—to indicate where in the overall sequence of key and focus questions in the unit this lesson fits.
6. *Generalisations*—the key ideas that you want students to take with them from the lesson.
7. *Objectives*—to be expressed in terms of the activities that students will be actually doing during the lesson in order to achieve an understanding of the generalisations.
8. *Resources*—a detailed list of the full range of resources that are necessary for the lesson so that the teacher can check before the lesson that everything has been prepared.
9. *Steps in the lesson*—set out in as much detail as individual teachers feel they need. It is always a good idea to detail the steps in a lesson in terms of teacher activities and student activities so that you have planned exactly what you and your students will be doing at each stage of the lesson. Add approximate time allocations to monitor the pacing of the lesson. Include a blackboard plan where appropriate.
10. *Lesson review*—to refocus attention on the focus question(s) that formed the basis of inquiry in the lesson and to consolidate the generalisations.
11. *Looking ahead*—to show how this lesson will lead onto the next one.
12. *Homework*—however small, homework should flow from every lesson to consolidate learning. It is best if homework provides an application of the ideas or skills learnt during the lesson and helps form a link between one lesson and the next.

Conclusion

One strategy for planning and teaching a curriculum unit has been described in this paper. It is based upon several preliminary safeguards which attempt to ensure that the teaching and learning of geography is well planned, inquiry based, and caters for the individual needs of students. In summary, the strategy is composed of two stages, unit planning and lesson planning. The steps suggested for unit planning are:

1. Ask the question: 'What about it?' to develop an *angle* for the unit.
2. Structure the topic according to a planned sequence of key and focus questions which reflect the angle of the unit.

3. Identify the ideas (generalisations) to be understood in order to answer these questions.
4. Prepare general objectives for the unit.
5. Identify appropriate case studies, resources, exercises and teaching strategies.
6. Develop the resources and learning experiences that are to be used.
7. Prepare the resources and activities in forms suitable for presentation to students.
8. Prepare an assessment program for the unit.

The steps suggested for lesson planning are:

1. Divide the unit into the required number of lessons.
2. Plan for the realities of school and classroom life.
3. Develop the aim of each lesson as the 'key ideas' or generalisations students will take with them from the lesson.
4. Organise the resources and exercises selected for each lesson into two or three different 10 to 15 minute activities.
5. Summarise these activities in terms of the things students will actually be doing in each lesson. These become the specific objectives of the lesson.
6. Write detailed procedural notes for conducting each of the activities in a lesson.

Note

1. Some of the examples used in this chapter are adapted from curriculum units written by geography teachers at unit writing workshops conducted for the Geography Teachers' Association of Queensland. The teachers used the same method of unit planning which is described here. My thanks to Robert Ritchie, Helen Sheriff, and Russell Smerdon for permission to adapt sections of the curriculum units they prepared at these workshops.

References

Bartlett, V.B. and Cox, G.B. (1982) *Learning to Teach Geography*, Brisbane: John Wiley and Sons.
Fien, J., Fossey, W. and Cox, B. (1984) 'Geography: A Medium for Education', *Queensland Geographer*, Vol. 19(2), 4–16.
Graves, N.J. (1980) *Geographical Education in Secondary Schools*, Sheffield: The Geographical Association.
Michaelis, J.U. (1980) *Social Studies for Children: A Guide to Basic Instruction*, 7th edition, Englewood Cliffs, NJ: Prentice Hall.
Ritchie, R. (1981) 'Refugees', in J.F. Fien (ed.) *Teaching Units in Geographical Education*, Volume 5, Unit 5, Brisbane: Geography Teachers' Association of Queensland.
Secondary Geography Education Project (1977) *SGEP-PAK*, Melbourne: Geography Teachers' Association of Victoria.
Sheriff, H. (1981) 'Transport in Nepal', in J.F. Fien (ed.) *Teaching Units in Geographical Education*, Volume 3, Unit 5, Brisbane: Geography Teachers' Association of Queensland.
Smerdon, R. (1981) 'Kangaroos: An Endangered Species or an Underdeveloped Resource?', in J.F. Fien (ed.) *Teaching Units in Geographical Education*, Volume 5, Unit 4, Brisbane: Geography Teachers' Association of Queensland.
Tolley, H. and Reynolds, J. (1977) *Geography 14–18: A Handbook for School-based Curriculum Development*, London: Macmillan.

27

School-based Curriculum Development in Geography

John Fien

This chapter is the first of four chapters designed to help geography departments and individual teachers in the process of school-based curriculum development and evaluation. This sequence of chapters is designed to aid in the process of developing courses that match the needs of particular students in particular school environments and thus foster the goals of modern geography teaching and the use of appropriate teaching methods outlined in earlier chapters. The decline in the importance of centralised syllabuses and external examinations now make both of these possible.

This chapter outlines six principles for effective school-based curriculum development, the most important two of which are the collective responsibility of the entire geography department and the relationship between curriculum development and teacher development. A case study of one school's experiences with school-based curriculum development is used to illustrate a three-step program for course planning. These steps include: the study of the context in which the course is to be taught, the selection of a course structure (concept-, principle-, theme-, and learner-centred course structures are explained), and selecting and implementing a course design strategy. Particular emphasis is given to thematic and learner-centred course structures and the process model of course design because of their potential for achieving a wide range of educational and personal benefits for students of geography.

'Curriculum' is a term that is defined by many writers in many different ways. Some see it as the total program of studies a school organises for all its students, and others as the total of all the experiences, planned and otherwise, that students have while in the care of a school. Some define 'curriculum' as the things that teachers do with students in the classroom, and others as the actual learnings students take away from their school experiences. More narrowly, some see it as the program of studies for one subject in a school, and narrower still, others see it as the program of studies for one subject for one group of students, for example a particular class level. All are correct, at least in part, for the term 'curriculum' is a contextual one and, hence, needs defining according to the context in which it is being used.

The context in this chapter is provided by the course planning activities of a geography department. Thus, the word 'curriculum' will be used to refer to a specific course of studies for a particular group or class of students in the subject of geography. This focuses the goal of the chapter on the provision of guidance on some of the issues and steps involved in the task of school-based curriculum development.

Curriculum development is both a decision making and a planning process. In curriculum development, decisions are sought to at least six questions:

1. What is to be taught?
2. Why is it to be taught?
3. How will it be organised for presentation to students?
4. How will it be taught and learned?
5. What resources will be needed?
6. How will the teaching and learning be evaluated?

The answers to the questions are organised into an overall plan for teaching a course of studies and written up in a course statement such as a work program or school curriculum document. Curriculum development is an on-going process, however, and changes need to be incorporated into such documents as a course is taught and evaluated.

The importance and value of cooperative school-based curriculum development has been well summarised by geography inspectors in England who recently wrote that a school-developed work program

> ... is of greatest value to those who design and subsequently review it, in that the process of constructing and revising such a statement can stimulate teachers to sort out their rationale and to focus their thinking on important curricular issues. The discussion which accompanies planning can bring assumptions to the surface to be questioned and reflected upon, and it can open the door to fresh ideas. However, a teaching ... [program] can also provide useful guidance for staff who did not contribute to its design, especially if it is regarded as a working document which is kept under review. A teaching ... [program] can also provide a basis for discussion and co-ordination with other staff within the school, such as the teachers of related subjects; the teachers providing remedial support for pupils or catering for special needs such as English as a second language; and the headteacher and senior staff responsible for broader curricular policies and decisions. A teaching ... [program] should reveal what sort of contribution a course makes to the total curriculum of a school and, therefore, should be of value when reviewing the overall breadth and balance of a curriculum; the treatment of cross-curricular themes (e.g. environmental education, political education); and the development of whole school curricular strategies (e.g. language across the curriculum, the development of learning skills, and educating pupils for living in a multiethnic society).
> (Department of Education and Science 1986)

Popular wisdom has it that school-based curriculum development is a recent trend. True, more and more responsibility to determine course content, teaching goals, learning experiences, resources and modes of assessment and evaluation has been placed on schools in the last decade. Such activities are not really new to geography teachers. However, their recent level of involvement in them does represent an expansion of the curriculum development role that they traditionally have had. Even under the most traditional of syllabuses and stringencies of external examinations, geography teachers have had the task of analysing the prescribed and inferred (from past examination papers and examiners' reports) requirements of a course of study, and determining the balance of knowledge, skill and value components to be in the course, and the case-study exemplars, textbooks, teaching methods and assessment procedures to be used.

Opportunities for teachers to engage in such tasks are increasing, however. Among the reasons for this expansion of the teacher's role in curriculum development are:

1. The decline in the importance and, sometimes, complete removal of external examinations in many geography syllabuses.[1]
2. The realisation by many central educational authorities that curricula need to be designed to suit local school conditions and particular groups of students.
3. The marked trend for geography syllabuses to be expressed in terms of aims and generalisations to be attained by students rather than as lists of content to be covered.
4. The increased freedom of teachers to select and sequence content exemplars and resources that these trends necessitate.
5. The marked decline in the use of a set textbook as the basis for a course of study and its replacement by class sets of various resources and multi-media approaches to teaching geography.
6. The increased professionalism of teachers brought about by better qualifications, membership of geography teachers' associations and the experience many teachers have gained from developing courses for non-examination students in lower secondary and lower ability classes.

School-based Curriculum Development: Six Principles

Two very good books on curriculum development in geography have been published since the mid-1970s.

They are Biddle's *Translating Curriculum Theory into Practice in Geographical Education: A Systems Approach* (1976) and Graves' *Curriculum Planning in Geography* (1979). Both are important guides for teachers involved in planning their own courses and provide more detailed advice than this short chapter can give. Other books, for example, Boden's *Developments in Geography Teaching* (1976) and Tolley and Reynolds' *Geography 14–18: A Handbook for School-based Curriculum Development* (1977) provide examples of course outlines and the experiences of teachers in designing school-based courses. However, none of these four (despite the title of the Tolley and Reynolds book) focuses on issues and principles involved in school-based curriculum development, *per se*

It is, therefore, necessary that the issues and principles that underlie the theory and practice of school-based curriculum development be explored. Such principles provide the basis for advice on teacher action in the design of their own courses. Six of the principles are:

1. Effective school-based curriculum development is not an activity to be undertaken alone. It rarely occurs when the head of a geography department plans a course of study for a particular class by him- or herself and presents it as a finished document to the staff. It is the collective responsibility of the entire geography department working as a course planning team.

2. School-based curriculum development does not require the abandonment of all the content, resources and strategies used in existing courses and their replacement by new ones. Rather, it is an evolutionary process and works best when it is seen and used as a way of establishing priorities and criteria for the planning, teaching and learning of geography in a school. Existing curricula can then be reviewed, long- and short-term needs identified, and courses of action to modify and, if necessary, replace existing practices be planned and undertaken.

3. School-based curriculum development does not deny the role of central authorities to review or construct syllabus documents for schools. Indeed, such authorities and their syllabus committees have a vital role to play as one of the *starting points* for teachers in school-based course planning. School-based curriculum development sees a school's geography department as an interpreter of centrally designed syllabus documents and the final arbiter on that which is best for its students.

4. School-based curriculum development demands that teachers be aware of all the factors which may influence the quality of their students' experiences in a geography course. These factors range from the demands of society and the school system for geographically educated people, the organisational structures and procedures of individual schools and the staffing and material resources of a geography department, to the nature, demands and problems of the local community and the socio-environmental needs, concerns and interests of their students.

5. School-based curriculum development in geography is a team enterprise, dependent upon good working relationships between department members. Good relationships depend upon each person having the opportunities and skills to work as part of a group towards the achievement of a common goal. The head of department's role is crucial here. In addition to the skills involved in preparing documents, organising meetings, locating resources and assigning tasks, he or she needs, above all, to have the confidence, sensitivity and competence to facilitate cooperative work in others.

6. School-based curriculum development and teacher development are parts of a single process, 'two sides of the one coin' so to speak. It requires the development of many skills in teachers including those involved in course design, group decision making and course evaluation. On the other hand, involvement in school-based curriculum development can do much to complement and enhance these and other professional skills and attitudes.

These principles highlight the importance of a team approach to curriculum development in geography departments. Little more will be said on the need for cooperative effort in this chapter. However, this principle should be seen as a necessary mode of operation for all the suggestions that follow.

The principles also indicate that examples of school-based curriculum development can be categorised into two broad groups. These are:

1. the evaluation and improvement of existing courses; and

2. the planning of new courses in geography.

This chapter develops these two forms of school-based curriculum development, with greater attention paid to the planning of new courses.

School-based Curriculum Development: Evaluating and Improving a Geography Course

Several questions can be asked in an evaluation of existing geography courses. The answers to such questions provide information for decision making on future modifications to course rationale, content, teaching strategies, resources and evaluation procedures. A very brief checklist of questions includes:

1. Is there an adequate *rationale* for the course explaining:
   (a) its purpose in respect of the needs of students?
   (b) the philosophy of geography upon which the course is based, and justifying its adoption in preference to others?
   (c) the knowledge, skills and values it is envisaged students will acquire from the course?

2. Does the *content* section of the course:
   (a) contain knowledge that is relevant to the needs of society and students?
   (b) build upon previous work?
   (c) contain a guide to the organisation of learning experiences through curriculum units or modules of study structured around the key questions of geography?
   (d) contain criteria for the selection of specific studies and the depth of coverage?
   (e) take account of student's levels of cognitive development?

3. Does the course statement recommend a variety of *teaching strategies* related to modern trends and new perspectives in geographical education, the rationale of the course, varied student preferences for ways of learning and the different teaching styles of members of your department?

4. Are the *resources* specified in the course statement suitable, relevant, stimulating, up to date and within students' ranges of comprehension? What additional resources are needed? How can they be obtained?

5. Does the course statement provide criteria for the selection of *evaluation* and *assessment* procedures? Are they consistent with the rationale of the course and the learning styles of students?

Such questions are the basic ones that could be used in evaluating a geography course. The information seeking and interpretation such questions generate should be seen as a team enterprise as should any decisions and actions that result from the evaluation. These ideas are developed more fully in Chapter 30, 'Evaluating Your Geography Courses'.

School-based Curriculum Development: Designing a New Geography Course

Many curriculum writers also believe a checklist of questions to be a good starting point for designing new courses. Among the questions suggested by Smith (1976) for geography course planning are:

1. What geographic knowledge and skills are important for students at a particular age and ability level to acquire?

2. What feelings, awarenesses and values are important for such students to experience and explore?

3. What constitute the most effective organisation and sequence of learning experiences to provide students with opportunities to obtain such knowledge, skills and values?

4. What content exemplars are most interesting and appropriate?

5. How will the learning experiences and content exemplars be organised and sequenced?

6. What resources will be necessary?

7. To what extent and how will the teaching and learning be evaluated?

It is common for such lists of curriculum questions to be provided for teachers. Less common, however, is practical guidance on ways of answering such questions and acting upon one's answers.

School-based Curriculum Development: The Baulkham Hills Experience

Douglas et al. (1981) have described their experiences with school-based curriculum development within the Geography Department of Baulkham Hills High School in Sydney. They went through five steps in formulating, implementing and evaluating a three-year lower secondary school geography program. These steps were:

1. *Getting motivated*—in which a lack of student interest in existing courses and the consequent decline in student numbers coupled with teacher perception of out-datedness in the methodology, content and philosophical basis of existing courses produced a desire for curriculum renewal.

2. *Information seeking*—in which they sought advice from other departments in their school, geography colleagues in other schools and geography advisers. They also studied alternative syllabus

documents, school and university geography textbooks, and the literature of geographical, social and environmental education.

3. *Course writing*—in which they applied the information they had gathered to the needs and interests of their students, the resources available and the organisational structure and procedures of their school to produce a new course. They based their three-year course on the theme of 'conflict' and developed teaching units to support and illustrate it. An overview of the three-year course is provided in Figure 27.1.

4. *Teaching the course*—The course was first taught in 1980 with a high level of enthusiasm for the new goals they were seeking and the new units. Department members cooperated in helping each other teach the units each had prepared.

5. *Evaluationg the course*—in which information was sought from teachers and students on the objectives, content, learning experiences, resources and assessment of each unit. The year convener of each course called meetings with teachers involved in the course at the completion of each unit to discuss this feedback and to plan modifications to the unit for future use. At the end of the year, similar group meetings evaluated the overall course and made modifications for the following year.

The Baulkham Hills experience provides encouraging insights into school-based curriculum development for others wishing to redesign their courses. Their five-step model is a sound one; but, as with all school-based curriculum development, it is highly dependent upon the experience and skill of the teachers concerned, the quality of available consultants and literature in geography and geographical education, and the time available for this information seeking. Much care, time and thought needs to go into the process of information seeking as the success of the course to be developed depends upon the quality of the information upon which it is based. This step, therefore, requires elaboration.

School-based Curriculum Development: Information, Decisions and Strategies

Information needs to be gathered in four major areas for successful school-based curriculum development. This allows:

1. planning to be made based upon the *antecedents* of the course;
2. a determination of the *course structure* to be used;
3. a determination of the *course design approach* to be used; and
4. the design and writing of the *course outline* to be finalised.

Studying Course Antecedents

The antecedents of a course are the factors that need to be considered before a course can be planned. They include: society's needs and expectations of geographical education; students' needs, abilities and interests; the school system and the infrastructure, the staffing and resources it provdes; the school itself and its curriculum policies and procedures in relation to timetabling, length of periods, expectations of teachers and students, etc.; and the established pattern of interactions in geography classrooms such as teacher–student relationships, other courses in geography, the layout of and facilities in

---

First year
Conflict—Nature vs. Nature
1. What is geography?
2. Earth-forming processes and results
3. Weather and climate
4. High altitude and latitudes
5. Interior lands
6. Coastal lands
7. Oceans

Second year
Conflict—People vs. Nature
1. Maritime usage and disturbances
2. Priorities and planning for coastal lands
3. Cities by the sea
4. Urban networks—city to farm
5. Land use and planning for mountain lands

Third year
Conflict—People vs. People
1. Introducing conflict
2. Black vs. white—racial conflict
3. Democracy vs. communism—political conflict
4. Conflicting beliefs—religious conflict
5. The 'haves' vs. the 'have-nots'—developmental conflict
6. The energy crisis—power conflicts
7. Developers vs. conservationists—environmental conflict
8. Revolution in Toraz—a simulation exercise, bringing the course content and methodology together

Figure 27.1 Three-year geography curriculum at Baulkham Hills High School, Sydney

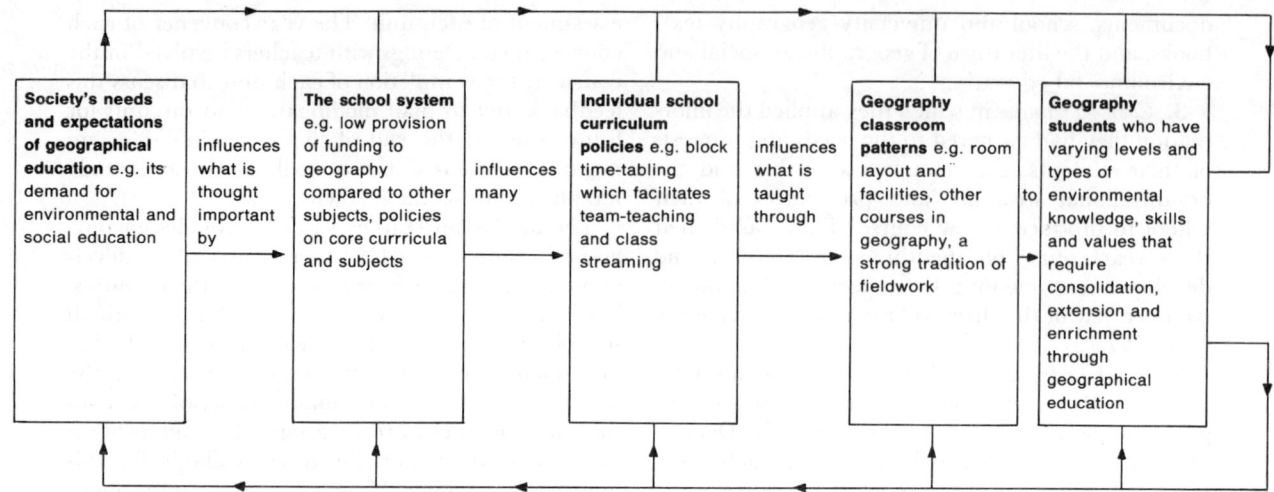

Figure 27.2 Some antecedents to course planning in geography

geography rooms, and the interests and skills of department members. Many of these antecedents to course planning are related. Figure 27.2 illustrates some of the interactions that exist between them.

Perhaps, the most important antecedent that requires attention is the students for whom the course is being planned. Information about their previous geographical learning and other studies, their range of ability levels in numeracy, verbalisation, reading, reasoning and graphicacy, their environmental needs, concerns and interests and the aspirations they have for themselves and their geographical studies needs to be collected and analysed in order to select objectives, content, resources and learning experiences to suit them.

Selecting a Course Structure

There are many different approaches that can be taken in determining the structure of a course of studies. In the 1960s, and before, most courses followed a simple structure based upon areal studies and topics. Global coverage, geographical facts and rote learning were the dominant themes in the teaching of such courses. The spatial and quantification revolutions that occurred in geography in the 1960s, the growth of insights from that branch of education known as curriculum studies and the application of both these developments to the teaching of geography have led to a multiplication of course structures. Four distinct structures that can be identified are:

1. concept-based courses;
2. principle-based courses;
3. thematic courses; and
4. learner-centred courses.

Each of these four approaches are described and illustrated with reference to representative geography textbooks. Readers may find it useful to have a copy of at least one of the books (or series) from each section at hand when they read through the following section in order to see examples of the points made in the text.

However, one note of warning is given before the four approaches are discussed—and this pertains to what has almost become the 'typical' approach to junior secondary geography textbooks in Australia—and it is a severe criticism of most of them. Far too many textbook series in Australia have no coherent structure. Most are heavily content-for-its-own-sake orientated and are still steeped in the landscape and sample study approaches that died out in Britain in the 1960s. While many of these Australian series (such as *A Geographer's World, Our Changing World, Explorations in Geography, Man and His World* (sic), etc. are 'best sellers' (but what competition have they really had locally ?), severe doubts must be raised over their lack of any obvious conceptual or thematic structure, organised scope and sequence, and clear criteria for the selection of content and case studies. None of these books are referred to below.

1. *Concept-based courses* focus on teaching the basic ideas upon which the discipline of geography is based. Textbooks such as Jones' (1980) *Understanding Places* series, the Blachford et al. *Going Places* series,

Out and About (1976), *Way Out* (1976) and *Out of Site* (1977) and Thomas' (1985) *Concept Geography* series provide resources for designing and teaching such courses. Each of the authors has identified several substantive concepts in geography (e.g. location, distance, distribution, association, interaction, movement, change, region and scale in the *Going Places* books) and written materials to illustrate and teach about the concepts. Some of these books have been severely criticised for the narrow spatial analysis view of geography assumed in the selected concepts, illogical selection of case studies and the lack of a sound educational theory of concept learning in their pedagogy. See Hall's (1986) review of the *Concept Geography* series for more details. However, the basic idea of structuring a course on the concepts of geography is a good one and many schools have in fact successfully developed and used such courses.

2. *Principle-based courses* generally focus on the major topic areas of geography (e.g. agriculture, manufacturing, trade, urbanisation, communications, etc.). A list of key generalisations for each area is identified for students to come to understand. Case-study exemplars for each generalisation can be identified and course outlines developed so that students gradually develop an understanding of the generalisations considered important in geography. Principle-based courses are considered an advance on concept-based ones because of the interrelationships sought between concepts and the application of concepts to real world problems that an emphasis on principles allows. Such courses are becoming increasingly popular in Britain through their advocation by Graves (1979) and Her Majesty's Inspectorate in Geography (Department of Education and Science 1978, 1986). Chapter 4 of the 1978 book, *The Teaching of Ideas in Geography*, is necessary reading for geography departments wishing to develop a principle-based course. This chapter considers the classroom implications of teaching for generalisations, and provides syllabus construction guidelines, including seven examples of syllabuses and teaching units developed through this approach. Textbook series such as *Harrap's Course in Reformed Geography* (Dinkele et al. 1976), *Harrap's Basic Geography* (Greasley et al. 1979), *People and Places* (Crisp 1974, 1975) and *Location and Links* (Walker and Wilson 1972, 1973) use this approach. Although material in them is now quite dated, use them as inspiration for activities you research yourself.

3. *Thematic courses* are becoming increasingly popular, with the Baulkham Hills High School course an early example of this approach to course structure in geography. The theme of 'conflict' provided the basis for their whole three-year junior geography program. Other themes that have been successfully used by schools include: 'space-ship earth', people–environment issues, natural hazards, world landscapes, global interdependence, human survival, Three Worlds, contemporary issues, leisure and recreation and cultural change. Thematic courses can provide opportunities for quite creative course structures and content organisation as themes can be selected to suit the problems or issues dominant in particular areas and the special interests of groups of students. The textbook series *Accent on Cities* (Oliver 1985), *A Sense of Place* (Beddis 1981, 1982), *Core Geography* (Martin and Whittle 1982, 1983 and 1985), *People and Environment* (Slater 1986) and *Worldwide Issues in Geography* (Hart et al. 1985) are excellent examples of the thematic approach to course structure. Of particular interest to schools wishing to develop courses based upon the themes of social justice and peace in geography is the *New Wave Geography* series (Stowell and Bentley 1988) sponsored by the Geography Teachers' Association of Victoria. This two-book series for Year 9–10 students provides exciting and well-designed units based upon the themes promoted in *Teaching Geography for a Better World*.

4. *Learner-centred courses* are the fourth possibility to be considered. They will be discussed in greater detail than the previous three because they are not commonly used in secondary school geography at the present time. There are several reasons for this, including the traditional discipline-of-knowledge orientation of most syllabuses and the false impression many teachers have of learner-centred education. It is commonly believed that such courses can only be taught to very small classes and that they involve allowing students to learn (or not learn) whatever it is they wish. Such beliefs are quite wrong but reflect the limited amount of available information on student-centred curriculum planning and the content of most pre-service teacher training courses. Elements of learner centredness can be seen in syllabuses and courses that involve individual studies to be completed by students and inquiry process based courses such as the Queensland one, Australian Geographical Inquiries, and the Geography 16–19 Project in Britain.

Learner-centred approaches to course planning in geography are based upon the principle that all education should be concerned to help students assign meaning to, and make meaning of, their own life experiences, their actions, and the many environments they encounter. Smith (1976) argues that geographical education, particularly, should seek to help students try out the concepts and inquiry methods used in geography and use them as a means of interpreting and mentally structuring their life experiences. This involves four philosophical principles:

(a) Students do not need to be *introduced* to geography. They are geographers already. Geography is really nothing more than the disciplined working of the human mind as it perceives and acts in the environment.

(b) The knowledge, skills and attitudes so gained constitute one's personal geography. Personal geographies contain many forms of knowledge not given credence in other approaches to geographical education. The learner-centred approach to geography courses stresses that the 'contents' of personal geographies should be utilised as the starting point in geography course planning because of their personal meaning and the integration of environmental experience, knowledge and values in students' lives.

(c) The goals of geographical education should focus on the refinement, extension and enrichment of personal geographies. What is taught and how it is taught should be designed to help students cope more effectively with the environment in which students live, their feelings and values in relation to it, and to empathise with the environmental situations and experiences of others.

(d) Learning experiences in geography should be designed to encourage experiential learning. This involves more than obtaining knowledge and skills necessary to obtain new knowledge. Experiential learning requires a regular and structured reflection on past experiences and active participation in new situations structured to allow students to find out more about their environment, themselves and the relations with it.

A curriculum model based upon a learner-centred approach to geography has been developed and is outlined in Fien (1980). Guidelines for the selection of content, learning experiences and evaluation strategies are also provided in that article. See also Hall (1979), Gilbert (1979), Romey and Elberty (1980), and Fien (1983), for ideas in relation to these. Examples of teaching materials designed for learner-centred courses in geography include *Essences I* (American Geological Institute 1971), *People in Places* (Farbstein and Kantrowitz 1978), books in the *Issues* series (Love and Edwards 1977, 1979) and the *Living Decisions* series (Menzies 1979).

Selecting and Implementing a Course Design Strategy

An understanding of two models of curriculum development, the *objectives model* and the *process model* are necessary at this point. One of these models needs to be selected and followed to prevent course design becoming merely a matter of selecting content topics to match the selected course structure. Unfortunately, such a selection appears to be rarely done. The Victorian *SGEP-PAK* (Secondary Geography Education Project 1977) guide to course planning ignores it altogether but, by default, leads teachers to use the objectives model of course design. The objectives model is also advocated in courses that require schools to state objectives explicitly and match their assessment items to the objects, as well as in such geography curriculum planning books as Biddle (1976) and Graves (1979).

The objectives model involves determining the general aims of a course and methodically translating them into statements of more specific goals or mediate objectives for each section of a course and then into quite specific objectives for each lesson. Content, teaching methods and assessment strategies are then selected to enable students to attain and to measure the achievement of the specific objectives, It is assumed that if these are attained then the mediate objectives and general aims will also be attained.

The objectives model has been severely criticised despite its popularity. The criticisms are of two types—practical and theoretical. The practical objections include the large number of objectives that need to be stated if the model is to be followed correctly, the time involved in objective writing, the inappropriateness of specific objectives to affective and many cognitive aims of teaching geography, and the way the model can limit students learning to only the objectives prescribed (see James 1968; Sockett 1976). The theoretical criticisms include doubts that the attainment of general aims is assured by the attainment of specific classroom objectives (that is, the process of education is more than merely the sum of its parts) and the fact that the learner's existing knowledge, beliefs, experiences and interests, rather than objectives, should be the starting point in curriculum planning. Indeed, it is only by studying each learner that a curriculum for that learner can be planned.

Stenhouse (1975) was the leading advocate of the alternative to the objective model of course design, the process model. It is especially suited to learner-centred course structures but can just as easily be used in concept, principle and thematic-based courses. Fien (1980) has outlined eight steps involved in a process model for course design in geography (Figure 27.3). The eight steps are:

1. An identification of the learners and learning groups within the class for whom the course is being planned.
2. An analysis of the personal geographies of students to ascertain their spatial and environmental needs, concerns and interests.
3. An analysis of the contribution geographical education can make to the environmental needs, concerns and interests of students.
4. An identification of the organising ideas of geography that can be used to provide structure to programs of geographical education based upon the personal geographies of students.
5. The selection and development of curriculum units, incorporating these main ideas.
6. An identification of the learning skills needed for, and/or that can be developed through the selected curriculum units.
7. A determination of teaching strategies that will allow teachers to link the skills and organising

Figure 27.3 A process model for curriculum development in geographical education (after Fien 1980)

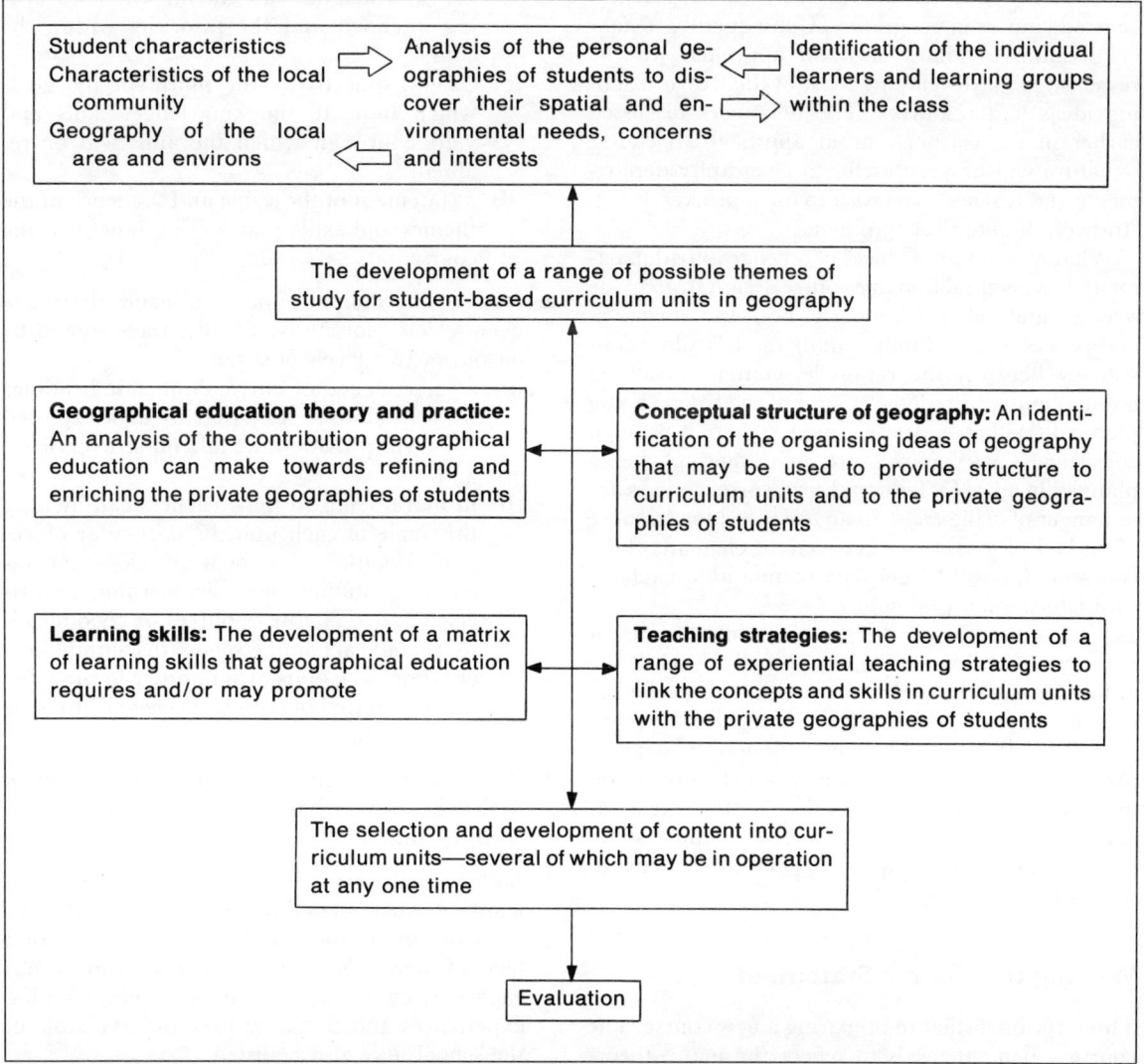

ideas in curriculum units with the personal geographies of their students.

8. A selection of approaches to evaluation that will allow student, peer and teacher decision making about the nature and value of the curriculum and their contributions to it.

There are problems with the process approach to course design. It is a new approach and many teachers received their training during the years when the objectives model was unchallenged. Many syllabus prescriptions are still based on the objectives model. Inservice education, much reading and thinking, and perhaps a change in attitude may be necessary in many cases. The learner-centred process approach requires greater sensitivity to student needs, concerns and interests than the objectives model. Few textbooks have been written for learner-centred courses in geography. Consequently, teachers wishing to follow such an approach will be responsible for developing most of their own teaching ideas and resources. The references discussed earlier in the learner-centred approach to course structure will be very useful to geography departments and teachers who wish to use a process design strategy, despite these problems.

Whatever the preferences of a geography department, it is essential that one course design strategy be selected and followed properly. Both the objectives and process curriculum planning models allow consistency between the rationale, content, teaching method and evaluation elements of the course being planned. Without a specific model, there is no such consistency and courses are in danger of being planned in an *ad hoc* fashion. Even worse, courses are in danger of being *taught* in an *ad hoc* fashion because of the lack of consistency between the elements of the course and, possibly, between them and a teacher's own educational philosophy.

This author prefers the process model because it closely matches his educational philosophy while the authors of some other chapters in this book prefer the objectives model. This difference illustrates the point that the selection of a course design strategy is problematic. It requires school-based curriculum development teams to examine their own educational philosophies and preferred teaching styles closely, and to select a course design strategy to match them.

Writing the Course Statement

This is the final stage in preparing a new course. The course statement needs to reflect the many theoretical and practical guidelines that have influenced the thinking of the curriculum developers This ensures that all the 'fine points' in the course are remembered and followed, and that the spirit, as well as the content of the course, is maintained.

Course statements will vary greatly depending upon the course structure and design strategy selected. However, it is important that course statements do include the following:

1. *Rationale*—a clear statement of the educational context and purpose of the course, which includes:
   (a) a summary of the school and community context in which students live and how this geography course relates to it;
   (b) educational aims—the long-term goals of the course which indicate the direction of learning intended and the priorities within the course;
   (c) general objectives—the more specific goals which indicate the knowledge, skills and values through which the aims will be realised;
   (d) a statement of the scope and sequence of the themes and skills that will be taught in the course.

2. *Content*—the themes, topics and issues that have been selected for study. Ideally, these should be set out in two levels of detail:
   (a) as a sequence of curriculum unit headings with the key questions that are to be explored (this will provide a useful course overview); and
   (b) in more detail so that teachers can identify the angle of each unit, its particular objectives, key questions, focus questions, suggested case studies, possible learning experiences and available resources. A good idea is to see that each unit is set out in columns on a chart or page of its own in order to show the detail required of both teachers and students at this second level.

3. *Time allocations*—the agreed amount or range of time to be spent on each curriculum unit together with the timing of fieldwork and assessment.

4. *Learning experiences*—suggestions on the types of learning experiences that are considered effective for achieving particular objectives. If the second level of detail (No. 2b) is being used to set out each curriculum unit, then particular learning experiences and prepared exercises available in the school may also be listed.

5. *Resources*—the relevant textbooks, class sets, field equipment, maps, computer software and audio-visual aids available for each curriculum unit.
6. *Differentiation*—guidance on how objectives, content, learning experience and resources may be differentiated to cater for the range of abilities of students in the course.
7. *Assessment plan*—the forms and frequency of assessment instruments to be used to monitor student progress. Advice on the weightings of each instrument and how records are to be kept for moderation and reporting purposes should also be given.
8. *Course evaluation*—the criteria and strategies to be used to evaluate the appropriateness of the sequence of curriculum units, learning experiences, resources and assessment instruments used in the course (adapted from Department of Education and Science 1986).

Conclusions

Two points require repetition by way of conclusion. Firstly, school-based curriculum development in geography is a cooperative enterprise in which all members of a geography department join resources to design an improved program of studies for students. Secondly, a role change for many teachers from syllabus interpreter to curriculum maker may be necessitated by school-based curriculum development. This will require the development of new awarenesses and skills, especially those of reflection on existing practices and group decision making. Fortunately, school-based curriculum development brings its own rewards, not only in improved courses for students, but also by providing teachers with the very skills they need for it. In this way, involvement in school-based curriculum development can provide on-the-job professional development for teachers who conscientiously engage in it.

Note

1. The word 'syllabus' is here used to refer to official prescriptions of courses of study in a subject issued by examination boards and education authorities.

References

Beddis, R. (1982, 1983) *A Sense of Place* (3 books), Oxford: Oxford University Press.
Biddle, D.S. (1976) *Translating Curriculum Theory into Practice in Geographical Education: A Systems Approach*, Geographical Education Monograph Series No. 1, Melbourne: Australian Geography Teachers' Association.
Blachford, K. et al. (1976, 1977) *Going Places* Series 3 (3 books), Adelaide: Rigby.
Boden, P. (1976) *Developments in Geography Teaching*, London: Open Books.
Crisp, T. (1974, 1975) *People and Places* Series (6 books), London: Thomas Nelson. Australian editions of books in this series are also available.
Department of Education and Science (1978) *The Teaching of Ideas in Geography: Some Suggestions for the Middle and Secondary Years of Education*, London: HMSO.
Department of Education and Science (1986) *Geography from 5-16*, Curriculum Matters 7, London: HMSO.
Dinkele, G., Cotterell, S. and Thorn, I. (1976) *Harrap's Course in Reformed Geography* (6 books), London: George G. Harrap and Co.
Douglas, J., Hirst, P. and Pettit, J. (1981) 'Programming for Real World Geography', *Geography Bulletin*, March, 35-53.
Farbstein, J. and Kantrowitz, M. (1978) *People in Places: Experiencing, Using and Changing the Built Environment*, Englewood Cliffs, New Jersey: Prentice Hall.
Fien, J.F. (1980) 'Operationalizing the Humanistic Perspective in Geographical Education', *Geographical Education*, Vol. 3(4), 507-532.
Fien, J.F. (1983) 'Humanistic Geography', in J.F. Huckle (ed.) *Geographical Education: Reflection and Action*, Oxford: Oxford University Press.
Fien, J.F. and Gerber, R. (eds) (1988) *Teaching Geography for a Better World*, 2nd edition, Edinburgh: Oliver and Boyd.
Gilbert, R.J. (1979) 'Image, Experience and Personal Geography: Some Implications for Education', *Geographical Education*, Vol. 3(3), 339-406.
Graves, N.J. (1979) *Curriculum Planning in Geography*, London: Heinemann.
Greasley, B. et al. (1979) *Harrap's Basic Geography*, London: George G. Harrap and Co.
Hall, S.R. (1979) 'Teaching Humanistic Geography', *Australian Geographer*, Vol. 14, 7-14.
Hall, S.R. (1986) Review of *Concept Geography* series, in *Geographical Education*, Vol. 5(2), 55.
Hart, C. et al. (1985) *Worldwide Issues in Geography*, London: Collins Educational.
James, C. (1968) *Young Lives at Stake*, London: Collins.
Jones, D.P. (1980) *Understanding Places*, London: Hodder and Stoughton.
Love, J. and Edwards, C. (1977, 1979) *Issues* Series (Books 1-3), London: George G. Harrap and Co.

Martin, F. and Whittle, A. (1982, 1983, 1985) *Core Geography* (4 books), London: Hutchinson.

Menzies, W. (1979) *Living Decisions* Series (2 books), London: George G. Harrap and Co.

New South Wales Directorate of Studies (1976) 'Curriculum-Making in Geography', *Curriculum Newsletter*, SG5.

Oliver, J. (1985) *Accent on Cities*, Melbourne: Thomas Nelson.

Romey, W. and Elberty, W. (1980) 'A "Person-Centred" Approach to Geography', *Journal of Geography in Higher Education*, Vol. 3(2), 42–50.

Secondary Geography Education Project (1977) *SGEP-PAK*, Melbourne: Geography Teachers' Association of Victoria.

Slater, F. (1986) *People and Environment*, London: Collins Educational.

Smith, D.L. (1976) 'Some Personal Thoughts on Developing Curriculum for Geography in Years 7 to 10 in N.S.W. High Schools', *Curriculum Newsletter*, SG5, 1–3.

Sockett, H. (1976) *Designing the Curriculum*, London: Open Books.

Stenhouse, L. (1975) *An Introduction to Curriculum Research and Development*, London: Heinemann.

Stowell, R. and Bentley, L. (1988) *New Wave Geography* (2 books), Brisbane: Jacaranda.

Thomas, S. (1985) *Concept Geography* Series, London: John Murray.

Tolley, H. and Reynolds, J. (1977) *Geography 14–18: A Handbook for School-based Curriculum Development*, London: Macmillan.

Walker, E. and M. and Wilson, T. (1972, 1973) *Location and Links* Series (5 books), Oxford: Blackwell.

28

Planning a School-based Assessment Program

Warren Halloway

The assessment of student learning is an important part of all teaching. One of the requirements of school-based curriculum development is that schools be responsible for planning appropriate programs of assessment for diagnostic, grading and reporting purposes. Three aspects of assessment are outlined in this chapter. These are determining what to assess, the test instruments to be used, and the integration of these into a balanced program of assessment. Warren Halloway explains that the focus of any assessment program should be the monitoring of the full range of objectives planned in a course, and shows how this may be achieved through the use of anecdotal records, objective tests, essays, data-response questions and work-based and decision-making assessment. The chapter concludes with a pro forma which could be used to plan the purposes, type, timing and weighting of test instruments in an assessment program. This chapter should be read in conjunction with Chapter 27 on school-based curriculum development and Chapter 30 on the evaluation of geography courses.

'How is Mary doing?' is a question teachers often ask of themselves in relation to each and every student they teach. It is a reasonable question also asked by many parents. Mary is also entitled to an answer.

The question has the added virtue that it puts the emphasis on the process of 'doing geography' rather than the product of knowing a body of information. Parents rarely ask 'What does Mary know?'.

Evaluation is an essential part of the teaching process which provides information to students, teachers, parents, education authorities and others such as tertiary institutions and employers on the appropriateness and adequacy of students' achievements in courses of study. Ultimately these interested parties will ascribe a value to the information provided. In Chapter 30 the broader aspects of course evaluation are discussed especially as they concern the worth of classroom processes. In this chapter some aspects of student assessment are considered as a way of providing information upon which judgements of worth will be made if the question 'How is Mary doing?' is to be answered.

It is useful briefly to distinguish between *assessment* and *evaluation*. In this chapter assessment refers to the value teachers place on the information they have about their students. Evaluation generally refers to the broader judgements made about courses of study and how they are taught. Clearly these overlap. The information we use to assess students' learning may also be useful in evaluating a course or the effectiveness of a teaching method. The difference is sufficiently important to use the two terms separately and it is a reminder that teaching needs to be examined as closely as students' learning.

School-based Assessment Program is One Aspect of School-based Curriculum Development

The increasing opportunities and even the requirement for teachers to engage in school-based curriculum development were explained in Chapters 26 and 27. Assessment is one aspect of school-based curriculum development and should be seen in that context. The reasons why teachers now carry a heavy responsibility for developing assessment programs include:

1. The decreasing frequency and importance of external examinations.
2. An increase in the need for school-based curriculum development to complement more general syllabuses issued by central authorities.
3. Increasing demands from education authorities, parents, employers and the community for accountability by schools.
4. Appropriate assessment procedures are required as school courses provide more adequately for local conditions and individual student needs and interests.
5. Increased professional expectations of teachers. The nature of modern school geography and techniques of assessment present new challenges and opportunities for teachers.

Six principles were stated in Chapter 27 to guide teachers in the design of school-based curriculum development. Because assessment is an aspect of any curriculum development it will be helpful to restate these principles with emphasis on the implications for planning a school-based assessment program:

1. Developing a school-based assessment program is a collective responsibility of the geography department, appropriately led by the head of department and in harmony with overall school assessment policy.
2. A school-based assessment program should evolve gradually from existing methods of assessment. Existing methods of assessment, even if uncoordinated and rudimentary, represent considerable effort and thinking by conscientious teachers. To reject these and adopt a 'clean slate' approach implies a lack of appreciation for work done and is likely to disrupt on-going courses. It is suggested that existing methods and policies should form the basis of any new programs. Using this strategy, a geography staff would need to codify existing practice, identify short- and long-term needs, develop new techniques and reject those which are found to be inadequate in terms of these newly identified needs.
3. Note must be taken of central authority assessment requirements by way of syllabus documents and external examinations. The geography department's role is to interpret such central assessment policies for local conditions.
4. Just as school-based curriculum development demands that teachers be aware of all the factors which impinge on the quality of students' experiences in their geography courses, so the planning of an assessment program requires an equal and sustained awareness and response. It is mainly for this reason that assessment policy should take cognizance of the *full* range of objectives developed for a course.
5. All school-based curriculum development is a team enterprise. Planning an assessment program is no exception. The development of a successful program will require good working relationships, leadership, knowledge and skills. If the necessary cooperation exists, communications will be maintained. Successful planning requires skill in designing assessment instruments, keeping records and using this information to make judgements about students learning and, eventually, about courses.
6. Planning an assessment program this way has the potential to enhance teacher development by calling forth knowledge and skills teachers have and enabling them to share these with their colleagues. Geography departments have a great variety of support services available to assist in areas where deficiencies may exist. The use of external consultants is a thoroughly professional practice and is strongly advocated.

Assessment Should be Tied to Course Objectives

Kevin Piper's *Evaluation in the Social Sciences for Secondary Schools* (Piper 1976) states the basic principle that all evaluation should be closely tied to the objectives of a course. These are conveniently listed as:

1. *Knowledge objectives*. Recall and recognition of factual content is usually expected although an understanding of concepts, generalisations and principles is regarded as more important.
2. *Skills objectives*. Three major types of skills are included in most courses:
    (a) *Information-gathering skills*. Locating, gathering, listing, grouping, organising, transform-

ing, utilising and communicating are commonly cited in course objectives.
(b) *Cognitive skills*. Forming hypotheses, critical analysis, thinking at abstract levels, inferring are among the objectives of this type. Bloom et al.'s (1956) taxonomy is utilised in many statements of cognitive objectives. Taken as a group such objectives promote rational, empirical and objective inquiry.
(c) *Social skills*. These objectives relate to social participation, interaction and group inquiry.
3. *Values objectives*. This is a complex and sensitive area in which feelings, attitudes and values overlap. For the teaching of geography to make any contribution to the moral growth of students, geography teachers must strive to orient their teaching towards exploration and development of such affective objectives as respect for life, excellence, integrity, honesty, tolerance, justice, social responsibility, compassion, adaptability, rationality and sensitivity (physical, aesthetic, emotional and spiritual). Additionally, we need to develop means of monitoring student attitudes, not only to know 'where they are at', but to serve as a guide in the selection of topics, themes, issues and case studies.

Another list of objectives is provided by Keith Cooper (1976) in *Evaluation, Assessment and Record Keeping in History, Geography and Social Science*. This list eschews a classification, but clearly covers a wide range of objectives in the cognitive and affective domains. The list was used by a teacher who wished to analyse the objectives he had covered in work-

Figure 28.1 An evaluative checklist of objectives for planning, teaching or assessment (after Cooper 1976)

Objective	Item 1	Item 2	Item 3	Item 4
1. Find information				
2. Show curiosity				
3. Sustain interest				
4. Communicate information				
5. Interpret information				
6. Evaluate information				
7. Organise information to form concepts and generalisations				
8. Formulate and test hypotheses				
9. Wariness of overcommitment				
10. Participate in small groups				
11. Awareness of groups in society				
12. Development of empathy				
13. Explore feelings and values				
14. Development of skill in use of equipment				
15. Development of expressive powers through communication and movement				

cards he prepared. Cooper reports that the teacher was horrified to find his workcards mainly asked his students to find information and communicate it in writing with an occasional drawing. Since the teacher knew that *all* the objectives in the course were important, he had to revise his workcards and develop new ones covering the full range of objectives.

A written list of objectives can help the teacher plan his or her course, whatever the approach taken, whether he or she uses workcards, lesson plans, or units and inquiry sequences. It is also invaluable in preparing an assessment program and in devising particular instruments such as a test.

Cooper (1976) provides useful advice for geography teachers under each of the objectives listed in Figure 28.1. Under the heading 'What to look for' he indicates specific ideas for formal and informal assessment of each objective. For example:

1. *Objective: Find information*
   What to look for:
   Student
   - uses content pages and index
   - looks for more than one source
   - looks for different kinds of sources (e.g. maps, books, atlases, pictures, resource people)
   - compares sources
   - uses sources from school, home and community

2. *Objective: Evaluate evidence*
   What to look for:
   Student
   - can detect probable bias (e.g. 'This Letter to the Editor says that there should be a total ban on wheat sales to the USSR. Perhaps he is not looking at it from the farmers point of view')
   - can distinguish between fact and opinion
   - comments on limitations of evidence (e.g. 'This only tells us about the actions of one settler'. 'This is a description of Sydney, not all of Australia')

3. *Objective: Organise information to form concepts and generalisations*
   What to look for:
   Student
   - tries to make groupings leading to concepts and generalisations (e.g. 'Those services are provided by government but these are provided by voluntary groups')
   - asks questions about groupings (e.g. 'Is sawmilling a primary or secondary industry?')
   - is prepared to change her conceptual framework (e.g. 'This boy is not an immigrant, even though his parents come from Italy. He was born here')
   - makes generalisations (e.g. 'I think all of these valleys are densely populated')
   - reviews his generalisations in the light of new evidence (e.g. 'The country has been prosperous most of the time; but then there was a severe drought and many people suffered heavy losses, so some people are not prosperous some time')

4. *Objective: Participate in small groups*
   What to look for:
   Student
   - understands the need for rules in a group
   - participates in deciding rules
   - shares tasks and roles in a group
   - accepts role of leader
   - appreciates qualities in other group members
   - can put aside personal goals for group cohesion

5. *Objective: Development of empathy*
   What to look for:
   Student
   - recognises significant similarities and differences between situations (e.g. communities, families, countries, places, periods)
   - is capable of understanding feelings and values other than her own or those of society
   - is able to tolerate other values
   - can infer the feelings and actions of others from knowledge of their situation

External examinations are notoriously poor in assessing student achievement of such research, thinking and attitudinal objectives. Hickman et al. (1973) say that because of this, external examinations have created a 'vicious cycle of curriculum underdevelopment'. We need to keep the full range of objectives for a course in mind to prevent school-based assessment programs falling into the same vicious cycle.

Assessment Should be Diagnostic, Formative and Summative

The previous section has dealt with the vital link between objectives and assessment. It was noted that Cooper (1976) provided guidance on 'what to look for' if particular objectives are to be assessed. The next section indicates how this can be done. It will be well to bear in mind the warning by Oliner (1976) that assessment should be directed at the progress of students and not their deficiencies.

Bloom et al. (1971) differentiate between three forms of assessment in terms of timing.

Diagnostic. This can be done prior to teaching to enable informed planning to be carried out in terms of appropriate learning objectives. Cooper (1976) illustrates how this can be done at a very basic level when first meeting a geography class. He says it is a simple matter to devise library assignments seeking specific items in order to measure reading ability and awareness and ability to use basic reference books, e.g. 'Find a book which tells you the states of USA' or 'Find a newspaper which tells you the weather conditions around Australia last Friday'. (See Chapter 24 for more ideas on diagnosing student learning.)

Formative. This is done throughout the course and is intended to indicate what has been learnt. It is useful because the teacher can change the 'form' of the course in the light of feedback obtained, to rectify deficiencies and improve students' performance.

Summative. This occurs at the end of teaching and is done to assign a grade, rank or to otherwise inform all concerned about the extent to which learning objectives have been achieved.

The purposes of diagnostic, formative and summative assessment overlap. In the real world of classroom teaching, all three forms of assessment may be incorporated into a test instrument.

Many writers in geographical education provide advice and sample items for assessment instruments. It is beyond the scope of the chapter to provide the detailed samples contained in these specialist sources. A very useful booklet which contains such detail is Kevin Blachford's 1972 book, *Evaluation*, which is Unit V in the series 'The Teaching of Geography'. Blachford describes three types of assessment instruments:

1. Classroom observations
2. Objective tests
3. Essays

Some techniques which may help the geography teacher in each category are described in the sections that follow. Techniques for a fourth type of assessment instrument, the data-response exercise, are also described.

Assessment Instruments Based on Classroom Observations

Teachers informally observe students in many situations and have opportunities to interview students as they work individually and in groups. The *anecdotal record* is a useful technique, provided entries are brief and kept regularly. It is highly recommended for assessment of students with learning difficulties. Try keeping a booklet in which you devote one page to each student over a year.

An anecdotal record on Jean, a Year 8 student, might read as follows:

29/4/88: Showed little interest in new unit on cities.

4/5/88: Assisted me with display on Sydney. Seemed to show interest in maps.

5/5/88: Borrowed book on Sydney's transport. Returned it saying it was too hard to read.

6/5/88: Have provided Jean with map exercise and will test for reading skill. Interview arranged for 11.30 a.m.

11/5/88: Records show frequent absences from school.

The *interview* is a valuable technique whether done with individuals or groups. When the course involves an inquiry into conditions which call for an individual stance of interpretation, e.g. the units 'How can we avoid the worst effects of drought?' or 'Why log rainforests?', the open-ended interview may be used.

A structured interview may also be used with students. For example, to find out their interest and understanding of mountain landscapes the teacher may show students a few pictures of mountains mixed up with city and farming landscapes and ask questions such as:

- Which of these pictures interests you most?
- Why do you find this picture of most interest?
- Which place interests you least?
- Would you like to know more about one of these areas?

The volume on *Evaluation Strategies* in the teachers' guides to 'Man: A Course of Study' (1970) contains extensive practical guidance on interviewing techniques with school students. The use of a tape recorder is strongly advocated, but the primary task of the teacher is to help students to give clear, relevant and complete answers.

The *checklist* is a useful device for recording in a systematic way the progress of each student. Perhaps, all members of a geography department could agree on a common set of criteria which will enable a continuity of records and, therefore, a comparison of progress over several years. Information collected regularly can make minimum demands on teachers' time and be of special value when advising students and parents. Extracts from two sample checklists are provided in Figures 28.2 and 28.3. Neither is intended to be definitive. Rather, use them as guides to the *format* of checklists you could develop in your geography department to monitor the progress of students.

A student's *self-evaluation checklist* may be useful, also. This can be completed by the student and used during an interview. Figure 28.4 is an extract from a sample instrument in the area of attitudes phrased in language a student may use. It is a profitable exercise to involve students in the development of the self-evaluation checklist. This will enable you to see the factors students consider important, and may increase student commitment to completing them on a regular basis.

Assessment Instruments Based on Objective Tests

Research studies show that the greatest weakness of current assessment practices is the overemphasis on the recall of information, that is, on knowledge objectives. Unfortunately, this frequently occurs to the near exclusion of questions which tap the higher cognitive levels (Banks 1974; Piper 1976). What is needed is the development of assessment programs which do not rely on a teacher's memory and hard

Figure 28.2 Extract from the attitudes and values part of a sample checklist to monitor student learning

Objective	Year 1			Year 2		
	Term			Term		
	1	2	3	1	2	3
1. Works independently	B					
2. Contributes to group work	A					
3. Shows consideration for others	C					
4. Developing empathy	C					
Key: A = Frequent and valuable B = Occasional C = Infrequent						

Objective	Year 1			Year 2	
5. Organises information to form concepts and generalisations	Term			Term	
	1	2	3	1	2
5.1 Makes groupings					
5.2 Asks questions about groupings					
5.3 Can change conceptual framework					
5.4 Makes generalisations					
5.5 Will review generalisations					

Key:	The student can do this	▩
	The student can partly do this	▨
	The student cannot do this	▭

Figure 28.3 Extract from the cognitive skills part of a sample checklist to monitor student learning

Figure 28.4 Extract from a self-evaluation checklist for students

How often do you do each of these:	Usually	Seldom
1. I concentrate well on my work.		
2. I keep to a job until it is finished.		
3. I am keen to try out new ideas.		
4. I am a good organiser of my own study.		
5. I work happily with others in class.		
6. I share equipment with others.		
7. I help to set up plans for group work.		
8. I consider the rights of others.		
9. I try to watch TV shows that help my work.		
10. I collect newspaper reports relevant to my work.		

work alone, but which systematically assess the full range of objectives planned for students, especially when those objectives call for learning at the higher levels of thinking and valuing. Objective tests are important in this regard as they can be written to assess specific pieces of knowledge, but also specific thinking and attitudinal objectives.

Blachford (1972) contains numerous sample objective questions which can be used in diagnostic, formative or summative assessment. Section 8 of this useful book is a bank of test items from which teachers can choose or design similar items. Another highly recommended source of sample items is in Cox (1966), which lists questions illustrating each part of Bloom's taxonomy of cognitive objectives (Bloom et al. 1956). Senathirajah and Weiss (1971), Salmon and Masterton (1974), Piper (1976) and Marsden (1976) also provide examples of objective test items based upon similar categories of knowledge, skill and thinking objectives in geography.

Samples of geography questions similar to those in such references include the following seven. The range of learnings being assessed are based upon Bloom's taxonomy of cognitive objectives.

1. *Knowledge of ways of dealing with specifics*
   Which of the following may be used to describe relief?
   - aspect
   - humidity
   - elevation
   - gradient

2. *Knowledge of methodology*
   Describe a system of fractional notation which may be used to record a land-use survey of your town.

3. *Comprehension—Translation*
   Study the table of population statistics and use them to draw a histogram.

4. *Comprehension—Interpretation*
   Describe the settlement pattern shown on the accompanying map sheet.

5. *Comprehension—Extrapolation*
   Study the weather charts for Australia from 20 to 24 October. Predict weather conditions in NSW on 25 October.

6. *Analysis*
   'The use of fertilisers and exotic pasture species during the 1950s and 1960s dramatically increased carrying capacity of livestock in the New England Region. The feasibility of improving pastures had been demonstrated by research organisations such as the N.S.W. Department of Agriculture and the C.S.I.R.O. Of equal importance however was the high wool prices during the Korean War in the 1950s which provided graziers with the surplus funds to invest in improving their pastures. The severe drought in recent years has affected all pastures in the region but the improved pastures have proved especially vulnerable to the dry conditions'.
   Read the passage and identify:
   (a) the physical, biotic and human elements of the environment described; and
   (b) the interactions between the environmental elements which led to changes occurring.

7 *Synthesis*
   (a) Provide students with a map showing numerous spot heights. Use contours to represent the relief.

(b) Refer to three field studies undertaken in the local area. Use these experiences to describe the patterns of land forms, settlement, land use and drainage in the areas visited.

In order to ensure that the range of objectives stated in the course plan are being assessed it is suggested that a matrix be prepared similar to Figure 28.1. The course objectives are set out along one axis and the questions in the instrument are set out on the other. The pattern will be revealed as all items are categorised. Omissions and areas over-emphasised will become apparent, and corrective action taken.

Objective tests have been described and illustrated to this point in terms of *when* they might be used and the educational *objectives* they tap. The next section will deal with the common types of objective test items.

1. *True–false test*. Also called alternate response questions. It is advisable to reduce guessing by including Uncertain (U) as a response besides True (T) and False (F). For example, the earth revolves around the sun once in 365 days. T.F.U.

2. *Completion test*. Also called a 'missing word' test. The usual form is a passage with missing words (Type 1). An alternative method is to supply a list of words in the space. The student underlines the correct word (Type 2).

   Type 1:
   The first Europeans who settled the New England Tableland in the 1830s were _____ who climbed the southern_____ near Tamworth in order to find natural_____ for their_____.

   Type 2:
   Many people have left (coastal, rural, mountainous) areas and moved to towns and cities as secondary and (primary, cottage, tertiary) industries have grown. (Fencing, Mechanisation, Labour) on farms has contributed to this drift of (population, settlement, livestock).

3. *Matching test*. Students seem to enjoy most objective test items including matching tests. They are very appropriate for checking recall and recognition. For example, in Figure 28.5 write the correct word from the list in the matching test in front of the definition it best matches.

4. *Multiple choice test*. These are difficult to design but among objective tests they are probably the most satisfactory way of measuring a range of educational objectives. Students are required to discriminate between alternatives. Well-designed items can test understanding of geographical relationships. For example, underline the most appropriate answer:
   Malaysia has become a major exporter of natural rubber because:
   A. the motor car industry demanded much rubber and the indigenous people were able to grow rubber trees on their farms.
   B. the climate of Malaysia was suitable for growing rubber trees.
   C. the environmental conditions were favourable and Western markets for rubber grew rapidly.
   D. rich soils, flat land and good transport services were available.

5. *Short-answer tests*. These items take many forms but most require only recall of facts. They include simple questions, definitions, identification and unfinished statements. In geography, much reading and interpretation of maps, pictures, diagrams, sketches and panoramas can commence with short-answer questions although testing for a wider range of skills and understandings is usually required in course objectives.

Assessment Instruments Based on Essays

The essay question is a valuable method for geography teachers because it helps to assess important objectives not satisfactorily assessed by other techniques.

The essay is pre-eminent in measuring originality, divergent thinking and students' ability to organise their thoughts in their own way. Essays are valued because they enable discussion of principles and points of view, interpretation of ideas, organisation of evidence and the presentation of a written statement clearly, convincingly and with appropriate style.

With such objectives in mind geography teachers have valid reasons for persevering with essay questions as an assessment technique despite difficulties of reliability. Piper (1976) describes these difficulties in some detail and offers advice to ensure greater reliability. This can be summarised as:

1. Providing favourable conditions for the students to do their best writing.

2. Sampling students' essays widely rather than at one sitting.

Answer	Definition	Select from list
	a drought-resistant plant	1. ratoon
		2. fazenda
	the coniferous forests in the Northern Hemisphere	3. xerophyte
		4. taiga
	a crop of sugar cane grown from roots left after a previous cutting	5. epiphyte
		6. ephemerals
	a Brazilian coffee plantation	7. liana
		8. tundra
	plants that complete their life cycle in a short space of time	9. sett.

Figure 28.5 A matching test (see text)

3. Setting the questions carefully. Specifically:
   (a) Clarify the objectives of the essay. If they are at the lower levels of the cognitive range consider other assessment instruments such as objective tests.
   (b) Tell the students exactly what is expected of them. If a particular structure is required state this in the question. For example, your essay should begin by defining transhumance and include a discussion of the factors which...
   (c) Spend time explaining and showing students what is meant by the *directive terms* used in essays such as:

   - *Discuss* with reference to...
   - *Describe* and account for...
   - Critically *examine*...
   - *Compare* and *contrast*...
   - Write a *regional description*...
   - What are the *geographical factors*...
   - *Evaluate* the contribution of...

   (d) Tell the students if there is a sectional marking scheme. It is helpful to set out the question with a line for each part carrying a particular weighting. This is not only fair to the student but helps with time allocation. For example, write an essay about a region of commercial agriculture in North America.
   In your essay *discuss*:
   (4 marks) (i) the factors influencing the nature of its agriculture;
   (4 marks) (ii) the settlement patterns in the region;
   (4 marks) (iii) the processing industries connected to its agriculture;
   (8 marks) (iv) recent changes in the region.
   (e) Reduce choice to a minimum consistent with freedom and flexibility in the course. Excessive choice contributes to unreliability.
   (f) Attempt to write a model answer to the question in less time than allocated to students. If it cannot be done change or discard the question.
   (g) Submit questions to fellow geography teachers for critical comment.
   (h) Trial test the questions. A time-saving and practical alternative is to build a bank of tried questions which have produced the desired responses.

4. Use an appropriate marking scheme. If there is considerable emphasis on skills, understanding and structure in an essay or library research assignment, it is possible to achieve a high level of reliability by using a detailed *analytical marking scheme*, such as:

   - Knowledge or recall of factual information — 5 marks
   - Understanding, interpretation and application of key concepts within the context of the question — 5 marks
   - The use of evidence and examples to support claims and conclusions — 5 marks
   - Graphic and written communication skills — 5 marks

   20 marks

For individual research projects involving the identification of a problem, collecting and analysing relevant data and applying it to solving the problem, an analytical marking scheme such as the following may be used:

- Identification and definition of the issue or problem   5 marks
- Selection, collection and presentation of data   5 marks
- Skills used in analysing and presenting data   5 marks
- Interpretation of data and application to the problem or issue being investigated   5 marks
- Graphic and written communication skills   5 marks

                               25 marks

Detailed analytical marking schemes have a number of disadvantages. They are time consuming and may overlook total quality of the essay. If the primary objectives of the essay are to encourage originality, divergent thinking and to have students express their ideas in their own way such schemes tend to inhibit their attainment. It is debatable if the quest for reliability is worth the loss of validity.

Piper (1976) reports that promising results have been obtained from holistic impression marking. In this system the marker reads the essay quickly and determines a mark on the basis of the whole impression it gives. This system has proved to be just as reliable as detailed analytical marking but has the advantage of saving teachers' time. To increase reliability with this system of marking, it is suggested that a second opinion be used, that students write their names on the *back* of their papers, that the marker reads each question through all papers (if several essays are involved as in an examination), and that the full marking scale be used to assist with discrimination.

Figure 28.6 An example of a data-response question for assessment (Jones 1979)

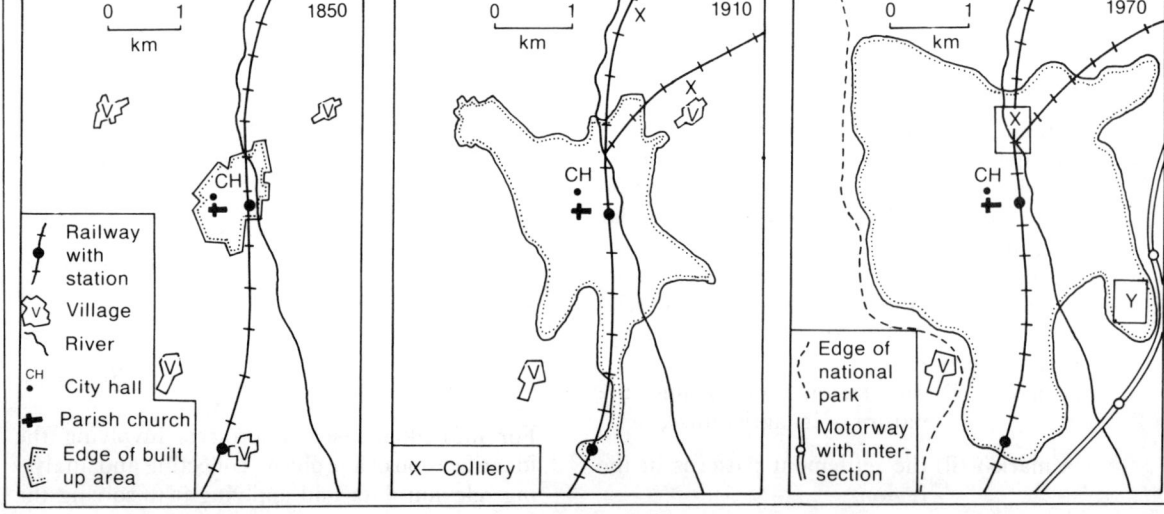

Study the maps below which show a large city in an industrialised country in 1850, 1910 and 1970.

1. Describe and account for the differences in the size and shape of the city in 1850 and 1910. (*5*)

2. Describe and account for the changes which have taken place in the size and shape of the city between 1910 and 1970. (*6*)

3. Areas X and Y are both industrial areas. Area X grew up between 1840 and 1890 and Area Y has grown up since 1960.
   (a) Account for the growth of the two areas.
   (b) Choose *one* of the areas and describe its probable appearance. (*9*)

Assessment Instruments Based on Data-response Questions

The changes in geography that replaced descriptive regional geography with analytical systematic studies in school geography programs also introduced a range of practical, statistical and decision-making techniques for students to master. These techniques are not amenable to assessment through objective tests or essays.

These techniques are best assessed by providing students with stimulus material or data in the form of maps, tables, graphs, photographs, cartoons or text, and asking them to make a response through a range of structured questions. Figure 28.6 contains an example of one such data-response assessment instrument. Others may be found in the many modern geography textbooks that incorporate an inquiry approach to learning, and in specialised assessment texts such as *Data Response Exercises in Physical and Human Geography* (Briggs et al. 1979) and *Assignment Geography: Structured Exercises in Human Geography* (Jones 1979).

Data-response questions are self-contained exercises that enable students to draw conclusions without recourse to too many remembered facts. What is important is the ability of students to comprehend, analyse, synthesise and evaluate the data (all the upper level cognitive processes in Bloom's taxonomy), manipulate or transform them using specific geographical skills, and present their results in a short report structured according to the subsections of the question. Jones (1979) recommends that the subsections be arranged in terms of increasing difficulty, the testing of a variety of abilities and skills, and increasing generalisation. The best data-response questions conclude with subsections that require students to relate the stimulus material to a case study or to apply their findings to a different situation, perhaps a second piece of geographical information.

Well-structured data-response questions thus combine the advantages of essay questions in requiring students to write clearly and concisely about a geographical issue, and of objective test questions which allow an assessment to be made of the full range of cognitive processes. Additionally, they provide teachers with a degree of flexibility in the scheduling of assessment as data-response questions may be written to be answered in a 35–45 minute period in-class test, as part of a longer examination, or as an independent study exercise as part of in-course assessment.

The use of data-response questions in an assessment program is an encouragement to teachers to include a wide range of learning experiences in their courses. In this way, the use of data-response questions can become a means of improving the scope of assessment and of improving the geography curriculum as well.

Decision-making Assessment

Geographers have an essential interest in human decision making because such behaviour has important spatial and environmental impacts. Conceptual and curricular developments since the mid-1960s have led to the prominence of decision making objectives in school geography courses. Inquiry into spatial distributions, patterns and flow depends on understanding the perceptions, culture, motivations and values of the decision makers. Naish (1986) describes how a decision-making exercise in The Geography 16–19 Project was developed as an assessment item for its advanced (matriculation) examination in the General Certificate of Education (UK). Students are presented with inquiry-based resources about a real world issue in the form of a case study. They are required to study the evidence and reach a decision about the problem. For example, students may be faced with the problem of alternative routes for a motorway or preserving a national park facing threats to its ecosystems. Students would be assessed on their ability to undertake a clear process of inquiry, appropriate analysis of data, appreciation of the values positions of those concerned as well as a clarification of their own values. An assessment could also require a clear presentation of alternative solutions and consequences and a judgement of the quality of the student report.

Work-based Assessment

Manuel (1986) describes a work-based assessment scheme which has been implemented at Elizabeth High School, South Australia. In this scheme, students and teachers negotiate a contract of how learning is to occur and be assessed in the course. The contract is settled at the commencement of the course or unit of work and may exist at various levels. Thus the broad scope of the topic may be determined by the teacher but some tasks may be selected by students. Manuel suggests that with experience of work-based assessment, teachers and students gain confidence to negotiate a range of learning activities, skills, assessment procedures and criteria. He also says that more negotiated work-based assessment occurs with senior classes.

The adoption of work-based assessment holds promise for accommodation of specific differences in interests and abilities in students. The approach requires flexibility and a change in teaching methodology for many teachers. It is important also for the completed tasks to be presented to a wide audience such as peers, other teachers and parents.

Nash (1985) compiled case studies and ideas from school practice aimed at increasing student participation through goal-based assessment and a negotiated curriculum. The negotiated curriculum enables students to contribute to what, why, how and when they learn and how their learning will be assessed. The aim is to have students committed to the learning process and to enhance the effectiveness of such learning. In many schools and school systems there would appear to be an uneasy contradiction between goal-based assessment and a negotiated curriculum on the one hand and an externally prescribed syllabus and examinations on the other. Many schools which design goal-based assessment and a negotiated curriculum find descriptive student profiling an effective means of assessment along with other procedures.

The Question of How Much Assessment

One of the most vexing problems for geography teachers besides *how* to assess is *when* and *with what frequency*. It has already been suggested that some diagnostic testing on basic reading skills and ability to find sources should be done at the beginning of a course. Formative assessment should be done throughout the course with a strong emphasis on remediation of deficiencies in students' learning, courses and teaching. Summative assessment is seemingly an expectation of students, parents, school authorities and the community regardless of professional misgivings about validity and reliability of our assessment instruments. It is suggested that the whole geography department should consider the variables of course content and structure, student capability and the demands of the system before deciding on the types and frequency of the assessment. Syllabus requirements will be a major factor in determining the amount, style and frequency of assessment in the upper years of secondary school. These requirements are more flexible in the lower years. It is suggested that the school-based assessment for such students include no more than four to six separate assessment items per semester. This will ensure that a sufficient range of course objectives is assessed, but avoid the excessive demands that could be placed on students especially in written assignments.

Figure 28.7 is a pro forma that could be used to integrate a testing program into a planned course. It has the advantages of matching assessment procedures to the concepts, skills and attitudes taught and the teaching strategies used. It enables these to be seen in the light of the assessment procedures used for grading and those used for diagnostic and other purposes.

Figure 28.7 Pro forma for planning a school-based assessment program

Week No.	Main concepts, skills and attitudes to be developed *or* Specific objectives (whichever appropriate)	Strategies and types of activities— e.g. small or large group work, fieldwork, simulation, library research, practical, workshops, etc.	Assessment procedure for grading		Assessment procedure not for grading
			Type of test	Mark weighting	

Conclusion

The following points need to be emphasised in concluding this brief guide to the geography teacher on assessment. Assessment program planning is a part of school-based curriculum development and is, similarly, a shared enterprise. A variety of forms of assessment are available which should be selected and geared to local needs. Diagnostic, formative and summative assessment each have their place. Techniques in designing items such as observation, objective tests and essay types should arise from the course objectives. While the quest for reliability is a worthy one it should not be followed at the expense of validity or teachers' and students' limited time. Finally, the purpose of all assessment should be to provide information to help promote student progress, rather than simply to find faults.

References

Banks, J.A. with Clegg, A.A.Jr (1974) *Teaching Strategies for the Social Studies*, Reading, Massachusetts: Addison-Wesley.

Blachford, K. (1972) *Evaluation*, Unit V, The Teaching of Geography, Melbourne: Education Department of Victoria.

Black, H.D. and Dockrell, W.B. (1980) *Diagnostic Assessment in Geography*, Edinburgh: The Scottish Council for Research in Education.

Bloom, B.S. et al. (eds) (1956) *A Taxonomy of Educational Objectives, Handbook 1: Cognitive Domain*, New York: David McKay.

Bloom, B.S. et al. (1971) *Handbook on Formative and Summative Evaluation of Student Learning*, New York: McGraw-Hill.

Briggs, K., Riley, D. and Tolley, H. (1979) *Data Response Exercise in Physical and Human Geography*, Oxford: Oxford University Press.

Cooper, K. (1976) *Evaluation, Assessment and Record Keeping in History, Geography and Social Science*, Place, Time and Society 8–13, Bristol: Collins for the Schools Council.

Cox, B. (1966) 'Test Items in Geography for a Taxonomy of Educational Objectives', *Monthly Bulletin*, No. 44. Reprinted in Biddle, D.S. (ed.) (1968) *Readings in Geographical Education*, volume 1, Sydney: Whitcombe and Tombs.

Fenton, E. (1967) *The New Social Studies*, New York: Holt, Rinehart and Winston.

Hickman, G., Reynolds, J. and Tolley, H. (1973) *A New Professionalism for a Changing Geography*, London: Schools Council.

Jones, M. (1979) *Assignment Geography: Structured Exercises in Human Geography*, London: Nelson.

Krathwohl, D.R., Bloom, B.S. and Masia, B.B. (1964) *Taxonomy of Educational Objectives, Handbook II: Affective Domain*, New York: David McKay.

'Man: A Course of Study' (1970) *Evaluation Strategies*, Washington: Curriculum Development Associates.

Manuel, M.D. (1986) 'Work Required Assessment in Geography', Workshop presented at the 10th Australian Geography Teachers' Association Conference, Brisbane.

Marsden, W.E. (1976) *Evaluating the Geography Curriculum*, Edinburgh: Oliver and Boyd.

Naish, M. (1986) 'Decisions, Decisions! Teaching and Assessing Environmental Thinking', *Geographical Education*, Vol. 5(2), 31–34.

Nash, J. (1985) 'Increasing Participation through Goal-Based Assessment and Negotiated Curriculum', Participation and Equity Program, Curriculum Services Branch, Division of Curriculum Services, Department of Education, Queensland.

Okunrotifa, P.A. (1977) *Evaluation in Geography*, Ibadan: Oxford University Press.

Okunrotifa, P.A. (1981) 'Evaluation in Social Studies', in H. D. Mehlinger (ed.) *UNESCO Handbook for the Teaching of Social Studies*, Paris: UNESCO.

Oliner, P. (1976) *Teaching Elementary Social Studies*, New York: Harcourt Brace Jovanovich Inc.

Piper, K. (1976) *Evaluation in the Social Sciences for Secondary Schools, Teachers' Handbook*, Canberra: Australian Council for Educational Research.

Salmon, R. and Masterton, T. (1974) *The Principles of Objective Testing in Geography*, London: Heinemann.

Senathirajah, N. and Weiss, J. (1971) *Evaluation in Geography*, Toronto: Ontario Institute for Studies in Education.

Social Studies K–10 Syllabus (1981) Perth: Curriculum Branch, Education Department of Western Australia.

29

Selecting and Evaluating Resources for Geography Teaching

Peter Maccoll

Geography teachers have to cope with an increasing variety of resources as they strive to make learning in geography both interesting and relevant. Peter Maccoll addresses the question of resource materials in geography from a process viewpoint. He attempts to equip geography teachers with an increased awareness of the variety of resources, their sources and factors which may influence the selection of these resources, before offering geography teachers a comprehensive set of criteria for evaluating resource materials. This process of resource selection and evaluation is based on a useful resource-based planning model. Relevant applications of this process of resource selection should provide teachers with practical benefits from reading this chapter.

It is not too far into the past that geography teaching in secondary schools was dominated by the desire for students to gain mastery of a wide range of factual information (almost solely for examination purposes) through traditional 'chalk and talk' approaches. Usually this knowledge was highly specialised and prescribed by tertiary institutions. This academic knowledge was 'simplified' and presented in set textbooks for students to absorb.

In these circumstances, the attitudes of students were generally attuned to 'conquering' mountains of facts. The few skills that were emphasised mainly related to mapping, the interpretation of tables, graphs or photographs and the reproduction of selected facts to answer specific examination essay questions.

To some readers, this may paint too gloomy a picture of past schooling, to others it may still be very familiar!

Thankfully, the focus of education is changing. Changes have taken place and are continuing: in the nature of the geography curriculum in schools, in the methods of teaching, in the resources being used, and also in the ways in which student performance is being monitored, assessed and evaluated. These changes are discussed and analysed in detail throughout this book. It is relevant, however, to briefly explore some of the changes which have taken place in relation to teaching and learning for they have significant implications for resource selection and management.

Changes in the Teaching of Geography

Reference was made earlier to the 'chalk, talk and textbook' approach to teaching. Certainly, this has not disappeared and, as Jones shows in 'Using Expository Methods Well' (Chapter 4), it is still a valid mode of teaching and learning. Yet it is only *one* of an increasing number of approaches to facilitate learning. Recent research into how children learn, developments in educational hardware and software, more efficient pre-service and inservice teacher education, and improved school facilities have all contributed to a vast improvement in the quality and diversity of learning experiences for students undertaking studies in geography. Such changes have seen an increased emphasis on inquiry learning and skill development, field studies being incorporated as integral components of the curriculum, and the use of case studies, games and simulation. These all reflect a movement towards student-centred teaching and learning. 'How do students learn?' is now as important a question as 'What should be taught?'.

Changes have also taken place in the nature of geography curriculum offerings in schools. A recognition of the various branches of geography has seen the proliferation of programs such as urban studies, economic and industrial geography, recreation geography and environmental studies. Accompanying this in many places has been the introduction of shorter programs based on a semester or even periods of five to nine weeks.

The Use of a Wide Range of Resources

Closely allied to changes in teaching methodology and curriculum development is the increasing variety of resources being used by both teachers and students. Although a single textbook may still be the focus in some programs, the tendency in recent times has been towards the use of multiple resources— both within and outside the classroom. The increased provision of audio-visual hardware has enabled the frequent use of films, filmstrips, slides, audio- and videotapes in the geography classroom. As well, the emphasis on student inquiry has seen the introduction of class sets of books, greatly increased use of libraries and has firmly cemented the place of newspapers, periodicals, pamphlets and other media as typical sources of data.

These moves have also been supported by the development of new and diverse curriculum materials by commercial publishers, government agencies and education authorities. For example, the materials produced by curriculum projects such as the Place, Time and Society 8–13 Project (UK), the Geography for the Young School Leaver Project (UK) and the Social Education Materials Project (Australia) have provided teachers with a wide and diverse range of contemporary data and student activities, and guidelines for implementation as well as vehicles for the development of the many student skills, attitudes and values.

Accompanying this growth in resource development, there has been the formal incorporation of field studies into most geography curricula and the recognition of the value of the community as a resource. As well as bringing the outside world into the classroom through the media and guest speakers, teachers and students are using both the urban and rural environments in more diverse ways and with more sophisticated techniques. The 'hop on the bus, look around and fill in the question sheets' type of field study is rapidly becoming a thing of the past. In its place we find such exercises as rural land-use studies, farm studies, social surveys, spatial interaction analyses, quantitative geomorphic studies and residential area analyses—all of these being a resource in their own right.

Against such a background, few would disagree that there is a plethora of resource materials readily available for use by geography teachers and students. Such a situation, however, is not without its problems.

Consider the following statements which might be heard in a staffroom, corridor or even in a meeting to decide on selecting resources for a geography program:

'That textbook we set is just too difficult for those kids to read and the exercises are almost at university level . . . '

'The statistics and newspaper extracts in that book are hopelessly out of date . . . '

'You'd think the author was being paid by industry the way he treats conservation . . . '

'The kit is very good in places, but you'll need other resources to treat the topic in depth . . . '

'That's almost two-thirds of our whole budget spent on materials which can only be used in the first two units of the course . . . '

'I've just found out that not only did my students use these materials we're using now in their course last year, they're also using them in science. No wonder they're not responding . . . '

The sentiments behind all these statements point to six problems associated with *selection* of resource materials:

1. readability and comprehension levels;
2. currency of content;
3. bias;
4. inconsistency and inadequacy of coverage;
5. value for money with regard to program needs; and
6. uncoordinated management and duplication of resources.

The Question of Resource Selection

How are textbooks and resource materials selected for use in schools? Certainly the answer in the past was much simpler than it is today. When there was a limited range, or even where a textbook was prescribed by a syllabus document, there was little choice to be made. Today the answer to that question is much more complex.

One is reminded of the choices facing the racegoer or punter when looking for an analogy to resource selection. This analogy reflects not only the disparate nature of recent research on resource selection and the great diversity of current practice, but also something of the role of the teacher responsible for resource selection. In his or her quest to choose a winner, the racegoer is faced with numerous variables to consider. For instance, such factors as which horses are in the field, the breeding of the horses, barrier positions, weights, jockeys, track reports and the recommendations of form guides must be taken into account if a rational choice is to be made (Jacobson and Maccoll 1976).

This is not entirely irrelevant to a consideration of the issues involved in the selection and management of educational resources, for it must be recognised that the actual process of selection is certainly a case of 'horses for courses'. Primarily, this has resulted from the increasing decentralisation of schooling which has meant the central educational authorities are, in many cases, no longer the selectors of resource materials as they often were in the past. The local school authority, the school geography department and the individual teacher have increasing control over the curriculum of the school and classroom and are making basic decisions about the resource materials that will be used. Hence, materials are often considered in relation to how well they meet the needs of specific teachers and students, and subsequently, the nature of resource materials used can vary quite significantly even between schools in close proximity.

Such diversity of resource selection and use does not mean, however, that those involved in the selection process do not have a 'form guide'. No one can complain about the lack of materials any longer, since there is the profusion of publishers' catalogues and brochures, displays of resources at conferences, as well as the reviews and advice offered by professional journals and newsletters. The major activity is to identify and select *appropriate* resource materials. Such a quest involves two tasks: firstly, the identification of what is currently available in the form of geography materials and, secondly, the evaluation of those materials so that an informed judgement can be made as to their appropriateness for teacher and learner.

The Nature of Resource Materials for the Teaching of Geography

As most geography teachers are aware, the nature of resource materials and sources of data can vary immensely. Piper (1976) suggests four major categories or modes in which information can be presented:

1. *Verbal*: Data presented mainly in words, from such sources as reference books or textbooks, newspapers, periodicals, documents and interviews.

2. *Pictorial*: Data presented in visual form, such as photographs, illustrations and cartoons.

3. *Quantitative*: Data in numerical form, especially statistical data as presented in groups and tables.

4. *Symbolic*: Data presented in representational forms, such as maps and diagrams.

Drawing on Piper's work, Hill (1977) suggests a more comprehensive categorisation of ways in which data can be found and presented and combines this with the elements of a basic information-processing model to produce a matrix which has significant value for planning programs of work based on inquiry learning. An adaptation of Hill's suggested grid is shown in Figure 29.1.

This model has four advantages for the geography teacher:

1. It allows ready recognition of the range of data sources represented by available materials which can be used in a unit of work or an entire program. This is important not only because of the diverse range of learning experiences which can result, but also because students are continually exposed to various forms and sources of information with which they have to work. By using the grid in planning, a teacher can ensure that students experience working with a range of resources.

2. It enables the skill objectives of any program to be related to the various forms of resources.

	Input →		Processing →		Output →		Further Research	
Skills Data Forms	Gathering and organising information	Interpreting information	Analysing information	Generalising	Making value judgements	Presenting the findings	Identifying issues and problems—Hypothesising	Generalising and testing hypotheses
Language • written • spoken								
Pictures • photographs • slides • cartoons								
Maps • orthophoto • atlas • topographic								
Diagrams • flow charts • time lines • block diagrams								
Graphs **Tables**								
Realia • artefacts • specimens • recording • sketching								
Field Studies • observation								
People • interviews • questionnaires								
Films • 16 mm • Video								

Figure 29.1 A resource-based planning model (Maccoll and Fennel 1982)

3. Teachers can easily identify areas in which resources are needed to support an inquiry approach.
4. It provides teachers with an opportunity to plan for resource usage in field studies and local area surveys.

Figure 29.2 illustrates how the planning matrix might be used in planning a study of a farm or farming locale. The data sources to be used are identified as are the student skills involved in extracting and using the information. Depending on the extent or purpose of the planned task the grid need not have every box completed; particular activities could be deleted or added as the study progresses.

Sources of Resource Materials

Geography resource materials are produced by a variety of sources. One major problem in identifying what is available is the tendency to overlook those resources already existing in the school. Most teachers can readily cite instances of previously purchased or developed resources which are lying idle in a cupboard, stockroom or library. In many cases, their neglect is not due to irrelevance or student indifference but teachers tiring of their use. Such instances beg the introduction of some form of resource

Figure 29.2 Resources for a study of a farm

	Input	Processing		
Skills / **Data Forms**	*Gathering and organising information*	*Interpreting information*	*Analysing information*	*Generalising*
Language • written • spoken	Research of farm development in area (Govt documents) • Identification and description of farming type			
Pictures • photographs • slides • cartoons	Photographing farm landscape	Drawing and labelling line sketches of photographs	Comparing photographs and line drawings with field sketches	
Maps • orthophoto • atlas • topographic	Constructing maps to indicate • farm layout • relationship to transport		Relating location of farm to regional networks of transport, land use and settlement.	Drawing conclusions about land use in a particular area
Diagrams • flow charts • time lines • block diagrams	Identification of production and processing stages from farm to factory			
Graphs **Tables**	Gathering statistics on farm production over time		Comparison of farm production figures with regional averages	Identifying trends in production
Realia • artefacts • specimens	Collecting soil samples	Analysing soil samples	Relating soil types to land use recorded in field sketches	
Field Studies • observation • recording • sketching	Field sketching of farm landscape; identifying features			
People • interviews • questionnaires	Interviews with farmer and Government Primary Industry officer	Identification of trends and problems		
Films • 16 mm • Video				

(Continued)

	Output		Further Research	
Skills Data Forms	Making value judgements	Presenting the findings	Identifying issues and problems; hypothesising	Generalising and testing hypotheses
Language • written • spoken		Written report of case study farm illustrated with sketches and photographs	Identification of problems facing rural producers in this area. Suggestions of possible solutions	Examining feasibility of hypotheses through other case studies and research
Pictures • photographs • slides • cartoons	Explanation of relationships between land use, terrain, water availability and transport networks			
Maps				
Diagrams • flow charts • time lines		Time line drawn to show the development of rural holdings in area and transport		
Graphs **Tables**		Constructing graphs showing trends in farm production		
Realia • artefacts • specimens				
Field Studies • observation				
People • interviews • questionnaires				
Movies • video				

identification and management scheme within the geography department across the whole school and between neighbouring schools. Quite often, what might be seen as a peripheral resource in one school or by one teacher, might be regarded as important by others.

In a similar fashion valuable resources can be neglected because they are not specifically labelled 'geography', or because they are held in another department or section of the library. In this sense, a full acquaintance with *existing resources* within the school can often pay handsome dividends. The possible sources of such materials are indicated in the two inner rings of Figure 29.3, which displays sources of resource materials which may be tapped by geography teachers.

Other readily available sources of resource materials are identified in the outer ring in Figure 29.3. Naturally, such sources can be further defined to identify specific departments, contact people or specific publications such as journals, catalogues and resource listings, and to explore means of access.

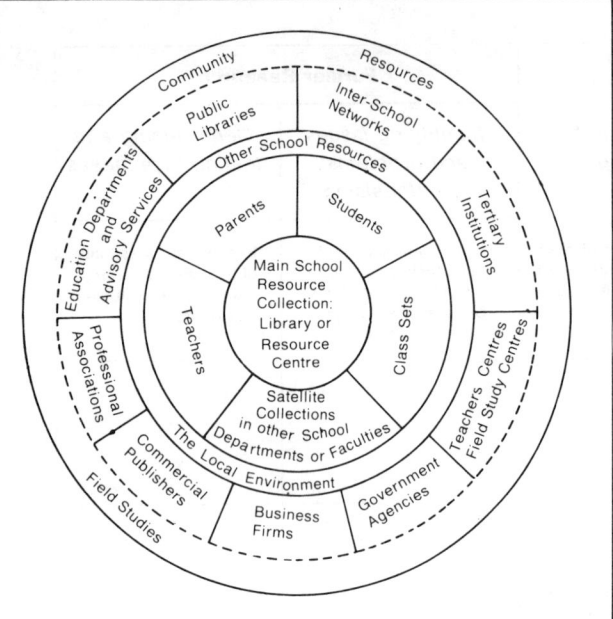

An example of how these numerous sources of information can be tapped to provide the basis for a unit of work on urban problems is illustrated in Figure 29.4.

The Selection of Resource Material

Indeed the plethora of resources that are available for many studies in geography is often as much a problem as a shortage of resources on other topics. Not only could there be a range of textbooks seemingly appropriate for a specific topic, but also other sources of data such as audio-visual kits, sets of activity cards, pictures or photographs, map extracts, previously collected field data, curriculum guidelines, 'how-to-do-it' booklets and corporate

Figure 29.3 Sources of teaching/learning materials (after Department of Education, Queensland 1980)

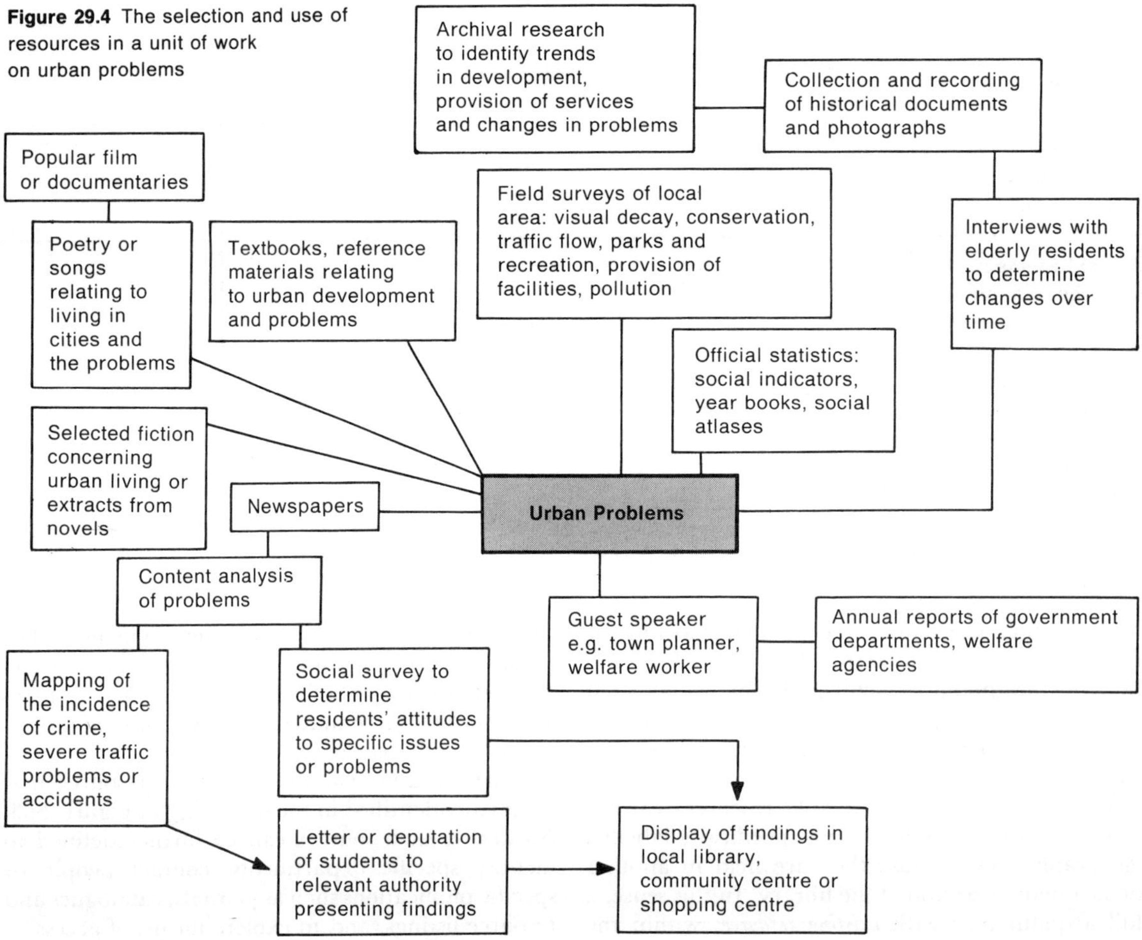

Figure 29.4 The selection and use of resources in a unit of work on urban problems

terials. These issues are very much interrelated and provide a context and a basis for the development of specific evaluation criteria. These issues include:

1. The use of school and department curriculum policies as a starting point.
2. An awareness of community attitudes towards education.
3. The need for group decision making.
4. The coordination of resources throughout the school.
5. The problems with available information about resources.

School Curriculum Policies as a Starting Point

The changes in teaching approaches referred to earlier in this chapter have developed hand in hand with changes in the nature of geography curricular offerings. As might be expected, these changes have been by no means uniform, and they reflect considerable differences in approach. These differences can be seen in the disagreement over such curricular questions as how to state behavioural objectives, or if they should be used at all in curriculum planning. (See Chapter 27, 'School-Based Curriculum Development in Geography', for more information and an elaboration of this debate.)

Such questions and differences are often reflected in materials published for schools. Where a significant mismatch arises between curriculum design and the design of resource materials, problems of use are inevitable. To overcome such problems, a sound starting point is an examination of the school curriculum policy. Most secondary schools and geography departments now have written policy statements which outline, among other aspects, particular principles of teaching and learning which should be promoted. Such statements can provide a sound initial basis for the geography teacher in deciding upon the nature of the resource materials to be used. The degree of emphasis on inquiry learning, data gathering and research skills, the development of attitudes and values, a concern for 'learning how to learn' and a concern for active student involvement in practical activities are the types of issues which can give a focus to resource selection procedures. For instance, where a school or department policy encourages the use of multiple resources, teacher attention can be directed to examining a wide range of kit and card materials, selecting class sets of booklets or integrating audio-visual and film resources into

---

An unfinished checklist of resources . . .*

activity cards	guest speakers	posters
archives	interviews	questionnaires
artefacts	item banks	radio
audio-tapes	jigsaw puzzle	readings
brochures	maps	recipes
bulletin boards	journals	recordings
cartoons	letters	reports
case studies	line sketches	reproductions
cassettes	logs	research findings
census findings	magazines	role plays
chalkboard	maps	sand tables
charts	mobiles	scrapbooks
collages	mock-ups	simulations
collections	mock newscasts	slides
computer-assisted	making models	sociodrama
learning packages	montages	specimens
cross-sections	murals	stamps
crossword puzzles	music	statistics
diaries	narratives	stencils
dioramas	newsletters	stories
documents	newspapers	surveys
excursions	novels	tables
exhibits	overhead projector	talks
field studies	transparencies	tape recordings
films	paintings	television
filmstrips	pamphlets	tests
flags	parents	textbooks
flannelboards	pen pals	timelines
flashcards	people	touring guides
flowcharts	periodicals	transparencies
games	photographs	trips
globes	pictures	video
graphs	poetry	workbooks . . .

* Although many of these might be viewed as activities, the results of these activities can provide valuable resources for other students.

Figure 29.5 Resources available to geography teachers

---

and government reports and handbooks could be available.

This situation can be made even more complex by the often conflicting information about resources that is available. Reviews in professional journals, publications by educational authorities, publishers' promotional brochures, and reports from other schools can often blur rather than refine the advantages and disadvantages of a particular text or slide set. It is here that a *definite thoughtful approach* is essential to ensure that the 'right horse' is selected for the 'right course'—especially when one considers what is available to geography teachers as illustrated in the unfinished list in Figure 29.5.

Teachers are at the centre of decision making with regard to purchasing or not purchasing new materials as they become available. Clear guidelines or evaluation criteria to assist in this decision making can be established. However, there are several broad issues which have to be considered before turning to procedures for appraising specific resource ma-

programs, rather than searching for the 'one next' to cover an entire program. On the other hand, where textbooks are seen as necessary, preference might be directed towards those containing student activities and suggested extension experiences.

Awareness of Community Attitudes Towards Education

Changes in curriculum have not occurred without criticism. In recent years increased public criticism has focused not only on what is being taught, but also on how it is being taught and the materials being used in that teaching. A concern for 'the basics', the handling of controversial issues, the discussion of values in the classroom, and the use of role play and simulation are some of the issues which have attracted vehement criticism—and which sometimes have resulted in government action. The withdrawal of 'Man: A Course of Study' (MACOS) and the materials of the Social Education Materials Project (SEMP) in Queensland, the discontinuation of funding of the dissemination of all National Science Foundation projects in the United States following controversy over MACOS, and the evolution and creation textbook dispute in Kanawha County, West Virginia, in the United States are recent examples of such action.

This certainly does not mean that the rights of schools to select resource materials have been abrogated and supplanted by forms of external censorship. In a number of cases such actions have resulted in definite guidelines being established for resource selection and use in schools—guidelines which move towards greater community involvement in the curriculum planning process. What it does mean for teachers, and geography teachers are no exception when one considers the changes that have taken place in geography teaching, is a need for a much greater awareness of, and sensitivity towards, community attitudes and expectations. Such awareness is often exhibited in school policy statements, but should never be lost by those involved in actual resource selection.

The Need for Group Decision Making

Although individual teachers are centrally involved in the selection process, certainly each teacher will not be able to evaluate in any systematic way all the materials he or she might like to use. The value of *group decision making* has become apparent, through the involvement of teachers, administrators and community members, in such activities as school policy development and school-based curriculum development. Such a concept, given time provisions for its conduct, can be applied very easily to the selection of specific resource materials. Because it is based on a broad range of expertise and experience it is more likely to meet the needs of a department much better than any one individual's decision making. In addition, the employment of a cooperative group or committee has the direct advantage of reducing the time necessary to complete the task whilst also providing valuable training for the less experienced teacher.

Coordination of Resources Throughout the School

The involvement of as many teachers as possible in the selection process can also enable the recognition of the need for coordination of geography resource management on a whole-school basis. This coordination can extend both *vertically*, through respective year levels, and *horizontally*, across each year level.

A recurring problem in many schools is the use of the same textbooks or pieces of resource materials in consecutive year levels (and often in different subject areas as well). Although teacher intent and the strategies adopted with the materials might be entirely different, student perception of the 'same materials being used again' can often result in lack of motivation, even boredom. Through coordination between year levels this problem can be avoided, particularly if members of other subject departments are involved in the resource management exercise which can be focused through the library or resource centre.

Horizontal coordination is more frequently a problem between subject areas, although it can often occur where different geography electives such as urban studies and environmental studies are offered in the same year level. The use of the same materials here can lead to significant repetition or overlap resulting in the same boredom. With careful examination of resource usage across a particular year level, it may be rationalised to some extent, and where overlap remains, such instances could be used to reinforce concepts or skills.

Problems of Available Information About Resources

Finally, one major issue which at present remains largely unresolved is the lack of opportunity for teachers to trial materials adequately before their purchase and implementation in the school. This places immense importance on the efficacy of the evaluation procedures developed and adopted by a school. There are, however, several sources of information which can initially assist these procedures:

1. Earlier mention was made of publishers' brochures and catalogues which might contain information relating to possible results of field trials, the possible use of the material/s in the classroom, particular areas of application and other factors such as flexibility and evaluation of student performance. If publishers do have information of this type, it usually applies to the total effectiveness of the product in relation to its perceived need rather than detailed information relating to different aspects and components of the materials. Such a source, however, should not be disregarded, for with increasing competitiveness in the market place, more publishers are using field trials to determine appropriateness of materials and it can be expected that such results will eventually filter through in promotional media.

2. Geography teachers' associations, educational authorities and education centres provide reviews of numerous textbooks and material, through the medium of their journals and newsletters. Unfortunately, many of these result from 'armchair evaluations' and may ignore many of the relevant practical considerations which might apply to a particular school program.

3. The major sources of potentially reliable information which *does* result from trialling are other teachers and students who have used a textbook or set of materials. Awaiting their advice might incur a time delay, and course requirements and environments will certainly differ between classrooms and between schools, but to the extent that teachers use the same evaluative procedures, questions and ways of responding to materials, they are in a good position to provide meaningful advice to their colleagues. Hence, there is a need to ensure that resource evaluations are not 'one-off' considerations, and that our recording of evaluation findings

Figure 29.6 Model of criteria and procedures for evaluating resources (after Harding 1978)

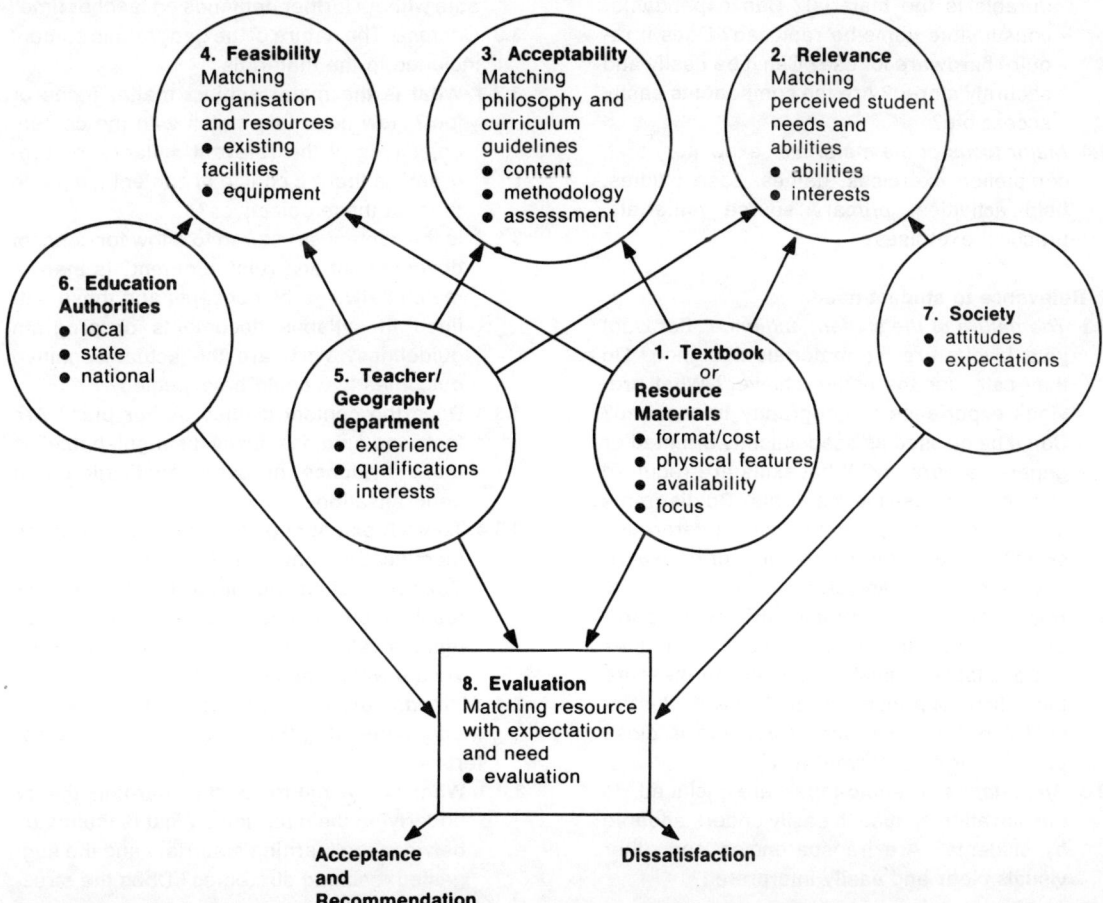

Figure 29.7 Criteria for evaluating educational materials

Criteria for evaluating educational materials

1. The teaching-learning materials
Basic identification data and the general characteristics of the materials can be organised as follows:
1.1 Title
1.2 Author(s)/Project
    Date of Publication
    Publisher/Distributor
    Cost
1.3 *General description:*
1.3.1 *Textbook:* Number of pages? Is it part of a series? Is it well indexed? Is there accompanying support material, e.g. teacher's manual? Is it well bound and durable?
1.3.2 *Kit materials:* What does the package contain? How can the materials be used? Individually? Whole class? Small groups? How durable is the material? Can expendable/consumable items be replaced? Does it require hardware for use? Can it be easily and securely stored? Are the components easily accessible?
1.4 *Major focus of the materials:* expository text, completion exercises, games, case studies, field activities, primary source materials, practical exercises?

2 Relevance to student needs
2.1 *The nature of the student audience:* For what year level/s are the materials designed? Do they cater for the underachiever? What previous experience in geography is required? Can it be classed as academically oriented or general geography? What skills are required of students to use the materials? Do diagrams and maps presuppose particular interpretive skills? Do the materials presume prior experience such as urban/rural living?
2.2 *Readability:* Is the readability level appropriate for the intended audience? (Techniques are available to enable teachers to measure the difficult of print material.) Is it well illustrated? Are the illustrations clear and in close proximity to the relevant text?
2.3 *Audio-tapes:* If audio-tapes are included, is the narration or speech easily understandable by students? Are transparencies and other visuals clear and easily interpreted?

3 Curriculum match
Background to the development of the materials and the objectives.
3.1 *Rationale:* Do the author's/project's educational assumptions match the philosophy of the geography syllabus? Are these assumptions and the goals of the materials clearly stated and consistent?
3.2 *Objectives:* Are the objectives clearly stated or are they implicit in the materials? Are they stated in behavioural terms and if so, to what degree of specificity? Are they too specific to be manageable without further teacher input?

Do the objectives relate to cognitive, skill and affective domains? In the cognitive domain, are they concerned with more than just recall and comprehension and require students to become involved in application, analysis, synthesis and evaluation? Do the skill objectives require previous levels of attainment? Do the skill objectives match the task provided for students? Are the affective objectives attainable without further demands on teacher time?
3.3 *Content:* The nature of the geographic content included in the materials.
3.3.1 What is the major subject matter focus or foci? How does this match with the content objectives of the relevant syllabus or program? Is there a choice of content available to meet these objectives?
3.3.2 Is the content structured to allow for concept development and reinforcement? Is there a match between the concepts and those outlined in syllabus documents or program guidelines? What are the actual cognitive outcomes that could be expected?
3.3.3 Does the content display author bias? Are facts and opinions easily distinguishable? Is there evidence of oversimplification and generalisation?
3.3.4 To what extent does the content encourage identification and clarification of values? Does the author attempt to teach values or teach about values? Do the materials attempt a value-free approach, or incorporate values within the content?
3.4 *Methodology:* An examination of the learning theory underlying the development of the materials.
3.4.1 What is the nature of the learning theory underlying the materials? What is the match between the learning materials and the suggested teaching strategies? Does the struc-

ture and form of knowledge presented match the structure and form of geographical knowledge? Is there a variety of teaching-learning strategies?

3.4.2 What are the major teaching forms encouraged by the materials? Teacher-to-student? Resource-to-student? Teacher–student interaction? Resource–student interaction? Are the materials rigidly sequenced or do they permit flexibility and choice?

3.5 *Assessment:* Assessment provisions inherent in the materials.

3.5.1 What provisions are made in the materials for student assessment? Are tests included? Does it provide for student self-evaluation?

4 School organisation and resources

4.1 *School policy:* Is there a school policy statement relating to resource selection and management? How could the library or resource centre assist in this process?

4.2 *Implementation requirements:* Do the materials and their inherent methods require special teaching situations or circumstances? Can they be used in traditional classroom settings or do they require flexible open space? Is special equipment or hardware necessary? Do they place undue demands on timetabling provisions? How much time is needed to prepare the materials for use?

5 Teacher characteristics and requirements

Can the materials be readily used by existing geography staff? Is special training necessary? Will teaching strategies and/or organisation be a major departure from normal practice? Are special skills needed to use the materials?

6 Education Authorities

What guidelines exist with regard to resource selection and implementation in schools? Are there specific directives relating to the handling of controversial issues, replacement of expendable materials, resource maintenance or safety in and outside the classroom? What funding provisions are available? Is there provision for sharing expensive materials between schools?

7 Societal attitudes and expectations

Are there any distinctive characteristics of the local community which could inhibit resource usage? Are there issues deemed controversial in the local community which would necessitate sensitive treatment in geography programs or which should not be treated at all? Are there sources of assistance and support in the local community which can be used as classroom resources to supplement the materials?

8 Evaluation: Matching resources with expectation and needs: an overall judgement

Categories 1–7 can provide a wealth of detailed information about a textbook or set of resource materials. According to purpose and constraints such as time and personnel, questions can be expanded upon or deleted where required. Often it will not be necessary to work through all categories if the qualities of a certain piece of material provides little match with perceived needs.

Where a piece of material is fully appraised according to the categories of the model, the information can help to provide an overall evaluative judgement. This in itself raises several questions:

8.1 *Recommendation:* What summary statements can be made about the overall appropriateness of the geography materials and the conditions under which they should or should not be used? Even though a certain textbook, for instance, will not satisfy the entire needs of a program, it could contain very appropriate components. Perhaps it would be fruitful to explore whether it also has relevance to other subject areas and if it could be used jointly.

8.2 *Presentation:* How might the final judgement be presented? Several options are available for reporting and disseminating evaluation findings. A relatively simple method of recording responses to focusing questions is by the use of a checklist or scale response. The former involves 'yes' or 'no' answers whilst the scale response requires the evaluator to make a definite decision about the particular aspects of the materials. Examples of this latter format are as follows:

Example A

Criterion: Currency of materials

Poor ⊢———┼———┼———⊣ Excellent

Includes content, pictures and techniques that are out of date.   Utilises current information, photography and experiences.

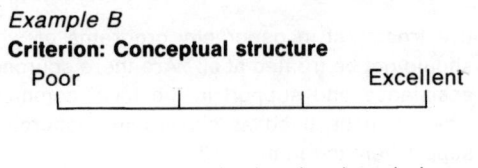

(both before and after use) are in a form that can provide assistance to other teachers and other schools.

These five broad issues may not be able to be resolved to everybody's satisfaction although they do provide foci for considering an organised and coordinated approach to resource selection and management, and certainly a context in which to establish specific criteria for evaluation.

Criteria for the Evaluation and Selection of Commercially Produced Resource Materials

Evaluation of resource materials for geography teaching is concerned with *appropriateness* or *match*. That is, evaluation procedures attempt to explore the degree of match between the demands and expectations as expressed in syllabus or program documents or by teachers, and the adequacy of the textbook or materials to satisfy those expectations.

To proceed well past the 'first impressions' or 'intuitive' basis for evaluation it is desirable to develop criteria which not only reflect the expectations and demands of the prospective purchaser but which also provide a methodical and comprehensive approach to evaluation. As discussed previously, the needs of individual teachers and schools could vary considerably and to meet these needs schools could develop their own set of specific criteria. To assist in just this task a variety of checklists, inventories, guidelines and handbooks has been developed. In addition, several very detailed analysis systems have been produced, perhaps the most notable being the Curriculum Materials Analysis System (CMAS) developed by Morrissett and associates (Morrissett and Stevens 1967; Morrissett et al. 1968). Although developed to facilitate evaluation of the materials produced by numerous significant curriculum projects, the CMAS categories of criteria provide a useful basis for establishing more specific forms of evaluation systems. Jones (1981) has provided a detailed analysis of CMAS applied specifically to geography materials to assist in this task.

For the practical purposes of an individual school or geography department, the extended use of a comprehensive system such as CMAS could be very time consuming and laborious (although these tasks could be shared between schools or within a network of schools). An alternative basis for developing categories of evaluation criteria is suggested by the model in Figure 29.6. The model indicates the major sources of issues, the areas in which 'match' is desirable, and the potential interaction in the resource evaluation and selection process. Figure 29.7 provides a guide to the sorts of focusing questions which could be used to evaluate the 'match' between any particular resource and its intended use and audience.

It must be pointed out that resource evaluation criteria systems are not 'tests', and 'scores' should not be the desired end results. If scores were used, two books could receive the same score, one possibly because it is visually attractive though its content is of doubtful validity, the other possibly because it is valid, well structured but unattractive in design and difficult to read.

In a quest for a rational system and to save time, the vital importance of brief comments is often overlooked. Certainly a lengthy and descriptive conclusion might suffer the fate of other reviews, whereas a series of short summary statements relating to each of the categories illustrated in Figures 29.6 and 29.7 provides an avenue to a concise yet comprehensive overview of the textbook or materials.

Conclusion

An exploration of resource provision, selection and management reveals a broad and complex area, which in terms of obtaining educational and economic 'value for money' can provide numerous pitfalls and problems for geography teachers. It is only through an increased awareness of the resources available to schools, and the use of coordinated and systematic procedure for evaluating resources, that these problems will at the worst be alleviated, and at the best, resolved. It is hoped that by identifying sources of relevant geography materials and outlining categories of criteria by which these may be evaluated, this chapter can also highlight the implications of such decision making for effective and efficient resource use by geography teachers.

References

Department of Education (1980) *Resource Management for Secondary Schools*, Brisbane: Library and Resource Services.

Harding, J. (1978) 'Curriculum Change: A Model of Teacher Decision-making', *Journal of Curriculum Studies*, Vol. 10(4), 354.

Hill, P.W. (1977) *Teaching of Skills in the Social Sciences: A Critical Survey of the Literature*, Canberra: Curriculum Development Centre.

Jacobson, E.J.P. and Maccoll, P.G. (1976) *So You Want a Winner? Workshops for Evaluating Resource Materials*, Brisbane: Department of Education, Queensland.

Jones, G. (1981) 'How can Geography Materials be Appraised?', *Queensland Geographer*, Vol. 16(2), 26–33.

Maccoll, P. and Fennell, P. (1982) *Teaching Social Science Skills*, Canberra: Curriculum Development Centre.

Morrissett, I. and Stevens, W.W. Jnr (1967) 'Curriculum Analysis', *Social Education*, Vol. 31(6), 483–487.

Morrissett, I., Stevens, W.W. Jnr and Woodley, C.P. (1968) 'A Model of Analysing Curriculum Materials and Classroom Transactions', in D. Fraser (ed.) *Social Studies Curriculum Development: Projects and Problems*, Washington DC: National Council for the Social Studies.

Piper, K. (1976) *Evaluation in the Social Sciences for Secondary Schools: Teachers' Handbook*, Canberra: Australian Council for Educational Research.

30

Evaluating Your Geography Courses

Barrie McElroy

This chapter assumes that the health and development of school geography courses depends on regular and effective evaluation. It briefly explains the nature of course evaluation and its role in curriculum development. A case-study approach to the evaluation of geography courses is recommended. Barrie McElroy describes this approach as an attempt to paint a picture of the course by assembling a montage of similar and disparate aspects of the course drawn from the widest range of appropriate sources. Suitable methods for collecting and 'displaying' information are suggested and discussed. Finally, the chapter explains several ways by which teachers can find meanings in the revealed pictures of their courses and thus enhance the process of curriculum redevelopment.

Do the following extracts from course evaluation files sound familiar?

Extract A:
(From a transcript of an interview with a teacher beginning to discuss his methods of teaching.)

Teacher: Well, I dictate notes or put them on the blackboard for them to copy. Then, of course, there's the good old 'chalk and talk.' And. . .
Interviewer: (waits silently, not prompting)
Teacher: . . .What other methods are there? I know I must do more than that.

Extract B:
(From a transcript of a group of students at the end of Year 9 discussing the frequency of fieldwork. Their teacher had previously, independently, put it at three or four times a year.)

Student 1: Did we go to that farm. . .
Student 2: . . .first term wasn't it?
Student 1: . . . No. . . may be . . .
Student 3: That was second term last year.
Student 1: Yes, last winter . . .
Interviewer: But what about this year? When have you been out of the classroom?
Student 3: She kept saying we'd go up to the shops. But we never. Did we?

Extract C:
(From a survey of 32 geography teachers when asked the question: 'How do you know when you have taught a topic well?'. They could check any items on a list that gave them this feedback. Percentages are expressed for each item against the total number of responses (viz. 73).)

	Approx. percentage
Does your Principal tell you?	$1\frac{1}{2}$
Does your Deputy Principal tell you?	—
Does your Head of Department tell you?	7
Do your colleagues tell you?	4
Do you simply *know* (a gut feeling)?	75
Do your pupils tell you?	7
Do pupil workbooks tell you?	4
Do pupil test results tell you?	$1\frac{1}{2}$
Other	—

These short examples highlight the need for an effective evaluation strategy that can be used to improve geography courses and teaching.

Course Evaluation: Rationale and Definitions

Who Wants It?

Generally, few teachers welcome evaluation of their own courses. Schools believe in publicly evaluating, grading and sometimes ranking students, but suggestions to assess or rank the courses or teachers within a school would not be welcomed by many teachers. It would be as threatening to teachers as current evaluation procedures are to students. In particular, teachers fear it could: lay them open to the scrutiny of administrators, parents and pupils; damage their self-concept by discovering they are not performing as well as they thought; and, maybe, reveal that their students do not like them.

All in all these are quite frightening possibilities that we teachers would rather not consider. Fortunately, this is *not* the style of course evaluation to be discussed in this chapter.

A broader view of evaluation is taken here, so that much more than the student or the teacher is probed. Much of the implied threat of evaluation is removed when its focus is the worth of classroom *processes* rather than classroom persons. Such a view of evaluation can create an atmosphere that supports the fundamental purpose of course improvement. A greater involvement of all interested parties in the planning as well as the evaluation of courses may help to minimise the threat of accountability, and to promote evaluation as a means of enhancing them. Similarly to allow the findings of a course evaluation to be used for anything other than the declared aims of course improvement betrays the trust of the participants and thereby most chances of such aims being achieved.

What is Evaluation?

A considerable literature on evaluation in general, and course evaluation in particular, exists. However, for the purpose of this chapter, evaluation is defined simply as the judgements and meanings ascribed to course assessment information. Thus evaluation is seen as a kind of 'value-added assessment'. That is, when some meaning, or value, is added to course assessment information, we have course evaluation. Expressed algebraically,

$$\text{Course evaluation} = \text{Course assessment} + \text{Value judgement.}$$

The main purpose of course assessment is to provide good-quality evidence for evaluation. Evaluation's major task is to inform course decision makers and planners as to the worth of a course, perhaps relative to another. Generally, evaluation is more concerned with the worthiness of various parts and activities of a course than with its overall worth.

Furthermore, while evaluation is interested in judging whether an assessed performance was good or bad, it is far more concerned with deciding how and why it happened that way.

Why Evaluate?

Course evaluation can be *summative* and occur at the end of a course or *formative* and take place as an on-going process throughout a course.

1. *Summative course evaluation*. Such *post hoc* evaluation has the role of judging the final success of a course and ascribing responsibility (blame?). However, often it is futile as it occurs too late to change the current course and is usually too remote to affect future offerings of the course.

2. *Formative course evaluation*. This form of evaluation is integral with the on-going development of courses and is undertaken to make decisions about:

(a) changes in the performance of students that can be related to specific aspects of the course; and

(b) the suitability of all aspects of the course including the intentions, teaching, resources and learning environment.

It is this kind of evaluation that is described in this chapter.

Principles of Sound Evaluation

Course developers and teachers need sound evidence about their courses in order to make decisions to improve them. This means that course evaluators need to:

(a) be sure of the evaluation's intent;

(b) select an appropriate variety of means of data collection and analysis;

(c) use criteria appropriate to the context of the course and intent of the evaluation; and

(d) regard the evaluation primarily as a means of improving the course in practice by informing decision makers.

Evaluate What?

Evaluation of geography courses should be concerned with assessing and judging:

(a) the quality and changes in students' learning;

(b) the effectiveness of teaching strategies; and

(c) the context of these activities.

It should seek meaning in this assessment of learning, teaching and context according to goals and criteria that acknowledge the needs of the students and society, and the nature of the discipline of geography.

How to Evaluate Your Geography Courses

Alternative Methodologies

Three alternative methodologies or models for course evaluation exist. They are:

1. *Experimental designs.* These seek to test the efficacy of a course, usually against another course or a previous one. Generally, school geography departments lack time and resources to set up sufficiently valid designs to 'prove' anything satisfactorily. Pre-test and post-test designs can provide useful evidence, but need the careful use of control groups which are usually hard to get in one school. The most common approach to experimental design evaluation seeks to match course objectives with outcomes. This often imposes a rigidity that stifles good curriculum development.

2. *Survey designs.* These are not very useful in evaluations of the kind conducted by a geography department because the purpose of a survey is to generalise from many examples. Generally, the number of students and teachers involved in a course is too small for this to be possible, and even when this is possible, great care needs to be taken in decision making because so many individual differences can be hidden by the research and generalisation process.

3. *Case study.* Case study and other ethnographic methods of evaluation provide a more appropriate approach to course evaluation as they are specific and sensitive enough to be able to inform school course developers in immediate and constructive ways.

The Case Study: A Recommended Approach

The major principle of case-study evaluation is that it must be individually designed and undertaken within, interactive with, and appropriate to, particular course contexts. It necessitates formative evaluation. The whole point and purpose of formative evaluation is the sympathetic interaction between the evidence and course development decisions.

As distinct from experimental and survey designs which, inappropriately for in-school evaluation, seek to generalise and to prove, *case-study evaluation seeks to particularise and improve.* The acceptability of case study comes not from a strength of explanation based on underlying, universal laws, but from its perceptions and understandings which derive from its deep and holistic examination of the processes at work in particular course contexts.

The purpose of a case study is to paint a picture of the unique situation that at once rings true for the participants and reveals new meanings from the familiar. It must stand as a mirror to expose the known, but hitherto unseen, dimensions of one's teaching, student learning and course contexts (Burgess 1985, 1986; Walker 1985).

Much of the value of the final case-study portrait is that it presents a montage of divergent images. Truth is not necessarily, nor likely to be, singular in any case studied. This complexity arises from the variety of perceptions of individual participants as well as the actual variety of elements in the curriculum system. A further value of case-study evaluation is that it speaks to curriculum decision makers of a world they know at a commonsense level and provides them with information of appropriate currency. The language of case-study evaluation is congruent with the language of classroom actions and decisions.

By Whom?

Chapter 27 suggests that school-based curriculum development should be a group activity, drawing upon the interest and skills of all geography department members. So should course evaluation. Indeed it should be impossible to see where course development stops and evaluation begins. Consequently, the group best placed to undertake such a study is the team of teacher-course developers. The big advantage of this as far as evaluation is concerned is that as teachers do the course development work of designing and teaching together, they gain the confidence in each other to expose themselves more willingly to the scrutiny implied by evaluation.

The inherent danger of the myopic view of course developers being the evaluators can be offset by inviting onto the evaluation team a specialist outsider such as a local geography adviser, or lecturer in geography curriculum from a nearby college or university. Ideally, this person should have no role in the school system and yet be able to identify and be identified readily with the classroom experience. This will probably involve some teaching with the group for a while.

As the group designs the evaluation it should examine the possibility of evaluation roles for the other participants in the course, such as students, parents and school administrators.

Choose your evaluation team from:

1. Geography department in the school (the core):
   (a) Head of department
   (b) Individual teachers
   (c) Pupil representatives

2. From the school outside the geography department:
   (a) Deputy Principal (curriculum)
   (b) Member of specialist committee (e.g. language across the curriculum, mixed-ability teaching, timetable. . .)

3. Outside the school:
   (a) Geography adviser/consultant
   (b) Lecturer from nearby college or university (e.g. geography curriculum studies, evaluation)
   (c) Parents
   (d) Employers

For Whom?

The definition of course evaluation used in this chapter indicates that the meanings revealed by evaluation are for the decision makers in the course development system. In practice each evaluation team must carefully, and specifically, identify the people who will have access to their findings. This must be clear at the start, and adhered to for the sake of the integrity of the evaluation.

Consider the possible users:

1. Outside the geography department:
   (a) Employers
      (i) Local Education Authority
      (ii) School Council
      (iii) State Education Department
   (b) School
      (i) Headmaster/mistress (Principal)
      (ii) Deputy Principal (curriculum)
      (iii) Bursar
      (iv) Timetable committee

2. Geography department in the school:
   (a) Head of department
   (b) Individual teachers
   (c) Department as a group
   (d) Students

When?

Ideally, evaluation should be continuing, pervasive and formative. Such a continuous evaluative effort is not always possible in the busy, practical setting of a school. Hence, it is appropriate that the microscope of evaluation be focused deliberately with greater acuity at specific times. The early trial stage in the development of a new course is an occasion when evaluation can be invaluable in helping adjustments. All courses also demand a critical review at least every two or three years to ensure that all aspects remain responsive to the needs of students and the context of the course.

Ways of Collecting Case-study Evidence

A variety of formal and informal methods exists for collecting the evidence to be used in a case study. As many techniques as time permits should be utilised so that decisions can be made on the widest range of cross-referenced data that has been gathered from as many participants in the evaluation as possible. Figure 30.1 is an evaluation planning matrix from which a selection of methods and evidence sources can be made. It should be noted that a lot of the evidence collected in a case study will be subjective as it depends on the values and perceptions of the people involved. This need not be a problem, however. These values can enhance the evidence and the evaluation that results from it provided one remains aware of the values and their source. Indeed, often it will be these values that will require probing, clarifying and sometimes changing if decisions to alter courses are to be carried through effectively.

Some of the methods of investigation outlined in Figure 30.1 warrant elaboration. *Informally gathered evidence* can be every bit as effective in a case study as more formally gathered evidence. Figure 30.2 illustrates a range of informal methods and provides a brief comment on each one. Figure 30.3 is an example of an *observation* schedule that could be used when a classroom is being observed to note the resources that are used in geography teaching. Observation schedules can also be used for more formal data collecting. For example, the Flanders' Interaction Analysis Categories (Flanders 1970) or modified versions of his schedules (e.g. Stones and Morris 1972), provide useful checklists for observing and analysing the type and direction of interpersonal interactions in a classroom. Teachers should write their own schedules to suit the particular focus of different evaluation goals. Figure 30.3 is an illustration of one of this kind. Do not be inflexible in the use of observation schedules, however, as they can sometimes cause the observer to note particular things and not be alert to unanticipated evidence.

Documents can be informative sources of evidence about courses also. Especially useful can be: statements of school philosophy and a geography department's policy, course outlines, budgets, timetables, committee minutes and examples of student work such as tests, assignments, practical exercises and other course work.

The *interview* is generally the best instrument for gathering data suitable for course evaluation. Especially if unstructured, it allows evaluators to collect data that are very little influenced by the research design or the values of research personnel. A brief interviewee profile, such as the teacher profile in Figure 30.4, developed at the start of the interview, is helpful in later analysis and judgement. The rest of the interview should proceed with as little intervention as possible by the interviewer.

Checklists that survey the context, teaching and learning aspects of the course may be used to help the interviewer prepare for the responses that may

Investigation methods (expand and specify)	Possible participants					
	Teachers	Students	Parents	Principal/Headmaster	Employers	Other
Informal methods Observation —participant —non-participant Unobtrusive						
Formal methods Observation schedule Documents Questionnaire Interview —unstructured —semistructured Reviews of resources Teacher diaries Exam performance Other						

Figure 30.1 Evaluation planning matrix. Select a range of investigation methods and participants to facilitate cross-referencing of data in a case study

1. Observation (a) participant —teacher —student —other	*Comments* This can be done more formally using an observation schedule (see Figure 30.3).
(b) non-participant	Usually the presence of such a person alters the event.
2. Unobtrusive measures (a) Attendance (b) Behaviour (c) Enrolment/participation in the course over several years. (d) Library borrowing on that topic for students of that course. (e) Questions that students ask, etc.	Little of this evidence on its own is worth much, but is very useful in conjunction with other data. For example, student statements of how much they enjoyed a particular topic can be checked by noting how many of them borrowed books from the library on it.

Figure 30.2 Some informal data gathering methods

Figure 30.3 Observation schedule for noting the resources used in geography teaching

The observer should record facts about the quantity and quality of things observed as objectively as possible. A few example observations are included in this figure. This evidence will be analysed with the rest later and then meanings can be sought.

(a) *School yard*
Stephenson screen weather station. Well equipped and located.
(b) *Library*
Well stocked regional section. (Latin America, Africa and the Pacific not represented.)
(c) *Text and/or book sets*
Book sets. Wide variety of recent books. Only one junior level physical geography title (1958).
(d) *Geography rooms/laboratories*
Laboratory. Flat-top work space for 20 students. Sink, drainboard, hot and cold running water. Plenty of cupboard and shelf storage. No map drawers. No light-table.
(e) *Classroom in which geography is taught*
Display of wall maps on current topic. Display of examples of student writing and mapping on current topic.
(f) *Non-book resources for geography*
About 25 large wall maps (all pre-1958).
3 aerial photographs (small scale).
6 local area topographic maps (1988).
1 large political globe (damaged).
(g) *Other* (e.g. school camp site)
School-owned bus for excursions (21 seats).

Figure 30.4 An unstructured interview schedule for geography teachers.

Teacher Profile		
Years of teaching		
Years in present school		
Status in present school		

Qualifications	Where obtained	When
Degree		
Teaching Qual.		
Diploma		
Other		

Current study

Recent in-service (short term as well as more formal courses)

Subject association membership

Journals subscribed to or read

Time	Open interview (see interview checklist)	Interviewer's comment

come. Interviewers should be cautious about seeking specific responses and so contravening the respondents' perceptual priorities, however. Later these checklists can be used to help in the analysis of the evidence that has been gathered. An example is given in Figure 30.5 of a checklist that could be used in developing an interview schedule to investigate the context in which a course is being taught.

Analysis of Evidence

The purpose of analysing the evidence that has been collected is to expose all the influential characteristics of the course. It is an attempt to probe and portray the *alternative perspectives* of the many participants and other sources of evidence that help to 'tell the story' about the course. Remember that the aim of a case study is to find this 'story', not to prove anything. Therefore, approach analysis with an open mind and with as few *a priori* notions of the outcome as you can.

While congruence of evidence on certain issues can be very revealing, contentious data may be equally useful. For instance, conflicting perceptions of an event may reveal that contrasting values are held by the participants and these, while unrecognised, may be hindering an effective team effort.

Variety of Display

Display is the main tool of analysis that provides insights into associations between the many elements

---

Context—Climate and resources for geography courses

1. Climate
- School ethos

  Students. What are they like? What are their needs?

  Staff–student relations. Describe. What philosophy underlies this?

  School-based curriculum decision making. Is it participatory? Who dominates? Comment on communication?

- Community

  What social influences are there on the geography curriculum? What does the community require of geography courses? What support resources are available?

- Local professional climate

  What sort of involvement does the geography faculty have with teacher centres, professional subject association, local college/university, subject adviser, etc.?

- Geography as a subject in the school

  How is its role and activities perceived by various other people in the school?

2. Resources
Teacher profiles (see Figure 30.4)
Total school curriculum
Timetable
Rooms
Facilities
Book resources
Non-book resources
Out of school resources/facilities (e.g. camp site)
Other

3. Major contexual constraints perceived (list)

4. Major contextual advantages perceived (list)

Figure 30.5 A sample checklist for developing an interview schedule

of curriculum that interest the course evaluator. Careful decisions concerning the appropriateness and variety of the display are essential in the analysis stage of evaluation. The intention is to cross-reference data from various sources.

Present the collected data in a suitable variety of ways to reveal and highlight important associations. For example, draw and complete sets of:

1. *Maps and plans* to reveal access to and the suitability of teaching places. Perhaps mental maps could be drawn to reveal teachers' perceptions of the school and its neighbourhood.

2. *Tables* to summarise and order collected data.

3. *Graphs* to compare performance with teaching methods.

4. *Diagrams* such as systems models to show the process of relationships between elements of the course.

5. *Grids* to cross-reference the elements of the course.

Very large sheets of paper and felt pens are invaluable during the early stages of analysis.

Grids are particularly valuable in helping identify associations between the three elements of course context, student learning and teaching. Most commonly, two-dimensional grids are employed to compare aspects of two of the three course elements. See the example in Figure 30.6 which associates context and learning. A more complex three-dimensional grid can be used to display associations between the three elements of the courses, when necessary. Figure 30.7 may be used as the basis for such an analytical display.

Making Judgements: Ascribing Meaning

Participation by the whole evaluation team in the vital stage of ascribing meaning to the collected and analysed information is essential.

The emphasis on portrayal in case study suggests a crucial interpretation role for the information user. If, as suggested earlier, the evaluation team is largely comprised of the course development and teaching team then the integrated processes of evaluation and

Figure 30.6 A two-dimensional analysis grid to display the associations between course content and student learning

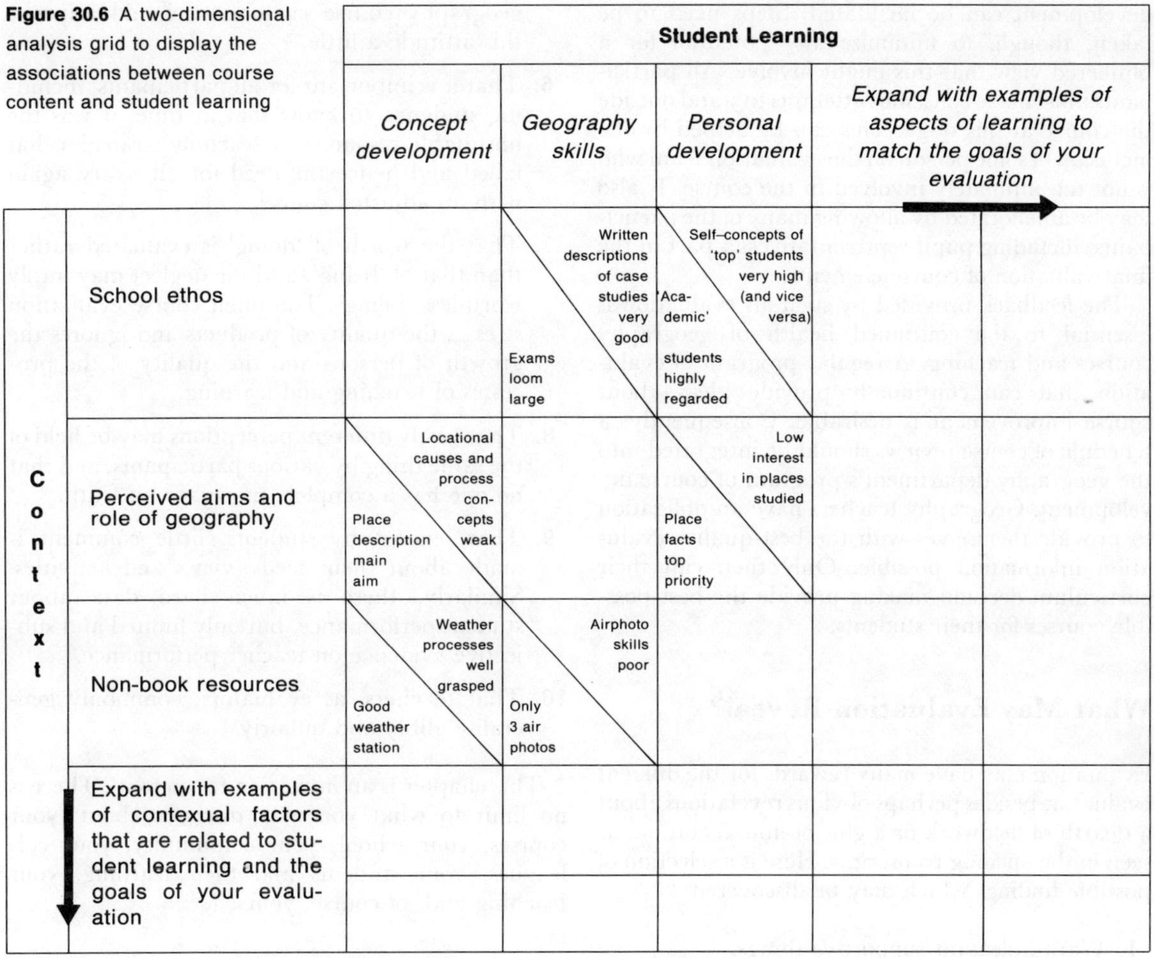

		Student Learning			
		Concept development	Geography skills	Personal development	Expand with examples of aspects of learning to match the goals of your evaluation →
C o n t e x t	School ethos		Written descriptions of case studies very good / Exams loom large	Self–concepts of 'top' students very high (and vice versa) / 'Academic' students highly regarded	
	Perceived aims and role of geography	Place description main aim / Locational causes and process concepts weak		Place facts top priority / Low interest in places studied	
	Non-book resources	Good weather station / Weather processes well grasped	Only 3 air photos / Airphoto skills poor		
	Expand with examples of contexual factors that are related to student learning and the goals of your evaluation ↓				

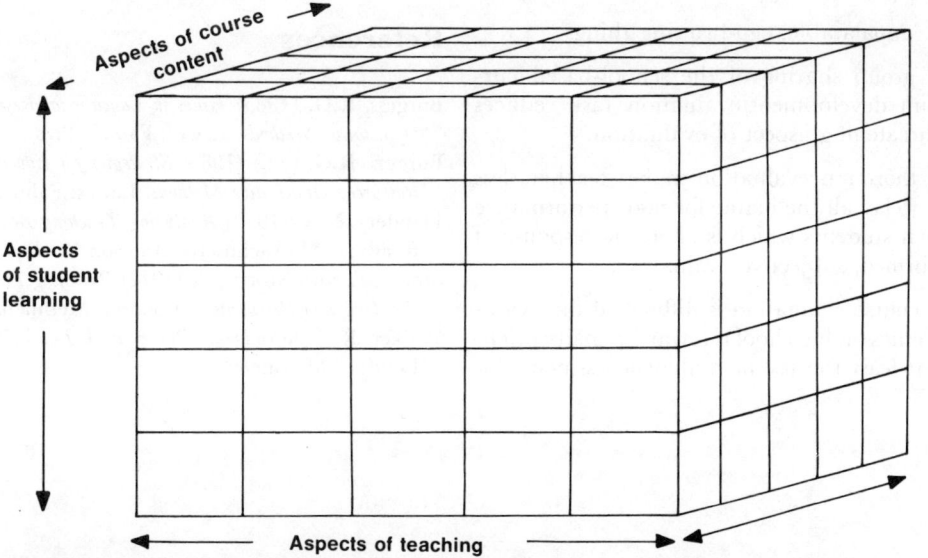

Figure 30.7 A three-dimensional analysis grid to display the associations between course context, student learning and teaching

development can be facilitated. Steps need to be taken, though, to minimise any potential for a blinkered view that this might involve. All participants must make conscious attempts to stand outside the course at this stage. This can be helped by the inclusion of some person on the evaluation team who is not too intimately involved in the course. It also may be ameliorated by allowing many of the participants, including pupil representatives, a part in the final valuation of course elements.

The feedback provided by such an evaluation is essential to the continued health of geography courses and teaching. A regular program of evaluation that can continually provide ideas about course improvement is desirable. Consequently, a schedule of course reviews should be integrated into the geography department's program of course development. Geography teachers have an obligation to provide themselves with the best-quality evaluation information possible. Only then can their curriculum decision making provide the best possible courses for their students.

What May Evaluation Reveal?

Evaluation may have many rewards for the diligent evaluator, besides perhaps obvious revelations about a dearth of fieldwork or a glut of note dictation, as seen in the opening transcripts. Here is a selection of possible findings which may be discovered:

1. Warm, pleasant supportive things.
2. Cold, unpalatable, ego-bruising things.
3. That group sharing of the school-based curriculum development/evaluation task reduces the threatening aspect of evaluation.
4. That thorough evaluation makes teachers less likely to lay all the blame for poor performance on their students which is often the response of uninformed, subjective evaluation.
5. That course evaluation is difficult if the evaluation ethos of the school is reward/penalty oriented. At least the use of student assessments for geography course evaluation should improve this attitude a little.
6. That it is important for all participants, including students, to know that at times it was the unsuitable resources or teaching strategies that failed and hence the need for all to try again with an adjusted course.
7. That the worth of 'doing' is evaluated rather than that of 'being', and the neglect may imply worthless 'being'. Too often course evaluation stresses the quality of products and ignores the growth of persons and the quality of the processes of teaching and learning.
8. That vastly different perceptions may be held of the same thing by various participants, and that no one has a complete mortgage on truth.
9. That, except by students, little comment is made about their needs, views and activities. Similarly, there is much hard data about student performance, but only limited and subjective evidence on teacher performance.
10. That teachers, as evaluators, commonly generalise glibly and unfairly.

This chapter is an invitation to 'go to it'. There is no limit to what you may discover about: your courses, your school, your community, your colleagues, your students and their learning, your teaching and, of course, yourself.

References

Burgess, R.G. (1985) *Issues in Educational Research: Qualitative Methods*, Lewes: Falmer Press.
Burgess, R.G. (ed.) (1986) *Strategies for Educational Research: Qualitative Methods*, Lewes: Falmer Press.
Flanders, N.A. (1970) *Analyzing Teaching Behavior*, Reading, Massachusetts: Addison-Wesley.
Stones, E. and Morris, S. (1972) *Teaching Practice: Problems and Perspectives*, London: Methuen.
Walker, R. (1985) *Doing Research: A Teacher's Handbook*, London: Methuen.

31
On Being a Geography Teacher in the 1990s and Beyond

William Romey and William Elberty Jnr

Every one of us is born as a geographer. The task of the geography teacher in the 1990s is to help students rediscover their 'geographic antennae' and to bring the geographic dimensions of all their activities and all events around them into conscious awareness. William Romey and William Elberty Jnr are American college geography teachers and many of the ideas they suggest may not match your school context exactly, but their suggestions for a variety of learner-centred approaches to teaching geography are worthy of consideration and implementation wherever possible in your course. Learner-centred approaches to geography teaching offer a means for recognising and including the wide range of subject matter geography touches upon and for helping students to develop and use their inborn geographer's potential to seek solutions to social and environmental problems. The authors use vignettes of examples of learner-centred approaches from their own experiences as pointers to geography teaching in the 1990s and beyond. Their chapter focuses on the personal and professional characteristics needed by geography teachers and thus provides a fitting conclusion to *The Geography Teacher's Guide to the Classroom*.

The symbol is not the thing symbolized.
The map is not the territory.
The word is not the thing. (Hayakawa 1978)

Too often we confuse the map for the territory, the report for the experience, the product for the process. We do this for a number of reasons including ease, convenience, conditioned unawareness and the semantic snares of our verbal and graphic languages. This confusion also stems from the fact that as humans we are inquisitive creatures, born with an intrinsic need to learn. Our inquisitiveness traps us. We want to believe that the map is the territory, for how else can we 'experience' places that we may never visit? We want to believe that a report is an accurate portrayal of an experience, for how else can we vicariously share it? As teachers, we want to believe that the product (test, lecturette or research report) is an accurate representation of the process (learning), for how else can our student assessment results have meaning?

We communicate via a complex symbolic system. As writers (communicators), we know the meaning of the symbols which make up this chapter. We do not know how you as a reader (communicatee) are going to interpret them for the message sent is often not the message received. Our hope in writing this chapter is that we can provide some perspectives that will make your life richer, your 'teaching' more rewarding and your 'students' more interested learners.

Twenty-five of us walked across the Icelandic plains in the centre of Krafla's caldera. In places, bright green grass, the only vegetation, covered the brown basalt. Every now and then we had to leap over a long, zipper-like gap in the ground where the hexagonal columns of basalt had pulled apart along a spreading fissure. Suddenly, we heard a cry of alarm. Looking around, we saw that Leslie had literally dropped out of sight. Hurrying back, we found a newly collapsed hole in the grassy blanket covering a gaping fissure. Ten feet down, Leslie stood forlorn and frightened but uninjured.

For eight weeks our group lived together. During the first two weeks on the campus of a small school in central New York we had classes on Icelandic culture, history, geography, politics, economic and social problems, and geology. We supplemented the program with local field trips to sharpen our powers of observation. We learned about each other and about living and working together and about sharing interests and expertise. We examined the dimensions of 'groupness' in our own culture before having to face an exotic foreign culture under field conditions.

On the four-week expedition in Iceland each person pursued a personally chosen topic in depth. Our itinerary served as the only 'structure'. As 'leaders', we led walks and talked a lot, but people were free to participate as they chose. Everyone was free to explore individually in each location we visited. Meals together for three or four weeks provided times for sharing information about individual activities. The group experience, shared in this way, exceeded the experience of single group members, creating a rich fabric of different viewpoints, perceptions, observations and information. Everyone became a guide for others at one time or another.

After the fieldwork in Iceland, an additional two-week session took place in central New York. Group members continued to live, eat and work together during this time set aside for analysis, reflection and synthesis. Colour slides trickled in from the processing laboratories; other people developed and printed their own black and white photographs; several people worked on sketches and paintings begun in Iceland; most people (males included) finished work on the Icelandic sweaters they had begun; and everyone ate, drank, played, and wrote a lot. On the last day, the group members participated in a process of group and individual self-evaluation. More than a year later students are still giving public presentations on their Icelandic experiences.

Every Wednesday afternoon 15 of us piled into a school van and drove off to explore a new area along the St Lawrence River within a few miles of school. People mapped their way along as the van passed through various communities and environments on our way to each trip's final destination. Once out of the van, group members wandered on their own. Each person listed or mapped personally chosen aspects of the urban, suburban, rural, commercial or wilderness environment we visited. Local citizens or community leaders sometimes met with us. An historian, meeting us in a preserved British fort from the War of 1812, discussed the military geography of that area as it affected the battles. At the St Regis Mohawk reservation three elected native American chiefs discussed their problems and hopes. An official of the St Lawrence Parks Commission (Canadian), meeting us at one of the parks, informed us about patterns of tourism. The manager of the international power dam across the river led us around the dam while introducing problems of power generation, politics and local geography. The plant manager of the Canada Starch Co. Ltd had his foreman conduct us through the plant to find out about transportation of goods on the St Lawrence, manufacture of a wide range of products from corn, sources of supplies, markets, innovation and testing in the plant, and the historical development of the industry in the 150-year-old company town.

As on the Iceland trip, *place* provided the structure for our studies. There was no need for the place to be an 'exotic' one. At each place, personal interest and curiosity guided the learning experiences of each person. In addition to participating in group excursions the 'common experience' for everyone was to write three individual field study reports, duplicate them, and distribute them to all members of the group. By the end of the term the students had created a textbook of regional geography based upon their own experiences and research.

The problem of waste disposal was brought up by a student in a physical geography class. We decided (spontaneously) to visit a 'sanitary' landfill site at the nearby town of Pyrites. An hour was spent watching the workers compact the garbage in 1.5 metre layers and then cover each layer with one metre of sand from the adjacent, ice-age delta. We then examined the leachate which drained out of the landfill into an adjacent stream.

The classroom program in this case led us to *place*, a particularly non-exotic one. Topics surveyed at the landfill site ranged from high-waste technology, the contents of municipal wastes, the physical structure of 'sanitary landfills' (high-rise rodent housing), infiltration rates and ground-water migration, anerobic systems, surface-water pollution, heavy-metal contamination, non-degradable plastics, and the deposition of the Pleistocene delta, to the possibility of a strong earthquake in the region of 10 000 years ago as evidenced by contorted strata in the upper portion of the deposit, and much more.

This encounter with one of the realities of a high-waste society left us aware of the cosmetic approach to an 'out-of-sight-out-of-mind' syndrome. We could see the symptoms of the problem in the form of ground- and surface-water pollution and environmental blight, and we increased our awareness of the geographic and geologic history of the region.

The weekly seminar provided a forum for sharing information and ideas about any aspect of geography, not restricted as to place or topic, that individual learners happened to be pursuing. Often people came in 'unprepared', but the conversation went on anyway, with people jumping from educational values through political activism, to questions of a conventional 'geographic' nature, and it included a strong component of personal problems and reactions to everyday life at school.

Seminars of this type serve as points of communication for people working on a variety of independent study projects in geography. Sometimes we attach one or another label to provide a focus, something to talk about when the 'real' conversation lags. Useful names have included 'Reality and Perception in the Sciences', 'Rural Social Geography', and other labels. Some seminars focus on regions or countries: the Soviet Union and eastern Europe; western Europe; Canada (just a few miles from us). Topically oriented seminars have dealt with 'Volcanoes', 'The World Ocean', 'Climate', 'Hazards', 'Acid Rain', 'Introductory Navigation' and 'Earth Resources'. The focus of individual sessions always comes back to topics the students choose to explore. When our turns come up in the rotation for presentations, we have an opportunity to add any balance we may feel is lacking. We also have ample opportunity to introduce what we think are important additional topics in short briefings and 'general discussion' periods that serve as 'warm-up' periods at the beginning of each seminar. The seminar also serves as a problem-solving group in which group members become 'consultants' to each other. Each person may pursue any topic or topics of personal interest. We have agreed to help train the group members in procedures which help everyone help each other. These have included the 'synectics' procedures of Gordon (1961); the 'lateral thinking' of de Bono (1970); the 'applied imagination' of Osborn (1953); the 'conceptual blockbusting' of Adams (1974); the 'visual thinking' of McKim (1972); and many others. We have also introduced the group-process facilitation of Rogers (1962) and others. These involve learning about how to listen. Out of a welter of seemingly unrelated topics introduced by students, a unified and interrelated picture of a multidimensional world sometimes emerges. Almost like real life.

Student: Ahhh?
Bill: Hello there. . .what's up?
Student: Hi. Are you Dr Romey?
Bill: No, I'm Bill Elberty; Bill Romey lives next door. Can I help you?
Student: Well. . . I don't know. I want to do an independent research project in geography.
Bill: Great! What do you want to do?
Student: I don't know. What do you want me to do?
Bill: If you're going to do a geography independent project sponsored by Bill or me, we would ask you to develop your own project . . . you know, something *you* are interested in.
Student: But I don't know anything about geography.
Bill: That's OK. . . Good reason for being here. The question is, what is it that you would like to do? We can take almost any topic and place it in a spatial or geographic perspective.
Student: Gee, I don't know what I'm interested in. . . .
Bill: What do you enjoy doing when you are not doing school?
Student: I like soccer (needlepoint, swimming, reading, travelling, hiking . . .).
Bill: Great! Do a project on soccer (needlepoint, language, photography, wine, Zen, transportation, navigation, maps, climate change, poverty, rural road maintenance, alternative energy sources, petroleum, legal services, volcanoes, earthquakes, waves, population, housing, your home, the park down the street. . .).
Student: Can I do that?
Bill: Sure—sounds great! Let's work up a project proposal. I think I know where you can find some initial resources. I have this book on sports geography (oceanography, shelter, energy, anthropology, demography, industrial location,

political geography, rural..., urban...). Why don't you look through it and get an idea of how geography relates to your area of interest. Then we can find some more specific resources. What an interesting topic! You may find that you have to limit it to some extent...

We have found that student-originated independent projects, outside the constraints imposed by seminars and classes, are invaluable in encouraging students to take responsibility for their learning. They generate skills for acquiring resources, and they encourage creativity. A problem is that many students are so conditioned to being told what is important for their education that they have lost the ability to think for themselves. As a consequence, the initial product of an independent project may appear rather unsophisticated. However, if self-directed exploration is encouraged, the results can be surprisingly rewarding.

Being a Geography Learner in the 1990s

These four vignettes illustrate learning experiences some of us have used to help people free the geographer that we believe is inborn in each of us. Unfortunately, that geographer is sometimes hidden and repressed. Every person is a born geographer. Newborn babies immediately determine their spatial relationship to 'mother'. The explorative tendency next leads them to explore the dimensions of the crib. Then they crawl toward the edge of whatever space they live in. Life becomes a process of exploring larger and larger spaces. The child explores the house and neighbourhood and then finds his or her way to playmates or school and home again. Mental maps are formulated even before attempts at verbalisation begin.

Our own earliest memories include mental images of spaces: the neighbourhoods we lived in, the frequent drives around town, trips to visit grandparents and relatives who lived several hours away from us, the long trips to vacation sites. The earliest stories we read vividly painted the places where events occurred. Think of the places described in the 'O_2' books, *Wind in the Willows* and *Winnie-the-Pooh*. Grandmother's attic contained piles of dusty old *National Geographics* which we read avidly on rainy Saturdays. Later came the travel books of Richard Haliburton and other adventurers and explorers. Tom Sawyer and Huckleberry Finn planned exotic voyages to distant places while exploring the Mississippi River. Then came the books of Jack London and others with their vivid images of space. Television documentaries took us all over the world under the care of guides such as David Attenborough and Richard Adams. Summers were spent on vacations many hundred kilometres from home. This often brought the experience of being 14 years old and having to find one's way, often alone, from one train station to another in a strange city. These trips later led to hitch-hiking and biking over vast distances. We grew up in a world of geography. Today's children do this even more.

Although this process of discovery may seem an uneconomical and wasteful approach to education, Jean Piaget's studies of the nature of learning long ago led to the conclusion that things are only understood to the extent that they are *reinvented* or discovered for oneself. Economical or not, the nature of the learning process in humans demands a strongly self-directed experiential component (and 'guided' inquiry will not do) no matter how time consuming or untidy it may seem. The idea of a 'walkabout' in which learners are put out into the world and forced to survive on their own represents a fascinating approach to this kind of experiential learning (Gibbons 1974). Every person has the inborn antennae of a geographer. *Our task as teachers is not to turn people into geographers but rather to free the geographer that already resides within each of them.*

We feel that learners in a person-centred geography program have an especially good opportunity for developing perceptual skills. Thus, much of our fieldwork and class activity is devoted to exploring the nature of perception. Tuan's (1986) *The Good Life* describes the kind of broadening experience that geography can provide, and elsewhere Tuan (1987) describes the kind of open-minded, experiential attentiveness that geography teachers can foster in their students.

Traditionally, the word 'geography' connotes the formal classification systems and procedures applied to places: knowing the locations of cities, rivers, national boundaries and ethnic groups; knowing the dimensions of economic production, transportation systems, etc. We by no means exclude these matters from the geography needed in the 1990s. We merely consider this an inadequate and incomplete idea of what geography has to offer. Every conceivable aspect of human life and interests involves a spatial or people–environment dimension and is thus an appropriate subject for application of the geographer's perspective, as pointed out long ago by Sauer (1956). We also agree with James and Martin (1981) that geography and geographical education need to discover the contribution they can make to the overriding problems facing humanity in the 1990s. We must help our students find solutions to the problems of poverty, hunger, injustice, violence and warfare. Hence we must refer you to two books to use as your daily guide: *Geographical Education: Reflection and Action* (Huckle 1983) and *Teaching Geography for a Better World* (Fien and Gerber 1988). Both these books are practical handbooks for educating for peace and justice through geography.

The Task of Geography 'Teachers' in the 1990s

The task ahead is to put these broader ideas into direct operation in our own learning environments in order to give geographic perspectives the integral role they deserve in every person's education. Perhaps, if geography is such an intrinsic part of every imaginable area of interest, we do not need to think of 'teaching' it is a separate 'subject'. Geographic content is inevitable, regardless of what we teach. Yet raising it into conscious awareness as a special subject can increase the power of the spatial perspective. In an overcrowded world beset by conflicting interests and territorial struggles at many political, social, economic and cultural levels, conscious awareness of spatial problems may have a lot to do with survival.

Our goal as geography teachers in the 1990s and beyond is to help each learner raise into conscious awareness the spatial and environmental factors in every human endeavour. We want all people to know about the existence of their geographers' antennae, to be consciously aware of the spatial and environmental dimensions of their lives, and to be able to seek well-conceived plans and responses to problems. From a theoretical point of view in education, this means that we must approach learners first at the awareness level of the taxonomy of affective educational objectives rather than at the content level of any taxonomy of cognitive objectives.

Geographic education needs to involve a balance between the inward-looking expedition of self-discovery and the more conventionally acknowledged outward-directed exploration of the world. Even the outward-directed exploration bears redefinition. In most education we devote our time to 'making the strange familiar', we need to spend more time 'making the familiar strange'. This means devoting more attention to the supposedly commonplace, everyday aspects of spatial experiences in order to see them freshly. The home town, the local neighbourhood, and the cultural factors of our own lives that we take for granted, being familiar, may pass unnoticed. These aspects of our lives are what E.T. Hall (1959) calls 'The Hidden Dimension'. We want our students to see 'old' things as if for the first time, with as much attention as they lavish on exotic, 'foreign' subjects.

The Necessity of a Person-centred Orientation

We see a 'person-centred' mode (Rogers 1980) of helping people learn about geography as a necessity in the 1990s. If self-exploration and self-discovery are necessary components of learning geography, we are forced to acknowledge and dignify the personal interests of each learner. This has important implications with respect to syllabuses, grades, the nature of common experience introduced, and the need to accept divergent thinking in schools. Let us look at each of these in turn.

On the Syllabus

In a person-centred mode, no firmly established conventional syllabus can be distributed at the beginning of a 'course'. Each learner arrives with a different set of learnings and understanding based on his or her own lifetime of experiences to date. The subject matter of the course has to unroll during the process of the learning and cannot be predicted in advance. Reports in newspapers, television, and other media will have a strong effect on the dimensions of subjects that arise (Romey 1977). It is necessary to work from a *process* syllabus which would suggest kinds of experience and modes of exploration for a period of time. See Chapter 27 for guidelines on developing such a process syllabus.

On Grades

A necessary corollary of a process approach of the person-centred type is that conventional assessment is not possible. The learner is the only person qualified to judge the importance and value of a particular experience to his or her own learning process. Unless we have been gifted with rare insight or are the recipients of very unusual trust on the part of a student, we have only a foggy notion of what a particular person knows or has learned.

If each learner travels a different path and no common 'coverage' can be expected, there is no basis for comparatively assessing the experiences of different learners any more than there is for determining whether an elephant is better than a grape. For a teacher to grade a student's personal experience would be presumptuous at the least. It certainly is not possible to grade a student on 'improvement' or 'growth' because it is not possible to know for sure what the student was really like upon entry; nor would it be possible to standardise judgements of growth in any fair way to either individuals or the group. Since the interest in person-centred geographical education is 'process' rather than 'products', the products, however 'excellent' or 'poor' they may seem, cannot be taken as accurate reflections of any learner's real growth. A recorded judgement in the form of a grade is merely irrelevant to the experience.

Ideally, to accommodate a person-centred approach, a simple narrative account of the learner's experience and learnings, prepared by the learner, ought to provide the final record for school files and the learner's personal records. Realistically, schools will probably continue to demand simplified grade indices of some sort to characterise complex learning experiences during the 1990s. An approach we now use, having previously experimented with blanket grades, pure student self-evaluation, and pass or not-yet-pass options (all of which caused us no end of misery with our school administration and with other departments) involves strong components of self-evaluation by the students, peer evaluation, and on-going dialogue within our groups about 'quality'. We work together to develop individual criteria by which this 'nebulous' thing called 'quality' can be assessed. We have come to use flexible contracts enabling students to set some personal goals, always open to change, and to get feedback about how we and their peers view what they do. Let students talk over their experiences and their work with other group members and with the 'teacher', and let them assign for themselves whatever comment they wish to have appear on their records. Self-evaluation, incidentally, is considered by Krathwohl et al. (1964) to be the highest level in the taxonomy of affective educational objectives.

On 'Common Experiences'

Person-centred courses need not be characterised by a lack of structure or standards. We can still spell out in advance a schedule of certain expectations and requirements for a set of activities, even if we do not wish to prescribe the actual content and case studies to be covered. We can demand participation and require written materials to be distributed to everyone in each group. For projects, students usually schedule special group meetings so that different students can share information about their work and ask for help and guidance beyond what we, as 'instructors' may be able to offer. We have begun now to schedule required 'projects seminars' in mid-semester. These are run like a professional meeting: students submit short abstracts describing their work, we publish these in a program that is widely distributed and then, at the seminar, each student presents a short, formal presentation to peers and visitors, with a discussion period following. Such communication is not intended as the conventional link between student and teacher, where the goal is usually accountability. Instead, it is intended as a link between one learner and other learners. The 'teacher' becomes just another co-learner. We take responsibility for planning some common experiences for the group, but we recognise that the experience will be shaped by individuals in the group. We can schedule a trip to a rural community and suggest possible approaches to exploration, but we ultimately have to work with whatever it is that the learners themselves derive from the experience. We have found that the less specific we make our instructions, the richer are the perspectives that students discover.

On Ambiguity and Divergent Thinking

We court ambiguity in the assignments we make. This is contrary to the usual view of education in which teachers seek specificity, accuracy, convergence and closure. A principal route to creative exploration involves divergent thinking. A specific assignment, with a narrow range of acceptable responses, trains people to think convergently. Ambiguous assignments almost guarantee divergent thinking. Instead of worrying about whether or not learners are finding right answers to our questions, we explore the nature of the questions they ask. The questions people ask determine the answers they will get. A focus on answers becomes largely trivial. A focus on the process of question development is more productive and useful.

Communicating About Person-centred Geography Courses

At the end of the term we arranged an old gymnasium as an exhibit hall. Each participant displayed something to represent his or her experience in the program, remained in the vicinity of the displays, and was available to talk with interested visitors about the work. Two girls baked a huge St Lawrence River cake decorated with a map (to scale) of our field trip areas. The punch they made was created from a recipe obtained from the Canada Starch Co. on a field trip, and its ingredients were almost all corn-based products. Two people who had studied fish populations and distribution along part of the river, in addition to their maps and papers, also produced a catch of 100 perch and bullheads which they fried up for visitors. One girl built a set of cardboard locks based on observations from the St Lawrence Seaway; visitors to the exhibit had to 'lock through' to get into the room. A girl interested in birds set up a large display on the biogeography of birds in the area and embellished it with over a dozen stuffed birds borrowed from the biology department. A student interested in the distribution of native American artefacts was able to procure materials from the County Historical Society and the anthropology department. An athlete mapped the competitive networks of local schools. Participants had tables, bulletin boards, floor space, slide projectors and screens—anything they needed to help in the organisation of the displays. After the exhibit a dozen of us went to dinner together at a Chinese restaurant on the Canadian side of the border, 55 kilometres away.

Part of the effort of geographers interested in person-centred education will be the search for better means of conveying their approach to a sceptical wider public. Most educators are not used to this mode of operation, do not understand its principles and may see it as a threat to traditional values, procedures and roles. Perhaps they are right, for the approach challenges many accepted ideas of societal organisation and responsibility. In its shunning of political hierarchy and policy and with its emphasis on trust, personal responsibility and autonomy it becomes a politically explosive power in its own right (Rogers 1977).

This year we introduced a 'walk-through', described in the vignette above. The idea came from an event one of us attended at a school of art and design. The organisation was more or less like a poster session now common at professional geographic meetings or, in some respects, like a 'science fair' (Romey 1984). Jenkins and Keene (1979) have described the use of posters as a final required product for geography courses, and we carried the concept through into a fully-fledged exhibition.

In addition to participation in the 'walk-through', each learner had already submitted three short progress reports during the term, and copies of these had been disseminated to all members of their classes. The requirement for interim reports helped many people avoid the temptation to put off their involvement until it was too late, to accomplish much, and, more important, it allowed people doing related work to find each other and to collaborate. Participants at the walk-through were free to prepare displays either related or unrelated to their earlier work. The 'walk-throughs' turned into *celebrations* of our experience. Many people took the business of 'going public' seriously. Posters and displays were imaginative, instructive, and creative. Administrators, members from other departments, friends and parents of our group members, and casual 'drop-ins' visited these events. Opportunities for 'walk-throughs' of projects in more public places such as school foyers and halls, civic centres and shopping plazas are really without limit. We are optimistic that such public 'showing' of academic work may hold promise for improving communication in a non-threatening way as well as creating a high-energy educational forum showing the compatibility of learning, personal fulfilment, recreation, and enjoyment.

In some classes we are requiring students seeking higher grades to submit and publish their independent research papers or essays in a departmental journal, *Geographical Essays*, which we issue in-house, with the students doing not only the writing, but also all of the layout, computer typesetting, printing, binding and distribution. To qualify for 'honours', students must prepare papers that pass peer-review and faculty-review stages. They may keep trying as long as they wish in order to complete a satisfactory, publishable paper. Our goals as editors are not to 'select' good work or carry out a rejection or grading function, but rather to help each writer bring his or her work up to a standard we all can agree makes it the best possible piece of work of which the student is capable, and work which will be to the department's long-term credit.

Personal Qualities of Person-centred Geography Teachers in the 1990s

The official from the National Science Foundation appeared at our residential workshop in environmental studies. After a full morning of going around to visit various groups and talking to people, he asked us to identify the people who were 'staff members'. We refused to tell him: his question acknowledged the nature of the relationships people had established within the program. You couldn't tell the students from the teachers.

A person-centred orientation requires redefinition of the teacher–student relationship. A view of 'teachers' and 'students' as co-equal learners replaces the view of the teacher as authority, knower and giver and of the student as learner and taker. The teacher, usually an older person with longer experience with life and supposedly deeper intellectual experience, may 'know' more about some things than any individual student. We must beware of the fallacy of suggesting that older people (instructors) are necessarily wiser than younger people (students). Young students often show great wisdom and older instructors often do not. Each student, however, certainly knows some things that the teacher does not. If we could accurately catalogue the aggregate knowledge of a group of 25 students and one teacher, we would find a rich bank of information and skills. The class that depends entirely upon the teacher's knowledge represents an impoverished learning environment. The group that opens itself to the contribution of all its members is maximally enriched.

Relinquishing the role of primary knowledge giver and becoming just one of the learners can produce a rough jolt for many adults. To survive this transition requires a self-assured, self-confident person with a strongly developed sense of personal worth. You have to believe in a philosophy of openness and equality in order to survive constant questioning and frequent criticism from colleagues and, harder still, from your own students. Many of them expect you to continue behaving according to the traditional rules of teacher behaviour. In a person-centred environment, the teacher stops being responsible for what the student learns. That responsibility is fully recognised to fall where it should be: right on the learner's own shoulders.

The person-centred geography teacher has to be a good listener. You have to be able to hear several different levels of messages in what people say. This includes not only the apparent message in the content of the words spoken but also the feelings, the intent and the hidden messages in the tone of voice and the body language students use (Fast 1971; Hall 1959). Significant cognitive learning is only possible when affective (emotional) factors are acknowledged. A person-centred teacher spends much time as a personal counsellor and adviser.

Among the most important personal characteristics of a person-centred geographer are the three qualities Carl Rogers (1962) identified for what he first labelled 'client-centred therapy'. These are:

1. *Congruence*—being as genuinely yourself as possible at all times; being in touch with your own real feelings of the moment.

2. *Empathy*—being able to get in touch with another person's inner world of meanings *as if* it were your own, but without adopting these feelings for yourself.

3. *Unconditional positive regard*—being able to regard the other person in an unconditionally positive light as a human being without judging his or her behaviour; accepting the person as he or she is without wanting to change him or her.

In addition, are requirements for a high level of personal energy, good humour, openness to new ideas, a personal delight in learning new things and savouring things already known, and a well-developed explorative curiosity. These qualities are inborn in all children. As adults, most of us need to recultivate them. Relearning how to 'have fun' rarely figures as a goal in a teacher-training program. This state of mind does not preclude serious and concerned involvement in problem-solving, intellectual activity in its most advanced forms, or in political and social change movements.

Professional Qualities of Person-centred Geography Teachers in the 1990s

He sat there in front of the television feeling vaguely uneasy about 'taking the evening off' instead of reading through all of those student essays and commenting on them. Then, as he took stock of his activities, he realised that he was watching a Cousteau film on undersea exploration and anthropology. Now he could allow himself 'credit' for an evening of professional development. Later, the novel from which he read a few pages before going to sleep, Michener's *Centennial*, turned out to be so full of geographic material that it almost seemed pedantic.

The breadth of geography poses huge difficulties for professional development. Traditionally, in a normal geography course based on a single textbook, a teacher could prepare for the next day by reading the assigned chapter in the text, some old notes and material from a couple of references the class members could be presumed *not* to have read, thus giving the teacher a slight edge over the students. The teacher controlled the content, and this seemed comforting.

A teacher in the 1990s, following our approach, might never even have heard of a topic introduced by a student. You have to look at your entire experience as an all-encompassing 'inservice course'. Your life is the course, and it is full to the brim of geographically relevant material. Recognition of this may not increase the volume of material that passes in front of you. It will increase your *attention* to a wider variety of things that surround you. It extends your appreciation of everyday events that relate to your professional life. Mowing the lawn may become as important to your professional training as going to a workshop. As a daily exercise, consider your activities, decide which one has been the 'least' relevant to your professional life as a geographer, and then redefine it in terms of geographic implications.

Professional Organisations

We find it important to maintain contact with appropriate professional organisations. Join your local geography teachers' association as a matter of urgency, but remain aware that really new perspectives may also come from community groups and meetings that would normally seem irrelevant. Go to such meetings with your antennae tuned for geographic content and ideas. If you do not find obviously relevant material, look for relevant metaphors. Force it; make it a game. Thus, the cultural geographer may find useful ideas at a meeting of organic chemists, music teachers, or at a drama festival. Each year, reach beyond your normal range of interests.

Inservice Workshops

> Phil reported to the workshop dutifully. After sitting through the first session without participating, he skipped the next one and went out to the trout stream nearby. No one criticised him: it has been made explict that each person was responsible for his or her own agenda. Many participants and 'leaders' even expressed interest in Phil's expedition. Once that agenda was satisfied, Phil became one of the most vigorous participants in the group.

Content-oriented or method-oriented workshops, with tightly organised agendas, do little to develop the kind of teachers we need. For the person who already has the perspective necessary to take information and adapt it to the needs of a person-centred educational environment, any workshop can be helpful. A workshop explicitly designed to engender helpful values and perspectives for person-centred work must create the same learning environment we want to see in the school. Talking and theorising about it has little useful effect. Only experiencing it can prepare anyone for functioning in it.

Over the years, we have been involved in many inservice education courses. We always opt for residential, open-format workshops. We often begin with small groups where participants can open their real concerns to each other. Not surprisingly, most participants are more concerned with their own lives and personal problems than they are about 'geography'. Just like their own students! The first agenda item is thus to let these concerns be listened to and acknowledged. This creates high levels of trust among the participants.

We place blackboards, bulletin boards, chalk, markers, newsprint pads, tacks and tape around our meeting places. Group members advertise and suggest times for making offerings to the group. They also list what they want to learn about, what their needs are. Staff members and guest specialists guarantee a certain minimum of diversity and 'coverage'. The staff must not monopolise by offering so much that other participants feel no pressure or need to offer their own options. As staff, we may consciously decline to schedule events in order to attend other sessions ourselves and to elicit the emergence of group members' own expertise: 'You want it; *you* do it or go find it!'. We schedule events without regard to time conflicts. Participants need to experience and deal with problems of choice, time conflicts, and even rejection, all inherent in a multidimensional learning environment. If the staff are suitably selected for their sensitivity to these problems and to the group, then staff members become 'invisible' as staff.

The Place of Controversial Issues in Geography Courses

With its interdisciplinary focus and attention to current events, geography lends itself naturally to the discussion of controversial issues. In a student-centred program, focused on the concerns students bring to the group, controversial topics become inevitable. Recently, we have begun making *The New York Times* a required 'text' for several courses. We also begin both large and small group meetings with information briefings given by group members, on a scheduled, rotational basis. We also issue a departmental newsletter, *The Geography Weekly*, which is widely distributed on campus, and which often deals with controversial issues. We have offered a course on 'Earth Stewardship' which includes environmental topics (with discussion not only on preservation and conservation, but also on benign use of earth resources), a semester course devoted entirely to problems of world hunger (featuring a symposium that generated highly controversial discussion presented by representatives from the World Bank, Industry and relief organisations), and another course on problems of world peace. Geography students have been strongly involved in campus controversies over problems in South Africa, Central America, the Middle East, US relationships with the Soviet Union, and, on a more local level, student problems with the administration, racial problems, poverty and other such general issues. When controversial topics arise naturally in seminar discussion, we generally encourage continuation of the discussion, even if previously scheduled events must be deferred. We regard geography as a dynamic, interdisciplinary forum for the analysis of controversial topics.

Conclusion: A Cautionary Note

Some readers may feel that some of the suggestions in this chapter imply that geography teachers have no need to learn 'traditional' content about geography. Not so! A firm, wide, deep and explicit background and a strong interest in 'geography' *per se* are invaluable and form a foundation for the other interests of the teacher of the 1990s. If you are not reading a lot of professional material in which the word 'geography' appears frequently, you may be neglecting a mainstream equally as important as 'looking for geography' in newspapers, general reading, music, sports, business and family life. At the same time, we *do* intend that a person 'untrained' in 'geography' (a 'musician', for example) may be as good a 'geography teacher' as a person with several degrees in 'geography'! Recently we have begun inviting members from widely different departments and members of the wider community to give talks on the connections they see between their own work or interests and geography. Often these people are apprehensive at first, but we are finding that people from biology, history, philosophy, the arts, music, literature and a wide range of fields of interest are willing to join us. When you pass the lecture hall where a large introductory geography group is meeting, you may just hear the university's resident string quartet giving a concert of geographically related music, followed by a spirited discussion. Neglecting any dimension decreases the potential for helping learning to occur in 'geography'.

Past successes also pose a danger to person-centred education in geography. Once something 'works' we tend to want to use the techniques over again in order to repeat the success. The next group, however, will be different. You never step in the same river twice, as the saying goes. If an approach works, rejoice, but then approach the next situation freshly, on its own terms and seek a new perspective. Abandon 'techniques' that get to feel like formulas, and search for freshness as if you have had no past experience. Mistakes? Yes, mistakes must continue to be made if progress is to continue. Failure to make mistakes generally means failure to grow. Teachers of the 1990s must join their students in exploring all possible paths, including what may appear to be dead ends, if better paths into the future are to be found. It's amazing how often a 'safe' path becomes a blind alley and an unlikely, overgrown trail leads to a previously unknown highway.

References

Adams, J.L. (1974) *Conceptual Blockbusting*, San Francisco: W.H. Freeman.
de Bono, E. (1970) *Lateral Thinking*, New York: Harper and Row.
Fast, J. (1971) *Body Language*, New York: Pocket Books.
Fien, J. and Gerber, R. (eds) (1988) *Teaching Geography for a Better World*, 2nd edition, Edinburgh: Oliver and Boyd.
Gibbons, M (1974). 'Walkabout: Searching for the Right Passage from Childhood and School', *Phi Delta Kappa*, Vol. 55, 594–602.
Gordon, W.J.J. (1961) *Synectics*, New York: Harper and Row.
Hall, E.T. (1959) *The Silent Language*, Garden City, New York: Doubleday.
Hall, E.T. (1966) *The Hidden Dimension*, Garden City, New York: Doubleday.
Hayakawa, S.I. (1978) *Language in Thought and Action*, fourth edition, New York: Harcourt, Brace, Jovanovich.
Huckle, J. (ed.) (1983) *Geographical Education: Reflection and Action*, Oxford: Oxford University Press.
James, P.E. and Martin, G.L. (1981) *All Possible Worlds: A History of Geographic Ideas*, New York: John Wiley and Sons.
Jenkins, A.M. and Keene, P. (1979) 'Use Posters in your Teaching', *Journal of Geography in Higher Education*, Vol. 3(1), 26–27.
Krathwohl, D.R., Bloom, B.S. and Masia, B.B. (1964) *Taxonomy of Educational Objectives. Affective Domain*, New York: McKay.

McKim, R. (1972) *Experience in Visual Thinking*, Monterey, California: Brooks/Cole.

Osborn, A.F. (1953) *Applied Imagination*, New York: Charles Scribners Son.

Rogers, C.R. (1962) 'The Interpersonal Relationship: the Core of Guidance', *Harvard Educational Review*, Vol. 32(4).

Rogers, C.R. (1977) *Carl Rogers on Personal Power*, New York: Delta.

Rogers, C.R. (1980) *A Way of Being*, Boston: Houghton Mifflin.

Romey, W.D. (1977) 'Introductory Geology from the Newspapers', *Journal of Geological Education*, Vol. 25, 111–114.

Romey, W.D. (1984) '"Walk-Through": An Alternative Approach to Final Examinations in Science', *Journal of College Science Teaching*, Vol. 13, 420–422.

Sauer, C. (1956) 'The Education of a Geographer', *Annals, Association of American Geographers*, Vol. 46, 287–299.

Tuan, Yi-fu (1986) *The Good Life*, Madison, Wisconsin: University of Wisconsin Press.

Tuan, Yi-fu (1987) 'Attention: Moral-Cognitive Geography', *Journal of Geography*, Vol. 86, 11–13.

Acknowledgements

The authors, editors and publishers are grateful to the following for permission to reproduce copyright material:

Chapter 7: Figure 7.2 is by John Mason from *Line-Up, Surfers Paradise*, June, 1982.
Chapter 9: Department of Lands, South Australia for figures 9.1 and 9.6.
Chapter 13: Mr John Berger for the quotation from *A Fortunate Man*.
Chapter 14: W H Auden, E. Mendelson, Random House Inc. and Faber and Faber Inc. for the extract from the poem *Bucolics*; Mr Rodney Moss for the ideas associated with figure 14.13; NSW Department of Tourism for the photograph with figure 14.3(a); Mrs O Streeton for permission to reproduce the painting 'Still glides the stream . . .' by Sir Arthur Streeton (figure 14.7) and the Art Gallery of NSW for the photographic print of the painting.
Chapter 16: Jacaranda Cartographics, a division of Jacaranda Wiley Ltd for figure 16.1; Longman Cheshire for figure 16.3 from *Geography Patterns* by W Farleigh & R Hinds; Nestle Australia Ltd for figure 16.12; New Internationalist for figures 16.5, 16.6, 16.7 & 16.11 from *New Internationalist Magazine*, PO Box 82, Fitzroy Vic. 3065 (annual subscription rates — $37.50 for an individual, $47.00 for an institution).
Chapter 20: The Rainforest Game (figure 20.1) was developed by D. Lergessner and D Cave, Department of Education, Queensland, with assistance from K Cordwell, Department of Education, Queensland.
Chapter 21: Active Learning Systems for figure 21.17; BP Educational Service for figure 21.13; Education Department of Western Australia for figure 21.10 from 'Climate' (1985) Wesoft Educational Software; Hutchinson Software for figure 21.4 from MEP/GAPE Project Geography Programs London, Hutchinson Software in association with the Geographical Association; Simon Jewell for figure 21.7; Longman Micro Software UK for figures 21.5, 21.6 and 21.11; Satchel Software, South Australia for figure 21.16; Science Education Software, UK for figure 21.13.
Chapter 25: Thomas Nelson and Sons Pty Ltd for figures 25.1 and 25.6 from *Going Places* Book 2 (page 29) and Book 3 (page 54) by M Renwick and W Pick.
Chapter 28: Thomas Nelson and Sons Pty Ltd for figure 28.5 from *Assignment Geography: Structure Exercises in Human Geography* by M Jones (1979).

While every care has been taken to trace and acknowledge copyright, the publishers tender their apologies for any accidental infringement where copyright has proved untraceable. They would be pleased to come to a suitable arrangement with the rightful owner in each case.